Cognitive Intelligence and Robotics

Cognitive Intelligence refers to the natural intelligence of humans and animals, it is considered that the brain performs intelligent activities. While establishing a hard boundary that distinguishes intelligent activities from others remains controversial, most common behaviors and activities of living organisms that cannot be fully synthesized using artificial means are regarded as intelligent. Thus the acts of sensing and perception, understanding the environment, and voluntary control of muscles, which can be performed by lower-level mammals, are indeed intelligent. Besides the above, advanced mammals can perform more sophisticated cognitive tasks, including logical reasoning, learning, recognition, and complex planning and coordination, none of which can yet be realized artificially to the level of a baby, and thus are regarded as cognitively intelligent.

This book series covers two important aspects of brain science. First, it attempts to uncover the mystery behind the biological basis of cognition, with a special emphasis on the decoding of stimulated brain signals or images. Topics in this area include the neural basis of sensory perception, motor control, sensory-motor coordination, and understanding the biological basis of higher-level cognition, including memory, learning, reasoning, and complex planning. The second objective of the series is to publish consolidated research on brain-inspired models of learning, perception, memory, and coordination, including results that can be realized on robots, enabling them to mimic the cognitive activities performed by living creatures. These brain-inspired models of machine intelligence complement the behavioral counterparts studied in traditional artificial intelligence.

The series publishes textbooks, monographs, and contributed volumes.

Maolong Lv • Boyang Zhang • Haibin Duan •
Yuyuan Shi • Chuhan Zhou

Unmanned Aerial Vehicles Formation Control and Safety Guarantee

Springer

Maolong Lv
Air Traffic Control and Navigation
Air Force Engineering University
Xi'an, Shaanxi, China

Haibin Duan
College of Automation Science and Electrical Engineering
Beihang University
Beijing, China

Chuhan Zhou
College of Air Traffic Control and Navigation
Air Force Engineering University
Xi'an, Shaanxi, China

Boyang Zhang
Beijing Blue Sky Science and Technology Innovation Center
Beijing, China

Yuyuan Shi
School of Mathematics and Statistics
Xidian University
Xi'an, Shaanxi, China

ISSN 2520-1956 ISSN 2520-1964 (electronic)
Cognitive Intelligence and Robotics
ISBN 978-981-96-5828-2 ISBN 978-981-96-5829-9 (eBook)
https://doi.org/10.1007/978-981-96-5829-9

This Springer imprint is published by the registered company Springer Nature Singapore Pte Ltd.
The registered company address is: 152 Beach Road, #21-01/04 Gateway East, Singapore 189721, Singapore

To our families for their love and support

Preface

With the rapid development of aerospace technology and information technology, cooperative multiple unmanned aerial vehicles (UAVs) formation demonstrates the excellent coordination and intelligence, as an important trend in the current field of UAVs. In response to complex flight environments and diverse dynamic tasks, cooperative multi-UAV flight exhibits well task scalability and fault tolerance, which are widely applied in several areas, such as remote sensing, search and rescue, surveillance, wireless relay, fire monitoring and detection, agricultural production management. Among these tasks, the cooperative control system plays an important role for ensuring the reliability and stability of UAV formation flight. It is typically divided into two parts: the guidance layer and the control layer. Regarding the guidance layer, the UAVs calculate the reference commands for speed and attitude based on the desired flight trajectory and formation configuration. On the other hand, the control layer stabilizes the reference tracking of speed and attitude by controlling the actuators of ailerons, rudders, elevators, and throttle.

Considering the requirements of flight missions on the control performance of UAVs, the various constraints that practical UAV systems are subject to, as well as the nonlinearity and uncertainty of UAV systems in complex flight environment, the cooperative control of multiple UAVs faces three challenges as follows:

(1) The flight mission always imposes strict requirements on convergence time of multi-UAV system. By enhancing the convergence speed, the multi-UAV system aims to complete command tracking and error convergence within the required time.
(2) Practical multi-UAV systems are subject to typical constraints on actuator response frequency, onboard computing resource, and communication resource, which greatly affects the control performance of multi-UAV system.
(3) Adverse factors in complex flight environments, such as gust perturbations, electromagnetic interference, and denial-of-service (DoS) attacks, significantly affect the control stability of multi-UAV system, resulting in the unsafe formation flight.

To address aforementioned challenges, the key technology is to design distributed cooperative control systems by utilizing some advanced control techniques for further guaranteeing the reliability, adaptability, and safety. The widely investigated cooperative multi-UAV control methods cannot be directly extended to handle the above challenges encountered by the UAV formation. This monograph focuses on systematical cooperative control design for multiple UAVs against faults, strongly unknown nonlinearities, rapid convergence, limited onboard computing resource, limited communication resource, strong disturbances, and DoS attacks. The main contents of this monograph are presented in the following aspects:

(1) To enhance the fixed-time convergence of six-degree-of-freedom multi-UAV system subject to actuator faults and state constraints, a distributed fixed-time fault-tolerant control scheme is proposed. To reduce the adverse influences caused by parameter uncertainties and actuator faults, the bound estimation approach and well-defined smooth functions are introduced, such that the tracking errors of velocity and attitude stably converge to the residual sets around origin within fixed time.
(2) To achieve the fixed-time synchronization tracking of six-degree-of-freedom multi-UAV system in the presence of actuator faults and saturation, a new adaptive-gain-based control design with fixed-time sliding mode surface was proposed. Moreover, model uncertainties and external disturbances could be effectively handled by using a bound estimation method and smooth functions.
(3) To address the specified-time synchronization tracking problem of six-degree-of-freedom multi-UAV system with actuator faults, the novel specified-time performance functions are designed in such a way that the desired performance bounds can be imposed on velocity and attitude tracking errors. Then a distributed specified-time control scheme is constructed with adaptive robustifying terms to enhance the fault-tolerant ability and compensate the modeling uncertainties.
(4) To further reduce the signal transmission burden of six-degree-of-freedom multi-UAV system subject to actuator faults and lumped disturbances, an event-triggered cooperative synchronization fault-tolerant control methodology is investigated. By integrating an event-triggering mechanism, the time instants of updating control signals are determined. The actuator faults as well as the network-induced errors are handled via the bound estimation approach and some well-defined smooth functions.
(5) To further reduce the communication burden of six-degree-of-freedom multi-UAV system in the presence of actuator faults and lumped disturbances, an event-based fault-tolerant coordinated control is utilized such that the consensus tracking of velocity and attitude is achieved. By introducing a dynamic event-triggered communication mechanism, the connected communications are scheduled, while avoiding the unnecessary information exchanges among UAVs. Regarding the actuator faults and lumped disturbances, the disturbance observer technique is incorporated into the control scheme to increase the composite approximation capability.

(6) To reduce the communication frequency among six-degree-of-freedom fixed-wing multi-UAVs, a topology-switching control method is proposed. The considered switching graph set includes non-connected graphs, thereby reducing communication links within the cluster. Additionally, an asynchronous switching observer is introduced to ensure that each UAV can estimate the lead-UAV's state information despite asynchronous switching caused by topology identification delay.
(7) To ensure the performance robustness and optimality of multi-UAV system in the face of strong disturbances, a feedforward-feedback learning-based optimal control methodology is established to facilitate cooperative formation tracking. By leveraging backstepping-based feedback control, the problem of formation tracking is transformed into an equivalent optimal regulation problem. Consequently, a learning-based feedforward control scheme is devised, wherein the critic-only echo state network is employed to approximate the optimal feedforward control policies.
(8) To address the resilient fixed-time security control problem for multiple UAVs under asynchronous DoS attacks, a novel resilient fixed-time observer is introduced to accurately estimate the leader's reference in spite of the attack signals. Then, an attack-oriented fixed-time security consensus controller is proposed to ensure the fast and stable convergence of the networked UAVs.

This monograph presents systematic and comprehensive descriptions of the cooperative control issues for multiple UAVs against faults, strongly unknown nonlinearities, rapid convergence, limited onboard computing resource, limited communication resource, strong disturbances, and DoS attacks. It offers readers with deep understanding and insights on reliability, adaptability, and safety of multi-UAV system and the corresponding control designs. The advanced cooperative control methods presented in this monograph can provide guidelines for engineers to improve the reliability, adaptability, and safety of aerospace engineering systems and some other related fields. This monograph is suitable for scientists and researchers, aerospace engineers, control engineers, lecturers and teachers, postgraduates, and undergraduates in the system and control community, especially those engaged in the field of UAV cooperation and multi agent systems.

Xi'an, China — Maolong Lv
Beijing, China — Boyang Zhang
Beijing, China — Haibin Duan
Xi'an, China — Yuyuan Shi
Xi'an, China — Chuhan Zhou

Acknowledgments

The contents included in this monograph are the outgrowth and summary of the authors' academic researches to address the cooperative control problems for multi-UAVs for guaranteeing the reliability, adaptability, and safety. It was partially supported by National Natural Science Foundation of China under grant 62303489, 62350048, 62403488, GKJJ24050502, and T2121003, Natural Science Basic Research Program of Shaanxi under grant 2024JC-YBQN-0668, Project for Science and Technology under grant 2022-JCJQ-QT-018, Young Talent Fund of Association for Science and Technology in Shaanxi under grant 20220101, Postdoctoral International Exchange Project under grant YJ20220347, Postdoctoral Science Foundation General Program under grant 2022M723877, and Postdoctoral Science Foundation Special Funding under grant 2023T160790. The authors greatly appreciate all above financial supports. We appreciate the permission from Springer Nature, Institute of Electrical and Electronics Engineers, and Elsevier to reuse the results published in the relevant journals, such as *IEEE Systems Journal* (Chap. 3), *IEEE Transactions on Aerospace and Electronic Systems* (Chap. 4), *ISA Transactions* (Chap. 5), *IEEE Transactions on Vehicular Technology* (Chaps. 6, 7, and 8), *IEEE Transactions on Automation Science and Engineering* (Chap. 9). In addition, the authors would like to thank the editors in Springer Nature for their help to accomplish the publication of this monograph.

Competing Interests The authors have no competing interests to declare that are relevant to the content of this manuscript.

Contents

Part I
Fundamentals and Theoretical Foundations

Chapter 1
Introduction to Cooperative UAV Control

This chapter provides a brief analysis of the research background and significance of cooperative unmanned aerial vehicle (UAV) control. It focuses on elaborating the current researches of the multi-UAV system model and control system, the UAV cooperative control methods with fast convergence, the UAV cooperative control methods based on event-triggered strategies, and the UAV cooperative control methods with safety theory. Additionally, it provides an overview of the structural arrangement of the book.

1.1 Background and Motivations

With the rapid advancement of aerospace and information technology, multi-UAV collaborative formation flight is emerging as a prominent trend in the field of UAV, showcasing the remarkable coordination and intelligence. In complex flight environment, multi-UAV formation with diverse dynamic tasks provides significant advantages in terms of scalability and fault tolerance, which makes it widely applicable in some areas such as agriculture [1], cargo delivery [2], and reconnaissance [3, 4]. UAVs can generally be categorized into fixed-wing and rotary-wing types. Compared with rotary-wing UAVs, fixed-wing UAVs are known for their long endurance, high flight speeds, and substantial payload capacity, which makes them well suited for military operations with long-range surveillance, high-speed penetration, and large-area coverage [5]. Accordingly, this monograph focuses on exploring collaborative formation challenges specific to fixed-wing UAVs.

For fixed-wing UAV formation, each UAV can be viewed as the underactuated control system, using ailerons, rudder, elevator and throttle with state variables including flight speed, roll angle, pitch angle, and yaw angle [6]. The collaborative guidance and control system, essential for achieving coordinated UAV formation flight, is generally divided into two layers: guidance layer and control layer. The guidance layer calculates speed and attitude reference commands based on

M. Lv et al., *Unmanned Aerial Vehicles Formation Control and Safety Guarantee*,
Cognitive Intelligence and Robotics, https://doi.org/10.1007/978-981-96-5829-9_1

Fig. 1.1 Fixed-wing UAV formation (*Excerpt from the Internet*)

the desired trajectory and formation configuration, to establish the framework for effective formation flight. Then, the control layer tracks the speed and attitude references by controlling the ailerons, rudder, elevator, and throttle [7]. Due to the nonlinearities and uncertainties inherent in UAV systems within complex flight environment, along with the constraints of convergence time and state error, the collaborative control problem for multiple UAVs presents the following challenges:

(1) Challenges of collaborative control within complex flight environment

In complex flight environments, multi-UAVs are affected by parameter uncertainties and external disturbances, with the added risk of unforeseen actuator failures during operations. These factors directly influence the accuracy of system modeling, which further affects the stability of collaborative control.

(2) Challenges of collaborative control related to flight task requirements

During multi-UAV formation flight, each UAV should respond promptly to reference commands and adjust flight states swiftly. Mission requirements impose stringent demands on the convergence speed of UAV states, while also necessitating that the states and tracking errors of each UAV adhere to specific constraints, which adds further challenges to collaborative control.

(3) Challenges of practical system constraints on collaborative control

For a practical multi-UAV system, collaborative control faces various realistic constraints, such as limited actuator response frequencies, restricted onboard computing resources, and constrained communication bandwidth. These limitations directly affect the effectiveness of task execution, posing significant challenges for the collaborative control of UAVs.

(4) Challenges of strong disturbances and communication attacks to safe control

In complex environment, the stability of UAV formation flight is significantly affected by external disturbances, such as strong gusts. Additionally, under adversarial conditions, UAVs face the threats of communication attacks, which pose substantial risks to flight safety. Considering the external disturbances and communication attacks, the issue of safe control for UAV formation necessitates further in-depth research.

In summary, this monograph focuses on multi-UAVs to investigate collaborative control issues in complex environment and under various constraints. To tackle the outlined challenges, the monograph develops effective algorithms for collaborative control of speed and attitude, as well as for formation control, enhancing the robustness, reliability, and safety of multi-UAV collaborative operations and improving the capabilities of UAV formation in task execution. The research presented in this monograph offers technical support for collaborative control and establishes a theoretical foundation for the engineering applications of multi-UAV systems.

1.2 Introduction to Cooperative Control

Multi-agent collaborative control effectively harnesses the capabilities of individual agents, enabling them to flexibly and efficiently adapt to ever-changing environments and task objectives [8]. In comparison to single-agent systems, multi-agent systems (MASs) offer significant advantages such as enhanced scalability and robustness, especially in specialized task scenarios including large-scale reconnaissance and collaborative transportation [9, 10]. Furthermore, when tackling complex tasks, MASs assign the simple tasks to individual agents to improve overall execution efficiency. Currently, the theoretical framework for multi-agent collaborative control is primarily categorized into two types: distributed tracking control and distributed formation control.

The issue of distributed tracking for MASs arises from the consensus problem, where a network of agents updates the states based on information from neighboring agents to ultimately converge on a common value. Achieving consensus is a prerequisite for most collaborative tasks of MASs. By designing controllers that enable all agents to track a common reference trajectory, distributed tracking control of MASs primarily focuses on first-order and second-order linear systems [11–13]. However, most practical systems exhibit nonlinear characteristics, such as inverted pendulum systems, robotic systems, and UAV system [14]. The nonlinear nature of these systems, combined with uncompensated model uncertainties and external disturbances, presents greater challenges for distributed tracking control of nonlinear MASs while also offering extensive application potential.

Distributed formation control of MASs involves designing formation topologies and distributed controllers to ensure that the networked agents can follow a desired formation while maintaining or adjusting the configuration over time to meet specific task requirements, such as reconnaissance, exploration, and obstacle avoidance. Common formation configurations are defined using relative positional information between agents. Do [15] proposed a potential function-based formation controller for a group of mobile robots with limited sensing ranges, facilitating the achievement of desired formation while ensuring collision avoidance. Ihle et al. [16]

presented an adaptive controller that guides agents to achieve the desired configuration using potential functions, combining favorable transient characteristics with the robustness of sliding mode control. Additionally, the method based on orientation constraints retains its invariance when the formation is translated or scaled. Zhao et al. [17] implemented formation control for MASs with double integrators in arbitrary dimensions, addressing practical issues such as external disturbances and obstacle avoidance.

Building on the theories of distributed tracking control and distributed formation control for MASs, the cooperative control problem for multi-UAVs can be categorized into two main fields: velocity and attitude tracking control and formation control for multi-UAVs. The following sections provide a detailed analysis of the literatures and methodologies related to abovementioned control issues.

1.3 Application of Cooperative Control in Multi-UAV

In recent years, the development of key technologies and related fields has made the issue of multi-UAV cooperative control as a prominent research topic. Both domestic and international researches have explored various aspects, including finite-time convergence, fixed-time convergence, event-triggered control, and event-triggered communication, and achieved significant progress in these fields [18–20]. Addressing above four challenges associated with multi-UAV cooperative control, this section elaborates on the current researches from the following four perspectives: modeling and control strategies for multi-UAVs, cooperative control methods with fast convergence for multi-UAVs, cooperative event-triggered control methods for multi-UAVs, and cooperative optimal control methods for multi-UAVs.

1.3.1 Modeling and Control Strategies of Multi-UAVs

The multi-UAV system consists of multiple single UAV systems, with each UAV representing a typical multi-input multi-output underactuated system characterized by nonlinearity and strong coupling [21]. Regarding UAV dynamic modeling, there are currently two approaches. The first approach separates the UAV system into an outer-loop guidance system and an inner-loop velocity and attitude control system and then establishes dynamic and kinematic models for each component [22–24]. The second approach considers the guidance and control systems as an integrated whole [25–27]. This modeling requires coupling multiple guidance and control variables. Due to the simpler structure and ease of implementation, the first approach has become the mainstream method in use. As illustrated in Fig. 1.2, the guidance system calculates velocity and attitude reference commands based on the expected trajectory and position information, which are then transmit to the velocity and

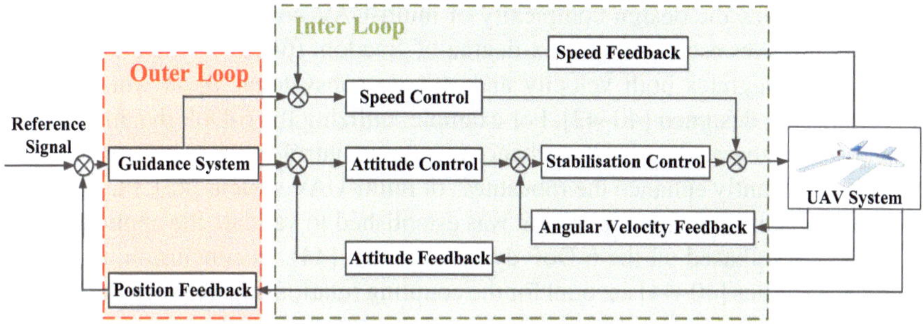

Fig. 1.2 UAV guidance and control system

attitude control system. The control system, in turn, ensures the stable tracking of velocity and attitude to the reference commands.

To tackle the guidance control of multiple UAVs, current primary methods can be divided into two categories. The first category employs a feedback linearized approach, which transforms the three-degree-of-freedom (3-DoF) UAV guidance model into a point mass model. This linearized model is utilized to design collaborative control algorithms for multiple UAVs [28, 29]. For instance, Wang et al. proposed an optimal control method for the multi-UAVs to achieve the trajectory tracking and collision avoidance for UAV formation [30]. Similarly, Zhang et al. presented a model predictive control method based on an extended state observer to enhance disturbance rejection for UAV formation [31]. The second category directly chooses the nonlinear guidance model and develops collaborative controllers for multiple UAVs [32, 33]. For example, Zhang et al. developed a formation control algorithm based on nonlinear model predictive control, to leverage intelligent algorithms to optimize control inputs online, which facilitates the reconfiguration of UAV formation [34]. To address the challenge of close formation control under wind disturbances, Zheng et al. proposed an adaptive control algorithm to achieve the finite time convergence for UAV formation reconfiguration [35].

The current approaches to address velocity and attitude control for multiple UAVs can be classified into two main categories. The first category designs the independent controllers for either the velocity or attitude subsystem. For instance, in the context of multi-UAV attitude system, a finite-time collaborative controller was developed to achieve finite-time tracking of attitude under input constraints and actuator faults [36]. Similarly, Lin designed an attitude collaborative controller that ensures the consistent tracking of attitude in spite of parameter uncertainties [37]. Additionally, Kuwata et al. tackled the consistency tracking problem for leader-follower UAV formation by implementing a two-layer control structure to guarantee the speed tracking [38]. Specifically for vertical motion and speed control, an adaptive control algorithm based on disturbance observers was introduced to achieve stable tracking of speed for each UAV [39]. However, in studies [36–39], the coupling relationship between the velocity and attitude subsystems is not addressed, which simplifies the design complexity of multi-UAV control system. The second category involves establishing a six-degree-of-freedom (6-DoF) dynamic model for UAVs that integrates both velocity and attitude subsystems, upon which control algorithms are designed [40–42]. For example, utilizing the 6-DoF dynamic model, Zhang et al. integrated active disturbance rejection control and dynamic surface control to significantly enhance the robustness of multi-UAV system [43]. Furthermore, a semi-physical experiment platform was established to validate the applicability of multiple UAVs based on the 6-DoF dynamic model [44]. In conclusion, the 6-DoF models in studies [40–44] account for the coupling relationship between the velocity and attitude subsystems, making it more representative of the characteristics of practical UAV systems.

1.3.2 Cooperative Control with Fast Convergence for Multi-UAVs

In recent years, various control methods, including backstepping control [45, 46], sliding mode control [47, 48], and robust control [49, 50], have been extensively applied in multi-UAV system. For instance, Tang et al. explored the cooperative control problem for multiple UAVs in spite of time delays and unknown disturbances and designed a distributed controller based on adaptive control techniques to achieve consensus of attitude [51]. To tackle the tracking control issue of UAV formation, an adaptive backstepping control method utilizing vector field techniques was proposed and validated through experiments [52]. Additionally, Yu et al. developed a distributed adaptive fault-tolerant control algorithm for close formation of UAVs affected by aerodynamic coupling, which enhances disturbance rejection capabilities through hybrid learning methods [53]. Generally, studies [45–53] successfully achieved the stable control of multi-UAV system. However, the methods primarily relied on the asymptotic stability convergence, which do not well satisfy the rapid convergence requirements of flight missions.

To enhance the convergence performance of UAV systems, researches have proposed the finite-time control methods, which offer the advantages of rapid convergence and strong robustness [54]. For example, Tian et al. developed an attitude control algorithm based on finite-time convergence for UAV attitude system and demonstrated its effectiveness through experiments [55, 56]. In the context of hierarchical UAV formation control, a finite-time control strategy was designed to achieve the desired configuration within finite time [57]. Similarly, to address UAV formation reconfiguration, a finite-time control method was proposed under input constraints to ensure that the formation error converges to a specified steady-state range within finite time [58]. Yu et al. developed a finite-time trajectory tracking control method utilizing finite-time disturbance observers, which improved both the rapid convergence and disturbance rejection capabilities of UAV systems [59]. In summary, finite-time control methods effectively enhance the convergence speed and control performance of UAV systems. However, the limits on convergence time of finite-time control are closely tied to the initial system states. When the initial state significantly deviates from the equilibrium state, guaranteeing convergence time becomes difficult, which restricts the practical application of finite-time control in UAV systems.

To address the challenges associated with finite-time control methods, Andrieu et al. introduced the concept of fixed-time convergence [60], which was later explored by Polyakov in the context of fixed-time Lyapunov stability [61]. Fixed-time control is a specialized form of finite-time control where the system state can converge to an equilibrium point within a finite time, regardless of initial conditions. This fixed time is independent of the initial states and relies solely on control parameters. Once the control parameters are established, the upper bound on convergence time can be predetermined. In recent years, researchers have proposed fixed-time control algorithms for various MASs, including first-order integrator systems [62, 63], second-order integrator systems [64, 65], and high-order systems [66, 67]. Concurrently, fixed-time control methods have been applied to UAV systems. For instance, Cui et al. developed a double-loop fixed-time control algorithm with fixed-time disturbance observer to tackle the attitude tracking problem of UAVs under external disturbances [68]. Similarly, Yu et al. designed a fixed-time controller that integrates nonsingular fixed-time sliding mode and adaptive control methods to address UAV attitude tracking in the presence of external disturbances and parameter uncertainties [69]. Further, considering the input constraints and actuator faults of UAV systems, studies [70, 71] designed an adaptive attitude controller based on an improved nonsingular fixed-time sliding mode. Additionally, to tackle the UAV control within autonomous ship landing task, Xia et al. employed a hierarchical architecture to achieve fixed-time convergence for both trajectory and attitude tracking [72]. In conclusion, fixed-time control significantly enhances the convergence performance of UAV systems and improves overall control efficacy. However, researches on collaborative control methods based on fixed-time control for 6-DoF multi-UAV system remains insufficient in both theoretical exploration and engineering practice, which highlights the need for further study.

1.3.3 Cooperative Control with Event-Triggered Strategy for Multi-UAVs

Multi-UAV system is a typical example of networked control systems. UAVs communicate state information over a network and compute control variables based on the states of neighboring UAVs. Traditional researches on cooperative control for multi-UAV systems typically assume the real-time continuous data transmission among UAVs, overlooking the limited communication resources [73–75]. As the number of UAVs increases, the required communication network bandwidth grows exponentially, leading to significant network loads, which results in congestion, communication delays, and data packet loss [64, 76, 77]. To address the limitations of communication resources of multi-UAV system, researches have proposed event-triggered communication strategies that substantially reduce unnecessary communication and ultimately conserve communication resources [78–81]. For example, Ji et al. [82] designed a formation control algorithm based on a triggering communication mechanism that effectively minimizes network communication among UAVs. Jang et al. combined event-triggered strategies with model predictive control to develop an event-triggered formation control method, which decreases the unnecessary network communication among UAVs [83]. In the context of UAV formation control with disturbances, Wei et al. designed an event-triggered control method utilizing disturbance observers to achieve stable control of UAV formation under triggering communication conditions [84]. Additionally, Xun systematically discussed the swarm control within communication-limited environment and developed a distributed control algorithm based on a triggering communication mechanism [85]. However, the aforementioned event-triggering mechanisms require UAVs to continuously monitor their state information to determine whether the triggering conditions are met in real time, leading to unnecessary dissipation of control and sensor resources. To alleviate this constraint, Anta et al. introduced a self-triggered mechanism that does not necessitate continuous monitoring of state information [86]. Instead, this mechanism only requires the current state information to compute the next sampling time. At state measurement point, control law computation, and actuator updates are performed. Following this idea, scholars began applying self-triggered mechanisms to the cooperative control problem in MASs. For instance, Wang et al. extended traditional stability results for self-triggered feedback systems, which predicts the next event based on sampled states, thereby relaxed the assumption that the process noise amplitude is limited by system states [87]. Wang et al. proposed a self-triggered control method capable of effectively compensating for input hysteresis, ensuring that tracking errors remain bounded by explicit functions of design parameters [88]. Xu et al. constructed a distributed self-triggered mechanism that schedules information transmission for each agent in real time [89]. Chen et al. developed a self-triggered consensus control scheme based on neural networks to address the sensor saturation conditions for nonlinear MASs [90].

On the other hand, traditional UAV control methods typically employ time-driven control strategies, where the UAV system samples states and updates control variables at fixed intervals, known as periodic sampling control [23, 24, 26, 29]. To ensure the control accuracy of UAV systems, these control methods always opt for smaller sampling periods. However, continuous high-frequency sampling and computation increase energy consumption, leading to unnecessary actuator wear, and reduce actuator lifespan. As a result, effective strategies to lower the controller update frequency and minimize actuator wear have garnered significant attention and promoted research into event-triggered cooperative control for multi-UAV system [91, 92]. The core idea is to design event-triggering thresholds that activate the controller to update control signals when the thresholds are met [93]. Compared with periodic sampling control, event-triggered control significantly reduces the number of controller updates and lowers actuator response frequencies [94]. Cao et al. developed an adaptive control method based on event-triggering strategy to achieve UAV formation tracking control under output constraints [95]. To tackle the collision avoidance problem for UAV formation, Wu et al. combined artificial potential fields and sliding mode methodologies to propose an event-driven formation control method [96]. Furthermore, regarding the encirclement control problem for UAV formation, Wei et al. studied an event-triggered control method using disturbance observers to lower actuator update frequency while enhancing the disturbance rejection capabilities [97].

In summary, as illustrated in Fig. 1.3, event triggering is a fundamental mechanism that does not correspond to a specific control method but rather integrates triggering update strategies and communication strategies with existing control methods. This integration aims to conserve communication and computation resources, providing significant theoretical and practical value. Although current researches on event-triggered control and communication for UAVs have made notable progress [81–97], most studies have employed point mass models of rotorcraft and fixed-wing UAVs. Consequently, research on event-triggered control and communication for 6-DoF UAVs remains insufficient and requires further exploration.

1.3.4 Cooperative Control with Safety Theory for Multi-UAVs

Ensuring the safe cooperative flight of multiple UAVs in complex environments remains a significant challenge for UAV formation control. On the one hand, the complex environment exposes UAV systems to various external disturbances, such as gusts of wind and electromagnetic interference, which jeopardize the safety and stability of multi-UAV cooperative control. On the other hand, the data transmission among UAVs with open communication networks is susceptible to

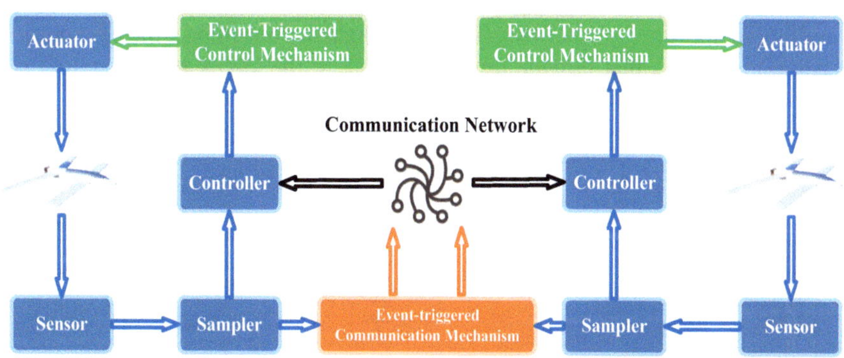

Fig. 1.3 Distributed multi-UAV system based on event-triggered strategies

malicious attacks, which can disrupt data transmission and lead to a loss of control actions [98, 99].

Firstly, to tackle safety issues of UAVs subject to multi-source disturbances, classical control methods such as robust control and disturbance observer control have been widely implemented in multi-UAV cooperative control [100, 101]. However, these disturbance-resistant methods fail to adequately address the optimality of control performance. Therefore, the need to adapt to complex environment and optimize performances is highlighted. Recently, reinforcement learning algorithms have shown considerable advantages in optimal control over comprehensive performance through autonomous learning [102, 103]. These algorithms can update control strategies based on interactions with the environment, without relying on precise system modeling [104], making it well suited to address optimal control problems for UAV systems in the presence of multisource disturbances.

Recent studies have applied reinforcement learning algorithms to cooperative control for UAVs. For instance, Hung et al. [105] investigated the formation flight of fixed-wing UAVs in disturbed environment using a model-free reinforcement learning algorithm with variable learning rates. Based on this work, Hung et al. further introduced a Q-learning algorithm for UAVs to achieve stable flight under disturbance conditions [106]. Additionally, to address the problem of multi-target tracking navigation for UAVs in complex environment, Zhou et al. proposed a distributed multi-agent reinforcement learning algorithm that employed maximum reciprocal reward principles to learn cooperative tracking strategies and combined with an experience-sharing reciprocity mechanism to enhance cooperation. Experimental results validated the effectiveness and superiority of this approach [107]. Furthermore, Liu et al. explored the optimal control problem for trajectory tracking of UAV formation and designed a control methodology that incorporated distributed observers and a model-free offline reinforcement learning algorithm [108].

Figure 1.4 illustrates a schematic of UAV cooperative control based on reinforcement learning algorithms. The learning process can be described as follows: (i) At each moment, the UAVs interact with the environment, collecting information and perceiving key state variables. (ii) Using these state variables, the UAV employs the reinforcement learning algorithm to compute and refine a control strategy, resulting in specific flight actions. (iii) These flight actions impact the environment, leading to changes that provide timely feedback to the UAVs and generating new

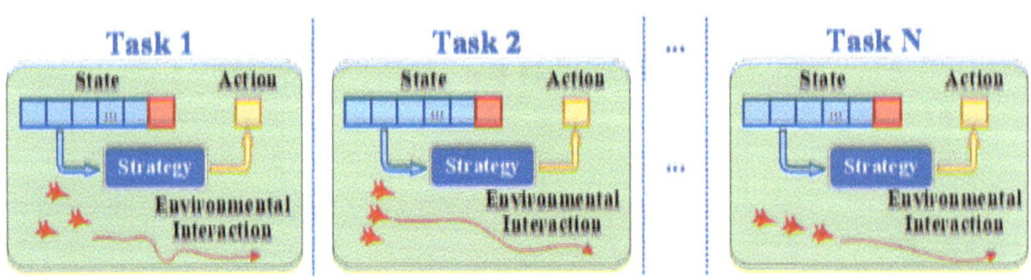

Fig. 1.4 Cooperative flight of UAVs based on reinforcement learning algorithms

state variables for the next interaction. This iterative process enables the UAVs to continuously update the control strategies, ultimately achieving optimal control of the UAV systems.

Secondly, considering the security issues related to the communication transmission, UAVs are vulnerable to two critical types of attacks: deception attacks and denial-of-service (DoS) attacks [109]. Deception attacks compromise the integrity of control data by injecting false information in place of actual transmission data [110]. In contrast, DoS attacks disrupt normal data transmission by severing communication links and preventing users from accessing distributed control data [111]. Compared with deception attacks, DoS attacks are generally easier to execute and own the broader applicability within various scenarios. Currently, most researches on safe control concerning DoS attacks have been theoretical, just focusing on MASs [110, 112, 113]. There exists limited exploration of safe control for UAV formation with DoS attacks. Therefore, investigating safe control strategies for UAV formation under DoS attacks is of significant theoretical and practical importance. This research not only enhances the robustness and security of UAV operations but also contributes to the broader understanding of MASs in the face of malicious interference, ultimately fostering the development of more resilient cooperative control frameworks for UAV networks.

1.4 Organization of the Monograph

This monograph comprises 12 chapters. Chapters 1 and 2 cover the fundamentals and theoretical foundations: Chap. 1 introduces the background, motivations, and cooperative control concepts, including their applications in UAVs, while Chap. 2 presents UAV dynamics, providing the basis for the chapters that follow.

Chapters 3, 4, and 5 focus on advanced control designs for UAVs with fixed-time and specified-time convergence. Specifically, Chap. 3 explores a distributed fixed-time fault-tolerant control scheme for multiple UAVs with actuator faults and state constraints. To mitigate adverse effects from parameter uncertainties and actuator faults, bound estimation methods and well-defined smooth functions are introduced. Chapter 4 presents a distributed fixed-time sliding mode control design with adaptive gains for multiple UAVs, addressing actuator faults and saturation. Model uncertainties and external disturbances are managed effectively through bound estimation and smooth functions. Chapter 5 tackles the specified-time synchronization tracking problem for multiple UAVs with actuator faults, constructing a distributed specified-time control scheme with adaptive robustifying terms to enhance fault tolerance and compensate for modeling uncertainties.

Chapters 6, 7, and 8 focus on event-triggered and communication-efficient control for UAVs. Chapter 6 examines a distributed event-triggered fault-tolerant control method for multiple UAVs subject to actuator faults and lumped disturbances, addressing actuator faults and network induced errors through bound estimation and well-defined smooth functions. Chapter 7 utilizes a distributed event-based fault-tolerant control with a dynamic event-triggered communication mechanism for multiple UAVs, incorporating a disturbance observer technique to enhance approximation capabilities when addressing actuator faults and lumped disturbances. In Chap. 8, a multi-UAV asynchronous topology switching control method is presented, and a switching observer is designed to ensure that the follower UAVs can track the state information of the leader UAV. In addition, a novel switching Lyapunov function is proposed to obtain more relaxed topology switching rules.

Chapters 9 and 10 focus on safety and security in multi-UAV operations. Chapter 9 presents a feedforward-feedback learning-based optimal control approach to facilitate cooperative formation tracking. Using backstepping-based feedback control, it introduces a learning-based feedforward control scheme that employs a critic-only echo state network to approximate optimal feedforward control policies. Chapter 10 addresses resilient fixed-time security control for multiple UAVs under asynchronous DoS attacks, proposing an attack-oriented fixed-time security consensus controller to achieve rapid and stable convergence of networked UAVs.

Chapter 11 concludes the monograph and discusses open problems and challenges in the cooperative control design for multiple UAVs.

This monograph addresses practical challenges in UAV cooperative control from three key perspectives: first, meeting the convergence time requirements for multi-UAV systems, second, accommodating the actual transmission, computation, and communication capabilities of multi-UAV systems, and third, ensuring safety in the presence of strong disturbances and DoS attacks. Additionally, the monograph presents cooperative control design, covering inner-loop velocity and attitude tracking to outer-loop position control for multiple UAVs.

References

1. L. Techy, I. Schmale, G. David, C.A. Woolsey, Coordinated aerobiological sampling of a plant pathogen in the lower atmosphere using two autonomous unmanned aerial vehicles. J. Field Rob. **27**(3), 335–343 (2010)
2. V. Spurný, T. Báča, M. Saska , R. Pěnička, T. Krajník, J. Thomas, D. Thakur, G. Loianno, V. Kumar, Cooperative autonomous search, grasping, and delivering in a treasure hunt scenario by a team of unmanned aerial vehicles. J. Field Rob. **36**(1), 125–148 (2019)
3. N. Nigam, S. Bieniawski, I. Kroo, J. Vian, Control of multiple UAVs for persistent surveillance: algorithm and flight test results. IEEE Trans. Control Syst. Technol. **20**(5), 1236–1251 (2012)
4. J. Keller , D. Thakur , M. Likhachev , J. Gallier, V. Kumar, Coordinated path planning for fixed-wing UAS conducting persistent surveillance missions. IEEE Trans. Autom. Sci. Eng. **14**(1), 17–24 (2017)
5. K. Li, N, Zhang, Design of multi-aircraft air combat simulation system based on layered architecture. J. Syst. Simul. **19**(22), 5160–5164 (2007)
6. F. Austin, Game theory for automated maneuvering during air-to-air combat. J. Guid. Control. Dyn. **13**(6), 1143–1147 (1990)
7. K. Wang, in *Research on Fixed-wing UAV Self-immunity Control and Autonomous Obstacle Avoidance Algorithm* (Nanjing University of Aeronautics and Astronautics, Nanjing, 2020)

8. Y. Su, J. Huang, Cooperative output regulation of linear multi-agent systems. IEEE Trans. Autom. Control **57**(4), 1062–1066 (2012)
9. J. Hu, L. Xie, K.Y. Lum, J. Xu, Multiagent information fusion and cooperative control in target search. IEEE Trans. Control Syst. Technol. **21**(4), 1223–1235 (2013)
10. A. Khan, B. Rinner, A. Cavallaro, Cooperative robots to observe moving targets: review. IEEE Trans. Cybern. **48**(1), 187–198 (2018)
11. R. Olfati-Saber, R.M. Murray, Consensus problems in networks of agents with switching topology and time-delays. IEEE Trans. Autom. Control **49**(9), 1520–1533 (2004)
12. Y. Hong, J. Hu, L. Gao, Tracking control for multi-agent consensus with an active leader and variable topology. Automatica **42**(7), 1177–1188 (2007)
13. J. Mei , W. Ren, J. Chen, Distributed consensus of second-order multi-agent systems with heterogeneous unknown inertias and control gains under a directed graph. IEEE Trans. Autom. Control **61**(8), 2019–2034 (2016)
14. S.S. Ge, C.C. Hang, T.H. Lee, T. Zhang, *Stable Adaptive Neural Network Control* (Springer Science & Business Media, New York, 2013)
15. K.D. Do, Bounded controllers for formation stabilization of mobile agents with limited sensing ranges. IEEE Trans. Autom. Control **52**(3), 569–576 (2007)
16. I.F. Ihle, J. Jouffroy, T.I. Fossen, Formation control of marine surface craft: a lagrangian approach. IEEE J. Ocean. Eng. **31**(4), 922–934 (2006)
17. S. Zhao, D. Zelazo, Translational and scaling formation maneuver control via a bearing-based approach. IEEE Trans. Control Network Syst. **4**(3), 429–438 (2017)
18. A. Budiyono, Advances in unmanned aerial vehicles technologies, in *International Symposium on Intelligent Unmanned System*, Nanjing, China (2008)
19. P.R. Rakesh, R. Padhi, Formation flying of UAVs with dynamic inversion based partial integrated guidance and control, in *AIAA Guidance, Navigation, and Control Conference*, Portland, Oregon (2011)
20. T. Yamasaki, S.N. Balakrishnan, H. Takano, Separate-channel integrated guidance and autopilot for automatic path-following. J. Guid. Control. Dyn. **36**(1), 25–34 (2012)
21. G. Campa, Y. Gu, B. Seanor, M.R. Napolitano, L. Pollini, M.L. Fravolini, Design and flight-testing of non-linear formation control laws. Control. Eng. Pract. **15**(9), 1077–1092 (2007)
22. Y. Gu, B. Seanor, G. Campa, M.R. Napolitano, L. Rowe, S. Gururajan, S. Wan, Design and flight testing evaluation of formation control laws. IEEE Trans. Control Syst. Technol. **14**(6), 1105–1112 (2006)
23. J. Zhang, J. Yan, A novel control approach for flight-stability of fixed-wing UAV formation with wind field. IEEE Syst. J. **15**(2), 2098–2108 (2021)
24. Y. Xie, L. Han, X. Dong , Q. Li, Z. Ren, Bio-inspired adaptive formation tracking control for swarm systems with application to UAV swarm systems. Neurocomputing **453**, 272–285 (2021)
25. Y. Xie, L. Han, X. Dong, Q. Li, Z. Ren, Distributed formation control of networked multi-agent systems using a dynamic event-triggered communication mechanism. Aerosp. Sci. Technol. **72**, 327–334 (2018)
26. Q. Luo, H. Duan, Distributed UAV flocking control based on homing pigeon hierarchical strategies. Aerosp. Sci. Technol. **70**, 257–264 (2017)
27. A. Tahir, J.M. Böling, M.H. Haghbayan, J. Plosila, Comparison of linear and nonlinear methods for distributed control of a hierarchical formation of UAVs. IEEE Access **8**, 95667–95680 (2020)
28. M.A. Lewis, K.H. Tan, High precision formation control of mobile robots using virtual structures. Auton. Robot. **4**(4), 387–403 (1997)
29. B. Zhang, X. Sun, S. Liu, X. Deng, Adaptive model predictive control with extended state observer for multi-UAV formation flight. Int. J. Adapt. Control Signal Process. **34**(10), 1341–1358 (2020)
30. M. Song, R.X Wei, M.L. Hu, UAV reconnaissance formation control method based on a virtual long aircraft. J. Syst. Eng. Electron. **32**(11), 2412–2415 (2010)

31. J. Zhang, J. Yan, P. Zhang, Multi-UAV formation control based on a novel back-stepping approach. IEEE Trans. Veh. Technol. **69**(3), 2437–2448 (2020)
32. X. Yu, L. Liu, Distributed formation control of nonholonomic vehicles subject to velocity constraints. IEEE Trans. Ind. Electron. **63**(2), 1289–1298 (2016)
33. A.S. Brandao, J.P. Barbosa, V. Mendoza, M. Sarcinelli-Filho, R. Carelli, A multi-layer control scheme for a centralized UAV formation, in *2014 International Conference on Unmanned Aircraft Systems*, Orlando, USA (2014)
34. B. Zhang, X. Sun, S. Liu, X. Deng, Adaptive differential evolution-based receding horizon control design for multi-UAV formation reconfiguration. Int. J. Control. Autom. Syst. **17**, 3009–3020 (2019)
35. W. Zheng, H. Wang, H. Ji, J. Wu, UAV formation flight and collision warning with centralized control of ground control station, *2019 IEEE International Conference on Unmanned Systems*, Beijing, China (2019)
36. C.H. Liu, Multiple UAV formations based on centralised control. J. Huazhong Univ. Sci. Technolog. (Nature Science Edition). **43**(1), 481–485 (2015)
37. W. Lin, Distributed UAV formation control using differential game approach. Aerosp. Sci. Technol. **35**, 54–62 (2014)
38. Y. Kuwata, J.P. How, Cooperative distributed robust trajectory optimization using receding horizon MILP. IEEE Trans. Control Syst. Technol. **19**(2), 423–431 (2011)
39. J. Zhang, P. Zhang, J. Yan, Distributed adaptive finite-time compensation control for UAV swarm with uncertain disturbances. IEEE Trans. Circuits Syst. I Regul. Pap. **68**(2), 829–841 (2021)
40. F. Liao, R. Teo, J.L. Wang, X. Dong, F. Lin, K. Peng, Distributed formation and reconfiguration control of VTOL UAVs. IEEE Trans. Control Syst. Technol. **25**(1), 270–277 (2017)
41. Z. Yu, Z. Liu, Y. Zhang, Y. Qu, C.Y. Su, Distributed finite-time fault-tolerant containment control for multiple unmanned aerial vehicles. IEEE Trans. Neural Networks Learn. Syst. **31**(6), 2077–2091 (2020)
42. Y.Q. Wen, R. Zhou, W.M. Wu, F.E. Kong, Decentralised optimal control of heterogeneous multi-UAV cooperative target tracking. J. Beijing University of Aeronautics and Astronautics **36**(5), 545–549 (2010)
43. H.B. Duan, Y.K. Shen, Y. Wang, D.L. Luo, N.F. Niu, A review of what's hot in UAVs in 2018. Sci. Technol. Rev. **37**(3), 82–90 (2019)
44. W. Xie, *PyeongChang Winter Olympics Opening Ceremony 1,218 drones performance system pre-recorded*. https://tech.huanqiu.com/article/9CaKrnK6FHm
45. T.Z. Muslimov, R.A. Munasypov, Consensus-based cooperative control of parallel fixed-wing UAV formations via adaptive backstepping. Aerosp. Sci. Technol. **109**, 106416 (2021)
46. F. Wang, Y. Gao, C. Zhou, Q. Zong, Disturbance observer-based backstepping formation control of multiple quadrotors with asymmetric output error constraints. Appl. Math. Comput. **415**, 126693 (2022)
47. J. Wang, L. Han, X. Dong, Q. Li, Z. Ren, Distributed sliding mode control for time-varying formation tracking of multi-UAV system with a dynamic leader. Aerosp. Sci. Technol. **111**, 106549 (2021)
48. X. Xiang, C. Liu, H. Su, Q. Zhang, On decentralized adaptive full-order sliding mode control of multiple UAVs. ISA Trans. **71**(2), 196–205 (2017)
49. B. Chen, C. Wang, M. Lee, Stochastic robust team tracking control of multi-UAV networked system under Wiener and Poisson random fluctuations. IEEE Trans. Cybern. **51**(12), 5786–5799 (2021)
50. S.N. Singh, R. Zhang, P. Chandler, S. Banda, Decentralized nonlinear robust control of UAVs in close formation. Int. J. Robust Nonlinear Control **13**(11), 1057–1078 (2003)
51. Y. Tang, Z. Xue, X. Liu, Q. Han, X. Tuo, Leader-following consensus control for multiple fixed-wing UAVs' attitude system with time delays and external disturbances. IEEE Access **7**, 169773–169781 (2019)
52. T.Z. Muslimov, R.A. Munasypov, Adaptive decentralized flocking control of multi-UAV circular formations based on vector fields and backstepping. ISA Trans. **107**, 143–159 (2020)

53. Z. Yu, Y. Zhang, B. Jiang, X. Yu, J. Fu, Y. Jin, T. Chai, Distributed adaptive fault-tolerant close formation flight control of multiple trailing fixed-wing UAVs. ISA Trans. **106**, 181–199 (2020)
54. L. Weiss, E.F. Infante, Finite time stability under perturbing forces and on product spaces. IEEE Trans. Autom. Control **12**(1), 54–59 (1967)
55. B. Tian, J. Cui, H. Lu, Z. Zuo, Q. Zong, Adaptive finite-time attitude tracking of quadrotors with experiments and comparisons. IEEE Trans. Ind. Electron. **66**(12), 9428–9438 (2019)
56. B. Tian, L. Liu, H. Lu, Z. Zuo, Q. Zong, Y. Zhang, Multivariable finite time attitude control for quadrotor UAV, theory and experimentation. IEEE Trans. Ind. Electron. **65**(3), 2567–2577 (2018)
57. M.A. Trujillo, H.M. Becerra, D.G. Gutiérrez, J. Ruiz-León, A. Ramírez-Trevino, Hierarchical task-based formation control and collision avoidance of UAVs in finite time. Eur. J. Control. **60**, 48–64 (2021)
58. B. Liu, A. Li, Y. Guo, C. Wang, Adaptive distributed finite-time formation control for multi-UAVs under input saturation without collisions. Aerosp. Sci. Technol. **120**, 107252 (2022)
59. X. Yu, J. Yang, S. Li, Finite-time path following control for small-scale fixed-wing UAVs under wind disturbances. J. Frankl. Inst. **357**(12), 7879–7903 (2020)
60. V. Andrieu, L. Praly, A. Astolfi. Homogeneous approximation, recursive observer design, and output feedback. SIAM J. Control. Optim. **47**(4), 1814–1850 (2008)
61. A. Polyakov, Nonlinear feedback design for fixed-time stabilization of linear control systems. IEEE Trans. Autom. Control **57**(8), 2106–2110 (2012)
62. Z. Zuo, L. Tie, A new class of finite-time nonlinear consensus protocols for multi-agent systems. Int. J. Control. **87**(2), 363–370 (2014)
63. S. Parsegov, A. Polyakov, P.S. Shcherbakov, Fixed-time consensus algorithm for multi-agent systems with integrator dynamics, in *Proceedings of the IFAC Workshop on Distributed Estimation and Control in Networked Systems*, Koblenz, Germany (2013)
64. Z. Zuo, Nonsingular fixed-time consensus tracking for second-order multi-agent networks. Automatica **54**, 305–309 (2015)
65. B. Tian, H. Lu, Z. Zuo, W. Yang, Fixed-time leader-follower output feedback consensus for second- order multiagent systems. IEEE Trans. Cybern. **49**(4), 1545–1550 (2019)
66. Z. Zuo, B. Tian, M. Defoort, Z. Ding, Fixed-time consensus tracking for multi-agent systems with high-order integrator dynamics. IEEE Trans. Autom. Control **63**(2), 563–570 (2017)
67. Z. Zuo, Q.L. Han, B. Ning, An explicit estimate for the upper bound of the settling time in fixed- time leader-following consensus of high-order multivariable multiagent systems. IEEE Trans. Ind. Electron. **66**(8), 6250–6259 (2019)
68. L. Cui, X. Hou, Z. Zuo, An adaptive fast super-twisting disturbance observer-based dual closed-loop attitude control with fixed-time convergence for UAV. J. Frankl. Inst. **359**(6), 2514–2540 (2022)
69. L. Yu, G. He, X. Wang, S. Zhao, Robust fixed-time sliding mode attitude control of tilt tri-rotor UAV in helicopter mode. IEEE Trans. Ind. Electron. **69**(10), 10322–10332 (2022)
70. Q. Chen, M. Tao, X. He, L. Tao, Fuzzy adaptive nonsingular fixed-time attitude tracking control of quadrotor UAVs. IEEE Trans. Aerosp. Electron. Syst. **57**(5), 2864–2877 (2021)
71. W. Gong, B. Li, Y Yang, H. Ban, B. Xiao, Fixed-time integral-type sliding mode control for the quadrotor UAV attitude stabilization under actuator failures. Aerosp. Sci. Technol. **95**, 105444 (2019)
72. K. Xia, H. Son, Adaptive fixed-time control of autonomous VTOL UAVs for ship landing operations. J. Frankl. Inst. **357**(10), 6175–6196 (2020)
73. K. Liu, R. Wang, S. Zheng, S. Dong, G. Sun, Fixed-time disturbance observer-based robust fault-tolerant tracking control for uncertain quadrotor UAV subject to input delay. Nonlinear Dyn. **107**, 2363–2390 (2022)
74. J. Chen, C. Hua, X. Guan, Image based fixed time visual servoing control for the quadrotor UAV. IET Control Theory Appl. **13**(18), 3117–3123 (2019)
75. Y. Song, Y. Wang, J. Holloway, M. Krstic, Time-varying feedback for robust regulation of normal- form nonlinear systems in prescribed finite time. Automatica **83**, 243–251 (2017)

76. Y. Wang, Y. Song, D.J. Hill, M. Krstic, Prescribed-time consensus and containment control of networked multiagent systems. IEEE Trans. Cybern. **49**(4), 1138–1147 (2019)
77. Y. Wang, Y. Song, Leader-following control of high-order multi-agent systems under directed graphs: Pre-specified finite time approach. Automatica **87**, 113–120 (2018)
78. J. Wang, X. Luo, W. Wong, X. Guan, Specified-time vehicular platoon control with flexible safe distance constraint. IEEE Trans. Veh. Technol. **68**(11), 10489–10503 (2019)
79. L. Xiong, S. Yang, P. Li, S. Huang, C. Wang, J. Wang, Discrete specified time consensus control of aggregated energy storage for load frequency regulation. Int. J. Electr. Power Energy Syst. **123**, 106224 (2020)
80. K. Liu, B. Liu, Optimization of smooth blasting parameters for mountain tunnel construction with specified control indices based on a GA and ISVR coupling algorithm. Tunn. Undergr. Space Technol. **70**, 363–374 (2017)
81. D. Xue, J. Yao, J. Wang, Y. Guo, X. Han, Formation control of multi-agent systems with stochastic switching topology and time-varying communication delays. IET Control Theory Appl. **7**(13), 1689–1698 (2013)
82. L. Ji, C. Fan, Y. Qiao, Formation control for multi-UAVs system based on event-triggered mechanism, in *2019 Chinese Control Conference*, Guangzhou, China (2019)
83. D. Jang, C.Y. Son, J. Yoo, H.J. Kim, K.H. Johansson, Efficient networked UAV control using event-triggered predictive control. IFAC-PapersOnLine **52**(15), 412–417 (2019)
84. L. Wei, M. Chen, T. Li, Dynamic event-triggered cooperative formation control for UAVs subject to time-varying disturbances. IET Control Theory Appl. **14**(17), 2514–2525 (2020)
85. W.W. Xun. *Research on UAV Group Distributed Control Technology Based on Triggered Communication Mechanism* (National University of Defense Technology, China, 2018)
86. A. Anta, P. Tabuada, To sample or not to sample: Self-triggered control for nonlinear systems. IEEE Trans. Autom. Control **55**(9), 2030–2042 (2010)
87. X. Wang, M.D. Lemmon, Self-triggering under state-independent disturbances. IEEE Trans. Automat. Contr. **55**(6), 1494–1500 (2010)
88. J. Wang, H. Zhang, K. Ma, Z. Liu, C.L.P. Chen, Neural adaptive self-triggered control for uncertain nonlinear systems with input hysteresis. IEEE Trans. Neural Networks Learn. Syst. **33**(11), 6206–6214 (2021)
89. W. Xu, S. Yang, J. Cao, Fully distributed self-triggered control for second-order consensus of multiagent systems. IEEE Trans. Syst. Man Cybern. Syst. **51**(6), 3541–3551 (2019)
90. D. Chen, X. Liu, W. Yu, L. Zhu, Q. Tang, Neural-network based adaptive self-triggered consensus of nonlinear multi-agent systems with sensor saturation. IEEE Trans. Network Sci. Eng. **8**(2), 1531–1541 (2021)
91. A. Cuenca, D.J. Antunes, A. Castillo, P. Garcia, B.A. Khashooei, W. Heemels, Periodic event-triggered sampling and dual-rate control for a wireless networked control system with applications to UAVs. IEEE Trans. Ind. Electron. **66**(4), 3157–3166 (2019)
92. S. Shao, M. Chen, J. Hou, Q. Zhao, Event-triggered-based discrete-time neural control for a quadrotor UAV using disturbance observer. IEEE/ASME Trans. Mechatron. **26**(2), 689–699 (2021)
93. V.T.J. Antonio, G. Adrien, A. Manuel, P. Jean-Christophe, C. Laurent, R. Damiano, T. Didier, Event-triggered leader-following formation control for multi-agent systems under communication faults: application to a fleet of unmanned aerial vehicles. J. Syst. Eng. Electron. **32**(5), 1014–1022 (2021)
94. J. Wang, C. Bi, D. Wang, Q. Kuang, C. Wang, Finite-time distributed event-triggered formation control for quadrotor UAVs with experimentation. ISA Trans. **126**, 585–596 (2022)
95. L. Cao, H. Ren, W. Meng, H. Li, R. Lu, Distributed event triggering control for six-rotor UAV systems with asymmetric time-varying output constraints. Sci. China Inf. Sci. **64**, 172213 (2021)
96. T. Wu, J. Wang, B. Tian, Periodic event-triggered formation control for multi-UAV systems with collision avoidance. Chin. J. Aeronaut. **35**(8), 193–203 (2022)

97. L. Wei, M. Chen, T. Li. Disturbance-observer-based formation-containment control for UAVs via distributed adaptive event-triggered mechanisms. J. Frankl. Inst. **358**(10), 5305–5333 (2021)
98. A. Cetinkaya, H. Ishii, T. Hatakawa, Effects of jamming attacks on wireless networked control systems under disturbance. IEEE Trans. Autom. Control **68**(2), 1223–1230 (2023)
99. Y.C. Chen, G.B. Liu, Z. Zhang, L. He, S. He, Improving physical layer security for multi-UAV systems against hybrid wireless attacks. IEEE Trans. Control Syst. Technol. **73**(5), 7034–17048 (2024)
100. B. Chen, C. Wang, M. Lee, Stochastic robust team tracking control of multi-UAV networked system under wiener and Poisson random fluctuations. IEEE Trans. Cybern. **51**(12), 5786–5799 (2021)
101. T. Yin, Z. Gu, X. Xie, Observer-based event-triggered sliding mode control for secure formation tracking of multi-UAV systems. IEEE Trans. Network Sci. Eng. **10**(2), 887–898 (2023)
102. D. Liu, S. Xue, B. Zhao, B. Luo, Q. Wei, Adaptive dynamic programming for control: A survey and recent advances. IEEE Trans. Syst. Man Cybern. Syst. **51**(1), 142–160 (2021)
103. P.J. Werbos, Z. Wang, Foreword-ADP: The key direction for future research in intelligent control and understanding brain intelligence. IEEE Trans. Syst. Man Cybern. B **38**(4), 898–900 (2008)
104. Q. Wei, D. Liu, G. Shi, Y. Liu, Multibattery optimal coordination control for home energy management systems via distributed iterative adaptive dynamic programming. IEEE Trans. Ind. Electron. **62**(7), 4203–4214 (2015)
105. S.M. Hung, S.N. Givigi, A. Noureldin, A Dyna-Q(λ) approach to flocking with fixed-wing UAVs in a stochastic environment, in *2015 IEEE International Conference on Systems, Man, and Cybernetics*, Hong Kong, China (2015)
106. S.M. Hung, S.N. Givigi, A Q-learning approach to flocking with UAVs in a stochastic environment. IEEE Trans. Cybern. **47**(1), 186–197 (2017)
107. W. Zhou, J. Li, Z. Liu, L.C. Shen, Improving multi-target cooperative tracking guidance for UAV swarms using multi-agent reinforcement learning. Chin. J. Aeronaut. **35**(7), 100–112 (2022)
108. H. Liu, Q. Meng, F. Peng, F.L. Lewis, Heterogeneous formation control of multiple UAVs with limited-input leader via reinforcement learning. Neurocomputing **412**, 63–71 (2020)
109. A. Nazarzadeh, M. Ghodrat, H.J. Marquez, Stability analysis for model-based event-triggered nonlinear control systems under DoS attacks. IEEE Trans. Control Network Syst. **10**(4), 2130–2140 (2023)
110. S.L. Hu, D. Yue, Q.L. Han, X. Xie, X. Chen, C. Dou, Observer-based event-triggered control for networked linear systems subject to denial-of-service attacks. IEEE Trans. Cybern. **50**(5), 1952–1964 (2020)
111. Z.Y. Wu, Y. Wang, J.L. Xiong, J. Xiong, M. Xie, QoS-dependent event-triggered control for UAVs on cognitive radio networks subject to deception attacks. IEEE Trans. Veh. Technol. **72**(9), 11389–11400 (2023)
112. R. Zhao, Z. Q. Zuo, Y.J. Wang, Event-triggered control for switched systems with denial-of-service attack. IEEE Trans. Autom. Control **67**(8), 4077–4090 (2022)
113. S.L. Hu, D. Yue, Z.H. Cheng, E. Tian, X. Xie, X. Chen, Co-design of dynamic event-triggered communication scheme and resilient observer-based control under aperiodic DoS attacks. IEEE Trans. Cybern. **51**(9), 4591–4601 (2021)

Chapter 2
UAV Dynamics Fundamentals

This chapter introduces the fixed-wing UAV dynamics for the subsequent cooperative control design. Since there exist several literatures containing fixed-wing UAV modeling, the corresponding UAV dynamics is directly demonstrated in this chapter. For more details about the modeling of fixed-wing UAV, the interested readers can refer to related literatures [1–3].

2.1 Definitions of Coordination Reference Frames for UAVs

To describe the six-degree-of-freedom (6-DoF) dynamics model of a fixed-wing UAV, this chapter introduces three coordinate frames: the inertial coordinate frame, the body coordinate frame, and the airflow coordinate frame [1], defined as follows:

(1) The inertial coordinate frame ($S_g - Ox_g y_g z_g$): The projection of the take-off point position of the UAV on the ground is selected as the coordinate origin O, with the Ox_g-axis oriented in a northward direction, the Oy_g-axis oriented in an eastward direction, and the Oz_g-axis determined according to the right-hand rule. Based on the geographic coordinate system, the velocity, attitude, and position of the UAV relative to the ground can be described.
(2) The body coordinate frame ($S_b - Ox_b y_b z_b$): The coordinate frame is firmly affixed to the UAV, with the origin situated at the center of mass of the UAV. The Ox_b-axis represents the longitudinal axis, which is parallel to the fuselage axis and is defined as forward. The $Oy_b g$-axis is perpendicular to the Ox_b-axis and points in the direction of the right flank. The Oz_b-axis is determined in accordance with the right-hand rule. The body coordinate frame allows for the description of the attitude of the UAV and the angular velocity of rotation relative to the ground.

M. Lv et al., *Unmanned Aerial Vehicles Formation Control and Safety Guarantee*, Cognitive Intelligence and Robotics, https://doi.org/10.1007/978-981-96-5829-9_2

(3) The airflow velocity coordinate frame ($S_a - Ox_a y_a z_a$): The origin is situated at the center of mass of the UAV. The Ox_a-axis is aligned with the airspeed vector, and forward is defined as positive. The Oz_a-axis is mutually perpendicular to the Ox_a-axis in the plane of symmetry of the UAV and is defined as pointing toward the belly of the aircraft as positive. The Oy_a-axis is perpendicular to the $Ox_a z_a$ plane in the direction determined by the right-hand rule.

In accordance with the aforementioned description of the three coordinate frames, we shall proceed to present the transformation relationship between said frames.

(1) Transformation between body coordinate frame and inertial coordinate frame: From the definition of the attitude angle of the UAV, it can be seen that the inertial coordinate frame is rotated along the Ox_g-, Oy_g- and Oz_g-axes by pitch angle θ, roll angle ϕ, and yaw angle ψ, respectively, in order to obtain the body coordinate frame. The transformation relationship between the two frames is given that

$$\boldsymbol{S}_b = \boldsymbol{R}_g^b \boldsymbol{S}_g, \tag{2.1}$$

where

$$\begin{aligned} \boldsymbol{R}_g^b &= \begin{bmatrix} 1 & 0 & 0 \\ 0 & \cos\phi & \sin\phi \\ 0 & -\sin\phi & \cos\phi \end{bmatrix} \begin{bmatrix} \cos\theta & 0 & -\sin\theta \\ 0 & 1 & 0 \\ \sin\theta & 0 & \cos\theta \end{bmatrix} \begin{bmatrix} \cos\psi & \sin\psi & 0 \\ -\sin\psi & \cos\psi & 0 \\ 0 & 0 & 1 \end{bmatrix} \\ &= \begin{bmatrix} \cos\theta\cos\psi & \cos\theta\sin\psi & -\sin\theta \\ \begin{pmatrix} \sin\theta\sin\phi\cos\psi \\ -\cos\phi\sin\psi \end{pmatrix} & \sin\theta\sin\phi\sin\psi + \cos\phi\cos\psi & \sin\phi\cos\theta \\ \begin{pmatrix} \sin\theta\cos\phi\cos\psi \\ +\sin\phi\sin\psi \end{pmatrix} & \sin\theta\cos\phi\sin\psi - \sin\phi\cos\psi & \cos\phi\cos\theta \end{bmatrix}. \end{aligned}$$

(2) Transformation between the body coordinate frame and the airflow coordinate frame: The definition of the airflow angle of the UAV reveals that the body coordinate frame is rotated by the attack angle α and the sideslip angle β, respectively, in order to obtain the airflow coordinate frame. The transformation relationship between the two is as follows:

$$\boldsymbol{S}_a = \boldsymbol{R}_b^a \boldsymbol{S}_b, \tag{2.2}$$

where

$$\begin{aligned} \boldsymbol{R}_b^a &= \begin{bmatrix} \cos\beta & \sin\beta & 0 \\ -\sin\beta & \cos\beta & 0 \\ 0 & 0 & 1 \end{bmatrix} \begin{bmatrix} \cos\alpha & 0 & \sin\alpha \\ 0 & 1 & 0 \\ -\sin\alpha & 0 & \cos\alpha \end{bmatrix} \\ &= \begin{bmatrix} \cos\alpha\cos\beta & \sin\beta & \sin\alpha\cos\beta \\ -\cos\alpha\sin\beta & \cos\beta & -\sin\alpha\sin\beta \\ -\sin\alpha & 0 & \cos\alpha \end{bmatrix}. \end{aligned}$$

2.2 Dynamics of UAVs

In a multi-UAV guidance control system, a distributed control strategy is commonly employed, with each UAV equipped with an independent guidance control system. This distributed guidance control system is typically divided into an outer-loop guidance system and an inner-loop velocity-attitude control system, as illustrated in references [2, 3]. The outer-loop guidance system calculates the necessary guidance commands based on real-time position and heading information, combined with a predefined reference trajectory, and transmits these commands to the inner-loop velocity-attitude control system. This allows the UAVs to follow the designated trajectory. The inner-loop control system stabilizes the flight attitude and ensures that the UAV tracks the specified speed. It is further divided into two subsystems: the velocity subsystem, which manages UAV speed by controlling engine thrust, and the attitude subsystem, which stabilizes the UAV's attitude by regulating the deflection angles of the elevator, aileron, and rudder. Figure 2.1 illustrates the 6-DoF model of the UAV.

2.2.1 *UAV Inner-Loop Velocity Dynamics*

For the ith UAV, $\boldsymbol{v}_i = [u_i, v_i, w_i]^{\mathrm{T}}$ denotes the flight velocity, i.e., the velocity vector in the body coordinate frame. By combining this with Newton's second law, one can obtain the advection equation [4, 5]:

$$\dot{\boldsymbol{v}}_i = -\boldsymbol{S}(\boldsymbol{\omega}_i)\,\boldsymbol{v}_i + \frac{\boldsymbol{T}_i}{m_i} + \boldsymbol{R}_1^{\mathrm{T}}(\boldsymbol{\varphi}_i)\,\boldsymbol{g} + \frac{\boldsymbol{F}_i}{m_i}, \tag{2.3}$$

Fig. 2.1 6-DoF UAV model

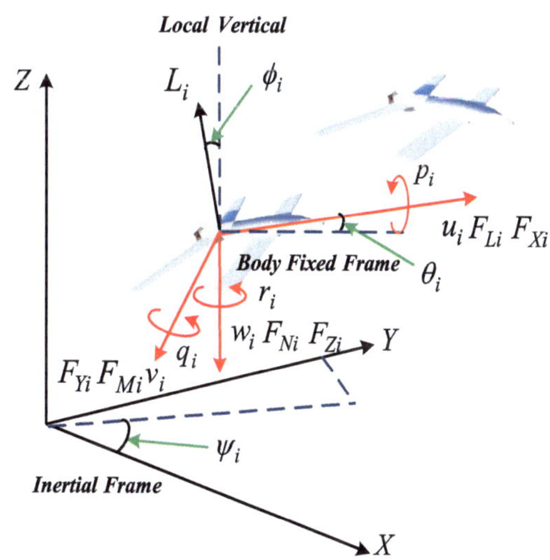

where m_l is the mass of the UAV, $\boldsymbol{\omega}_l = [p_l, q_l, r_l]^{\mathrm{T}}$ represents the angular velocity in the body coordinate frame, respectively. $\boldsymbol{T}_l = [T_{xl}, 0, 0]^{\mathrm{T}}$ is the thrust vector with T_{xl} being the actual engine thrust control quantity. Moreover, m_l is the mass of the UAV, and $\boldsymbol{g} = [0, 0, g]^{\mathrm{T}}$ is the gravity acceleration in inertial coordinate frame. $\boldsymbol{R}_1(\boldsymbol{\varphi}_l)$ denotes the matrix of transformation from the body coordinate frame to inertial coordinate frame, which is expressed as

$$\boldsymbol{R}_1 = \begin{bmatrix} \cos\psi_l\cos\theta_l & \begin{pmatrix} -\sin\psi_l\sin\phi_l \\ +\cos\psi_l\sin\theta_l\sin\phi_l \end{pmatrix} & \sin\psi_l\sin\phi_l + \cos\psi_l\sin\theta_l\cos\phi_l \\ \sin\psi_l\cos\theta_l & \begin{pmatrix} \cos\psi_l\cos\phi_l \\ +\sin\psi_l\sin\theta_l\sin\phi_l \end{pmatrix} & -\cos\psi_l\sin\phi_l + \sin\psi_l\sin\theta_l\cos\phi_l \\ -\sin\theta_l & \cos\theta_l\sin\phi_l & \cos\theta_l\cos\phi_l \end{bmatrix}. \tag{2.4}$$

$\boldsymbol{F}_l$ denotes the aerodynamic force on the UAV in the airflow coordinate system, expressed as

$$\begin{aligned} F_l &= \bar{q}_l S_l R_3(\alpha_l, \beta_l)^{-1}[-C_{Dl}, C_{Yl}, -C_{Ll}]^{\mathrm{T}} \\ &= \bar{q}_l S_l R_3(\alpha_l, \beta_l)^{-1} \begin{bmatrix} -\left(C_{D0l} + k_l C_{Ll}^2\right) \\ C_{Y\beta l}\beta_l \\ -\left(C_{L0l} + C_{L\alpha l}\alpha_l\right) \end{bmatrix}, \end{aligned} \tag{2.5}$$

where S_l is the wing area, and $\bar{q}_l$ is the dynamic pressure with $\bar{q}_l = \frac{1}{2}\bar{\rho}V_l^2$. $\bar{\rho}$ is the air density, and V_l is the scalar velocity corresponding to the velocity vector $\boldsymbol{v}_l$, which satisfies $V_l = \|\boldsymbol{v}_l\| = \sqrt{\boldsymbol{v}_l^{\mathrm{T}}\boldsymbol{v}_l}$. $\alpha_l = \arctan\left(\frac{w_l}{u_l}\right)$ and $\beta_l = \arcsin\left(\frac{v_l}{V_l}\right)$ denote the attack and sideslip angle, respectively. $\boldsymbol{C}_{(\cdot)}$ is the aerodynamic parameter. $\boldsymbol{R}_3(\alpha_l, \beta_l)$ indicates that the transformation matrix between the body coordinate system and the airflow coordinate system is expressed as follows:

$$\boldsymbol{R}_3(\alpha_l, \beta_l) = \begin{bmatrix} \cos\alpha_l\cos\beta_l & \sin\beta_l & \sin\alpha_l\cos\beta_i \\ -\cos\alpha_l\cos\beta_l & \cos\beta_l & -\sin\alpha_l\sin\beta_i \\ -\sin\alpha_l & 0 & \cos\alpha_i \end{bmatrix}. \tag{2.6}$$

According to (2.3), the derivation of the scalar velocity V_i yields that

$$\dot{V}_l = \frac{1}{V_l}\boldsymbol{v}_l^{\mathrm{T}}\dot{\boldsymbol{v}}_l = \frac{1}{V_l}\boldsymbol{v}_l^{\mathrm{T}}\left(-\boldsymbol{S}(\boldsymbol{\omega}_l)\boldsymbol{v}_l + \frac{\boldsymbol{T}_l}{m_l} + \boldsymbol{R}_1^{\mathrm{T}}(\boldsymbol{\varphi}_l)\,\boldsymbol{g} + \frac{\boldsymbol{F}_l}{m_l}\right). \tag{2.7}$$

In order to facilitate the resolution of the form of the controller, it is necessary to separate T_{xl} from $\boldsymbol{T}_l$. It is further obtained that

$$\dot{V}_l = \frac{1}{V_l}\boldsymbol{v}_l^{\mathrm{T}}\dot{\boldsymbol{v}}_l = \frac{u_l T_{xl}}{m_l V_l} + \frac{1}{V_l}\boldsymbol{v}_l^{\mathrm{T}}\left(\boldsymbol{R}_1^{\mathrm{T}}(\boldsymbol{\varphi}_l)\,\boldsymbol{g} + \frac{\boldsymbol{F}_i}{m_l}\right), \tag{2.8}$$

where $-\boldsymbol{v}_l^{\mathrm{T}}\boldsymbol{S}(\boldsymbol{\omega}_l)\boldsymbol{v}_l = 0$. In the case of actual UAV system, it is challenging to ascertain precise values for the aerodynamic parameters due to inherent uncertainties. Therefore, a reasonable approximation of the parameter uncertainty terms is essential. The aerodynamic parameter $\boldsymbol{F}_l$ can be decomposed into known term $\boldsymbol{F}_{l0}$ and uncertain term $\boldsymbol{\Delta F}_l$ [6].

Furthermore, actuator failure as an important factor affecting the stable flight of UAVs has been extensively investigated [7, 8]. The engine failure is considered as

$$T_{xl} = \rho_{Tl} T_{xl0} + T_{xlf}, \tag{2.9}$$

where T_{xl0} is the desired thrust, $\rho_{Tl} \in [0, 1]$ is an unknown actuator efficiency factor, and T_{xlf} is the bounded unknown stuck fault or bias fault. Note that (2.9) implies that following four cases:

(i) $\rho_{Tl} = 1$ and $T_{xlf} = 0$. This means the fault-free case.
(ii) $0 < \rho_{Tl} < 1$ and $T_{xlf} = 0$. This indicates the partial loss of effectiveness.
(iii) $\rho_{Tl} = 1$ and $T_{xlf} \neq 0$. This indicates the bias fault.
(iv) $\rho_{Tl} = 0$ and $T_{xlf} \neq 0$. This means that T_{xl} is stuck at the bounded time-varying function T_{xlf}.

Thus, with the actuator faults and modeling uncertainties, the velocity kinematics is reformulated as

$$\dot{V}_l = \frac{u_l \rho_{Tl} T_{xl0}}{m_l V_l} + \frac{u_l T_{xlf}}{m_l V_l} + \frac{1}{V_l}\boldsymbol{v}_l^{\mathrm{T}}\left(\boldsymbol{R}_1^{\mathrm{T}}(\boldsymbol{\varphi}_l)\boldsymbol{g} + \frac{\boldsymbol{F}_{l0}}{m_l} + \boldsymbol{\Delta}_{vl} + \boldsymbol{d}_{vl}\right), \tag{2.10}$$

where $\boldsymbol{d}_{vl}$ is the unknown bounded external disturbances, $\boldsymbol{\Delta}_{vl} = \frac{\boldsymbol{\Delta F}_l}{m_l}$.

Assumption 2.1 The unknown bounded external disturbances, aerodynamic parameter uncertainty, and its derivatives are bounded, i.e., $\|\boldsymbol{d}_{vl}\| \leq \bar{d}_{vl}$, $\|\boldsymbol{\Delta}_{vl}\| \leq \bar{\Delta}_{vl}$, and $\|\dot{\boldsymbol{\Delta}}_{vl}\| \leq \bar{\Delta}'_{vl}$, where $\bar{d}_{vl}$, $\bar{\Delta}_{vl}$, $\bar{\Delta}'_{vl}$ are the positive constants.

2.2.2 UAV Inner-Loop Attitude Dynamics

The equation governing the rotational kinematics of the attitude angle is given by [4, 5]

$$\dot{\boldsymbol{\varphi}}_l = \boldsymbol{R}_2^{-1}(\boldsymbol{\varphi}_l)\,\boldsymbol{\omega}_l, \tag{2.11}$$

where the matrix $\boldsymbol{R}_2^{-1}(\boldsymbol{\varphi}_l)$ maps the time derivative of the Euler angles to the noninertial expression of the angular velocity, and it is given as follows:

$$\boldsymbol{R}_2\left(\boldsymbol{\varphi}_i\right)=\begin{bmatrix}1 & 0 & -\sin\psi_i\\ 0 & \cos\psi_i & \sin\psi_i\cos\theta_i\\ 0 & -\sin\theta_i & \cos\psi_i\cos\theta_i\end{bmatrix}.$$

The angular acceleration of the UAV is derived from Euler's momentum equation as

$$\boldsymbol{J}_i\dot{\boldsymbol{\omega}}_i=-\boldsymbol{\omega}_i\times\boldsymbol{J}_i\boldsymbol{\omega}_i+\boldsymbol{M}_i, \tag{2.12}$$

where $\boldsymbol{M}_i$ is the aerodynamic moment of the UAV in the body coordinate frame, and the inertia tensor $\boldsymbol{J}_i$ is symmetric about the $x-z$ plane and can be represented as

$$\boldsymbol{J}_i=\begin{bmatrix}J_{xi} & 0 & -J_{xzi}\\ 0 & J_{yi} & 0\\ -J_{xzi} & 0 & J_{zi}\end{bmatrix},$$

where J_{xi}, J_{yi}, J_{zi}, and J_{xzi} denote constant value inertia coefficients.

To facilitate the solution of the rudder controller $\boldsymbol{\delta}_i=[\delta_{ai},\delta_{ei},\delta ri]^{\mathrm{T}}$, the aerodynamic moment $\boldsymbol{M}_i$ can be further expressed as

$$\begin{aligned}\boldsymbol{M}_i&=\boldsymbol{N}_i\left(V_i,\boldsymbol{\omega}_i,\alpha_i,\beta_i\right)+\boldsymbol{C}\left(V_i,\alpha_i,\beta_i\right)\delta_i\\ &=\boldsymbol{L}_i\left(V_i,\alpha_i,\beta_i\right)-\boldsymbol{D}\left(V_i,\alpha_i,\beta_i\right)\boldsymbol{\omega}_i+\boldsymbol{C}\left(V_i,\alpha_i,\beta_i\right)\boldsymbol{\delta}_i,\end{aligned} \tag{2.13}$$

where the dynamics vector $\boldsymbol{L}_i\left(V_i,\alpha_i,\beta_i\right)$ is defined as

$$L_i\left(V_i,\alpha_i,\beta_i\right)=\bar{q}_iS_i\begin{bmatrix}b_i\left(C_{l0i}+C_{l\beta i}\beta_i\right)\\ \bar{c}_i\left(C_{m0i}+C_{m\alpha i}\alpha_i\right)\\ b_i\left(C_{n0i}+C_{n\beta i}\beta_i\right)\end{bmatrix}, \tag{2.14}$$

where wingspan b_i and mean aerodynamic chord $\bar{c}_i$ are constant parameters. The damping matrix $\boldsymbol{D}\left(V_i,\alpha_i,\beta_i\right)$ is defined as

$$\boldsymbol{D}\left(V_i,\alpha_i,\beta_i\right)=-\bar{q}_iS_i\begin{bmatrix}\frac{b_i^2}{2V_i}C_{lpi} & 0 & \frac{b_i^2}{2V_i}C_{lri}\\ 0 & \frac{\bar{c}_i^2}{2V_i}C_{mqi} & 0\\ \frac{b_i^2}{2V_i}C_{npi} & 0 & \frac{b_i^2}{2V_i}C_{nri}\end{bmatrix}. \tag{2.15}$$

The control gain matrix $\boldsymbol{C}_i\left(V_i,\alpha_i,\beta_i\right)$ is defined by

$$\boldsymbol{C}_i=\bar{q}_iS_i\begin{bmatrix}b_iC_{l\delta ai} & 0 & b_iC_{l\delta ri}\\ 0 & \bar{c}_iC_{m\delta ei} & 0\\ b_iC_{n\delta ai} & 0 & b_iC_{n\delta ri}\end{bmatrix}. \tag{2.16}$$

According to (2.12)–(2.16), it obtains that

$$\dot{\boldsymbol{\omega}}_i = \boldsymbol{J}_i^{-1}\boldsymbol{S}(\boldsymbol{J}_i\boldsymbol{\omega}_i) + \boldsymbol{J}_i^{-1}\boldsymbol{N}_i + \boldsymbol{J}_i^{-1}\boldsymbol{C}\boldsymbol{\delta}_i. \tag{2.17}$$

Considering the coefficient uncertainties, $\boldsymbol{J}_i^{-1}\boldsymbol{S}(\boldsymbol{J}_i\boldsymbol{\omega}_i)$ is decomposed into a known component $\boldsymbol{J}_{i0}^{-1}\boldsymbol{S}(\boldsymbol{J}_{i0}\boldsymbol{\omega}_i)$ and an uncertain one $\boldsymbol{\Delta}\boldsymbol{J}_i^{-1}\boldsymbol{S}(\boldsymbol{J}_i\boldsymbol{\omega}_i)$. Similarly, $\boldsymbol{J}_i^{-1}\boldsymbol{N}_i$ is composed of a known part $\boldsymbol{J}_{i0}^{-1}\boldsymbol{N}_{i0}$ and an uncertain one $\boldsymbol{\Delta}\boldsymbol{J}_i^{-1}\boldsymbol{N}_i$. Also, $\boldsymbol{J}_i^{-1}\boldsymbol{C}$ consists of a known component $\boldsymbol{J}_{i0}^{-1}\boldsymbol{C}_0$ and an uncertain one $\boldsymbol{\Delta}\boldsymbol{J}_i^{-1}\boldsymbol{C}$.

Meanwhile, the flight control surfaces inevitably suffer from actuator faults, which are considered as

$$\delta_{mi} = \rho_{\delta mi}\delta_{mi0} + \delta_{mif}, \tag{2.18}$$

where δ_{mi0} is the mth applied control input with $m = \{a, e, r\}$, $\rho_{\delta mi}$ is the unknown actuator efficiency factor satisfying $0 \le \rho_{\delta mi} \le 1$, and δ_{mif} is the bounded unknown stuck fault or bias fault. Consider that (2.18) satisfies the similar cases as (2.9) in the following:

(i) $\rho_{\delta mi} = 1$ and $\delta_{mif} = 0$. This means the fault-free case.
(ii) $0 < \rho_{\delta mi} < 1$ and $\delta_{mif} = 0$. This indicates the partial loss of effectiveness.
(iii) $\rho_{\delta mi} = 1$ and $\delta_{mif} \neq 0$. This indicates the bias fault.
(iv) $\rho_{\delta mi} = 0$ and $\delta_{mif} \neq 0$. This means that δ_{mi} is stuck at the bounded time-varying function δ_{mif}.

According to (2.18), the rudder fault model is obtained as

$$\boldsymbol{\delta}_i = \boldsymbol{\rho}_{\delta i}\boldsymbol{\delta}_{i0} + \boldsymbol{\delta}_{if}, \tag{2.19}$$

where $\boldsymbol{\delta}_{i0} = [\delta_{ai0}, \delta_{ei0}, \delta_{ri0}]^{\mathrm{T}}$, $\boldsymbol{\rho}_{\delta i} = \mathrm{diag}\{\rho_{\delta ai}, \rho_{\delta ei}, \rho_{\delta ri}\}$, $\boldsymbol{\delta}_{if} = \left[\delta_{aif}, \delta_{eif}, \delta_{rif}\right]^{\mathrm{T}}$.

Assumption 2.2 The fault bias disturbance signal, T_{xif} and δ_{mif}, is bounded, and there exist two positive constants $\bar{T}_{xif}$, $\bar{\delta}_{mif}$ such that $|T_{xif}| \le \bar{T}_{xif}$, $|\delta_{mif}| \le \bar{\delta}_{mif}$.

Accordingly, in the presence of modeling uncertainties and actuator faults, the dynamic model of the angular velocity can be formulated by

$$\dot{\boldsymbol{\omega}}_i = \boldsymbol{J}_{i0}^{-1}\boldsymbol{S}(\boldsymbol{J}_{i0}\boldsymbol{\omega}_i) + \boldsymbol{J}_{i0}^{-1}\boldsymbol{N}_{i0} + \boldsymbol{J}_{i0}^{-1}\boldsymbol{C}_0\boldsymbol{\rho}_{\delta i}\boldsymbol{\delta}_{i0} + \boldsymbol{\Delta}_{\omega i} + \boldsymbol{d}_{\omega i}, \tag{2.20}$$

where $\boldsymbol{\Delta}_{\omega i} = \boldsymbol{\Delta}\boldsymbol{J}_i^{-1}\boldsymbol{S}(\boldsymbol{J}_i\boldsymbol{\omega}_i) + \boldsymbol{\Delta}\boldsymbol{J}_i^{-1}\boldsymbol{N}_i + \boldsymbol{\Delta}\boldsymbol{J}_i^{-1}\boldsymbol{C}\boldsymbol{\delta}_i + \boldsymbol{J}_{i0}^{-1}\boldsymbol{C}_0\boldsymbol{\delta}_{if}$, $\boldsymbol{d}_{\omega i}$ is the unknown bounded external disturbance.

Assumption 2.3 The unknown bounded external disturbances, aerodynamic parameter uncertainty, and its derivatives are bounded, i.e., $\|\boldsymbol{d}_{\omega i}\| \le \bar{d}_{\omega i}$, $\|\boldsymbol{\Delta}_{\omega i}\| \le \bar{\Delta}_{\omega i}$, and $\|\dot{\boldsymbol{\Delta}}_{\omega i}\| \le \bar{\Delta}'_{\omega i}$, where $\bar{d}_{\omega i}$, $\bar{\Delta}_{\omega i}$, $\bar{\Delta}'_{\omega i}$ are the positive constants.

2.2.3 *UAV Outer-Loop Position Dynamics*

In order to analyze the dynamics of the outer-loop position of the UAV, it is necessary to establish the positional motion equations which represent the equation relationship between the three-dimensional position of the UAV in the inertial coordinate frame and the flight speed, attitude angle, and flight path angle. The kinematics equations of the UAV's outer loop can be derived as [9]

$$\begin{cases} \dot{x}_l = V_l \cos\gamma_l \cos\psi_l + w_{xl}, \\ \dot{y}_l = V_l \cos\gamma_l \sin\psi_l + w_{yl}, \\ \dot{z}_l = V_l \sin\gamma_l + w_{zl}, \end{cases} \tag{2.21}$$

where x_l, y_l, and z_l denote the UAV three-dimensional position coordinates in the inertial coordinate frame, and w_{xl}, w_{yl}, and w_{zl} denote the disturbance components along the Ox_g-, Oy_g-, and Oz_g-axes of the combined disturbances including gust disturbances and electromagnetic disturbances.

2.3 Conclusions

This chapter has introduced the UAV reference coordinate frame, along with the inner-loop velocity and attitude model and the outer-loop position model. These foundational concepts will serve as a basis for addressing the UAV cooperative control problem in the following chapters.

References

1. B.L. Stevens, F.L. Lewis, E.N. Johnson, *Aircraft Control and Simulation: Dynamics, Controls Design, and Autonomous Systems* (Wiley, New York, 2015)
2. S.U. Ali, R. Samar, M.Z. Shah, A.I. Bhatti, K. Munawar, U.M. Al-Sggaf, Lateral guidance and control of UAVs using second-order sliding modes. Aerosp. Sci. Technol. **49**, 88–100 (2016)
3. R. Rysdyk, Unmanned aerial vehicle path following for target observation in wind. J. Guid. Control. Dyn. **29**(5), 1092–1100 (2006)
4. E. Oland, R. Schlanbusch, R. Kristiansen, Underactuated waypoint tracking of a fixed-wing UAV. IFAC Proceedings Volumes **46**(30), 126–133 (2013)
5. H. Castañeda, O.S. Salas-Peña, J. de León-Morales, Extended observer based on adaptive second order sliding mode control for a fixed wing UAV. ISA Trans. **66**, 226–232 (2017)
6. Y.X. Wang, T. Zhang, Z.H. Cai, Multi-UAV coordination control by chaotic grey wolf optimization based distributed MPC with event-triggered strategy. Chin. J. Aeronaut. **33**(11), 2877–2897 (2020)
7. H. Wang, P.X. Liu, X. Zhao, X. Liu, Adaptive fuzzy finite-time control of nonlinear systems with actuator faults. IEEE Trans. Cybern. **50**(5), 1786–1797 (2019)
8. X.Yu, P. Li, Y. Zhang, The design of fixed-time observer and finite-time fault-tolerant control for hypersonic gliding vehicles. IEEE Trans. Ind. Electron. **65**(5), 4135–4144 (2017)
9. J. Yang, C. Liu, M. Coombes, Y. Yan, W.H. Chen. Optimal path following for small fixed-wing UAVs under wind disturbances. IEEE Trans. Control Syst. Technol. **29**(3), 996–1008 (2020)

Part II
Advanced Control Systems for UAVs

Chapter 3
Fixed-Time Fault-Tolerant Control

In the process of executing tasks for multiple UAVs, the flight missions impose strict requirements on the convergence time, as an important indicator of control performance for UAV system. Currently, most control algorithms of UAVs are based on asymptotic stable tracking [1–3], which makes it difficult to satisfy the rapid convergence for UAV systems. Therefore, this chapter focuses on enhancing convergence speed of multi-UAV systems with fixed-time control method.

3.1 Introduction

Coordinated flight of multiple fixed-wing unmanned aerial vehicles (UAVs) has been extensively studied over the past decades, due to its important role in achieving flexibility and cost-effectiveness of mission cooperation [4, 5]. In general, coordinated control problem has been investigated in several aspects, including formulating and maintaining a particular formation shape for UAVs in a group [6] and synchronous tracking of velocity and attitude for UAV formation [7]. Fixed-wing UAVs are peculiar in terms of six-degree-of-freedom (6-DoF) dynamics, and these peculiarities should be borne in mind when formulating a control law.

As technologies advance, the distributed control strategy has been attached tremendous attentions and is desirable for UAV groups due to the fact that the global information is no longer required in control design while providing the scalability and tackling the vulnerability of centralized UAVs tasked to control a group [8, 9]. Driven by the emerging distributed cooperative control research on multiple UAVs, some results have appeared in [10–15]. To list a few, Yu et al. [10] presented a distributed cooperative control design for networked UAVs against actuator faults and model uncertainties. In [11], for longitudinal synchronization tracking of multiple UAVs, a distributed cooperative fault-tolerant controller was proposed in the presence of input saturation. Based on the swarm intelligence,

M. Lv et al., *Unmanned Aerial Vehicles Formation Control and Safety Guarantee*,
Cognitive Intelligence and Robotics, https://doi.org/10.1007/978-981-96-5829-9_3

Wang et al. [12] proposed a distributed model predictive approach for coordination control of multiple UAVs. In [13], with path-following vector fields, the UAV group achieves a circular motion around the target. Notably, most existing studies on the problem analyze UAV dynamics models as material points [12–15] and focus on the attitude or longitudinal dynamics [10, 11], which is an excessive and unnecessary simplification in practice. Therefore, the motivation of investigation on 6-DoF UAV dynamics naturally arises.

In ideal conditions, the aforementioned control methods can achieve good control results with asymptotic convergence. However, in real flight conditions, there is a strict requirement for convergence time, which mainly influences the system performances. The advanced algorithm to improve the efficiency of tracking convergence is finite-time control design. This technique has been successfully applied to UAV flights in [1–3]. But the main disadvantages of finite-time controls are as follows: (i) The convergence time strongly relies on the initial states, and (ii) considering the settling time increases with the initial states, the finite-time controls will be infeasible with large initial states. In [16], the fixed-time control is firstly developed to achieve the finite-time convergence independent of initial condition. This feature successfully promotes the fixed-time stability for spacecrafts [17], surface vessels [18], surface vehicles [19], and robot manipulators [20]. However, the crucial question of designing the fixed-time control scheme for multiple UAVs still remains open.

In practical cases, state constraints are commonly found in UAV system in the form of physical stoppage performance. The constraint consideration has been significantly important for UAVs because it ensures the avoidance of collision hazards and stability simultaneously. Some significant achievements on constrained control problem have been obtained in recent years, including model predictive control [21], use of set invariance [22], and Barrier Lyapunov Function [23]. Beyond these, the non-Barrier Lyapunov Function was proposed to handle constraints in [24, 25], where the constraints are guaranteed not transgressed by ensuring the boundedness of the proposed state-dependent function. Additionally, for UAVs during flight, negative factors including actuator faults, external disturbances, together with model uncertainties will degrade the control performance or even make the system unstable if being ignored. Therefore, there is an urgent need to take into account all negative factors when designing flight controllers to ensure the stability and tracking accuracy for UAV system. Some literatures pay attention to fault-tolerant control and disturbance rejection control for UAVs and gain some achievements in [26–29].

Lots of works deal with the control of fixed-wing UAV in terms of the linearized model or separate attitude and longitudinal dynamics [10–15]. On the one hand, this linearized model cannot accurately reflect the motion of a UAV and the anti-disturbance and fault-tolerant ability is limited; on the other hand, there can appear problems related to the transient tracking errors which can further lead to the failure of flight. Besides, due to the presence of partially unknown nonlinearities in aircraft dynamics and the nonlinear behavior of actuators depending on different flight conditions, the adaptive controllers with anti-disturbance and fault-tolerant ability

Table 3.1 Control models and control methods for existing works

Control model	Existing works	Control method	Existing works
2-DOF model	[6, 12–15, 21]	Fault-tolerant control	[1, 3, 7, 10, 11, 26–29]
3-DOF model	[2, 30]	Constrained control	[1, 3, 6, 11–15, 21, 29, 30]
Longitudinal model	[3, 11]	Adaptive control	[1, 2, 7, 10, 11, 13, 15, 28–30]
Attitude model	[1, 7, 10, 26, 29]	Finite-time control	[1–3, 29]
6-DOF model	[27, 28, 31, 32]	Fixed-time control	[16–20]

are better choices than conventional controllers. Although extensive studies have been conducted for multiple UAVs and achieved accurate tracking performance in ideal flight, there are some difficulties that need to be addressed: (i) Many studies in the research area of fixed-wing UAVs merely deal with the asymptotic convergence regardless of the requirement of convergence time, and (ii) the time-varying state-constrained problem of multi-UAV system is rarely considered so far.

Motivated by the aforementioned existing works whose control models and control methods have been summarized in Table 3.1, the main contributions of this chapter are stated as follows:

- Differently from asymptotic control for 6-DoF dynamics [27, 28, 31, 32] and finite-time control for 3-DoF model [2], longitudinal dynamics [3], and attitude dynamics [1, 29], our scheme extends the design to adaptive fixed-time fault-tolerant control scheme for 6-DoF UAVs.
- Unlike most nonadaptive controls of UAV flight [6, 12–14, 21, 31, 32], an adaptive backstepping control is proposed with the help of bound estimation approach and well-defined smooth functions to effectively compensate for the actuator faults, external disturbances, and model uncertainties in the same time with the guarantee of global stability.
- The scaling functions are designed to address the full-state constraints with respect to the system transformation obtained for translational and rotational dynamics, respectively. The fixed-time properties are obtained in such transformed unconstrained subsystems.

The rest of the chapter is organized as follows. In Sect. 3.2, the problem statements and preliminaries are presented. Section 3.3 gives the cooperative distributed fixed-time control design and stability analysis. Section 3.4 shows the validation results, and Sect. 3.5 draws the conclusions.

Remark The main control schemes and contents of this chapter are from the published journal paper "B. Y. Zhang, X. X. Sun, M. L. Lv, S. G. Liu, L. Li, Distributed adaptive fixed-time fault-tolerant control for multiple 6-DOF UAVs with full-state constraints guarantee. IEEE Systems Journal. 2021, 16(3): 4792–4803." The authors appreciate the permission from Institute of Electrical and Electronics Engineers to reuse the results published in the relevant journal.

3.2 Problem Formulation

3.2.1 Basic Graph Theory

Considering a multi-UAV system consisting of N fixed-wing UAVs, an undirected graph $\mathcal{G} = (\mathcal{V}, \mathcal{E}, \mathcal{A})$ is concerned. The set of all UAVs is denoted by $\mathcal{V} = \{v_1, \cdots, v_N\}$ with the edges $\mathcal{E} = \{(i, j),\ i, j \in \mathcal{V}, \text{and } i \neq j\}$. Let $\mathcal{A} = [a_{ij}] \in \Re^{N \times N}$ denote the weighted adjacency matrix of $\mathcal{G}$. The Laplacian matrix is defined by $\mathcal{L} = \mathcal{D} - \mathcal{A}$, where $\mathcal{D} = \text{diag}\{d_1, \cdots, d_N\}$ with $d_i = \sum_{j=1}^{N} a_{ij}$. Given the leader-following structure, let $\mathcal{H} = \text{diag}\{h_1, \cdots, h_N\}$ denote the leader adjacency matrix of the graph $\mathcal{G}$. If the jth follower receives the information from the leader UAV, then $h_j > 0$, and otherwise $h_j = 0$. The neighbors of each UAV are denoted by $\mathcal{N}_i = \{v_j : (v_j, v_i) \in \mathcal{E}\}$.

Assumption 3.1 The topology of considered multi-UAV system is an undirected graph with at least one node having access to the leading node.

3.2.2 UAV Dynamic Model

Using Newton-Euler convention, the full 6-DoF dynamics of each UAV is given by [31, 32]

$$\dot{\boldsymbol{p}}_i = \boldsymbol{R}_1 \left(\boldsymbol{\varphi}_i\right) \boldsymbol{v}_i \tag{3.1}$$

$$\dot{\boldsymbol{v}}_i = - \boldsymbol{S} \left(\boldsymbol{\omega}_i\right) \boldsymbol{v}_i + \frac{\boldsymbol{T}_i}{m_i} + \boldsymbol{R}_1^{\mathrm{T}} \left(\boldsymbol{\varphi}_i\right) \boldsymbol{g} + \frac{\boldsymbol{F}_i}{m_i} + \boldsymbol{d}_{vi} \tag{3.2}$$

$$\dot{\boldsymbol{\varphi}}_i = \boldsymbol{R}_2^{-1} \left(\boldsymbol{\varphi}_i\right) \boldsymbol{\omega}_i \tag{3.3}$$

$$\boldsymbol{J}_i \dot{\boldsymbol{\omega}}_i = - \boldsymbol{\omega}_i \times \boldsymbol{J}_i \boldsymbol{\omega}_i + \boldsymbol{N}_i + \boldsymbol{C} \left(\boldsymbol{\delta}_i\right) \boldsymbol{\delta}_i + \boldsymbol{J}_i \boldsymbol{d}_{\omega i}, \tag{3.4}$$

where $\boldsymbol{p}_i = [x_i, y_i, z_i]^{\mathrm{T}}$ denotes the inertial position, $\boldsymbol{v}_i = [u_i, v_i, w_i]^{\mathrm{T}}$ is the linear velocity in body fixed frame, $\boldsymbol{\varphi}_i = [\phi_i, \theta_i, \psi_i]^{\mathrm{T}}$ represents the attitude described by Euler angles, and $\boldsymbol{\omega}_i = [p_i, q_i, r_i]^{\mathrm{T}}$ is the non-inertial expression of the angular velocity. Moreover, $\boldsymbol{T}_i = [T_{xi}, 0, 0]^{\mathrm{T}}$ denotes the thrust vector along x body axis, $\boldsymbol{\delta}_i = [\delta_{ai}, \delta_{ei}, \delta_{ri}]^{\mathrm{T}}$ is the control input, where δ_{ei}, δ_{ai}, δ_{ri} represent the deflection of elevator, ailerons, and rudder, respectively, $\boldsymbol{g} = [0, 0, g_z]^{\mathrm{T}}$ is the gravity acceleration in inertia frame, $\boldsymbol{d}_{vi} \in \Re^3$ and $\boldsymbol{d}_{\omega i} \in \Re^3$ are the unknown bounded external disturbances, and the inertia tensor $\boldsymbol{J}_i$ is symmetric about the $x - z$ plane and can be represented as

$$\boldsymbol{J}_i = \begin{bmatrix} J_{xi} & 0 & -J_{xzi} \\ 0 & J_{yi} & 0 \\ -J_{xzi} & 0 & J_{zi}. \end{bmatrix} \tag{3.5}$$

Besides, $\boldsymbol{R}_1(\boldsymbol{\varphi}_i) \in \mathrm{SO}(3)$ is the rotation matrix transforming the body frame coordinates to inertial axis coordinates, and the matrix $\boldsymbol{R}_2(\boldsymbol{\varphi}_i) \in \Re^{3\times 3}$ maps the time derivative of the Euler angles to the non-inertial expression of the angular velocity. Both matrices are given as follows:

$$\boldsymbol{R}_1(\boldsymbol{\varphi}_i) = \begin{bmatrix} c_{\psi_i} c_{\theta_i} & -s_{\psi_i} s_{\phi_i} + c_{\psi_i} s_{\theta_i} s_{\phi_i} & s_{\psi_i} s_{\phi_i} + c_{\psi_i} s_{\theta_i} c_{\phi_i} \\ s_{\psi_i} c_{\theta_i} & c_{\psi_i} c_{\phi_i} + s_{\psi_i} s_{\theta_i} s_{\phi_i} & -c_{\psi_i} s_{\phi_i} + s_{\psi_i} s_{\theta_i} c_{\phi_i} \\ -s_{\theta_i} & c_{\theta_i} s_{\phi_i} & c_{\theta_i} c_{\phi_i} \end{bmatrix}$$

$$\boldsymbol{R}_2(\boldsymbol{\varphi}_i) = \begin{bmatrix} 1 & 0 & -s_{\psi_i} \\ 0 & c_{\psi_i} & s_{\psi_i} c_{\theta_i} \\ 0 & -s_{\theta_i} & c_{\psi_i} c_{\theta_i} \end{bmatrix}, \tag{3.6}$$

where s_a and c_a stand for $\sin(a)$ and $\cos(a)$ functions, respectively. The aerodynamics force $\boldsymbol{F}_i = [F_{Xi}, F_{Yi}, F_{Zi}]^{\mathrm{T}}$ and moment $\boldsymbol{N}_i = [F_{Li}, F_{Mi}, F_{Ni}]^{\mathrm{T}}$ are calculated by means of aerodynamic coefficients:

$$\boldsymbol{F}_i = \bar{q}_i S_i \boldsymbol{R}_3^{-1}(\alpha_i, \beta_i)[-C_{Di}, C_{Yi}, -C_{Li}]^{\mathrm{T}} \tag{3.7}$$

$$\boldsymbol{N}_i = \bar{q}_i S_i \left[b_i C'_{li}, \bar{c}_i C'_{Mi}, b_i C'_{ni}\right]^{\mathrm{T}}, \tag{3.8}$$

where $\alpha_i = \arctan(w_i/u_i)$ and $\beta_i = \arcsin(u_i/v_i)$ denote the attack angle and sideslip angle, respectively. The dynamic pressure $\bar{q}_i = \frac{1}{2}\rho V_i^2$ is a function of total velocity $V_i = \sqrt{u_i^2 + v_i^2 + w_i^2}$. Air density ρ, wingspan b_i, wing surface area S_i, and mean aerodynamic chord $\bar{c}_i$ are constant parameters. The transformation matrix $\boldsymbol{R}_3(\alpha_i, \beta_i)$ is

$$\boldsymbol{R}_3(\alpha_i, \beta_i) = \begin{bmatrix} c_{\alpha_i} c_{\beta_i} & s_{\beta_i} & s_{\alpha_i} c_{\beta_i} \\ -c_{\alpha_i} s_{\beta_i} & c_{\beta_i} & -s_{\alpha_i} s_{\beta_i} \\ -s_{\alpha_i} & 0 & c_{\alpha_i}, \end{bmatrix} \tag{3.9}$$

and C_{Di}, C_{Yi}, C_{Li}, C'_{li}, C'_{Mi}, and C'_{ni} are the dimensionless coefficients in force/moment expressions, of which the corresponding detailed descriptions are shown in [32]. The control effectiveness matrix is defined as

$$\boldsymbol{C}(\boldsymbol{\delta}_i) = \begin{bmatrix} \bar{q}_i S_i b_i c_{l\delta_{ai}} & 0 & \bar{q}_i S_i b_i c_{l\delta_{ri}} \\ 0 & \bar{q}_i S_i \bar{c}_i c_{m\delta_{ei}} & 0 \\ \bar{q}_i S_i b_i c_{n\delta_{ai}} & 0 & \bar{q}_i S_i b_i c_{n\delta_{ri}} \end{bmatrix}. \tag{3.10}$$

3.2.3 Control-Oriented Model

According to studies [31, 32], the whole system can be decomposed into two interconnected subsystems, i.e., translational kinematics (3.2) and rotational one

(3.3)–(3.4). From (3.2), the thrust is extracted, and using the property $\boldsymbol{v}_l^{\mathrm{T}} \boldsymbol{S}(\boldsymbol{\omega}_l)\, \boldsymbol{v}_l = 0$, it holds that

$$\dot{V}_l = \frac{\boldsymbol{v}_l^{\mathrm{T}} \dot{\boldsymbol{v}}_l}{V_l} = \frac{u_l T_{xl}}{m_l V_l} + \frac{\boldsymbol{v}_l^{\mathrm{T}}}{V_l}\left(\boldsymbol{R}_1^{\mathrm{T}}(\boldsymbol{\varphi}_l)\, \boldsymbol{g} + \frac{\boldsymbol{F}_l}{m_l} + \boldsymbol{d}_{vl}\right), \tag{3.11}$$

which satisfies $V_{l,\min}(t) < V_l(t) < V_{l,\max}(t)$ for $\forall t \geq 0$.

During operation, the thrust possibly suffers from actuator faults, modeled by [33]

$$T_{xl} = \rho_{T_l} T_{xl0} + T_{xlf}, \tag{3.12}$$

where T_{xl0} is the designed control input, ρ_{T_l} is the unknown actuator efficiency factor satisfying $0 \leq \rho_{T_l} \leq 1$, and T_{xlf} is the bounded unknown stuck fault or bias fault. Note that (3.12) implies the following four cases:

(i) $\rho_{T_l} = 1$ and $T_{xlf} = 0$. This means the fault-free case.
(ii) $0 < \rho_{T_l} < 1$ and $T_{xlf} = 0$. This indicates the partial loss of effectiveness.
(iii) $\rho_{T_l} = 1$ and $T_{xlf} \neq 0$. This indicates the bias fault.
(iv) $\rho_{T_l} = 0$ and $T_{xlf} \neq 0$. This means that T_{xl} is stuck at bounded time-varying function T_{xlf}.

Moreover, note that the accurate information of $\boldsymbol{F}_l$ cannot be known a prior due to the coefficient uncertainties. In this sense, $\boldsymbol{F}_l$ is decomposed into a known component $\boldsymbol{F}_{l0}$ and an uncertain one $\Delta \boldsymbol{F}_l$ [12]. Thus, with the actuator faults and modeling uncertainties, the velocity kinematics is reformulated as

$$\dot{V}_l = \frac{u_l \rho_{T_l} T_{xl0}}{m_l V_l} + \frac{u_l T_{xlf}}{m_l V_l} + \frac{\boldsymbol{v}_l^{\mathrm{T}}}{V_l}\left(\boldsymbol{R}_1^{\mathrm{T}}(\boldsymbol{\varphi}_l)\, \boldsymbol{g} + \frac{\boldsymbol{F}_{l0}}{m_l} + \frac{\Delta \boldsymbol{F}_i}{m_i} + \boldsymbol{d}_{vl}\right). \tag{3.13}$$

From (3.3) and (3.4), the dynamical attitude motion is rewritten as

$$\dot{\boldsymbol{\varphi}}_l = \boldsymbol{R}_2^{-1}(\boldsymbol{\varphi}_l)\, \boldsymbol{\omega}_l \tag{3.14}$$

$$\dot{\boldsymbol{\omega}}_l = \boldsymbol{J}_l^{-1} \boldsymbol{S}(\boldsymbol{J}_l \boldsymbol{\omega}_l) + \boldsymbol{J}_l^{-1} \boldsymbol{N}_l + \boldsymbol{J}_l^{-1} \boldsymbol{C}(\boldsymbol{\delta}_l)\, \boldsymbol{\delta}_l + \boldsymbol{d}_{\omega i}, \tag{3.15}$$

which satisfies $-\boldsymbol{\varphi}_{l,\min}(t) < \boldsymbol{\varphi}_l(t) < \boldsymbol{\varphi}_{l,\max}(t)$ and $-\boldsymbol{\omega}_{i,\min}(t) < \boldsymbol{\omega}_l(t) < \boldsymbol{\omega}_{l,\max}(t)$ for $\forall t \geq 0$.

During flight, the flight control surfaces inevitably suffer from actuator faults, which are considered as [1]

$$\delta_{ml} = \rho_{\delta_{ml}} \delta_{ml0} + \delta_{mlf}, \tag{3.16}$$

where δ_{ml0} is the mth applied control input with $m = \{a, e, r\}$, $\rho_{\delta_{ml}}$ is the unknown actuator efficiency factor satisfying $0 \leq \rho_{\delta_{ml}} \leq 1$, and δ_{mlf} is the bounded unknown stuck fault or bias fault. Consider that (3.16) satisfies the similar cases as (3.12) in the following.

(i) $\rho_{\delta_{ml}} = 1$ and $\delta_{mlf} = 0$. This refers to the fault-free case.
(ii) $0 < \rho_{\delta_{ml}} < 1$ and $\delta_{mlf} = 0$. This means the partial loss of effectiveness.
(iii) $\rho_{\delta_{ml}} = 1$ and $\delta_{mlf} \neq 0$. This indicates the bias fault.
(iv) $\rho_{\delta_{ml}} = 0$ and $\delta_{mlf} \neq 0$. This means that δ_{ml} is stuck at bounded time-varying function δ_{mlf}.

For the simplicity of presentation, the actuator fault model is formulated by

$$\boldsymbol{\delta}_l = \boldsymbol{\rho}_{\delta_l}\boldsymbol{\delta}_{l0} + \boldsymbol{\delta}_{lf}, \tag{3.17}$$

where $\boldsymbol{\delta}_{l0} = [\delta_{al0}, \delta_{el0}, \delta_{rl0}]^{\mathrm{T}}$, $\boldsymbol{\rho}_{\delta_l} = \mathrm{diag}\left\{\rho_{\delta_{al}}, \rho_{\delta_{el}}, \rho_{\delta_{rl}}\right\}$, $\boldsymbol{\delta}_{lf} = \left[\delta_{alf}, \delta_{elf}, \delta_{rlf}\right]^{\mathrm{T}}$.

Meanwhile, $\boldsymbol{J}_l^{-1}\boldsymbol{S}(\boldsymbol{J}_l\boldsymbol{\omega}_l)$ is decomposed into a known component $\boldsymbol{J}_{l0}^{-1}\boldsymbol{S}(\boldsymbol{J}_{l0}\boldsymbol{\omega}_l)$ and an uncertain one $\Delta\boldsymbol{J}_l^{-1}\boldsymbol{S}(\boldsymbol{J}_l\boldsymbol{\omega}_l)$. Similarly, $\boldsymbol{J}_l^{-1}\boldsymbol{N}_l$ is composed of a known part $\boldsymbol{J}_{l0}^{-1}\boldsymbol{N}_{l0}$ and an uncertain one $\Delta\boldsymbol{J}_l^{-1}\boldsymbol{N}_l$. Also, $\boldsymbol{J}_l^{-1}\boldsymbol{C}(\boldsymbol{\delta}_l)$ consists of a known component $\boldsymbol{J}_{l0}^{-1}\boldsymbol{C}_0(\boldsymbol{\delta}_l)$ and an uncertain one $\Delta\boldsymbol{J}_l^{-1}\boldsymbol{C}(\boldsymbol{\delta}_l)$. Accordingly, in the presence of actuator faults and modeling uncertainties, the dynamic model of angular velocity can be formulated by

$$\dot{\boldsymbol{\omega}}_l = \boldsymbol{J}_{l0}^{-1}\boldsymbol{S}(\boldsymbol{J}_{l0}\boldsymbol{\omega}_l) + \boldsymbol{J}_{l0}^{-1}\boldsymbol{N}_{l0} + \boldsymbol{J}_{l0}^{-1}\boldsymbol{C}_0(\boldsymbol{\delta}_l)\boldsymbol{\rho}_{\delta_l}\boldsymbol{\delta}_{l0} + \boldsymbol{\Delta}_{\omega l} + \boldsymbol{d}_{\omega l}, \tag{3.18}$$

where $\boldsymbol{\Delta}_{\omega l} = \Delta\boldsymbol{J}_l^{-1}\boldsymbol{S}(\boldsymbol{J}_l\boldsymbol{\omega}_l) + \Delta\boldsymbol{J}_l^{-1}\boldsymbol{N}_l + \Delta\boldsymbol{J}_l^{-1}\boldsymbol{C}(\boldsymbol{\delta}_l)\boldsymbol{\delta}_l + \boldsymbol{J}_{l0}^{-1}\boldsymbol{C}_0(\boldsymbol{\delta}_l)\boldsymbol{\delta}_{lf}$ is the lumped uncertainties induced by actuator faults and modeling uncertainties.

Assumption 3.2 For all $t > 0$, it is supposed that $\|\Delta\boldsymbol{J}_l^{-1}\boldsymbol{C}(\boldsymbol{\delta}_l)\boldsymbol{\rho}_{\delta_l}\left(\boldsymbol{J}_{l0}^{-1}\boldsymbol{C}_0(\boldsymbol{\delta}_l)\right)^{-1}\|_\infty < 1$ and $\partial\boldsymbol{\Delta}_{\omega l}/\partial\boldsymbol{\delta}_{l0} + \boldsymbol{J}_{l0}^{-1}\boldsymbol{C}_0(\boldsymbol{\delta}_l)\boldsymbol{\rho}_{\delta_l} \neq 0$.

Assumption 3.3 There exists a constant $g_{l,\min}$ such that $\lambda_{\min}(\boldsymbol{J}_{l0}^{-1}\boldsymbol{C}_0(\boldsymbol{\delta}_l)\boldsymbol{\rho}_{\delta_l}\left(\boldsymbol{J}_{l0}^{-1}\boldsymbol{C}_0(\boldsymbol{\delta}_l)\right)^{\mathrm{T}}) \geqq g_{l,\min} > 0$.

Remark 3.1 Assumption 3.2 imposes the controllability condition to the model (3.18), which guarantees the existence of controller $\boldsymbol{\delta}_{l0}$, while Assumption 3.2 is essential so that the control signal $\boldsymbol{\delta}_{l0}$ dominates the uncertain vector $\left(\boldsymbol{J}_{l0}^{-1}\boldsymbol{C}_{l0}(\boldsymbol{\delta}_l)\boldsymbol{\rho}_{\delta_l} + \Delta\boldsymbol{J}_l^{-1}\boldsymbol{C}(\boldsymbol{\delta}_l)\right)\boldsymbol{\delta}_{l0}$. Similar assumptions can be found in [1, 10, 34]. Moreover, from Assumption 3.3, $g_{l,\min}$ is just used for analysis and not required a priori.

Remark 3.2 Considering that $\boldsymbol{\Delta}_{\omega l}$ involves the control signal $\boldsymbol{\delta}_{l0}$, if neural networks are used to approximate $\boldsymbol{\Delta}_{\omega l}$, the algebraic loops inevitably exist. To break the algebraic loops, the low-pass filter technique is employed to filter $\boldsymbol{\delta}_{l0}$ [33]. We obtain $\boldsymbol{\Delta}_{\omega l} = \boldsymbol{\Delta}'_{\omega l} + \boldsymbol{l}_{\omega l}$, where $\boldsymbol{l}_{\omega l} \in \Re^3$ is the filtering error.

3.2.4 Neural Network Approximation

As the universal approximation property, the RBFNNs are employed in this chapter to compensate the modeling uncertainties [35, 36]. Denote $\boldsymbol{\Theta}_{vi} = \left[\boldsymbol{v}_i^{\mathrm{T}}, \boldsymbol{\varphi}_i^{\mathrm{T}}, \boldsymbol{\omega}_i^{\mathrm{T}}, \boldsymbol{\delta}_{i0}^{\mathrm{T}}\right]^{\mathrm{T}} \in \Re^{12}$ and $\boldsymbol{\Theta}_{\omega i} = \left[\boldsymbol{v}_i^{\mathrm{T}}, \boldsymbol{\varphi}_i^{\mathrm{T}}, \boldsymbol{\omega}_i^{\mathrm{T}}, \boldsymbol{\delta}_{i0}^{\mathrm{T}}\right]^{\mathrm{T}} \in \Re^{12}$. For unknown continuous terms $\boldsymbol{\Delta}_{vi}\left(\boldsymbol{\Theta}_{vi}\right) = \left[\Delta_{vi1}, \Delta_{vi2}, \Delta_{vi3}\right]^{\mathrm{T}}$ and $\boldsymbol{\Delta}'_{\omega i}\left(\boldsymbol{\Theta}_{\omega i}\right) = \left[\Delta'_{\omega i1}, \Delta'_{\omega i2}, \Delta'_{\omega i3}\right]^{\mathrm{T}}$, there exist RBFNNs such that

$$\begin{aligned}\Delta_{vil}\left(\boldsymbol{\Theta}_{vi}\right) &= \boldsymbol{W}_{vil}^{*\mathrm{T}}\boldsymbol{\varsigma}_{vil}\left(\boldsymbol{\Theta}_{vi}\right) + \tau_{vil}\left(\boldsymbol{\Theta}_{vi}\right),\\ \Delta'_{\omega il}\left(\boldsymbol{\Theta}_{\omega i}\right) &= \boldsymbol{W}_{\omega il}^{*\mathrm{T}}\boldsymbol{\varsigma}_{\omega il}\left(\boldsymbol{\Theta}_{\omega i}\right) + \tau_{\omega il}\left(\boldsymbol{\Theta}_{\omega i}\right),\end{aligned} \tag{3.19}$$

where $l = 1, 2, 3$, and $\tau_{vil}\left(\boldsymbol{\Theta}_{vi}\right) \in \Re$ and $\tau_{\omega il}\left(\boldsymbol{\Theta}_{\omega i}\right) \in \Re$ are the bounded approximation errors. $\boldsymbol{W}_{vil}^{*} \in \Re^{M\times 1}$ and $\boldsymbol{W}_{\omega il}^{*} \in \Re^{M\times 1}$ are ideal weight matrixes with M nodes. $\boldsymbol{\varsigma}_{vil}\left(\boldsymbol{\Theta}_{vi}\right) = \left[\varsigma_{vil}^{1}\left(\boldsymbol{\Theta}_{vi}\right), \ldots, \varsigma_{vil}^{M}\left(\boldsymbol{\Theta}_{vi}\right)\right]^{\mathrm{T}} \in \Re^{M}$ and $\boldsymbol{\varsigma}_{\omega il}\left(\boldsymbol{\Theta}_{\omega i}\right) = \left[\varsigma_{\omega il}^{1}\left(\boldsymbol{\Theta}_{\omega i}\right), \ldots, \varsigma_{\omega il}^{M}\left(\boldsymbol{\Theta}_{\omega i}\right)\right]^{\mathrm{T}} \in \Re^{M}$ are known vectors consisted of Gaussian basis functions $\varsigma_{vil}^{n}\left(\boldsymbol{\Theta}_{vi}\right)$ and $\varsigma_{\omega il}^{n}\left(\boldsymbol{\Theta}_{\omega i}\right)$, $n = 1, \ldots, M$, commonly selected as the following exponential form:

$$\begin{aligned}\varsigma_{vij}^{n}\left(\boldsymbol{\Theta}_{vi}\right) &= \exp\left[-\frac{\left(\boldsymbol{\Theta}_{vi} - \boldsymbol{\Upsilon}_{vi}\right)^{\mathrm{T}}\left(\boldsymbol{\Theta}_{vi} - \boldsymbol{\Upsilon}_{vi}\right)}{\kappa_{vi}^{2}}\right],\\ \varsigma_{\omega ij}^{n}\left(\boldsymbol{\Theta}_{\omega i}\right) &= \exp\left[-\frac{\left(\boldsymbol{\Theta}_{\omega i} - \boldsymbol{\Upsilon}_{\omega i}\right)^{\mathrm{T}}\left(\boldsymbol{\Theta}_{\omega i} - \boldsymbol{\Upsilon}_{\omega i}\right)}{\kappa_{\omega i}^{2}}\right],\end{aligned} \tag{3.20}$$

where $\boldsymbol{\Upsilon}_{vi} \in \Re^{12}$ and $\boldsymbol{\Upsilon}_{\omega i} \in \Re^{12}$ are the centers of the receptive fields, and κ_{vi} and $\kappa_{\omega i}$ are the width of the Gaussian basis functions.

3.2.5 Control Objective

Given the velocity reference V_r and attitude reference $\boldsymbol{\varphi}_r$, the goal is to design the cooperative distributed fixed-time tracking controller such that:

- All the closed-loop signals remain bounded and the velocity and attitude tracking errors for each UAV are SGUUB within fixed time in spite of actuator faults, modeling uncertainties, and external disturbances.
- The performances of velocity and attitude preserve certain constraints at all time.

Assumption 3.4 The designed parameters $V_{i,\min}$, $V_{i,\max}$, $\boldsymbol{\varphi}_{i,\min}$, $\boldsymbol{\varphi}_{i,\max}$, $\boldsymbol{\omega}_{i,\min}$, and $\boldsymbol{\omega}_{i,\max}$ and their time-derivatives up to second-order are continuous and bounded.

For the convenience of derivation, the following lemmas are needed.

Lemma 3.1 ([37]) *Consider the system*

$$\dot{\boldsymbol{x}}(t) = f(\boldsymbol{x}(t)), \ \boldsymbol{x}(0) = \boldsymbol{x}_0. \tag{3.21}$$

Suppose that there exist constants $\alpha > 0$, $\beta > 0$, $p > 1$, $0 < q < 1$, $0 < \eta < \infty$ *such that*

$$\dot{V}(\boldsymbol{x}) \leq -\alpha V^p(\boldsymbol{x}) - \beta V^q(\boldsymbol{x}) + \eta. \tag{3.22}$$

Then the system (3.21) is practically fixed-time stable. The settling time T *is presented as* $T \leq T_{\max} := \frac{1}{\alpha\phi(p-1)} + \frac{1}{\beta\phi(1-q)}$ *with* $0 < \phi < 1$. *And the convergence neighborhood is given by* $\boldsymbol{x} \in \{V(\boldsymbol{x}) \leq \min\{(\frac{\eta}{(1-\phi)\alpha})^{\frac{1}{p}}, (\frac{\eta}{(1-\phi)\beta})^{\frac{1}{q}}\}\}$.

Lemma 3.2 ([38]) *For any scalars* $\varepsilon > 0$, $z \in \Re$, *the following equation holds:* $0 \leq |z| - \frac{z^2}{\sqrt{z^2+\varepsilon}} \leq \sqrt{\varepsilon}$.

Lemma 3.3 ([39]) *For any scalars* $\varepsilon > 0$, $\vartheta \in \Re$, $0 \leq |\vartheta| - \vartheta \tanh\left(\frac{\vartheta}{\varepsilon}\right) \leq 0.2785\varepsilon$ *holds.*

Lemma 3.4 ([40]) *For positive constants* b_1, b_2, *and* b_3, *it follows that*

$$|x|^{b_1}|y|^{b_2} \leq \frac{b_1}{b_1+b_2} b_3 |x|^{b_1+b_2} + \frac{b_2}{b_1+b_2} b_3^{-\frac{b_1}{b_2}} |y|^{b_1+b_2},$$

where x *and* y *are real variables.*

3.3 Main Results

3.3.1 State Transformation

The system transformation is developed to transform the original constrained system (3.13), (3.14), and (3.18) into the unconstrained ones, whose stability can guarantee the full-state constraints.

Let $\tilde{V}_i = V_i - V_r$ represent the tracking error of velocity for each UAV with V_r piecewise continuously differentiable.

Definition 3.1 Velocity tracking is said to guarantee $-F_{vi1}(t) < \tilde{V}_i(t) < F_{vi2}(t)$ if any given bounded initial condition satisfies $-F_{vi1}(0) < \tilde{V}_i(0) < F_{vi2}(0)$. If there exists

$$\xi_{vi} = \frac{\tilde{V}_i}{\left(F_{vi1} + \tilde{V}_i\right)\left(F_{vi2} - \tilde{V}_i\right)}, \tag{3.23}$$

the constraint of $\tilde{V}_i$ can be ensured as ξ_{vi} is bounded $\forall t \geq 0$.

From (3.23), taking the derivative of ξ_{vl} gives

$$\dot{\xi}_{vl} = \eta_{vl}\dot{\tilde{V}}_l(t) + \mu_{vl}, \tag{3.24}$$

where

$$\eta_{vl} = \frac{F_{vl1}F_{vl2} + \tilde{V}_l^2}{(F_{vl1} + \tilde{V}_l)^2(F_{vl2} - \tilde{V}_l)^2}$$

$$\mu_{vl} = \frac{\left[\dot{F}_{vl1}F_{vl2} + F_{vl1}\dot{F}_{vl2} + (\dot{F}_{vl2} - \dot{F}_{vl1})\tilde{V}_l\right]\tilde{V}_l}{(F_{vl1} + \tilde{V}_l)^2(F_{vl2} - \tilde{V}_l)^2}.$$

Let $\boldsymbol{e}_{\varphi l}$ denote the error vector that specifies synchronization attitude for each UAV with $\boldsymbol{\varphi}_r$ piecewise continuously differentiable

$$\boldsymbol{e}_{\varphi l} = \lambda_{l1}\tilde{\boldsymbol{\varphi}}_l + \lambda_{l2}\sum_{j\in\mathcal{N}_l} a_{lj}\left(\tilde{\boldsymbol{\varphi}}_l - \tilde{\boldsymbol{\varphi}}_j\right), \tag{3.25}$$

where $\tilde{\boldsymbol{\varphi}}_l = \boldsymbol{\varphi}_l - \boldsymbol{\varphi}_r$, $\tilde{\boldsymbol{\varphi}}_j = \boldsymbol{\varphi}_j - \boldsymbol{\varphi}_r$, and $\lambda_{l1}, \lambda_{l2}$ are positive design parameters.

Definition 3.2 Synchronization attitude tracking is identified with $-F_{\varphi ll1}(t) < e_{\varphi ll}(t) < F_{\varphi ll2}(t)$ for $l = 1, 2, 3$ if for any given bounded initial condition satisfies $-F_{\varphi ll1}(0) < e_{\varphi ll}(0) < F_{\varphi ll2}(0)$. If there exists

$$\xi_{\varphi ll} = \frac{e_{\varphi ll}}{\left(F_{\varphi ll1} + e_{\varphi ll}\right)\left(F_{\varphi ll2} - e_{\varphi ll}\right)}, \tag{3.26}$$

the constraint on $e_{\varphi ll}$ can be guaranteed if $\xi_{\varphi ll}$ is bounded $\forall t \geq 0$.

Take the derivative of $\xi_{\varphi ll}$ along (3.26) as

$$\dot{\xi}_{\varphi ll} = \eta_{\varphi ll}\dot{e}_{\varphi ll} + \mu_{\varphi ll}, \tag{3.27}$$

where

$$\eta_{\varphi ll} = \frac{F_{\varphi ll1}F_{\varphi ll2} + e_{\varphi ll}{}^2}{(F_{\varphi ll1} + e_{\varphi ll})^2(F_{\varphi ll2} - e_{\varphi ll})^2}$$

$$\mu_{\varphi ll} = \frac{\left[\dot{F}_{\varphi ll1}F_{\varphi ll2} + F_{\varphi ll1}\dot{F}_{\varphi ll2} + (\dot{F}_{\varphi ll2} - \dot{F}_{\varphi ll1})e_{\varphi ll}\right]e_{\varphi ll}}{(F_{\varphi ll1} + e_{\varphi ll})^2(F_{\varphi ll2} - e_{\varphi ll})^2}.$$

We have $\boldsymbol{\xi}_{\varphi l} = \left[\xi_{\varphi l1}, \xi_{\varphi l2}, \xi_{\varphi l3}\right]^{\mathrm{T}}$, $\boldsymbol{\mu}_{\varphi l} = \left[\mu_{\varphi l1}, \mu_{\varphi l2}, \mu_{\varphi l3}\right]^{\mathrm{T}}$, and $\boldsymbol{\eta}_{\varphi l} = \mathrm{diag}\left\{\eta_{\varphi l1}, \eta_{\varphi l2}, \eta_{\varphi l3}\right\}$.

Definition 3.3 The time-varying constraint on $\boldsymbol{\omega}_l$ is never transgressed, that is, $-F_{\omega l l1}(t) < \omega_{ll}(t) < F_{\omega l l2}(t)$ for $l = 1, 2, 3$ if for any given bounded initial state satisfies $-F_{\omega l l1}(0) < \omega_{ll}(0) < F_{\omega l l2}(0)$. If there exists

$$\xi_{\omega l l} = \frac{\omega_{ll}}{(F_{\omega l l1} + \omega_{ll})(F_{\omega l l2} - \omega_{ll})}, \tag{3.28}$$

the constrained problem involving ω_{ll} is equivalent to ensuring the boundedness of $\xi_{\omega l l}$.

From (3.28), taking the derivative of $\xi_{\omega l l}$ gives

$$\dot{\xi}_{\omega l l} = \eta_{\omega l l}\dot{\omega}_{ll} + \mu_{\omega l l}, \tag{3.29}$$

where

$$\eta_{\omega l l} = \frac{F_{\omega l l1}F_{\omega l l2} + \omega_{ll}^2}{(F_{\omega l l1} + \omega_{ll})^2(F_{\omega l l2} - \omega_{ll})^2}$$

$$\mu_{\omega l l} = \frac{\left[\dot{F}_{\omega l l1}F_{\omega l l2} + F_{\omega l l1}\dot{F}_{\omega l l2} + (\dot{F}_{\omega l l2} - \dot{F}_{\omega l l1})\omega_{ll}\right]\omega_{ll}}{(F_{\omega l l1} + \omega_{ll})^2(F_{\omega l l2} - \omega_{ll})^2}.$$

We define $\boldsymbol{\xi}_{\omega l} = [\xi_{\omega l 1}, \xi_{\omega l 2}, \xi_{\omega l 3}]^{\mathrm{T}}$, $\boldsymbol{\mu}_{\omega l} = [\mu_{\omega l 1}, \mu_{\omega l 2}, \mu_{\omega l 3}]^{\mathrm{T}}$, and $\boldsymbol{\eta}_{\omega l} = \mathrm{diag}\{\eta_{\omega l 1}, \eta_{\omega l 2}, \eta_{\omega l 3}\}$.

Remark 3.3 $F_{vl1}(t)$, $F_{vl2}(t)$, $F_{\varphi l l1}(t)$, $F_{\varphi l l2}(t)$, $F_{\omega l l1}(t)$, and $F_{\omega l l2}(t)$ are utilized to describe the time-varying constraints for $\tilde{V}_l$, $e_{\varphi l l}$, and ω_{ll}, respectively. Note that $F_{vl1}(t)$, $F_{vl2}(t)$, $F_{\varphi l l1}(t)$, $F_{\varphi l l2}(t)$, $F_{\omega l l1}(t)$, and $F_{\omega l l2}(t)$ are strictly positive smooth functions. More details on the properties of the constrained problem described by (3.23), (3.26), and (3.28) can refer to the illustration example in [25]. Based on Definitions 3.1, 3.2, and 3.3, new unconstrained subsystems (3.23), (3.26), and (3.28) for translational and rotational dynamics are obtained, which are guaranteed to be bounded.

3.3.2 Cooperative Distributed Fixed-Time Controller Design for Velocity Subsystem

Following the double-layer control structure, the translational and rotational controller designs are proposed. Hereafter, the design of T_{xl0} and δ_{l0} is given.

To achieve the control objective, we employ the dynamic controller T_{xl0} (see Algorithm 3.1)

$$T_{xl0} = -\frac{\alpha_{vl}\hat{\beta}_{vl}}{\zeta_{vl}}\tanh\left(\frac{\xi_{vl}\eta_{vl}\alpha_{vl}\hat{\beta}_{vl}}{\varepsilon_{vl}}\right)$$

$$-\hat{g}_{vi}\tanh\left(\frac{\xi_{vi}\eta_{vi}\hat{g}_{vi}\zeta_{vi}}{\varepsilon_{vi}}\right), \tag{3.30}$$

where $\alpha_{vi} = \frac{c_{vi1}}{\eta_{vi}}\text{sig}(\xi_{vi})^{2p_v-1} + \frac{c_{vi2}}{\eta_{vi}}\text{sig}(\xi_{vi})^{2q_v-1} + \frac{\hat{\Phi}_{vi}\boldsymbol{v}_i^{\mathrm{T}}\boldsymbol{\Xi}_{vi}}{2h_{vi1}^2 V_i{}^2} + \frac{\boldsymbol{v}_i^{\mathrm{T}}}{V_i}(\boldsymbol{R}_1^{\mathrm{T}}\boldsymbol{g} + \frac{\boldsymbol{F}_{i0}}{m_i}) + \frac{\xi_{vi}\eta_{vi}\boldsymbol{v}_i^{\mathrm{T}}\boldsymbol{v}_i}{2h_{vi2}^2 V_i^2} - \dot{V}_r + \frac{\mu_{vi}}{\eta_{vi}}$ with $p_v > 1$, $0 < q_v < 1$, $\boldsymbol{\Xi}_{vi} = \xi_{vi}\eta_{vi}[v_{i1}\boldsymbol{\varsigma}_{vi1}^{\mathrm{T}}\boldsymbol{\varsigma}_{vi1}, v_{i2}\boldsymbol{\varsigma}_{vi2}^{\mathrm{T}}\boldsymbol{\varsigma}_{vi2}, v_{i3}\boldsymbol{\varsigma}_{vi3}^{\mathrm{T}}\boldsymbol{\varsigma}_{vi3}]^{\mathrm{T}}$. The adaptive parameters $\hat{\Phi}_{vi}$, $\hat{g}_{vi}$, and $\hat{\beta}_{vi}$ are updated by

$$\dot{\hat{\Phi}}_{vi} = \frac{\gamma_{vi1}\xi_{vi}\eta_{vi}\boldsymbol{v}_i^{\mathrm{T}}\boldsymbol{\Xi}_{vi}}{2h_{vi1}^2 V_i} - \gamma_{vi1}\sigma_{vi1}\hat{\Phi}_{vi} \tag{3.31}$$

$$\dot{\hat{g}}_{vi} = \gamma_{vi2}\left|\xi_{vi}\eta_{vi}\right|\zeta_{vi} - \gamma_{vi2}\sigma_{vi2}\hat{g}_{vi} \tag{3.32}$$

$$\dot{\hat{\beta}}_{vi} = \gamma_{vi3}\xi_{vi}\eta_{vi}\alpha_{vi} - \gamma_{vi3}\sigma_{vi3}\hat{\beta}_{vi}, \tag{3.33}$$

where $h_{vi1}, h_{vi2}, \gamma_{vi1}, \gamma_{vi2}, \gamma_{vi3}, \sigma_{vi1}, \sigma_{vi2}, \sigma_{vi3}, \varepsilon_{vi}$ are positive design parameters.

3.3.3 Cooperative Distributed Fixed-Time Controller Design for Attitude Subsystem

To obtain the control objective, we propose the dynamic controller $\boldsymbol{\delta}_{i0}$ (see Algorithm 3.2):

$$\boldsymbol{\delta}_{i0} = -\left(\boldsymbol{J}_{i0}^{-1}\boldsymbol{C}_0\right)^{\mathrm{T}}\boldsymbol{\eta}_{\omega i}^{-1}\frac{\boldsymbol{e}_{\omega i}\hat{\beta}_{\omega i}^2\boldsymbol{\zeta}_{\omega i}^{\mathrm{T}}\boldsymbol{\zeta}_{\omega i}}{\sqrt{\hat{\beta}_{\omega i}^2\boldsymbol{e}_{\omega i}^{\mathrm{T}}\boldsymbol{e}_{\omega i}\boldsymbol{\zeta}_{\omega i}^{\mathrm{T}}\boldsymbol{\zeta}_{\omega i} + \varepsilon_{\omega i}}}, \tag{3.34}$$

Algorithm 3.1 Design the distributed control protocol T_{xi0}

Input: The parameters F_{vi1} and F_{vi2} in state transformation (3.23); the parameters $\gamma_{vi1}, \gamma_{vi2}, \gamma_{vi3}, \sigma_{vi1}, \sigma_{vi2}, \sigma_{vi3}$ and h_{vi1} in adaptive laws (3.31)–(3.33); the parameters $c_{vi1}, c_{vi2}, h_{vi2}$ and ε_{vi} in control protocol (3.30).

Output: The distributed control protocol (3.30).

1: **Step 1:** Solve the velocity constraints to obtain the state transformation (3.23).

2: **Step 2:** Use RBFNN (3.19) to approximate the modeling uncertainties, and acquire the adaptive law (3.31).

3: **Step 3:** Introduce $\beta_{vi} = 1/\rho_{T_i}$ and $g_{vi} = \beta_{vi}\sup_{t\geq 0} T_{xif}$ to compensate for the actuator faults, and obtain the adaptive laws (3.32) and (3.33).

4: **Step 4:** Select the parameters $c_{vi1}, c_{vi2}, h_{vi2}$ and ε_{vi}, and integrate the adaptive laws (3.31)–(3.33) into the control protocol (3.30).

---

where $\boldsymbol{\zeta}_{\omega i} = \boldsymbol{K}_{4i}\boldsymbol{e}_{\omega i}\|\boldsymbol{e}_{\omega i}\|^{2p_\varphi-2} + \boldsymbol{K}_{5i}\boldsymbol{e}_{\omega i}\|\boldsymbol{e}_{\omega i}\|^{2q_\varphi-2} + \lambda_i\boldsymbol{K}_{1i}^{\mathrm{T}}\left(\boldsymbol{R}_2^{-1}\right)^{\mathrm{T}}\boldsymbol{\eta}_{\varphi i}^{\mathrm{T}}\boldsymbol{\xi}_{\varphi i} + \boldsymbol{\eta}_{\omega i}\boldsymbol{J}_{i0}^{-1}$ $\boldsymbol{S}(\boldsymbol{J}_{i0}\boldsymbol{\omega}_i) + \boldsymbol{\eta}_{\omega i}\boldsymbol{J}_{i0}^{-1}\boldsymbol{N}_{i0} + \frac{\hat{\Phi}_{\omega i}\boldsymbol{\eta}_{\omega i}\boldsymbol{\Xi}_{\omega i}}{2h_{\omega i1}^2} + \frac{\boldsymbol{\eta}_{\omega i}\boldsymbol{\eta}_{\omega i}^{\mathrm{T}}\boldsymbol{e}_{\omega i}}{2h_{\omega i2}^2} + \boldsymbol{\mu}_{\omega i} - \dot{\boldsymbol{\xi}}_{\omega di}$ with $p_\varphi > 1$, $0 < q_\varphi < 1$, $\boldsymbol{\Xi}_{\omega i} = \left[e_{\omega i1}\eta_{\omega i1}\varsigma_{\omega i1}^{\mathrm{T}}\varsigma_{\omega i1}, e_{\omega i2}\eta_{\omega i2}\varsigma_{\omega i2}^{\mathrm{T}}\varsigma_{\omega i2}, e_{\omega i3}\eta_{\omega i3}\varsigma_{\omega i3}^{\mathrm{T}}\varsigma_{\omega i3}\right]^{\mathrm{T}}$, $\lambda_i = \lambda_{i1} + \lambda_{i2}\sum_{j\in\mathcal{N}_i} a_{ij}$. The adaptive parameters $\hat{\Phi}_{\omega i}$ and $\hat{\beta}_{\omega i}$ are updated by

$$\dot{\hat{\Phi}}_{\omega i} = \gamma_{\omega i1}\frac{\boldsymbol{e}_{\omega i}^{\mathrm{T}}\boldsymbol{\eta}_{\omega i}\boldsymbol{\Xi}_{\omega i}}{2h_{\omega i1}^2} - \sigma_{\omega i1}\gamma_{\omega i1}\hat{\Phi}_{\omega i} \tag{3.35}$$

$$\dot{\hat{\beta}}_{\omega i} = \gamma_{\omega i2}\boldsymbol{e}_{\omega i}^{\mathrm{T}}\boldsymbol{\zeta}_{\omega i} - \sigma_{\omega i2}\gamma_{\omega i2}\hat{\beta}_{\omega i}, \tag{3.36}$$

where $h_{\omega i1}, h_{\omega i2}, \gamma_{\omega i1}, \gamma_{\omega i2}, \sigma_{\omega i1}, \sigma_{\omega i2}, \varepsilon_{\omega i}$ are positive design parameters.

3.3.4 *Stability Analysis*

Theorem 3.1 *Consider the translational kinematics (3.13) composed of the adaptive controller (3.30) and parameter adaptation laws (3.31)–(3.33). Let Assumptions 3.4 hold. There exist positive parameters $c_{vi1}, c_{vi2}, \varepsilon_{vi}, h_{vi1}, h_{vi2}, \gamma_{vi1}, \gamma_{vi2}, \gamma_{vi3}, \sigma_{vi1}, \sigma_{vi2}$, and σ_{vi3} ($i = 1, 2, \ldots, N$) such that:*

(1) All signals of the translational subsystem are SGUUB in the presence of actuator faults, modeling uncertainties, and external disturbances.
(2) Velocity tracking error $\tilde{V}_i$ satisfies $\lim_{t\to T_{\max,v}}|\tilde{V}_i| \le \mu_{vi}$ with $\mu_{vi} > 0$ a constant.

Algorithm 3.2 Design the distributed control protocol $\boldsymbol{\delta}_{i0}$

Input: The parameters $F_{\varphi il1}$, $F_{\varphi il2}$, $F_{\omega il1}$ and $F_{\omega il2}$ in state transformations (3.26) and (3.28); the parameters $\gamma_{\omega i1}, \gamma_{\omega i2}, \sigma_{\omega i1}, \sigma_{\omega i2}$ and $h_{\omega i1}$ in adaptive laws (3.35) and (3.36); the parameters $\boldsymbol{K}_{1i}$, $\boldsymbol{K}_{2i}$ and $\boldsymbol{K}_{3i}$ in virtual control (3.54); the parameters $\boldsymbol{K}_{4i}$, $\boldsymbol{K}_{5i}$, $h_{\omega i2}$ and $\varepsilon_{\omega i}$ in control protocol (3.34).
Output: The distributed control protocol (3.34).
1: **Step 1:** Solve the attitude and angular velocity constraints to obtain the state transformations (3.26) and (3.28).
2: **Step 2:** Choose the parameters $\boldsymbol{K}_{1i}$, $\boldsymbol{K}_{2i}$ and $\boldsymbol{K}_{3i}$, and propose the virtual control (3.54).
3: **Step 3:** Use RBFNN (3.19) to approximate the modeling uncertainties, and acquire the adaptive law (3.35).
4: **Step 4:** Introduce $\beta_{\omega i} = 1/g_{\omega i}$ and $g_{\omega i} = \inf_{t\ge 0}\lambda_{\min}\left(\boldsymbol{J}_{i0}^{-1}\boldsymbol{C}_0(\boldsymbol{\delta}_i)\boldsymbol{\rho}_{\delta_i}(\boldsymbol{J}_{i0}^{-1}\boldsymbol{C}_0(\boldsymbol{\delta}_i))^{\mathrm{T}}\right)$ to compensate for the actuator faults, and obtain the adaptive laws (3.36).
5: **Step 5:** Choose the parameters $\boldsymbol{K}_{4i}$, $\boldsymbol{K}_{5i}$, $h_{\omega i2}$ and $\varepsilon_{\omega i}$, and integrate the adaptive laws (3.35) and (3.36) into the control protocol (3.34).

---

(3) The required velocity constraint for translational subsystem is guaranteed for $\forall t \geq 0$.

Proof Combined with (3.13), (3.24) further becomes

$$\begin{aligned} \dot{\xi}_{vl} =& \eta_{vl}\left[\zeta_{vl}\rho_{T_l}T_{xl0} + \zeta_{vl}T_{xlf} + \frac{\boldsymbol{v}_l^{\mathrm{T}}}{V_l}\left(\boldsymbol{R}_1^{\mathrm{T}}\boldsymbol{g} + \frac{\boldsymbol{F}_{l0}}{m_l}\right)\right] \\ &+ \eta_{vl}\left[\frac{\boldsymbol{v}_l^{\mathrm{T}}\left(\boldsymbol{\Delta}_{vl} + \boldsymbol{d}_{vl}\right)}{V_l} - \dot{V}_r\right] + \mu_{vl}, \end{aligned} \tag{3.37}$$

where $\zeta_{vl} = \frac{u_l}{m_l V_l}$ and $\boldsymbol{\Delta}_{vl} = \frac{\Delta \boldsymbol{F}_l}{m_l}$.

Using Young's inequality, we obtain from (3.19) that

$$\begin{aligned} \frac{\xi_{vl}\eta_{vl}}{V_l}\boldsymbol{v}_l^{\mathrm{T}}\left(\boldsymbol{\Delta}_{vl} + \boldsymbol{d}_{vl}\right) \leq \sum_{l=1}^{3}\Bigg(&\frac{\xi_{vl}^2\eta_{vl}^2 v_{ll}^2 \Phi_{vl}\boldsymbol{S}_{vll}^{\mathrm{T}}\boldsymbol{S}_{vll}}{2h_{vl1}^2 V_l^2} + \frac{\xi_{vl}^2\eta_{vl}^2 v_{ll}^2}{2h_{vl2}^2 V_l^2} \\ &+ \frac{h_{vl1}^2}{2} + \frac{h_{vl2}^2\bar{\tau}_{vll}^2}{2}\Bigg), \end{aligned} \tag{3.38}$$

where $\Phi_{vl} = \max\left\{\boldsymbol{W}_{vl1}^{*\mathrm{T}}\boldsymbol{W}_{vl1}^{*}, \boldsymbol{W}_{vl2}^{*\mathrm{T}}\boldsymbol{W}_{vl2}^{*}, \boldsymbol{W}_{vl3}^{*\mathrm{T}}\boldsymbol{W}_{vl3}^{*}\right\}$ and $\tau_{vll} + d_{vll}$ satisfies $|\tau_{vll} + d_{vll}| \leq \bar{\tau}_{vll}$.

From (3.37) and (3.38), taking the time derivative of $\frac{1}{2}\xi_{vl}^2$ derives

$$\begin{aligned} \xi_{vl}\dot{\xi}_{vl} \leq& \xi_{vl}\eta_{vl}\left[\frac{\boldsymbol{v}_l^{\mathrm{T}}}{V_l}\left(\boldsymbol{R}_1^{\mathrm{T}}\boldsymbol{g} + \frac{\boldsymbol{F}_{l0}}{m_l}\right) + \frac{\Phi_{vl}\boldsymbol{v}_l^{\mathrm{T}}\boldsymbol{\Xi}_{vl}}{2h_{vl1}^2 V_l^2} + \frac{\xi_{vl}\eta_{vl}\boldsymbol{v}_l^{\mathrm{T}}\boldsymbol{v}_l}{2h_{vl2}^2 V_l^2}\right] \\ &+ \xi_{vl}\eta_{vl}\left(\zeta_{vl}\rho_{T_l}T_{xl0} + \zeta_{vl}T_{xlf} - \dot{V}_{rl} + \frac{\mu_{vl}}{\eta_{vl}}\right) \\ &+ \sum_{l=1}^{3}\left(\frac{h_{vl1}^2}{2} + \frac{h_{vl2}^2\bar{\tau}_{vll}^2}{2}\right). \end{aligned} \tag{3.39}$$

Let $\beta_{vl} = 1/\rho_{T_l}$ and $g_{vl} = \beta_{vl}\sup_{t\geq 0} T_{xlf}$. And define $\tilde{\Phi}_{vl} = \hat{\Phi}_{vl} - \Phi_{vl}$, $\tilde{\beta}_{vl} = \hat{\beta}_{vl} - \beta_{vl}$, and $\tilde{g}_{vl} = \hat{g}_{vl} - g_{vl}$, where $\hat{\Phi}_{vl}$, $\hat{\beta}_{vl}$, and $\hat{g}_{vl}$ are the estimations of Φ_{vl}, β_{vl}, and g_{vl}, respectively. Then it follows that

$$\xi_{vl}\eta_{vl}\zeta_{vl}T_{xlf} \leq \rho_{T_l}\hat{g}_{vl}\left|\xi_{vl}\eta_{vl}\right|\zeta_{vl} - \rho_{T_l}\tilde{g}_{vl}\left|\xi_{vl}\eta_{vl}\right|\zeta_{vl}. \tag{3.40}$$

According to Lemma 3.3, $\rho_{T_i}\hat{g}_{vl}\,|\xi_{vl}\eta_{vl}|\,\zeta_{vl}$ satisfies

$$\rho_{T_i}\hat{g}_{vl}\,|\xi_{vl}\eta_{vl}|\,\zeta_{vl}\leq\rho_{T_i}\xi_{vl}\eta_{vl}\hat{g}_{vl}\zeta_{vl}\tanh\left(\frac{\xi_{vl}\eta_{vl}\hat{g}_{vl}\zeta_{vl}}{\varepsilon_{vl}}\right)+0.2785\rho_{T_i}\varepsilon_{vl}. \tag{3.41}$$

Choose the following Lyapunov function candidate:

$$L_1=\frac{1}{2}\xi_{vl}^2+\frac{1}{2\gamma_{vl1}}\tilde{\Phi}_{vl}^2+\frac{\rho_{T_i}}{2\gamma_{vl2}}\tilde{g}_{vl}^2+\frac{\rho_{T_i}}{2\gamma_{vl3}}\tilde{\beta}_{vl}^2. \tag{3.42}$$

Taking the time derivative of (3.42) along (3.37)–(3.41) gives

$$\begin{aligned}\dot{L}_1=&\,\xi_{vl}\eta_{vl}\left[-\frac{c_{vl1}}{\eta_{vl}}\mathrm{sig}(\xi_{vl})^{2p_v-1}-\frac{c_{vl2}}{\eta_{vl}}\mathrm{sig}(\xi_{vl})^{2q_v-1}+\alpha_{vl}\right]\\&+\xi_{vl}\eta_{vl}\left[\rho_{T_i}\hat{g}_{vl}\zeta_{vl}\tanh\left(\frac{\xi_{vl}\eta_{vl}\hat{g}_{vl}\zeta_{vl}}{\varepsilon_{vl}}\right)+\zeta_{vl}\rho_{T_i}T_{xl0}\right]\\&+\frac{1}{\gamma_{vl1}}\tilde{\Phi}_{vl}\left(\dot{\hat{\Phi}}_{vl}-\frac{\gamma_{vl1}\xi_{vl}\eta_{vl}\boldsymbol{v}_l^{\mathrm{T}}\boldsymbol{\Xi}_{vl}}{2h_{vl1}^2V_l}\right)+\frac{\rho_{T_i}}{\gamma_{vl3}}\tilde{\beta}_{vl}\dot{\hat{\beta}}_{vl}\\&+\frac{\rho_{T_i}}{\gamma_{vl2}}\tilde{g}_{vl}\left(\dot{\hat{g}}_{vl}-\gamma_{vl2}\,|\xi_{vl}\eta_{vl}|\,\zeta_{vl}\right)+0.2785\rho_{T_i}\varepsilon_{vl}\\&+\sum_{l=1}^{3}\left(\frac{h_{vl1}^2}{2}+\frac{h_{vl2}^2\bar{\tau}_{vll}^2}{2}\right).\end{aligned} \tag{3.43}$$

Invoking Lemma 3.3, it follows along (3.43) that

$$\xi_{vl}\eta_{vl}\zeta_{vl}\rho_{T_i}T_{xl0}\leq-\xi_{vl}\eta_{vl}{}_{l}\zeta_{vl}\rho_{T_i}\hat{g}_{vl}\tanh\left(\frac{\xi_{vl}\eta_{vl}\hat{g}_{vl}\zeta_{vl}}{\varepsilon_{vl}}\right)-\rho_{T_i}\xi_{vl}\eta_{vl}\alpha_{vl}\hat{\beta}_{vl}+0.2785\rho_{T_i}\varepsilon_{vl}. \tag{3.44}$$

Noting that $-\rho_{T_i}\xi_{vl}\eta_{vl}\alpha_{vl}\hat{\beta}_{vl}=-\rho_{T_i}\xi_{vl}\eta_{vl}\alpha_{vl}\tilde{\beta}_{vl}-\xi_{vl}\eta_{vl}\alpha_{vl}$ and substituting (3.43)–(3.44) into (3.42) give

$$\begin{aligned}\dot{L}_1\leq&-c_{vl1}|\xi_{vl}|^{2p_v}-c_{vl2}|\xi_{vl}|^{2q_v}-\sigma_{vl1}\tilde{\Phi}_{vl}\hat{\Phi}_{vl}\\&-\rho_{T_i}\sigma_{vl2}\tilde{g}_{vl}\hat{g}_{vl}-\rho_{T_i}\sigma_{vl3}\tilde{\beta}_{vl}\hat{\beta}_{vl}+0.557\varepsilon_{vl}\rho_{T_i}\\&+\sum_{l=1}^{3}\left(\frac{h_{vl1}^2}{2}+\frac{h_{vl2}^2\bar{\tau}_{vll}^2}{2}\right).\end{aligned} \tag{3.45}$$

Choose $\gamma_{vi1} = \frac{2\vartheta_{vi1}}{2\vartheta_{vi1}-1}$, $\gamma_{vi2} = \frac{2\vartheta_{vi2}}{2\vartheta_{vi2}-1}$, and $\gamma_{vi3} = \frac{2\vartheta_{vi3}}{2\vartheta_{vi3}-1}$ with $\vartheta_{vi1} > \frac{1}{2}$, $\vartheta_{vi2} > \frac{1}{2}$, and $\vartheta_{vi3} > \frac{1}{2}$. It holds that

$$
\begin{aligned}
-\sigma_{vi1}\tilde{\Phi}_{vi}\hat{\Phi}_{vi} &\le -\frac{\sigma_{vi1}}{\gamma_{vi1}}\tilde{\Phi}_{vi}^2 + \frac{\vartheta_{vi1}\sigma_{vi1}}{2}\Phi_{vi}^2\\
-\rho_{T_i}\sigma_{vi2}\tilde{g}_{vi}\hat{g}_{vi} &\le -\frac{\rho_{T_i}\sigma_{vi2}}{\gamma_{vi2}}\tilde{g}_{vi}^2 + \frac{\vartheta_{vi2}\rho_{T_i}\sigma_{vi2}}{2}g_{vi}^2\\
-\rho_{T_i}\sigma_{vi3}\tilde{\beta}_{vi}\hat{\beta}_{vi} &\le -\frac{\rho_{T_i}\sigma_{vi3}}{\gamma_{vi3}}\tilde{\beta}_{vi}^2 + \frac{\vartheta_{vi3}\rho_{T_i}\sigma_{vi3}}{2}\beta_{vi}^2.
\end{aligned}
\tag{3.46}
$$

By adding and subtracting the terms of $\left(\frac{\tilde{\Phi}_{vi}^2}{2\gamma_{vi1}}\right)^{p_v}$, $\left(\frac{\tilde{\Phi}_{vi}^2}{2\gamma_{vi1}}\right)^{q_v}$, $\left(\frac{\rho_{T_i}\tilde{g}_{vi}^2}{2\gamma_{vi2}}\right)^{p_v}$, $\left(\frac{\rho_{T_i}\tilde{g}_{vi}^2}{2\gamma_{vi2}}\right)^{q_v}$, $\left(\frac{\rho_{T_i}\tilde{\beta}_{vi}^2}{2\gamma_{vi3}}\right)^{p_v}$, and $\left(\frac{\rho_{T_i}\tilde{\beta}_{vi}^2}{2\gamma_{vi3}}\right)^{q_v}$ and substituting (3.46) into (3.45), it follows that

$$
\begin{aligned}
\dot{L}_1 \le & -c_{vi1}|\xi_{vi}|^{2p_v} - c_{vi2}|\xi_{vi}|^{2q_v} - \sigma_{vi1}\left(\frac{\tilde{\Phi}_{vi}^2}{2\gamma_{vi1}}\right)^{p_v}\\
& - \sigma_{vi1}\left(\frac{\tilde{\Phi}_{vi}^2}{2\gamma_{vi1}}\right)^{q_v} - \sigma_{vi2}\left(\frac{\rho_{T_i}\tilde{g}_{vi}^2}{2\gamma_{vi2}}\right)^{p_v} - \sigma_{vi2}\left(\frac{\rho_{T_i}\tilde{g}_{vi}^2}{2\gamma_{vi2}}\right)^{q_v}\\
& - \sigma_{vi3}\left(\frac{\rho_{T_i}\tilde{\beta}_{vi}^2}{2\gamma_{vi3}}\right)^{p_v} - \sigma_{vi3}\left(\frac{\rho_{T_i}\tilde{\beta}_{vi}^2}{2\gamma_{vi3}}\right)^{q_v} + \sigma_{vi1}\left(\frac{\tilde{\Phi}_{vi}^2}{2\gamma_{vi1}}\right)^{p_v}\\
& + \sigma_{vi2}\left(\frac{\rho_{T_i}\tilde{g}_{vi}^2}{2\gamma_{vi2}}\right)^{p_v} + \sigma_{vi3}\left(\frac{\rho_{T_i}\tilde{\beta}_{vi}^2}{2\gamma_{vi3}}\right)^{p_v} - \frac{\sigma_{vi1}}{2\gamma_{vi1}}\tilde{\Phi}_{vi}^2\\
& - \frac{\rho_{T_i}\sigma_{vi2}}{2\gamma_{vi2}}\tilde{g}_{vi}^2 - \frac{\rho_{T_i}\sigma_{vi3}}{2\gamma_{vi3}}\tilde{\beta}_{vi}^2 + \frac{\vartheta_{vi1}\sigma_{vi1}}{2}\Phi_{vi}^2\\
& + \frac{\vartheta_{vi2}\rho_{T_i}\sigma_{vi2}}{2}g_{vi}^2 + \frac{\vartheta_{vi3}\rho_{T_i}\sigma_{vi3}}{2}\beta_{vi}^2 + 0.557\varepsilon_{vi}\rho_{T_i}\\
& + \sum_{l=1}^{3}\left(\frac{h_{vi1}^2}{2} + \frac{h_{vi2}^2\bar{\tau}_{vil}^2}{2} + \sigma_{vil}(1-q_v)\,q_v^{\frac{q_v}{1-q_v}}\right),
\end{aligned}
\tag{3.47}
$$

where the following inequalities are used from Lemma 3.4:

$$
\left(\frac{\tilde{\Phi}_{vi}^2}{2\gamma_{vi1}}\right)^{q_v} \le \frac{\tilde{\Phi}_{vi}^2}{2\gamma_{vi1}} + (1-q_v)\,q_v^{\frac{q_v}{1-q_v}}
$$

$$\left(\frac{\rho_{T_i}\tilde{g}_{vi}^2}{2\gamma_{vi2}}\right)^{q_v} \le \frac{\rho_{T_i}\tilde{g}_{vi}^2}{2\gamma_{vi2}} + (1-q_v)\, q_v{}^{\frac{q_v}{1-q_v}}$$

$$\left(\frac{\rho_{T_i}\tilde{\beta}_{vi}^2}{2\gamma_{vi3}}\right)^{q_v} \le \frac{\rho_{T_i}\tilde{\beta}_{vi}^2}{2\gamma_{vi3}} + (1-q_v)\, q_v{}^{\frac{q_v}{1-q_v}}. \tag{3.48}$$

Suppose that there exist unknown constants Δ_{vi1}, Δ_{vi2}, Δ_{vi3} such that $|\tilde{\Phi}_{vi}| \le \Delta_{vi1}$, $\sqrt{\rho_{T_i}}|\tilde{g}_{vi}| \le \Delta_{vi2}$, $\sqrt{\rho_{T_i}}|\tilde{\beta}_{vi}| \le \Delta_{vi3}$. The following two cases are discussed:

Case 1 If $\Delta_{vi1} < \sqrt{2\gamma_{vi1}}$ or $\Delta_{vi2} < \sqrt{2\gamma_{vi2}}$ or $\Delta_{vi3} < \sqrt{2\gamma_{vi3}}$, then

$$\sigma_{vi1}\left(\frac{\tilde{\Phi}_{vi}^2}{2\gamma_{vi1}}\right)^{p_v} - \frac{\sigma_{vi1}}{2\gamma_{vi1}}\tilde{\Phi}_{vi}^2 < 0$$

$$\sigma_{vi2}\left(\frac{\rho_{T_i}\tilde{g}_{vi}^2}{2\gamma_{vi2}}\right)^{p_v} - \frac{\rho_{T_i}\sigma_{vi2}}{2\gamma_{vi2}}\tilde{g}_{vi}^2 < 0$$

$$\sigma_{vi3}\left(\frac{\rho_{T_i}\tilde{\beta}_{vi}^2}{2\gamma_{vi3}}\right)^{p_v} - \frac{\rho_{T_i}\sigma_{vi3}}{2\gamma_{vi3}}\tilde{\beta}_{vi}^2 < 0. \tag{3.49}$$

Case 2 If $\Delta_{vi1} \ge \sqrt{2\gamma_{vi1}}$ or $\Delta_{vi2} \ge \sqrt{2\gamma_{vi2}}$ or $\Delta_{vi3} \ge \sqrt{2\gamma_{vi3}}$, then

$$\sigma_{vi1}\left(\frac{\tilde{\Phi}_{vi}^2}{2\gamma_{vi1}}\right)^{p_v} - \frac{\sigma_{vi1}}{2\gamma_{vi1}}\tilde{\Phi}_{vi}^2 \le \sigma_{vi1}\left(\frac{\Delta_{vi1}^2}{2\gamma_{vi1}}\right)^{p_v} - \frac{\sigma_{vi1}}{2\gamma_{vi1}}\Delta_{vi1}^2$$

$$\sigma_{vi2}\left(\frac{\rho_{T_i}\tilde{g}_{vi}^2}{2\gamma_{vi2}}\right)^{p_v} - \frac{\rho_{T_i}\sigma_{vi2}}{2\gamma_{vi2}}\tilde{g}_{vi}^2 \le \sigma_{vi2}\left(\frac{\Delta_{vi2}^2}{2\gamma_{vi2}}\right)^{p_v} - \frac{\sigma_{vi2}}{2\gamma_{vi2}}\Delta_{vi2}^2$$

$$\sigma_{vi3}\left(\frac{\rho_{T_i}\tilde{\beta}_{vi}^2}{2\gamma_{vi3}}\right)^{p_v} - \frac{\rho_{T_i}\sigma_{vi3}}{2\gamma_{vi3}}\tilde{\beta}_{vi}^2 < \sigma_{vi3}\left(\frac{\Delta_{vi3}^2}{2\gamma_{vi3}}\right)^{p_v} - \frac{\sigma_{vi3}}{2\gamma_{vi3}}\Delta_{vi3}^2. \tag{3.50}$$

Summarizing case 1 and case 2, it can be concluded that

$$\upsilon_{vi1} = \begin{cases} 0, & \text{if } \Delta_{vi1} < \sqrt{2\gamma_{vi1}} \\ \sigma_{vi1}\left(\frac{\Delta_{vi1}^2}{2\gamma_{vi1}}\right)^{p_v} - \frac{\sigma_{vi1}}{2\gamma_{vi1}}\Delta_{vi1}^2, & \text{if } \Delta_{vi1} \ge \sqrt{2\gamma_{vi1}} \end{cases}$$

$$\upsilon_{vi2} = \begin{cases} 0, & \text{if } \Delta_{vi2} < \sqrt{2\gamma_{vi2}} \\ \sigma_{vi2}\left(\frac{\Delta_{vi2}^2}{2\gamma_{vi2}}\right)^{p_v} - \frac{\sigma_{vi2}}{2\gamma_{vi2}}\Delta_{vi2}^2, & \text{if } \Delta_{vi2} \ge \sqrt{2\gamma_{vi2}} \end{cases}$$

$$\upsilon_{\upsilon l3} = \begin{cases} 0, & \text{if } \Delta_{\upsilon l3} < \sqrt{2\gamma_{\upsilon l3}} \\ \sigma_{\upsilon l3}\left(\frac{\Delta_{\upsilon l3}^2}{2\gamma_{\upsilon l3}}\right)^{p_\upsilon} - \frac{\sigma_{\upsilon l3}}{2\gamma_{\upsilon l3}}\Delta_{\upsilon l3}^2, & \text{if } \Delta_{\upsilon l3} \geq \sqrt{2\gamma_{\upsilon l3}}. \end{cases} \tag{3.51}$$

From (3.47) and (3.51), we see that

$$\dot{L}_1 \leq -\chi_{\upsilon l1} L_1{}^{p_\upsilon} - \chi_{\upsilon l2} L_1{}^{q_\upsilon} + \chi_{\upsilon i3}, \tag{3.52}$$

where $\chi_{\upsilon l1} = \min\{c_{\upsilon l1}2^{p_\upsilon}, \sigma_{\upsilon l1}, \sigma_{\upsilon l2}, \sigma_{\upsilon l3}\}$, $\chi_{\upsilon l2} = \min\{c_{\upsilon i2}2^{q_\upsilon}, \sigma_{\upsilon l1}, \sigma_{\upsilon l2}, \sigma_{\upsilon i3}\}$, and $\chi_{\upsilon l3} = \upsilon_{\upsilon l1} + \upsilon_{\upsilon l2} + \upsilon_{\upsilon l3} + \frac{\vartheta_{\upsilon l1}\sigma_{\upsilon l1}}{2}\Phi_{\upsilon l}^2 + \frac{\vartheta_{\upsilon l2}\rho_{T_l}\sigma_{\upsilon l2}}{2}g_{\upsilon l}^2 + \frac{\vartheta_{\upsilon i3}\rho_{T_l}\sigma_{\upsilon i3}}{2}\beta_{\upsilon l}^2 + 0.557\rho_{T_l}\varepsilon_{\upsilon l} + \sum_{l=1}^{3}\left(\frac{h_{\upsilon l1}^2}{2} + \frac{h_{\upsilon l2}^2\bar{\tau}_{\upsilon ll}^2}{2} + \sigma_{\upsilon ll}(1-q_\upsilon)q_\upsilon{}^{\frac{q_\upsilon}{1-q_\upsilon}}\right)$.

In accordance with Lemma 3.1, all signals of translational subsystem are semi-globally practically fixed-time bounded and converge to the following compact set: $L_1 \leq \min\{(\frac{\chi_{\upsilon l3}}{\chi_{\upsilon l1}(1-\sigma)})^{\frac{1}{p_\upsilon}}, (\frac{\chi_{\upsilon l3}}{\chi_{\upsilon l2}(1-\sigma)})^{\frac{1}{q_\upsilon}}\}$, and the setting time is $T \leq T_{\max,\upsilon} := \frac{1}{\chi_{\upsilon l1}\sigma(p_\upsilon - 1)} + \frac{1}{\chi_{\upsilon l2}\sigma(1-q_\upsilon)}$. It should be noticed that $|\tilde{V}_l|$ can be made arbitrarily small by increasing $c_{\upsilon l1}$, $c_{\upsilon l2}$, $\gamma_{\upsilon l1}$, $\gamma_{\upsilon l2}$, and $\gamma_{\upsilon l3}$ and meanwhile decreasing $h_{\upsilon l1}$, $h_{\upsilon l2}$, $\sigma_{\upsilon l1}$, $\sigma_{\upsilon l2}$, $\sigma_{\upsilon l3}$, and $\varepsilon_{\upsilon l}$. The proper choice of p_υ and q_υ helps to reduce convergence time and improve convergence accuracy. This will be shown in the following numerical example. This completes the proof of Theorem 3.1.

Theorem 3.2 *Consider the rotational kinematics (3.14) and (3.18) composed of the controller (3.34) and adaptation laws (3.35) and (3.36). Let Assumptions 3.1–3.4 hold. There exist positive parameters* $\boldsymbol{K}_{2i}$, $\boldsymbol{K}_{3l}$, $\boldsymbol{K}_{4l}$, $\boldsymbol{K}_{5i}$, $\varepsilon_{\omega l}$, $h_{\omega l1}$, $h_{\omega l2}$, $\gamma_{\omega l1}$, $\gamma_{\omega l2}$, $\sigma_{\omega l1}$, *and* $\sigma_{\omega l2}$ $(i = 1, 2, \ldots, N)$ *such that:*

(1) All signals of the rotational subsystem are SGUUB in spite of actuator faults, modeling uncertainties, and external disturbances.
(2) Attitude synchronization tracking error $\boldsymbol{e}_{\varphi l}$ *achieves that* $\lim_{t\to T_{\max,\varphi}} \|\boldsymbol{e}_{\varphi l}\| \leq \mu_{\varphi l}$ *with* $\mu_{\varphi l} > 0$ *a constant.*
(3) The involved attitude constraints for rotational subsystem are guaranteed for $\forall t \geq 0$.

Proof The following proof comprises two steps:
Step 1: Using (3.14), (3.25), and (3.27) yields

$$\dot{\boldsymbol{\xi}}_{\varphi l} = \eta_{\varphi l}\left[\lambda_l\left(\boldsymbol{R}_2^{-1}\boldsymbol{K}_{1l}\boldsymbol{\xi}_{\omega l} - \dot{\boldsymbol{\varphi}}_r\right) - \lambda_{l2}\sum_{j\in\mathcal{N}_l} a_{lj}\dot{\tilde{\boldsymbol{\varphi}}}_j\right] + \boldsymbol{\mu}_{\varphi l}, \tag{3.53}$$

where $\boldsymbol{K}_{1l} = \text{diag}\{(F_{\omega l11} + \omega_{l1})(F_{\omega l12} - \omega_{l1}), (F_{\omega i21} + \omega_{l2})(F_{\omega l22} - \omega_{l2}), (F_{\omega l31} + \omega_{l3})(F_{\omega l32} - \omega_{l3})\}$.

Based on (3.53), we design the virtual controller $\boldsymbol{\xi}_{\omega cl}$ as

$$\boldsymbol{\xi}_{\omega cl} = -\frac{\boldsymbol{K}_{1l}^{-1}\boldsymbol{R}_2\boldsymbol{\eta}_{\varphi l}^{-1}}{\lambda_l}\left(\boldsymbol{K}_{2l}\boldsymbol{\xi}_{\varphi l}\left\|\boldsymbol{\xi}_{\varphi l}\right\|^{2p_\varphi-2} + \boldsymbol{K}_{3l}\boldsymbol{\xi}_{\varphi l}\left\|\boldsymbol{\xi}_{\varphi l}\right\|^{2q_\varphi-2}\right)$$
$$-\frac{\boldsymbol{K}_{1l}^{-1}\boldsymbol{R}_2}{\lambda_l}\left(\boldsymbol{\eta}_{\varphi l}^{-1}\boldsymbol{\mu}_{\varphi l} - \lambda_l\dot{\boldsymbol{\varphi}}_r - \lambda_{l2}\sum_{j\in\mathcal{N}_l}a_{lj}\dot{\boldsymbol{\varphi}}_j\right). \tag{3.54}$$

To avoid the direct differentiation of complex $\boldsymbol{\xi}_{\omega cl}$, we introduce a new variable $\boldsymbol{\xi}_{\omega dl}$ using dynamic surface control [36]. The following nonlinear filter is exploited to guarantee the overall fixed-time convergence.

$$\dot{\boldsymbol{\xi}}_{\omega dl} = -\kappa_{\omega l}\left(\boldsymbol{y}_{\omega l}\left\|\boldsymbol{y}_{\omega l}\right\|^{r_{\varphi 1}-1} + \boldsymbol{y}_{\omega l}\left\|\boldsymbol{y}_{\omega l}\right\|^{r_{\varphi 2}-1}\right). \tag{3.55}$$

Let $\boldsymbol{e}_{\omega l} = \boldsymbol{\xi}_{\omega l} - \boldsymbol{\xi}_{\omega dl}$. Substituting (3.54) and (3.55) into (3.53), the derivative of $\frac{1}{2}\boldsymbol{\xi}_{\varphi l}^{\mathrm{T}}\boldsymbol{\xi}_{\varphi l}$ denotes

$$\boldsymbol{\xi}_{\varphi l}^{\mathrm{T}}\dot{\boldsymbol{\xi}}_{\varphi l} = -\boldsymbol{\xi}_{\varphi l}^{\mathrm{T}}\left(\boldsymbol{K}_{2l}\boldsymbol{\xi}_{\varphi l}\left\|\boldsymbol{\xi}_{\varphi l}\right\|^{2p_\varphi-2} + \boldsymbol{K}_{3l}\boldsymbol{\xi}_{\varphi l}\left\|\boldsymbol{\xi}_{\varphi l}\right\|^{2q_\varphi-2}\right)$$
$$+\lambda_l\boldsymbol{\xi}_{\varphi l}^{\mathrm{T}}\boldsymbol{\eta}_{\varphi l}\boldsymbol{R}_2^{-1}\boldsymbol{K}_{1l}\left(\boldsymbol{e}_{\omega l} + \boldsymbol{y}_{\omega l}\right). \tag{3.56}$$

Step 2: Differentiating $\frac{1}{2}\boldsymbol{e}_{\omega l}^{\mathrm{T}}\boldsymbol{e}_{\omega l}$ along (3.18) and (3.29) gives

$$\boldsymbol{e}_{\omega l}^{\mathrm{T}}\dot{\boldsymbol{e}}_{\omega l} = \boldsymbol{e}_{\omega l}^{\mathrm{T}}\boldsymbol{\eta}_{\omega l}\boldsymbol{J}_{l0}^{-1}\boldsymbol{C}_0\boldsymbol{\rho}_{\delta_l}\boldsymbol{\delta}_{l0} + \boldsymbol{J}_{l0}^{-1}\boldsymbol{S}\left(\boldsymbol{J}_{l0}\boldsymbol{\omega}_l\right) + \boldsymbol{J}_{l0}^{-1}\boldsymbol{N}_{l0}$$
$$+\boldsymbol{e}_{\omega l}^{\mathrm{T}}\left[\boldsymbol{\eta}_{\omega l}\left(\boldsymbol{\Delta}_{\omega l}' + \boldsymbol{l}_{\omega l} + \boldsymbol{d}_{\omega l}\right) + \boldsymbol{\mu}_{\omega l} - \dot{\boldsymbol{\xi}}_{\omega dl}\right]. \tag{3.57}$$

Using Young's inequality, it holds along (3.19) that

$$\boldsymbol{e}_{\omega l}^{\mathrm{T}}\boldsymbol{\eta}_{\omega l}\left(\boldsymbol{\Delta}_{\omega l}' + \boldsymbol{l}_{\omega l} + \boldsymbol{d}_{\omega l}\right) \le \boldsymbol{e}_{\omega l}^{\mathrm{T}}\boldsymbol{\eta}_{\omega l}\left(\frac{\Phi_{\omega l}\boldsymbol{\Xi}_{\omega l}}{2h_{\omega l1}^2} + \frac{\boldsymbol{\eta}_{\omega l}^{\mathrm{T}}\boldsymbol{e}_{\omega l}}{2h_{\omega l2}^2}\right)$$
$$+\sum_{l=1}^{3}\left(\frac{h_{\omega l1}^2}{2} + \frac{h_{\omega l2}^2\bar{\tau}_{\omega ll}^2}{2}\right), \tag{3.58}$$

where $\Phi_{\omega l} = \max\left\{\boldsymbol{W}_{\omega l1}^{*\mathrm{T}}\boldsymbol{W}_{\omega l1}^*, \boldsymbol{W}_{\omega l2}^{*\mathrm{T}}\boldsymbol{W}_{\omega l2}^*, \boldsymbol{W}_{\omega l3}^{*\mathrm{T}}\boldsymbol{W}_{\omega l3}^*\right\}$, and there exists $|\tau_{\omega ll} + l_{\omega ll} + d_{\omega ll}| \le \bar{\tau}_{\omega ll}$.

According to Assumption 3.2, we define $g_{\omega l} = \inf_{t\ge 0}\lambda_{\min}(\boldsymbol{J}_{l0}^{-1}\boldsymbol{C}_0(\boldsymbol{\delta}_l)\boldsymbol{\rho}_{\delta_l}(\boldsymbol{J}_{l0}^{-1}\boldsymbol{C}_0(\boldsymbol{\delta}_l))^{\mathrm{T}})$ and $\beta_{\omega l} = 1/g_{\omega l}$. Let $\tilde{\Phi}_{\omega l} = \hat{\Phi}_{\omega l} - \Phi_{\omega l}$ and $\tilde{\beta}_{\omega l} = \hat{\beta}_{\omega l} - \beta_{\omega l}$ with $\hat{\Phi}_{\omega l}$ and $\hat{\beta}_{\omega l}$ being the estimations of $\Phi_{\omega l}$ and $\beta_{\omega l}$, respectively.

Consider the Lyapunov function candidate as

$$L_2 = \frac{1}{2}\boldsymbol{\xi}_{\varphi i}^{\mathrm{T}}\boldsymbol{\xi}_{\varphi i} + \frac{1}{2}\boldsymbol{e}_{\omega i}^{\mathrm{T}}\boldsymbol{e}_{\omega i} + \frac{1}{2}\boldsymbol{y}_{\omega i}^{\mathrm{T}}\boldsymbol{y}_{\omega i} + \frac{1}{2\gamma_{\omega i1}}\tilde{\Phi}_{\omega i}^{2} + \frac{g_{\omega i}}{2\gamma_{\omega i2}}\tilde{\beta}_{\omega i}^{2}. \tag{3.59}$$

From (3.56)–(3.58), taking the time derivative of (3.59) gives

$$\begin{aligned}
\dot{L}_2 \le & -\boldsymbol{\xi}_{\varphi i}^{\mathrm{T}}\left(\boldsymbol{K}_{2i}\boldsymbol{\xi}_{\varphi i}\|\boldsymbol{\xi}_{\varphi i}\|^{2p_\varphi-2}+\boldsymbol{K}_{3i}\boldsymbol{\xi}_{\varphi i}\|\boldsymbol{\xi}_{\varphi i}\|^{2q_\varphi-2}\right)+\boldsymbol{e}_{\omega i}^{\mathrm{T}}\boldsymbol{\zeta}_{\omega i}\\
& - \boldsymbol{e}_{\omega i}^{\mathrm{T}}\left(\boldsymbol{K}_{4i}\boldsymbol{e}_{\omega i}\|\boldsymbol{e}_{\omega i}\|^{2p_\varphi-2} + \boldsymbol{K}_{5i}\boldsymbol{e}_{\omega i}\|\boldsymbol{e}_{\omega i}\|^{2q_\varphi-2}\right)\\
& + \boldsymbol{e}_{\omega i}^{\mathrm{T}}\boldsymbol{\eta}_{\omega i}\boldsymbol{J}_{i0}^{-1}\boldsymbol{C}_0\boldsymbol{\rho}_{\delta_i}\boldsymbol{\delta}_{i0} - \boldsymbol{e}_{\omega i}^{\mathrm{T}}\frac{\tilde{\Phi}_{\omega i}\boldsymbol{\eta}_{\omega i}\boldsymbol{\Xi}_{\omega i}}{2h_{\omega i1}^{2}}\\
& + \boldsymbol{y}_{\omega i}^{\mathrm{T}}\left[\dot{\boldsymbol{y}}_{\omega i} + \lambda_i\boldsymbol{K}_{1i}^{\mathrm{T}}\left(\boldsymbol{R}_2^{-1}\right)^{\mathrm{T}}\boldsymbol{\eta}_{\varphi i}^{\mathrm{T}}\boldsymbol{\xi}_{\varphi i}\right]\\
& + \frac{1}{\gamma_{\omega i1}}\tilde{\Phi}_{\omega i}\dot{\hat{\Phi}}_{\omega i} + \frac{g_{\omega i}}{\gamma_{\omega l2}}\tilde{\beta}_{\omega i}\dot{\hat{\beta}}_{\omega i}\\
& + \sum_{l=1}^{3}\left(\frac{h_{\omega i1}^{2}}{2} + \frac{h_{\omega i2}^{2}\bar{\tau}_{\omega il}^{2}}{2}\right).
\end{aligned} \tag{3.60}$$

In (3.60), define $\boldsymbol{\alpha}_{\varphi i} = \lambda_i\boldsymbol{K}_{1i}^{\mathrm{T}}\left(\boldsymbol{R}_2^{-1}\right)^{\mathrm{T}}\boldsymbol{\eta}_{\varphi i}^{\mathrm{T}}\boldsymbol{\xi}_{\varphi i}$. Noting that the inequality holds: $x \le x^m + x^n$, where $x \ge 0$, $0 < m < 1$, and $n > 1$, it follows that

$$\boldsymbol{y}_{\omega i}^{\mathrm{T}}\left(\dot{\boldsymbol{y}}_{\omega i} + \boldsymbol{\alpha}_{\varphi i}\right) \le -\kappa_{\omega i}\left(\|\boldsymbol{y}_{\omega i}\|^{r_{\varphi 1}+1} + \|\boldsymbol{y}_{\omega i}\|^{r_{\varphi 2}+1}\right) + \frac{\varpi_{\omega i}^{2}}{2}, \tag{3.61}$$

where $\kappa_{\omega i} = \lambda_{\min}(\boldsymbol{\kappa}_{\omega i}) - 1/2$, and there exists a positive constant $\varpi_{\omega i}$ such that $\|\boldsymbol{\alpha}_{\varphi i} - \dot{\boldsymbol{\xi}}_{\omega ci}\| \le \varpi_{\omega i}$.

Using (3.34) and invoking Lemma 3.2, the term of $\boldsymbol{e}_{\omega i}^{\mathrm{T}}\boldsymbol{\eta}_{\omega i}\boldsymbol{J}_{i0}^{-1}\boldsymbol{C}_0\boldsymbol{\rho}_{\delta_i}\boldsymbol{\delta}_{i0}$ renders

$$\boldsymbol{e}_{\omega i}^{\mathrm{T}}\boldsymbol{\eta}_{\omega i}\boldsymbol{J}_{i0}^{-1}\boldsymbol{C}_0\boldsymbol{\rho}_{\delta_i}\boldsymbol{\delta}_{i0} = -\boldsymbol{e}_{\omega i}^{\mathrm{T}}\boldsymbol{\eta}_{\omega i}\boldsymbol{J}_{i0}^{-1}\boldsymbol{C}_0\boldsymbol{\rho}_{\delta_i}\left(\boldsymbol{J}_{i0}^{-1}\boldsymbol{C}_0\right)^{\mathrm{T}}\boldsymbol{\eta}_{\omega i}^{-1}.$$

$$\begin{aligned}
\frac{\boldsymbol{e}_{\omega i}\hat{\beta}_{\omega i}^{2}\boldsymbol{\zeta}_{\omega i}^{\mathrm{T}}\boldsymbol{\zeta}_{\omega i}}{\sqrt{\hat{\beta}_{\omega i}^{2}\boldsymbol{e}_{\omega i}^{\mathrm{T}}\boldsymbol{e}_{\omega i}\boldsymbol{\zeta}_{\omega i}^{\mathrm{T}}\boldsymbol{\zeta}_{\omega i} + \varepsilon_{\omega i}}} & \le -\frac{g_{\omega i}\hat{\beta}_{\omega i}^{2}\boldsymbol{e}_{\omega i}^{\mathrm{T}}\boldsymbol{e}_{\omega i}\boldsymbol{\zeta}_{\omega i}^{\mathrm{T}}\boldsymbol{\zeta}_{\omega i}}{\sqrt{\hat{\beta}_{\omega i}^{2}\boldsymbol{e}_{\omega i}^{\mathrm{T}}\boldsymbol{e}_{\omega i}\boldsymbol{\zeta}_{\omega i}^{\mathrm{T}}\boldsymbol{\zeta}_{\omega i} + \varepsilon_{\omega i}}}\\
& \le g_{\omega i}\sqrt{\varepsilon_{\omega i}} - g_{\omega i}|\hat{\beta}_{\omega i}|\,\|\boldsymbol{e}_{\omega i}\|\,\|\boldsymbol{\zeta}_{\omega i}\|\\
& \le g_{\omega i}\sqrt{\varepsilon_{\omega i}} - g_{\omega i}\hat{\beta}_{\omega i}\boldsymbol{e}_{\omega i}^{\mathrm{T}}\boldsymbol{\zeta}_{\omega i}.
\end{aligned} \tag{3.62}$$

Based on $g_{\omega i}(\hat{\beta}_{\omega i} - \tilde{\beta}_{\omega i}) = g_{\omega i}\beta_{\omega i} = 1$, substitute (3.35), (3.36), (3.61), and (3.62) into (3.60). With the similar operations of (3.46)–(3.48), it holds that

$$\begin{aligned}
\dot{L}_2 \le & -\lambda_{\min}(\boldsymbol{K}_{2i})\left\|\boldsymbol{\xi}_{\varphi i}\right\|^{2p_\varphi} - \lambda_{\min}(\boldsymbol{K}_{3i})\left\|\boldsymbol{\xi}_{\varphi i}\right\|^{2q_\varphi} \\
& - \lambda_{\min}(\boldsymbol{K}_{4i})\left\|\boldsymbol{e}_{\omega i}\right\|^{2p_\varphi} - \lambda_{\min}(\boldsymbol{K}_{5i})\left\|\boldsymbol{e}_{\omega i}\right\|^{2q_\varphi} \\
& - \kappa_{\omega i}\left\|\boldsymbol{y}_{\omega i}\right\|^{r_{\varphi 1}+1} - \kappa_{\omega i}\left\|\boldsymbol{y}_{\omega i}\right\|^{r_{\varphi 2}+1} + g_{\omega i}\sqrt{\varepsilon_{\omega i}} \\
& - \sigma_{\omega i1}\left(\frac{\tilde{\Phi}_{\omega i}^2}{2\gamma_{\omega i1}}\right)^{p_\varphi} - \sigma_{\omega i1}\left(\frac{\tilde{\Phi}_{\omega i}^2}{2\gamma_{\omega i1}}\right)^{q_\varphi} - \sigma_{\omega i2}\left(\frac{g_{\omega i}\tilde{\beta}_{\omega i}^2}{2\gamma_{\omega i2}}\right)^{p_\varphi} \\
& - \sigma_{\omega i2}\left(\frac{g_{\omega i}\tilde{\beta}_{\omega i}^2}{2\gamma_{\omega i2}}\right)^{q_\varphi} + \sigma_{\omega i1}\left(\frac{\tilde{\Phi}_{\omega i}^2}{2\gamma_{\omega i1}}\right)^{p} + \sigma_{\omega i2}\left(\frac{g_{\omega i}\tilde{\beta}_{\omega i}^2}{2\gamma_{\omega i2}}\right)^{p_\varphi} \\
& - \frac{\sigma_{\omega i1}}{2\gamma_{\omega i1}}\tilde{\Phi}_{\omega i}^2 - \frac{g_{\omega i}\sigma_{\omega i2}}{2\gamma_{\omega i2}}\tilde{\beta}_{\omega i}^2 + \frac{\vartheta_{\omega i1}\sigma_{\omega i1}}{2}\Phi_{\omega i}^2 \\
& + \frac{\vartheta_{\omega i2}g_{\omega i}\sigma_{\omega i2}}{2}\beta_{\omega i}^2 + \sum_{l=1}^{2}\sigma_{\omega il}\left(1-q_\varphi\right)q_\varphi^{\frac{q_\varphi}{1-q_\varphi}} \\
& + \sum_{l=1}^{3}\left(\frac{h_{\omega i1}^2}{2} + \frac{h_{\omega i2}^2\bar{\tau}_{\omega il}^2}{2}\right) + \frac{\varpi_{\omega i}^2}{2}.
\end{aligned} \tag{3.63}$$

Suppose that there exist unknown constants $\Delta_{\omega i1}$, $\Delta_{\omega i2}$ such that $|\tilde{\Phi}_{\omega i}| \le \Delta_{\omega i1}$, $\sqrt{g_{\omega i}}|\tilde{\beta}_{\omega i}| \le \Delta_{\omega i2}$. Following case 1 and case 2 in (3.49) and (3.50), one has

$$\begin{aligned}
\upsilon_{\omega i1} &= \begin{cases} 0, & \text{if } \Delta_{\omega i1} < \sqrt{2\gamma_{\omega i1}} \\ \sigma_{\omega i1}\left(\frac{\Delta_{\omega i1}^2}{2\gamma_{\omega i1}}\right)^{p_\varphi} - \frac{\sigma_{\omega i1}}{2\gamma_{\omega i1}}\Delta_{\omega i1}^2, & \text{if } \Delta_{\omega i1} \ge \sqrt{2\gamma_{\omega i1}} \end{cases} \\
\upsilon_{\omega i2} &= \begin{cases} 0, & \text{if } \Delta_{\omega i2} < \sqrt{2\gamma_{\omega i2}} \\ \sigma_{\omega i2}\left(\frac{\Delta_{\omega i2}^2}{2\gamma_{\omega i2}}\right)^{p_\varphi} - \frac{\sigma_{\omega i2}}{2\gamma_{\omega i2}}\Delta_{\omega i2}^2, & \text{if } \Delta_{\omega i2} \ge \sqrt{2\gamma_{\omega i2}}. \end{cases}
\end{aligned} \tag{3.64}$$

Choosing $\frac{r_{\varphi 1}+1}{2} = q_\varphi$ and $\frac{r_{\varphi 2}+1}{2} = p_\varphi$ and substituting (3.64) into (3.63), it can be checked that

$$\dot{L}_2 \le -\chi_{\omega i1}L_2{}^{p_\varphi} - \chi_{\omega i2}L_2{}^{q_\varphi} + \chi_{\omega i3}, \tag{3.65}$$

where $\chi_{\omega i1} = \min\{\lambda_{\min}(\boldsymbol{K}_{2i})2^{p_\varphi}, \lambda_{\min}(\boldsymbol{K}_{4i})2^{p_\varphi}, \kappa_{\omega i}2^{p_\varphi}, \sigma_{\omega i1}, \sigma_{\omega i2}\}$, $\chi_{\omega i2} = \min\{\lambda_{\min}(\boldsymbol{K}_{3i})2^{q_\varphi}, \lambda_{\min}(\boldsymbol{K}_{5i})2^{q_\varphi}, \kappa_{\omega i}2^{q_\varphi}, \sigma_{\omega i1}, \sigma_{\omega i2}\}$, $\chi_{vi3} = \upsilon_{\omega i1} + \upsilon_{\omega i2} +$

$\frac{\vartheta_{\omega l1}\sigma_{\omega l1}}{2}\Phi_{\omega l}^2 + \frac{\vartheta_{\omega l2}g_{\omega l}\sigma_{\omega l2}}{2}\beta_{\omega l}^2 + g_{\omega l}\sqrt{\varepsilon_{\omega l}} + \sum_{l=1}^{3}(\frac{h_{\omega l1}^2}{2} + \frac{h_{\omega l2}^2\bar{\tau}_{\omega ll}^2}{2}) + \frac{\varpi_{\omega l}^2}{2} + \sum_{l=1}^{2}\sigma_{\omega ll}(1-$
$q_{\varphi})\, q_{\varphi}^{\frac{q_{\varphi}}{1-q_{\varphi}}}$.

Along similar lines to Lemma 3.1, all signals of rotational subsystem are semi-globally practically fixed-time bounded and converge to the following compact set $L_2 \leq \min\{(\frac{\chi_{\omega l3}}{\chi_{\omega l1}(1-\sigma)})^{\frac{1}{p_{\varphi}}}, (\frac{\chi_{\omega l3}}{\chi_{\omega l2}(1-\sigma)})^{\frac{1}{q_{\varphi}}}\}$, and the setting time is $T \leq T_{\max,\varphi} := \frac{1}{\chi_{\omega l1}\sigma(p_{\varphi}-1)} + \frac{1}{\chi_{\omega l2}\sigma(1-q_{\varphi})}$. Note that the convergence region of $\boldsymbol{e}_{\varphi l}$ can be made smaller by increasing $\boldsymbol{K}_{2l}$, $\boldsymbol{K}_{3l}$, $\boldsymbol{K}_{4l}$, $\boldsymbol{K}_{5l}$, $\gamma_{\omega l1}$, $\gamma_{\omega l2}$ and meanwhile decreasing $h_{\omega l1}$, $h_{\omega l2}$, $\sigma_{\omega l1}$, $\sigma_{\omega l2}$, and $\varepsilon_{\omega l}$. Thus attitude synchronization tracking error can be made as small as desired by an appropriate choice of the design parameters. Moreover, the choice of p_{φ} and q_{φ} requires to balance the control amplitude and convergence time simultaneously. This completes the proof of Theorem 3.2.

Remark 3.4 By running the distributed control algorithm in parallel for each UAV, the proposed distributed control structure optimizes the time complexity, compared with the centralized control with an integral control algorithm for all UAVs [5]. Besides, note that only three scalar parameter adaptation laws (3.31)–(3.33) for translational dynamics and two scalar parameter adaptation laws (3.35), (3.36) for rotational dynamics are involved in our control design to deal with the actuator faults, modeling uncertainties, and disturbances, which makes it simpler than vector-based adaptation laws in backstepping control for UAVs [10]. In addition, our proposed fixed-time control owns the superiority in terms of convergence time, where the time convergence is independent of initial condition. That is, the proposed fixed-time control can effectively shorten the execution time, which is comparable to time complexity of asymptotic controls and finite-time controls proposed for UAVs [1–3, 27–29, 31, 32].

Remark 3.5 The network induced errors and actuator faults are effectively handled with the help of bound estimations for parameters $\beta_{vl} = 1/\rho_{T_l}$, $g_{vl} = \beta_{vl}\sup_{t\geq 0} T_{xlf}$ in translational control protocol and $\beta_{\omega l} = 1/g_{\omega l}$, $g_{\omega l} = \inf_{t\geq 0}\lambda_{\min}(\boldsymbol{J}_{l0}^{-1}\boldsymbol{C}_0(\boldsymbol{\delta}_l)\boldsymbol{\rho}_{\delta l}\left(\boldsymbol{J}_{l0}^{-1}\boldsymbol{C}_0(\boldsymbol{\delta}_l)\right)^{\mathrm{T}})$ in rotational control design. Further, we employ the smooth function $\vartheta\tanh\left(\frac{\vartheta}{\varepsilon}\right)$ in design of T_{xl0} in (3.30) and function $\frac{z^2}{\sqrt{z^2+\varepsilon}}$ when designing $\boldsymbol{\delta}_{l0}$ in (3.34).

3.4 Semi-Physical Experiments and Analyses

To demonstrate the feasibility of proposed controller, the experimental validation is carried out based on the Links-RT UAV Platform (supported by Beijing Links Co., Ltd.). The experiment setup and relationship between functional units are depicted

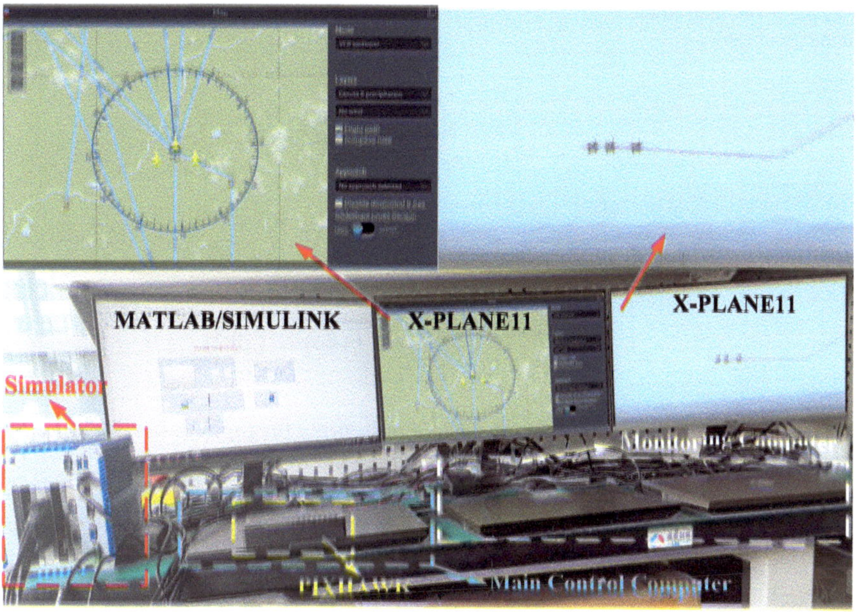

Fig. 3.1 Experiment prototype of the Links-RT UAV Platform

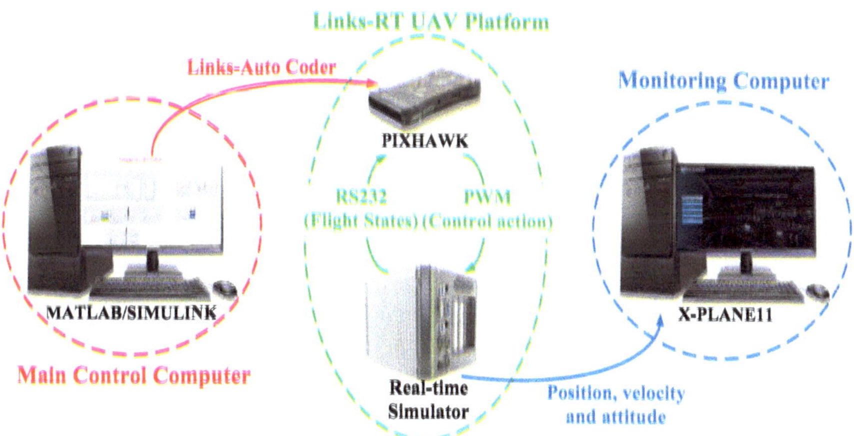

Fig. 3.2 Relationship between functional units

in Figs. 3.1 and 3.2. The proposed controller is first implemented in MATLAB. Then the code generation by Links-Auto Coder is used to convert the MATLAB language into C code, which is downloaded by Pixhawk. The Pixhawk is responsible for executing the proposed control algorithm with a sampling time of 2 ms and generating pulse-width modulation (PWM) signals, which are sent to the real-time simulator, such that the velocity and attitude motion are calculated and transmitted

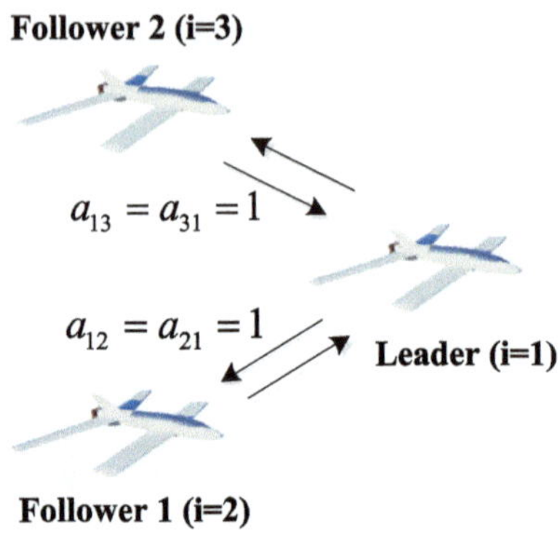

Fig. 3.3 Communication topology of UAVs

to Pixhawk and monitoring computer while presenting the flight scene in X-PLANE 11 and plotting the velocity and attitude tracking curves on MATLAB.

In this part, three networked 6-DoF fixed-wing UAVs are considered. The communication topology is illustrated in Fig. 3.3. The initial states are selected as $\boldsymbol{p}_1 = [20, 0, 150]^{\mathrm{T}}\mathrm{m}$, $\boldsymbol{p}_2 = [0, 20, 150]^{\mathrm{T}}\mathrm{m}$, $\boldsymbol{p}_3 = [0, -20, 150]^{\mathrm{T}}\mathrm{m}$, $V_1 = V_2 = 40\mathrm{m/s}$, $V_3 = 39.5\mathrm{m/s}$, $\boldsymbol{\varphi}_1 = [1, 0, 0]^{\mathrm{T}}\,\mathrm{deg}$, $\boldsymbol{\varphi}_2 = \boldsymbol{\varphi}_3 = [0, 0, 0]^{\mathrm{T}}\,\mathrm{deg}$, $\boldsymbol{\omega}_1 = \boldsymbol{\omega}_2 = \boldsymbol{\omega}_3 = [0, 0, 0]^{\mathrm{T}}\,\mathrm{deg/s}$. The control objective for experimental test is twofold:

(1) Task of low altitude penetration: Velocity tracks the trajectory $V_r = 60\mathrm{m/s}$, and the desired attitude trajectory is denoted as

$$[\phi_r, \theta_r, \psi_r]^{\mathrm{T}} = \begin{cases} [0, 0, 0]^{\mathrm{T}}\,\mathrm{deg}, & 0\mathrm{s} \le t \le 5\mathrm{s} \\ [0, -10, 0]^{\mathrm{T}}\,\mathrm{deg}, & 5\mathrm{s} < t \le 12\mathrm{s} \\ [0, 0, 0]^{\mathrm{T}}\,\mathrm{deg}, & 12\mathrm{s} < t \le 18\mathrm{s}. \end{cases}$$

The external disturbances in velocity channel are supposed as $\boldsymbol{d}_{v3} = [0.5\sin(3t), 0.3\cos(2t), 0.6\sin(2t)]^{\mathrm{T}}$ during $t \in (6, 15]\mathrm{s}$.

(2) The following full-state constraints are achieved: $0\mathrm{m/s} < V_i < 100\mathrm{m/s}$, $-60^\circ < \phi_i < 60^\circ$, $-60^\circ < \theta_i < 60^\circ$, $-180^\circ < \psi_i < 180^\circ$, $-10\,\mathrm{deg/s} < p_i < 10\,\mathrm{deg/s}$, $-10\,\mathrm{deg/s} < q_i < 10\,\mathrm{deg/s}$, $-5\,\mathrm{deg/s} < r_i < 5\,\mathrm{deg/s}$.

The control law and adaptation law are provided with design parameters as $p_v = 1.4$, $p_\varphi = 1.2$, $q_v = q_\varphi = 0.8$, $c_{vi1} = c_{vi2} = 300$, $\varepsilon_{vi} = \varepsilon_{\omega i} = 0.01$, $h_{v32} = 0.001$, $\kappa_{\omega i1} = \kappa_{\omega i2} = \kappa_{\omega i3} = 0.01$, $\gamma_{v22} = \sigma_{v22} = 20$, $\boldsymbol{K}_{2i} = \boldsymbol{K}_{3i} = \mathrm{diag}\{3 \times 10^{-4}, 4 \times 10^{-4}, 5 \times 10^{-4}\}$, $\boldsymbol{K}_{4i} = \boldsymbol{K}_{5i} = \mathrm{diag}\{50, 50, 50\}$ $(t \in (0, 5], (7, 12], (14, 18])$, and $\boldsymbol{K}_{4i} = \boldsymbol{K}_{5i} = \mathrm{diag}\{2, 2, 2\}(t \in (5, 7], (12, 14])$.

Experimental results are exhibited in Fig. 3.4a–d. Apparently, the proposed method can produce a rapid and accurate velocity and attitude tracking behavior because of the contribution of disturbance rejection and fixed-time property. In view of Fig. 3.4e–h, the tracking errors of velocity and attitude remain in a small

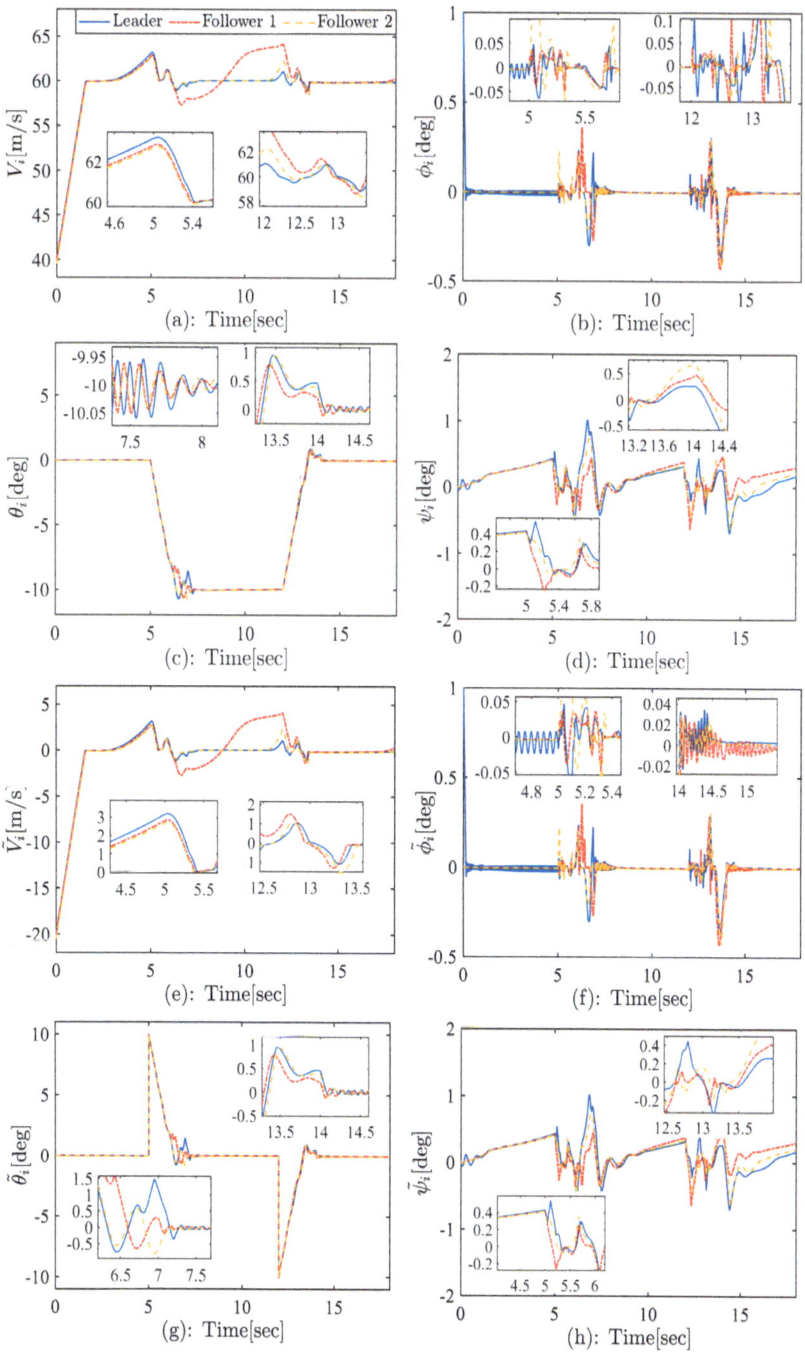

Fig. 3.4 Velocity and attitude tracking performance under proposed controller for experimental test. (**a**) V_i response. (**b**) ϕ_i response. (**c**) θ_i response. (**d**) ψ_i response. (**e**) Tracking error of V_i. (**f**) Tracking error of ϕ_i. (**g**) Tracking error of θ_i. (**h**) Tracking error of ψ_i

neighborhood of zero within fixed time. During the operation, the states strictly keep within corresponding constraints.

To highlight the superiority of suggested control scheme, the related traditional controller in [31], FTC in [26], and finite-time FTC in [1] are taken as comparative objects under the same conditions. Taking follower 1 for example, the actuator faults are introduced as $\rho_{\delta 2} = \text{diag}\{1,\ 0.9,\ 1\}$ during $t \in (15, 18]$. Task of coordinated turn and climb is considered: The velocity is supposed to be kept as 40m/s, and the attitude tracks the trajectory

$$[\phi_r, \theta_r, \psi_r]^T = \begin{cases} [0, 0, 0]^T \text{deg}, & 0s \leq t \leq 5s \\ [10, 10, 10]^T \text{deg}, & 5s < t \leq 15s \\ [0, 0, 10]^T \text{deg}, & 15s < t \leq 24s. \end{cases}$$

Comparative simulation results are shown in Fig. 3.5a–c. One can evidently find that although both FTC and finite-time FTC can accurately track the desired attitude commands regardless of actuator faults and disturbances, a slight rapid dynamic response and decreased tracking deviation can be discerned by the proposed scheme. What is more, it is distinct to see that no matter in transient and static

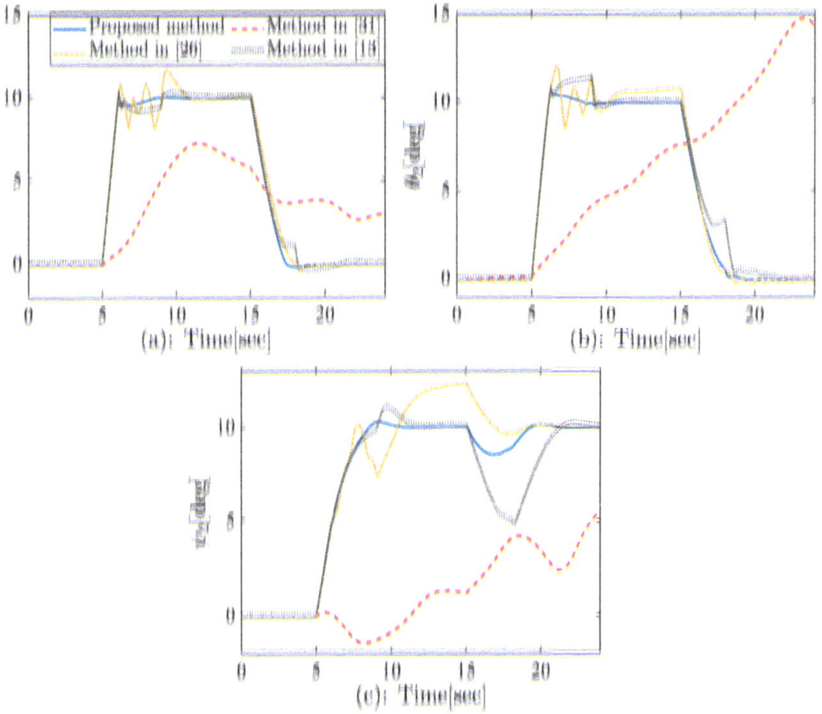

Fig. 3.5 Attitude tracking performances for comparison under the proposed method and methods in [1, 26, 31]. **(a)** ϕ_2 response. **(b)** θ_2 response. **(c)** ψ_2 response

phase, in contrast to severe fluctuations appearing in traditional controller [31] and larger overshoots resulting in FTC [26], controlling oscillations are effectively accommodated with the help of FTC and fixed-time property, naturally yielding the faster convergence performance and better robustness to the actuator faults and disturbances.

3.5 Conclusions

An adaptive fixed-time fault-tolerant control methodology has been designed for multiple 6-DoF UAVs with actuator faults and constraint guarantees. Specifically, system transformation with scaling function is introduced to guarantee the states within involved constraints. The fixed-time convergence is ensured based on the transformed unconstrained system. The actuator faults and the network induced errors are dealt with by the bound estimation approach and some well-defined smooth functions. All closed-loop signals are SGUUB, and meanwhile the velocity and attitude tracking are achieved by fixed time. This work lays down the groundwork for other coordinated control problems of UAVs including motion re-coordination and trajectory planning. This extension of coordinated control to guidance system for UAVs deserves further investigation. In practical flight, the communication among UAVs may be switching and directed. Future works can focus on the distributed fixed-time fault-tolerant control for multi-UAVs within a switching and directed communication. Despite that the proposed strategy is capable of tolerating actuator faults, the issues of actuator saturation and sensor fault diagnosis are not yet considered. Investigation of above factors is our future work.

References

1. Z. Yu, Y. Zhang, Z. Liu, Y. Qu, C.Y. Su, B. Jiang, Decentralized finite-time adaptive fault-tolerant synchronization tracking control for multiple UAVs with prescribed performance. J. Frankl. Inst. **357**(16), 11830–11862 (2020)
2. X. Yu, J. Yang, S. Li, Finite-time path following control for small-scale fixed-wing UAVs under wind disturbances. J. Frankl. Inst. **357**(12), 7879–7903 (2020)
3. Z. Yu, Z. Liu, Y. Zhang, Y. Qu, C.Y. Su, Distributed finite-time fault-tolerant containment control for multiple unmanned aerial vehicles. IEEE Trans. Neural Networks Learn. Syst. **31**(6), 2077–2091 (2020)
4. J. Lwowski, A. Majumdar, P. Benavidez, J.J. Prevost, M. Jamshidi, Bird flocking inspired formation control for unmanned aerial vehicles using stereo camera. IEEE Syst. J. **13**(3), 3580–3589 (2019)
5. J. Zhang, J. Yan, A novel control approach for flight-stability of fixed-wing UAV formation with wind field. IEEE Syst. J. **15**(2), 2098–2108 (2021)

6. Y. Wang, M. Shan, D. Wang, Motion capability analysis for multiple fixed-wing UAV formations with speed and heading rate constraints. IEEE Trans. Control Network Syst. **7**(2), 977–989 (2020)
7. Z. Yu, Y. Zhang, B. Jiang, C.Y. Su, J. Fu, Y. Jin, T. Chai, Fractional-order adaptive fault-tolerant synchronization tracking control of networked fixed-wing UAVs against actuator-sensor faults via intelligent learning mechanism. IEEE Trans. Neural Networks Learn. Syst. **32**(12), 5539–5553 (2021)
8. F. Mohseni, A. Doustmohammadi, M.B. Menhaj, Distributed receding horizon coverage control for multiple mobile robots. IEEE Syst. J. **10**(1), 198–207 (2016)
9. M. Lv, B.D. Schutter, C. Shi, S. Baldi, Logic-based distributed switching control for agents in power chained form with multiple unknown control directions. Automatica **137**, 110143 (2022)
10. Z. Yu, Y. Zhang, Z. Liu, Y. Qu, C.Y. Su, Distributed adaptive fractional-order fault tolerant cooperative control of networked unmanned aerial vehicles via fuzzy neural networks. IET Control Theory Appl. **13**(17), 2917–2929 (2019)
11. Z. Yu, Y. Qu, Y. Zhang, Distributed fault-tolerant cooperative control for multi-UAVs under actuator fault and input saturation. IEEE Trans. Control Syst. Technol. **27**(6), 2417–2429 (2019)
12. Y. Wang, T. Zhang, Z. Cai, J. Zhao, K. Wu, Multi-UAV coordination control by chaotic grey wolf optimization based distributed MPC with event-triggered strategy. Chin. J. Aeronaut. **33**(11), 2877–2897 (2020)
13. T.Z. Muslimov, R.A. Munasypov, Adaptive decentralized flocking control of multi-UAV circular formations based on vector fields and backstepping. ISA Trans. **107**, 143–159 (2020)
14. H. Chen, X. Wang, L. Shen, Y. Cong, Formation flight of fixed-wing UAV swarms: A group-based hierarchical approach. Chin. J. Aeronaut. **34**(2), 504–515 (2021)
15. T.Z. Muslimov, R.A. Munasypov, Consensus-based cooperative control of parallel fixed-wing UAV formations via adaptive backstepping. Aerosp. Sci. Technol. **109**, 106416 (2021)
16. A. Polyakov, Nonlinear feedback design for fixed-time stabilization of linear control systems. IEEE Trans. Autom. Control **57**(8), 2106–2110 (2012)
17. A.M. Zou, K.D. Kumar, A.H. Ruiter, Fixed-time attitude tracking control for rigid spacecraft. Automatica **113**, 108792 (2020)
18. J. Zhang, S. Yu, Y. Yan, D. Wu, Fixed-time output feedback sliding mode tracking control of marine surface vessels under actuator faults with disturbance cancellation. Appl. Ocean Res. **104**, 102378 (2020)
19. N. Wang, H. Li, Leader-follower formation control of surface vehicles: A fixed-time control approach. ISA Trans. **124**, 356–364 (2022)
20. M. Van, D. Ceglarek, Robust fault tolerant control of robot manipulators with global fixed-time convergence. J. Frankl. Inst. **358**(1), 699–722 (2021)
21. Z. Cai, L. Wang, J. Zhao, K. Wu, Y. Wang, Virtual target guidance-based distributed model predictive control for formation control of multiple UAVs. Chin. J. Aeronaut. **33**(3), 1037–1056 (2020)
22. M. Burger, M. Guay, Robust constraint satisfaction for continuous-time nonlinear systems in strict feedback form. IEEE Trans. Autom. Control **55**(11), 2597–2601 (2010)
23. W. He, Z. Yin, C. Sun, Adaptive neural network control of a marine vessel with constraints using the asymmetric barrier Lyapunov function. IEEE Trans. Cybern. **47**(7), 1641–1651 (2017)
24. K. Zhao, Y.D. Song, Removing the feasibility conditions imposed on tracking control designs for state-constrained strict-feedback systems. IEEE Trans. Autom. Control **64**(3), 1265–1272 (2019)
25. Y, Cao, Y. Song, C. Wen, Practical tracking control of perturbed uncertain nonaffine systems with full state constraints. Automatica **110**, 108608 (2019)
26. Z. Yu, Y. Zhang, B. Jiang, C.Y. Su, J. Fu, Y. Jin, T. Chai, Decentralized fractional-order backstepping fault-tolerant control of multi-UAVs against actuator faults and wind effects. Aerosp. Sci. Technol. **104**, 105939 (2020)

27. S. Zeghlache, H. Mekki, A. Bougueera, A. Djerioui, Actuator fault tolerant control using adaptive RBFNN fuzzy sliding mode controller for coaxial octorotor UAV. ISA Trans. **8**, 267–278 (2018)
28. Z. Yu, Y. Zhang, B. Jiang, X. Yu, J. Fu, Y. Jin, T. Chai, Distributed adaptive fault-tolerant close formation flight control of multiple trailing fixed-wing UAVs. ISA Trans. **106**, 181–199 (2020)
29. Z. Yu, Y. Zhang, B. Jiang, C.Y. Su, J. Fu, Y. Jin, T. Chai, Nussbaum-based finite-time fractional-order backstepping fault-tolerant flight control of fixed-wing UAV against input saturation with hardware-in-the-loop validation. Mech. Syst. Signal Process. **153**, 107406 (2021)
30. H. Yang, B. Jiang, H. Yang, H.H. Liu, Synchronization of multiple 3-DoF helicopters under actuator faults and saturations with prescribed performance. ISA Trans. **75**, 118–126 (2018)
31. O. Espen, S. Rune, K. Raymond, Underactuated waypoint tracking of a fixed-wing UAV. IFAC Proc. **46**(30), 126–133 (2015)
32. H. Castaneda, O.S. Salas-Pena, J. Leon-Morales, Extended observer based on adaptive second order sliding mode control for a fixed wing UAV. ISA Trans. **66**, 226–232 (2017)
33. H. Wang, P.X. Liu, X. Zhao, X. Liu, Adaptive fuzzy finite-time control of nonlinear systems with actuator faults. IEEE Trans. Cybern. **50**(5), 1786–1797(2020)
34. X. Yu, P. Li, Y. Zhang, The design of fixed-time observer and finite-time fault-tolerant control for hypersonic gliding vehicles. IEEE Trans. Ind. Electron. **65**(5), 4135–4144 (2018)
35. M. Lv, W. Yu, J. Cao, S. Baldi, A separation-based methodology to consensus tracking of switched high-order nonlinear multi-agent systems. IEEE Trans. Neural Networks Learn. Syst. **33**(10), 5467–5479 (2021)
36. M. Lv, W. Yu, J. Cao, S. Baldi, Consensus in high-power multi-agent systems with mixed unknown control directions via hybrid Nussbaum-based control. IEEE Trans. Cybern. **52**(6), 5184–5196 (2020)
37. D. Ba, Y.X. Li, S. Tong, Fixed-time adaptive neural tracking control for a class of uncertain nonstrict nonlinear systems. Neurocomputing **363**, 273–280 (2019)
38. C. Wang, Y. Lin, Decentralized adaptive tracking control for a class of interconnected nonlinear time-varying systems. Automatica **54**, 16–24 (2015)
39. S.S. Ge, C. Wang, Adaptive NN control of uncertain nonlinear pure-feedback systems. Automatica **38**(4), 671–682 (2002)
40. H. Wang, Q. Zhu, Adaptive output feedback control of stochastic nonholonomic systems with nonlinear parameterization. Automatica **98**, 247–255 (2018)

Chapter 4
Fixed-Time Control Against Saturation

Chapter 3 presents the fixed-time cooperative control algorithm for the velocity and attitude tracking of multiple UAVs. This method effectively enhances the convergence speed while addressing the issue of state constraints. In addition to state constraints, the phenomenon of input saturation is commonly encountered in UAV systems due to the physical limitations of actuators, which plays a crucial role in the safety and stability of multi-UAV flight. Therefore, on the basis of the fixed-time control strategy in Chap. 3, this chapter investigates the fixed-time control problem for the velocity and attitude tracking of multiple UAVs in the presence of input saturation.

4.1 Introduction

Research on the coordinated flight of networked unmanned aerial vehicles (UAVs) has obtained significant attention over the past few decades, owing to its importance in providing high flexibility and cost-effectiveness in several mission operations [1–5]. In this context, precise cooperative tracking of networked UAVs is essential. In terms of emerging distributed coordinated control [6–8], several control approaches for networked UAVs appear in the literature. Among this literature, Yu et al. [9] proposed a distributed coordinated control framework to solve the attitude synchronization problem for networked UAVs. By considering actuator saturation, Yu et al. [10] devised a distributed coordinated fault-tolerant control methodology for longitudinal synchronization of multiple UAVs. By using swarm intelligence, a fully distributed model predictive control method was designed for networked UAVs in [11]. In [12], with the aid of vector field-based control, the problem of path following for networked UAVs was investigated in which the angles and speeds converged to some user-specified values. Most studies on distributed control of multi-UAVs always simplify the involved UAV model as the three-degree-of-

M. Lv et al., *Unmanned Aerial Vehicles Formation Control and Safety Guarantee*,
Cognitive Intelligence and Robotics, https://doi.org/10.1007/978-981-96-5829-9_4

freedom (3-DoF) dynamics and even linear dynamics. Few works actually chose the six-degree-of-freedom (6-DoF) dynamics of UAVs as the control model.

Unlike traditional asymptotic controllers such as those studied in [9–12], controllers with finite-time design are capable of improving the convergence speed, and they have been successfully used in UAV systems [13–16]. In finite-time control methods, the upper bound of the convergence time is strongly dependent on the initial states. Such a fact limits the implementation of synthesis algorithms. To overcome such deficiencies, researchers have further developed fixed-time control and successfully made the upper bound of the convergence time independent of the initial state values [17]. The resulting control method has been used to obtain a feasible solution for improving convergence for spacecraft [18, 19], surface vessels [20], surface vehicles [21, 22], and robot manipulators [23] and has high potential in practical scenarios. However, few attempts to design the fixed-time control method for networked 6-DoF UAVs have already been made.

There is another challenge for coordinated control of networked UAVs: Actuator saturation and actuator faults are ubiquitous during flight, and they may cause performance degradation and even system instability [24, 25]. Several studies have been conducted to tackle this issue. A few studies are briefly described here. In studies [26, 27], dynamic surface control involving a disturbance observer was used to devise a distributed fault-tolerant control scheme for networked UAVs in the presence of actuator faults. In [28], an attitude stabilization fault-tolerant control approach was developed along with a fault estimation observer for multiple UAVs. Using the fractional-order backstepping iterative control design, Chen et al. [29] investigated a finite-time fault-tolerant control strategy for UAVs against actuator faults and saturation. Notably, most existing fault-tolerant control methods such as those considered in [24–29] are applicable only to asymptotic and finite-time stability of UAV systems. Thus, the development of a fixed-time-based distributed adaptive control approach for networked 6-DoF UAV systems with actuator faults and saturation remains an open problem.

This study aimed to solve the aforementioned problems, and its main contributions are threefold:

- Unlike most existing control results for multi-UAV systems [9–12], a distributed fixed-time anti-saturation fault-tolerant control method was developed for multi-UAV systems whose dynamics are subject to actuator failures and input saturation.
- To overcome the incapability of existing control designs to guarantee fixed-time convergence [13–16], we developed a new fixed-time sliding mode surface to make attitude-tracking errors converge to a residual set within a fixed time.
- Unlike most existing fault-tolerant control schemes that do not consider the effect of actuator saturation [26–28], our proposed adaptation-gain-based control method can handle actuator saturation and actuator faults through online tuning of well-defined adaptation parameters.

The remainder of this chapter is organized as follows: Problem statements and preliminaries are presented in Sect. 4.2, and the control design and stability analysis are described in Sect. 4.3. The simulation results are presented and discussed in Sects. 4.4, and 4.5 summarizes the conclusion.

Remark The main control schemes and contents of this chapter are from the published journal paper "M. L. Lv, C. K. Ahn, B. Y. Zhang, A. Q. Fu, Fixed-Time Antisaturation Cooperative Control for Networked Fixed-Wing Unmanned Aerial Vehicles Considering Actuator Failures, IEEE Transactions on Aerospace and Electronic Systems, 2023, 59(6), 8812–8825." The authors appreciate the permission from Institute of Electrical and Electronics Engineers to reuse the results published in the relevant journal.

4.2 Problem Formulation

4.2.1 Basic Graph Theory

The multi-UAV system is assumed to have N UAVs. The communication topology among UAVs is characterized via a graph $\mathcal{G} = (\mathcal{V}, \mathcal{E}, \mathcal{A})$, in which $\mathcal{V} = \{v_1, \cdots, v_n\}$ represents the set of vertices, $\mathcal{E} = \{(i, j),\ i, j \in \mathcal{V}, \text{and } i \neq j\}$ denotes the set of edges, and $\mathcal{A} = [a_{ij}] \in \Re^{n\times n}$ denotes the weighted adjacency matrix of $\mathcal{G}$. In case there is an edge between nodes i and j, then, it holds that $a_{ij} = a_{ji} \neq 0$, and otherwise $a_{ij} = a_{ji} = 0$. Moreover, a_{ii} becomes 0 for all $i \in \mathcal{V}$. The neighbors of the node i can be represented by $\mathcal{N}_i = \{v_j : (v_j, v_i) \in \mathcal{E}\}$. The Laplacian matrix is defined in the form of $\mathcal{L} = \mathcal{D} - \mathcal{A}$, in which $\mathcal{D} = \text{diag}\{d_1, \cdots, d_n\}$ with $d_i = \sum_{j=1}^{n} a_{ij}$. Note that $\mathcal{G}$ is connected when there exists a path from every UAV to every other UAV.

Assumption 4.1 The communication topology of multiple UAVs is connected in the sense that there exists at least one path from a leader to a follower.

4.2.2 UAV Dynamic Model

Let us consider $N(N \geq 2)$ identical fixed-wing UAVs. The 6-DoF dynamics of the ith UAV can be described as follows according to [30, 31]:

$$\dot{\boldsymbol{p}}_i = \boldsymbol{R}_1(\boldsymbol{\varphi}_i)\,\boldsymbol{v}_i \tag{4.1}$$

$$\dot{\boldsymbol{v}}_i = -\boldsymbol{S}(\boldsymbol{\omega}_i)\,\boldsymbol{v}_i + \frac{\boldsymbol{T}_i}{m_i} + \boldsymbol{R}_1^{\mathrm{T}}(\boldsymbol{\varphi}_i)\,\boldsymbol{g} + \frac{\boldsymbol{F}_i}{m_i} + \boldsymbol{d}_{vi} \tag{4.2}$$

$$\dot{\boldsymbol{\varphi}}_i = \boldsymbol{R}_2^{-1}(\boldsymbol{\varphi}_i)\,\boldsymbol{\omega}_i \tag{4.3}$$

$$\boldsymbol{J}_i\dot{\boldsymbol{\omega}}_i = -\boldsymbol{\omega}_i \times \boldsymbol{J}_i\boldsymbol{\omega}_i + \boldsymbol{N}_i + \boldsymbol{C}_i(\boldsymbol{\delta}_i)\,\boldsymbol{\delta}_i + \boldsymbol{J}_i\boldsymbol{d}_{\omega i}, \tag{4.4}$$

where $\boldsymbol{p}_i = [x_i, y_i, z_i]^{\mathrm{T}}$ and $\boldsymbol{v}_i = [u_i, v_i, w_i]^{\mathrm{T}}$ are the inertial position and body fixed velocity, respectively, $\boldsymbol{\varphi}_i = [\phi_i, \theta_i, \psi_i]^{\mathrm{T}}$, $\boldsymbol{\omega}_i = [p_i, q_i, r_i]^{\mathrm{T}}$, and

$\boldsymbol{T}_l = [T_{xl}, 0, 0]^{\mathrm{T}}$, respectively, mean attitude angle, angular velocity, and thrust vector, $\boldsymbol{\delta}_l = [\delta_{al}, \delta_{el}, \delta_{rl}]^{\mathrm{T}}$ stands for control surfaces with δ_{el}, δ_{al}, and δ_{rl} being the deflection of elevator, ailerons, and rudder, respectively, $\boldsymbol{g} = [0, 0, g_z]^{\mathrm{T}}$ represents gravity acceleration, $\boldsymbol{d}_{vl} \in \Re^3$ and $\boldsymbol{d}_{\omega l} \in \Re^3$ are disturbances, and $\boldsymbol{J}_l$ denotes inertia tensor whose expression is given by

$$\boldsymbol{J}_l = \begin{bmatrix} J_{xl} & 0 & -J_{xzl} \\ 0 & J_{yl} & 0 \\ -J_{xzl} & 0 & J_{zl} \end{bmatrix}. \tag{4.5}$$

The detailed descriptions of $\boldsymbol{R}_1(\boldsymbol{\varphi}_l)$, $\boldsymbol{R}_2(\boldsymbol{\varphi}_l)$, $\boldsymbol{F}_l$, $\boldsymbol{N}_l$, and $\boldsymbol{C}_i(\boldsymbol{\delta}_l)$ are shown in our previous work [31].

Inspired by studies [30, 31], the dynamics of the considered UAV system can be decomposed into translational kinematics and rotational kinematics as follows.

By extracting the thrust and using the fact that $\boldsymbol{v}_l^{\mathrm{T}} \boldsymbol{S}(\boldsymbol{w}_i) \boldsymbol{v}_i = 0$, it follows from (4.2) that

$$\dot{V}_l = \frac{u_l T_{xl}}{m_l V_l} + \frac{\boldsymbol{v}_l^{\mathrm{T}}}{V_l}\left(\boldsymbol{R}_1^{\mathrm{T}}(\boldsymbol{\varphi}_l)\boldsymbol{g} + \frac{\boldsymbol{F}_l}{m_l} + \boldsymbol{d}_{vl}\right). \tag{4.6}$$

For a practical UAV system, actuator faults of the thrust are common, and they may cause flight performance degradation. Therefore, the faults, including loss of effectiveness and bias faults, can be modeled as [32]

$$T_{xl} = \rho_{T_l} T_{xlc} + T_{xlf}, \tag{4.7}$$

where T_{xlc} stands for the designed control signal with saturation, ρ_{T_l} is an unknown actuator effectiveness coefficient satisfying $\rho_{T_l} \in (0, 1]$, and T_{xlf} is an unknown bounded bias fault.

Further, considering the coefficient uncertainties of $\boldsymbol{F}_l$, $\boldsymbol{F}_l$ can be decomposed into $\boldsymbol{F}_{l0}$ and $\Delta\boldsymbol{F}_l$ according to [13]. Consequently, (4.6) can be reformulated as

$$\dot{V}_l = \frac{u_l \rho_{Tl} T_{xlc}}{m_l V_l} + \frac{u_l T_{xlf}}{m_l V_l} + \frac{\boldsymbol{v}_l^{\mathrm{T}}}{V_l}\left(\boldsymbol{R}_1^{\mathrm{T}}(\boldsymbol{\varphi}_l)\boldsymbol{g} + \frac{\boldsymbol{F}_{l0}}{m_l} + \frac{\Delta\boldsymbol{F}_l}{m_l} + \boldsymbol{d}_{vl}\right), \tag{4.8}$$

where the thrust is constrained as $0\mathrm{N} \le T_{xlc} \le T_{xlc,\max}$, with $T_{xlc,\max}$ being a positive constant.

On the basis of (4.3) and (4.4), the actuator faults of control surfaces can be given by Xu et al. [8]

$$\delta_{ml} = \rho_{\delta_{ml}} \delta_{mlc} + \delta_{mlf}, \tag{4.9}$$

where δ_{mlc} denotes the designed control signal with saturation, $\rho_{\delta_{ml}}$ is an unknown actuator effectiveness coefficient satisfying $\rho_{\delta_{ml}} \in (0, 1]$, and δ_{mlf} represents an unknown bounded bias fault, with $m = \{a, e, r\}$. The corresponding actuator fault model can be described as

$$\boldsymbol{\delta}_l = \boldsymbol{\rho}_{\delta_l}\boldsymbol{\delta}_{lc} + \boldsymbol{\delta}_{lf}, \tag{4.10}$$

where $\boldsymbol{\delta}_{lc} = [\delta_{alc}, \delta_{elc}, \delta_{rlc}]^{\mathrm{T}}$, $\boldsymbol{\rho}_{\delta_l} = \mathrm{diag}\left\{\rho_{\delta_{al}}, \rho_{\delta_{el}}, \rho_{\delta_{rl}}\right\}$, and $\boldsymbol{\delta}_{lf} = \left[\delta_{alf}, \delta_{elf}, \delta_{rlf}\right]^{\mathrm{T}}$.

Following [30] and [31], we can express the angular velocity as

$$\boldsymbol{J}_l\dot{\boldsymbol{\omega}}_l = -\boldsymbol{\omega}_l \times \boldsymbol{J}_l\boldsymbol{\omega}_l + \boldsymbol{N}_l + \boldsymbol{C}_l\boldsymbol{\rho}_{\delta l}\boldsymbol{\delta}_{lc} + \boldsymbol{C}_l\boldsymbol{\delta}_{lf} + \boldsymbol{d}_{\omega l}, \tag{4.11}$$

where $\boldsymbol{\delta}_{lc}$ is saturated by $|\delta_{mlc}| \le \delta_{mlc,\max}$, with $\delta_{mlc,\max}$ being a positive constant.

Remark 4.1 Note that $\boldsymbol{J}_l$ is always affected by the onboard payload and fuel consumption, and hence, it always changes during flight. Therefore, it is challenging to identify $\boldsymbol{J}_l$ under complex circumstances. Since $\boldsymbol{J}_l$ is positive definite and bounded, the following assumption is reasonable.

Assumption 4.2 ([33]) The value of $\|\boldsymbol{J}_l\|$ and the derivative of $\boldsymbol{J}_l$ are bounded, and they satisfy $J_{\min} \leqq \|\boldsymbol{J}_l\| \le J_{\max}$ and $0 \le \left\|d\boldsymbol{J}_l/dt\right\| \le J_d$, with $J_{\min}$, $J_{\max}$, and J_d being positive constants.

Assumption 4.3 ([34]) The matrix $\boldsymbol{C}_l\boldsymbol{\rho}_{\delta l}\boldsymbol{C}_l^{\mathrm{T}}$ is positive definite, and $\lambda_{\min}\left(\boldsymbol{C}_l\boldsymbol{\rho}_{\delta l}\boldsymbol{C}_l^{\mathrm{T}}\right) \ge g_l$, where g_l is a positive constant.

Remark 4.2 Assumption 4.3 ensures the controllability of UAVs despite the presence of actuator faults. Otherwise, the matrix $\boldsymbol{C}_l\boldsymbol{\rho}_{\delta l}\boldsymbol{C}_l^{\mathrm{T}}$ may become singular.

4.2.3 Control Objective

The objective of this study is to derive T_{xlc} and $\boldsymbol{\delta}_{lc}$ such that synchronous velocity and attitude tracking of networked UAVs could be performed within a fixed time in spite of actuator faults and actuator input saturation.

To achieve the control objective, the following lemmas are presented.

Lemma 4.1 ([35]) *Let us consider the system below:*

$$\dot{\boldsymbol{x}}(t) = f(\boldsymbol{x}(t)), \ \boldsymbol{x}(0) = \boldsymbol{x}_0. \tag{4.12}$$

Let us assume that there exists a Lyapunov function $V(\boldsymbol{x})$ satisfying

$$\dot{V}(\boldsymbol{x}) \le -\alpha V^p(\boldsymbol{x}) - \beta V^q(\boldsymbol{x}) + \eta, \tag{4.13}$$

where $\alpha > 0$, $\beta > 0$, $p > 1$, $0 < q < 1$, and $0 < \eta < \infty$. Then, the system in (4.12) is practical and fixed-time stable. The settling time T is bounded as $T \le$

$T_{\max} := \frac{1}{\alpha\phi(p-1)} + \frac{1}{\beta\phi(1-q)}$ *with* $0 < \phi < 1$. *And the residual set of state is* $\boldsymbol{x} \in \{V(\boldsymbol{x}) \leq \min\{(\frac{\eta}{(1-\phi)\alpha})^{\frac{1}{p}}, (\frac{\eta}{(1-\phi)\beta})^{\frac{1}{q}}\}\}$.

Lemma 4.2 ([36]) *For* $z \in \Re$, $0 \leq |z| - \frac{z^2}{\sqrt{z^2+\varepsilon}} \leq \sqrt{\varepsilon}$ *holds, with* $\varepsilon > 0$.

Lemma 4.3 ([37]) *If* $\varepsilon > 0$, *then, for* $\vartheta \in \Re$, *we have* $0 \leq |\vartheta| - \vartheta \tanh\left(\frac{\vartheta}{\varepsilon}\right) \leq 0.2785\varepsilon$.

Lemma 4.4 ([37]) *For* $x \in \Re$ *and* $y \in \Re$, *there exist positive constants* b_1, b_2, *and* b_3 *such that*

$$|x|^{b_1}|y|^{b_2} \leq \frac{b_1}{b_1+b_2} b_3 |x|^{b_1+b_2} + \frac{b_2}{b_1+b_2} b_3^{-\frac{b_1}{b_2}} |y|^{b_1+b_2}.$$

4.3 Main Results

This section is devoted to developing a sliding mode-based control scheme for the purpose of guaranteeing fixed-time velocity and attitude tracking in the presence of actuator faults and actuator input saturation. The proposed controller framework is sketched in Fig. 4.1.

4.3.1 Fixed-Time Fault-Tolerant Anti-Saturation Controller Design for Velocity Subsystem

Given the velocity reference V_r, an error variable is firstly constructed by using the velocity tracking errors $\tilde{V}_i = V_i - V_r$ and $\tilde{V}_j = V_j - V_r$

$$s_{vi} = \lambda_{i1}\tilde{V}_i + \lambda_{i2}\sum_{j\in\mathcal{N}_i} a_{ij}\left(\tilde{V}_i - \tilde{V}_j\right), \tag{4.14}$$

where λ_{i1} and λ_{i2} are positive design parameters.

Substituting (4.8) into (4.14) yields

$$\begin{aligned} \dot{s}_{vi} = {} & \lambda_i\left[\frac{u_i\rho_{Ti}T_{xic}}{m_iV_i} + \frac{\boldsymbol{v}_i^{\mathrm{T}}}{V_i}\left(\boldsymbol{R}_1^{\mathrm{T}}\boldsymbol{g} + \frac{\boldsymbol{F}_{i0}}{m_i}\right) - \dot{V}_r\right] \\ & + \lambda_i\left[\frac{u_iT_{xif}}{m_iV_i} + \frac{\boldsymbol{v}_i^{\mathrm{T}}}{V_i}\left(\frac{\Delta\boldsymbol{F}_i}{m_i} + \boldsymbol{d}_{vi}\right)\right] - \lambda_{i2}\sum_{j\in N_i} a_{ij}\dot{\tilde{V}}_j, \end{aligned} \tag{4.15}$$

where $\lambda_i = \lambda_{i1} + \lambda_{i2}\sum_{j\in\mathcal{N}_i} a_{ij}$.

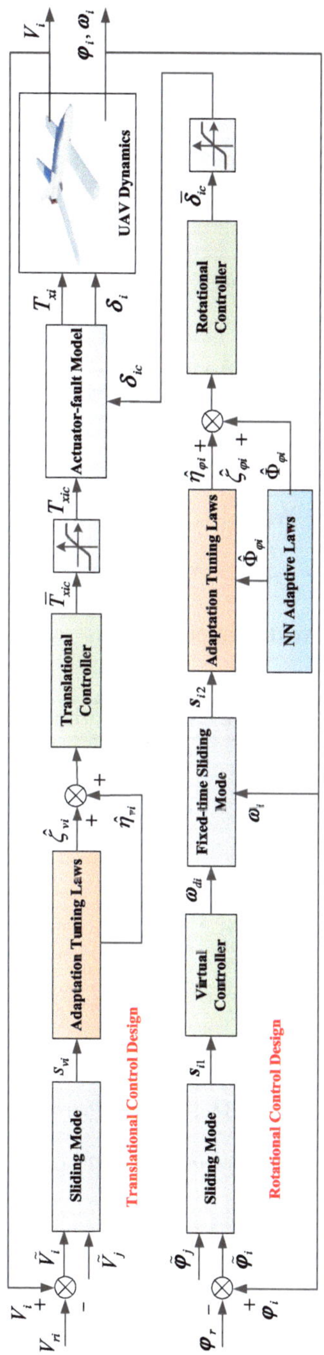

Fig. 4.1 Proposed framework of the controller design

Remark 4.3 Explicitly considering the velocity tracking error information s_{vl}, a fixed-time sliding variable is proposed for the velocity tracking of multi-UAVs. Specifically, the fixed-time sliding variable is given as $S_{vl} = \frac{c_{l1}}{\lambda_l}\text{sig}(s_{vl})^{r_{v1}} + \frac{c_{l2}}{\lambda_l}\text{sig}(s_{vl})^{r_{v2}} + \frac{c_{l3}}{\lambda_l}s_{vl}$, which fully integrates the features of fixed-time convergence lemma, i.e., Lemma 4.1.

In order to achieve the fixed-time stability of velocity dynamic by using Lemma 4.3, we invoke the tanh(·) function to design the control input $\bar{T}_{xlc}$ as

$$\begin{aligned}\bar{T}_{xlc} = &-\frac{m_l V_l}{u_l}\alpha_{vl}\hat{\xi}_{vl}\tanh\left(\frac{s_{vl}\alpha_{vl}\hat{\xi}_{vl}}{\varepsilon_{vl}}\right)\\&-\frac{m_l V_l}{u_l}\hat{\eta}_{vl}\tanh\left(\frac{s_{vl}\hat{\eta}_{vl}}{\varepsilon_{vl}}\right)\end{aligned}\tag{4.16}$$

and the adaptation tuning laws as

$$\dot{\hat{\eta}}_{vl} = \gamma_{vl1}\lambda_l\left|s_{vl}\right| - \gamma_{vl1}\sigma_{vl1}\hat{\eta}_{vl}\tag{4.17}$$

$$\dot{\hat{\xi}}_{vl} = \gamma_{vl2}s_{vl}\lambda_l\alpha_{vl} - \gamma_{vl2}\sigma_{vl2}\hat{\xi}_{vi},\tag{4.18}$$

where σ_{vl1} and σ_{vl2} are positive design parameters.

Algorithm 4.1 Fixed-time cooperative control algorithm for velocity dynamic

Let $l = 1$;
Repeat
For UAV l in the network:
Compute the velocity controller

$$\begin{aligned}\bar{T}_{xlc} = &-\frac{m_l V_l}{u_l}\alpha_{vl}\hat{\xi}_{vl}\tanh\left(\frac{s_{vl}\alpha_{vl}\hat{\xi}_{vl}}{\varepsilon_{vl}}\right)\\&-\frac{m_l V_l}{u_l}\hat{\eta}_{vl}\tanh\left(\frac{s_{vl}\hat{\eta}_{vl}}{\varepsilon_{vl}}\right)\end{aligned}$$

Compute the adaption law

$$\dot{\hat{\eta}}_{vl} = \gamma_{vl1}\lambda_l\left|s_{vl}\right| - \gamma_{vl1}\sigma_{vl1}\hat{\eta}_{vl}$$

$$\dot{\hat{\xi}}_{vl} = \gamma_{vl2}s_{vl}\lambda_l\alpha_{vl} - \gamma_{vl2}\sigma_{vl2}\hat{\xi}_{vl}$$

Let $l = l + 1$, go to step 3 and continue;
Until $l = N$

Remark 4.4 In fact, this form of adaptive law has been widely used in the control design of nonlinear systems and acrobats like unmanned aerial vehicles. However, we need to point out that the core research objective of this chapter is to achieve the fixed-time stability of the UAVs' speed and attitude dynamics. Therefore, we put more emphasis on the convergence time in the design process, and the UUB is adopted. It is true that UUB cannot achieve better asymptotic convergence, while the tracking errors can be smaller by properly adjusting the control parameters.

Remark 4.5 As we know, the tanh(·) function can be used to limit the parameters and reduce the occurrence of chatter. However, as formulated in (4.16), $\bar{T}_{xlc}$ is also affected by other parameters besides tanh(·)-terms $\tanh\left(\frac{s_{vl}\alpha_{vl}\hat{\xi}_{vl}}{\varepsilon_{vl}}\right)$ and $\tanh\left(\frac{s_{vl}\hat{\eta}_{vl}}{\varepsilon_{vl}}\right)$. Therefore, the occurrence of chatter in T_{x3} is the result of the joint action of multiple parameter items. To be sure, after using the tanh(·) function, a chatter will be weaker than when it is not used.

4.3.2 *Fixed-Time Fault-Tolerant Anti-Saturation Controller Design for Attitude Subsystem*

Given the attitude reference φ_r, let us define $\tilde{\varphi}_l = \varphi_l - \varphi_r$, $\tilde{\varphi}_j = \varphi_j - \varphi_r$, and $\tilde{\omega}_l = \omega_l - \omega_{dl}$. The sliding variables are proposed as follows:

$$s_{l1} = \lambda_{l3}\tilde{\varphi}_l + \lambda_{l4}\sum_{j\in\mathcal{N}_l} a_{lj}\left(\tilde{\varphi}_l - \tilde{\varphi}_j\right) \tag{4.19}$$

$$s_{l2} = \tilde{\omega}_l + \frac{R_2}{\lambda_{l3} + \lambda_{l4}\sum_{j\in\mathcal{N}_l} a_{lj}}\alpha_l, \tag{4.20}$$

where λ_{l3} and λ_{l4} are positive design parameters, and α_l is defined as

$$\alpha_l = k_{l1}\|s_{l1}\|^{r_{\varphi1}-1}s_{l1} + k_{l2}\|s_{l1}\|^{r_{\varphi2}-1}s_{l1}, \tag{4.21}$$

where k_{l1} and k_{l2} are positive design parameters, $0 < r_{\varphi1} < 1$, and $r_{\varphi2} > 1$.

Remark 4.6 Fully considering the Euler angle tracking error and angular velocity tracking error information s_{l1}, $\tilde{\omega}_l$, two fixed-time sliding variables are proposed for the attitude tracking of multi-UAVs. Specifically, two fixed-time sliding variables are presented as $S_{\varphi l1} = k_{l1}\|s_{l1}\|^{r_{\varphi1}-1}s_{l1} + k_{l2}\|s_{l1}\|^{r_{\varphi2}-1}s_{l1}$ and $S_{\varphi l2} = \tilde{\omega}_l + \frac{R_2}{\lambda_{l3}+\lambda_{l4}\sum_{j\in\mathcal{N}_l} a_{lj}}(k_{l1}\|s_{l1}\|^{r_{\varphi1}-1}s_{l1} + k_{l2}\|s_{l1}\|^{r_{\varphi2}-1}s_{l1})$, which fully integrates the features of fixed-time convergence lemma, i.e., Lemma 4.1.

To achieve the control objective, we propose an adaptive fixed-time-based rotational controller in the following:

$$\begin{aligned}\bar{\boldsymbol{\delta}}_{ic} =& - \boldsymbol{C}_i^{\mathrm{T}} \left(k_{\phi i} \|\boldsymbol{s}_{i2}\|^2 + k_{si} + \frac{\hat{\Phi}_{\varphi i} \|\boldsymbol{\varphi}_i\|^2}{k_{\Phi i}} \right) \boldsymbol{s}_{i2} \\ & - \boldsymbol{C}_i^{\mathrm{T}} \xi_i \hat{\zeta}_{\varphi i} \left(k_{pi} + \hat{\eta}_{\varphi i}\right) \|\boldsymbol{s}_{i2}\|^{r_{\varphi 1}-1} \boldsymbol{s}_{i2} \\ & - \boldsymbol{C}_i^{\mathrm{T}} \xi_i \hat{\zeta}_{\varphi i} \left(k_{hi} + \hat{\eta}_{\varphi i}\right) \|\boldsymbol{s}_{i2}\|^{r_{\varphi 2}-1} \boldsymbol{s}_{i2}\end{aligned} \tag{4.22}$$

and update it as

$$\dot{\hat{\Phi}}_{\varphi i} = \frac{\|\boldsymbol{\varphi}_i\|^2}{k_{\Phi i}} \|\boldsymbol{s}_{i2}\|^2 - \gamma_{\Phi i} \hat{\Phi}_{\varphi i} \tag{4.23}$$

$$\dot{\hat{\eta}}_{\varphi i} = \|\boldsymbol{s}_{i2}\|^{r_{\varphi 1}+1} + \|\boldsymbol{s}_{i2}\|^{r_{\varphi 2}+1} - \gamma_{\eta i} \hat{\eta}_{\varphi i} \tag{4.24}$$

$$\dot{\hat{\zeta}}_{\varphi i} = \begin{cases} 0, & \text{if } \hat{\zeta}_{\varphi i} = 1 \text{ and } \alpha_{\varphi i} < 0 \\ \alpha_{\varphi i}, & \text{otherwise, } \hat{\zeta}_{\varphi i}(0) > 1 \end{cases}$$

$$\begin{aligned}\alpha_{\varphi i} =& \beta_i \xi_i \hat{\zeta}_{\varphi i}^3 \left[\left(k_{pi} + \hat{\eta}_{\varphi i}\right) \|\boldsymbol{s}_{i2}\|^{r_{\varphi 1}+1} + \left(k_{hi} + \hat{\eta}_{\varphi i}\right) \|\boldsymbol{s}_{i2}\|^{r_{\varphi 2}+1} \right] \\ & - \beta_i \xi_i \gamma_{\zeta i} \hat{\zeta}_{\varphi i}^5,\end{aligned} \tag{4.25}$$

where $k_{\Phi i}, k_{si}, k_{pi}, k_{hi}, \gamma_{\Phi i}, \gamma_{\eta i}$, and $\gamma_{\zeta i}$ are positive design parameters.

4.3.3 Stability Analysis

Theorem 4.1 *Consider the velocity dynamics in (4.8). Let the thrust controller be given by (4.16), and let the adaptation tuning laws be given by (4.17) and (4.18). With Assumption 4.1, the sliding mode surface s_{vi} is uniformly ultimately bounded (UUB) within a fixed time despite the presence of actuator faults and actuator input saturation.*

Proof The thrust with saturation is modeled as

$$T_{xic} = \mathrm{sat}(\bar{T}_{xic}) = h_{Ti} \bar{T}_{xic}, \tag{4.26}$$

where $\bar{T}_{xic}$ is designed in (4.16) and h_{Ti} is defined as

$$h_{Ti} = \begin{cases} \frac{T_{xic,\max}}{\bar{T}_{xic}}, & \text{if } \bar{T}_{xic} > T_{xic,\max} \\ 1, & \text{if } 0\mathrm{N} \le \bar{T}_{xic} \le T_{xic,\max} \\ 0, & \text{if } \bar{T}_{xic} < 0\mathrm{N}. \end{cases} \tag{4.27}$$

Algorithm 4.2 Fixed-time cooperative control algorithm for attitude dynamic

1: Let $i = 1$;
2: **Repeat**
3: For UAV i in the network:
4: Compute the attitude controller

$$\begin{aligned}\bar{\boldsymbol{\delta}}_{ic} = & -\boldsymbol{C}_i^{\mathrm{T}}\left(k_{\phi i}\|\boldsymbol{s}_{i2}\|^2 + k_{si} + \frac{\hat{\Phi}_{\varphi i}\|\boldsymbol{\varphi}_i\|^2}{k_{\Phi i}}\right)\boldsymbol{s}_{i2} \\ & -\boldsymbol{C}_i^{\mathrm{T}}\xi_i\hat{\zeta}_{\varphi i}\left(k_{pi} + \hat{\eta}_{\varphi i}\right)\|\boldsymbol{s}_{i2}\|^{r_{\varphi 1}-1}\boldsymbol{s}_{i2} \\ & -\boldsymbol{C}_i^{\mathrm{T}}\xi_i\hat{\zeta}_{\varphi i}\left(k_{hi} + \hat{\eta}_{\varphi i}\right)\|\boldsymbol{s}_{i2}\|^{r_{\varphi 2}-1}\boldsymbol{s}_{i2}\end{aligned}$$

5: Compute the adaption law

$$\begin{aligned}\dot{\hat{\Phi}}_{\varphi i} = & \frac{\|\boldsymbol{\varphi}_i\|^2}{k_{\Phi i}}\|\boldsymbol{s}_{i2}\|^2 - \gamma_{\Phi i}\hat{\Phi}_{\varphi i} \\ \dot{\hat{\eta}}_{\varphi i} = & \|\boldsymbol{s}_{i2}\|^{r_{\varphi 1}+1} + \|\boldsymbol{s}_{i2}\|^{r_{\varphi 2}+1} - \gamma_{\eta i}\hat{\eta}_{\varphi i} \\ \dot{\hat{\zeta}}_{\varphi i} = & \begin{cases} 0, & \text{if } \hat{\zeta}_{\varphi i} = 1 \text{ and } \alpha_{\varphi i} < 0 \\ \alpha_{\varphi i}, & \text{otherwise, } \hat{\zeta}_{\varphi i}(0) > 1 \end{cases} \\ \alpha_{\varphi i} = & \beta_i\xi_i\hat{\zeta}_{\varphi i}^3\left[\left(k_{pi} + \hat{\eta}_{\varphi i}\right)\|\boldsymbol{s}_{i2}\|^{r_{\varphi 1}+1}\right. \\ & \left. + \left(k_{hi} + \hat{\eta}_{\varphi i}\right)\|\boldsymbol{s}_{i2}\|^{r_{\varphi 2}+1}\right] - \beta_i\xi_i\gamma_{\zeta i}\hat{\zeta}_{\varphi i}^5\end{aligned}$$

6: Let $i = i + 1$, go to step 3 and continue;
7: **Until** $i = N$

Let us define $\underline{\zeta}_{vi} = \rho_{Ti}h_{Ti}$ and $\zeta_{vi} = 1/\underline{\zeta}_{vi}$. The time derivative of $\frac{1}{2}s_{vi}^2$ is

$$\begin{aligned}s_{vi}\dot{s}_{vi} \leq & s_{vi}\lambda_i\left[\frac{u_i\rho_{Ti}T_{xic}}{m_iV_i} + \frac{\boldsymbol{v}_i^{\mathrm{T}}}{V_i}\left(\boldsymbol{R}_1^{\mathrm{T}}\boldsymbol{g} + \frac{\boldsymbol{F}_{i0}}{m_i}\right) - \dot{V}_r\right] \\ & - s_{vi}\lambda_{i2}\sum_{j\in\mathcal{N}_i}a_{ij}\dot{\hat{V}}_j + |s_{vi}|\lambda_i\eta_{vi}\underline{\zeta}_{vi},\end{aligned} \tag{4.28}$$

where $\eta_{vi}\underline{\zeta}_{vi} = \sup_{t\geq 0}|\frac{u_iT_{xif}}{m_iV_i} + \frac{\boldsymbol{v}_i^{\mathrm{T}}}{V_i}(\frac{\Delta\boldsymbol{F}_i}{m_i} + \boldsymbol{d}_{vi})|$.

Furthermore, let us define $\tilde{\zeta}_{vi} = \hat{\zeta}_{vi} - \zeta_{vi}$ and $\tilde{\eta}_{vi} = \hat{\eta}_{vi} - \eta_{vi}$, and let $\hat{\zeta}_{vi}$ and $\hat{\eta}_{vi}$ be the estimations of ζ_{vi} and η_{vi}, respectively. From (4.18), it follows that

$$|s_{vi}|\lambda_i\eta_{vi}\underline{\zeta}_{vi} = |s_{vi}|\lambda_i\hat{\eta}_{vi}\underline{\zeta}_{vi} - |s_{vi}|\lambda_i\tilde{\eta}_{vi}\underline{\zeta}_{vi}. \tag{4.29}$$

Let us choose the following Lyapunov function candidate:

$$L_{vl} = \frac{1}{2}s_{vl}^2 + \frac{\underline{\zeta}_{vl}}{2\gamma_{vl1}}\tilde{\eta}_{vl}^2 + \frac{\underline{\zeta}_{vl}}{2\gamma_{vl2}}\tilde{\zeta}_{vl}^2, \tag{4.30}$$

where γ_{vl1} and γ_{vl2} are positive design parameters.

Using (4.26), (4.28), and (4.29) and taking the time derivative of (4.30), we obtain

$$\begin{aligned}\dot{L}_{vl} \leq & s_{vl}\lambda_l\left[-\frac{c_{l1}}{\lambda_l}\mathrm{sig}(s_{vl})^{r_{v1}} - \frac{c_{l2}}{\lambda_l}\mathrm{sig}(s_{vl})^{r_{v2}} - \frac{c_{l3}}{\lambda_l}s_{vl} + \alpha_{vl}\right] \\ & + s_{vl}\lambda_l\frac{u_l\rho_{Tl}h_{Tl}\bar{T}_{xlc}}{m_l V_l} + |s_{vl}|\,\lambda_l\hat{\eta}_{vl}\underline{\zeta}_{vl} + \frac{\underline{\zeta}_{vl}}{\gamma_{vl2}}\tilde{\zeta}_{vl}\dot{\hat{\zeta}}_{vl} \\ & + \frac{\underline{\zeta}_{vl}}{\gamma_{vl1}}\tilde{\eta}_{vl}\left(\dot{\hat{\eta}}_{vl} - \gamma_{vl1}\lambda_l\,|s_{vl}|\right),\end{aligned} \tag{4.31}$$

where $\alpha_{vl} = \frac{c_{l1}}{\lambda_l}\mathrm{sig}(s_{vl})^{r_{v1}} + \frac{c_{l2}}{\lambda_l}\mathrm{sig}(s_{vl})^{r_{v2}} + \frac{c_{l3}}{\lambda_l}s_{vl} + \frac{\boldsymbol{v}_l^{\mathrm{T}}}{V_l}\left(\boldsymbol{R}_1^{\mathrm{T}}\boldsymbol{g} + \frac{\boldsymbol{F}_{l0}}{m_l}\right) - \dot{V}_r - \frac{\lambda_{l2}}{\lambda_l}\sum_{j\in\mathcal{N}_l} a_{lj}\dot{\hat{V}}_j$.

From (4.16), it follows that

$$\begin{aligned}s_{vl}\lambda_l\frac{u_l\rho_{Tl}h_{Tl}\bar{T}_{xlc}}{m_l V_l} = & -\underline{\zeta}_{vl}s_{vl}\lambda_l\alpha_{vl}\hat{\zeta}_{vl}\tanh\left(\frac{s_{vl}\alpha_{vl}\hat{\zeta}_{vl}}{\varepsilon_{vl}}\right) \\ & -\underline{\zeta}_{vl}s_{vl}\lambda_l\hat{\eta}_{vl}\tanh\left(\frac{s_{vl}\hat{\eta}_{vl}}{\varepsilon_{vl}}\right).\end{aligned} \tag{4.32}$$

Using Lemma 4.3, we obtain

$$\begin{aligned}& -\underline{\zeta}_{vl}s_{vl}\lambda_l\alpha_{vl}\hat{\zeta}_{vl}\tanh\left(\frac{s_{vl}\alpha_{vl}\hat{\zeta}_{vl}}{\varepsilon_{vl}}\right) \\ & \leq -\underline{\zeta}_{vl}\lambda_l\left|s_{vl}\alpha_{vl}\hat{\zeta}_{vl}\right| + 0.2785\underline{\zeta}_{vl}\lambda_l\varepsilon_{vl} \\ & \leq -\underline{\zeta}_{vl}s_{vl}\lambda_l\alpha_{vl}\tilde{\zeta}_{vl} - s_{vl}\lambda_l\alpha_{vl} + 0.2785\underline{\zeta}_{vi}\lambda_i\varepsilon_{vi}\end{aligned} \tag{4.33}$$

and

$$\begin{aligned}& -\underline{\zeta}_{vl}s_{vl}\lambda_l\hat{\eta}_{vl}\tanh\left(\frac{s_{vl}\hat{\eta}_{vl}}{\varepsilon_{vl}}\right) \\ & \qquad \leq -\underline{\zeta}_{vl}\lambda_l\,|s_{vl}|\,\hat{\eta}_{vl} + 0.2785\underline{\zeta}_{vi}\lambda_i\varepsilon_{vi}.\end{aligned} \tag{4.34}$$

Substituting (4.17)–(4.34) into (4.31) gives

$$\begin{aligned}\dot{L}_{vl} \leq & - c_{l3}s_{vl}^2 - \frac{\sigma_{vl1}\underline{\zeta}_{vl}}{2}\tilde{\eta}_{vl}^2 - \frac{\sigma_{vl2}\underline{\zeta}_{vl}}{2}\tilde{\zeta}_{vl}^2 + \frac{\sigma_{vl1}\underline{\zeta}_{vl}}{2}\eta_{vl}^2 \\ & + \frac{\sigma_{vl2}\underline{\zeta}_{vl}}{2}\zeta_{vl}^2 + 0.557\underline{\zeta}_{vl}\lambda_l\varepsilon_{vl} \\ \leq & - \chi_{vl1}L_{vl} + \chi_{vl2},\end{aligned} \tag{4.35}$$

where $\chi_{vl1} = \min\{2c_{l3}, \sigma_{vl1}\gamma_{vl1}, \sigma_{vl2}\gamma_{vl2}\}$ and $\chi_{vl2} = \frac{\sigma_{vl1vl}}{2}\eta_{vl}^2 + \frac{\sigma_{vl2vl}}{2}\zeta_{vl}^2 + 0.557_{vl}\lambda_l\varepsilon_{vl}$.

It can be concluded that for any $s_{vl}(0), \hat{\eta}_{vl}(0), \hat{\zeta}_{vl}(0) \in \Re \times \Re \times \Re$, there exist $|s_{vl}| \leq \mu_{sl}$, $|\tilde{\eta}_{vl}| \leq \mu_{\eta l}$, and $|\tilde{\zeta}_{vl}| \leq \mu_{\zeta l}$ with μ_{sl}, $\mu_{\eta l}$, and $\mu_{\zeta l}$ being positive constants.

In order to show the fixed-time convergence of s_{vl}, we again consider (4.31)–(4.34). Then, it holds that

$$\begin{aligned}\dot{L}_{vl} \leq & -c_{l1}|s_{vl}|^{2q_v} - c_{l2}|s_{vl}|^{2p_v} - \sigma_{vl1}\left(\frac{\underline{\zeta}_{vl}\tilde{\eta}_{vl}^2}{2\gamma_{vl1}}\right)^{q_v} \\ & - \sigma_{vl1}\left(\frac{\underline{\zeta}_{vl}\tilde{\eta}_{vl}^2}{2\gamma_{vl1}}\right)^{p_v} - \sigma_{vl2}\left(\frac{\underline{\zeta}_{vl}\tilde{\zeta}_{vl}^2}{2\gamma_{vl2}}\right)^{q_v} - \sigma_{vl2}\left(\frac{\underline{\zeta}_{vl}\tilde{\zeta}_{vl}^2}{2\gamma_{vl2}}\right)^{p_v} \\ & + \sigma_{vl1}\left(\frac{\underline{\zeta}_{vl}\tilde{\eta}_{vl}^2}{2\gamma_{vl1}}\right)^{q_v} + \sigma_{vl1}\left(\frac{\underline{\zeta}_{vl}\tilde{\eta}_{vl}^2}{2\gamma_{vl1}}\right)^{p_v} + \sigma_{vl2}\left(\frac{\underline{\zeta}_{vl}\tilde{\zeta}_{vl}^2}{2\gamma_{vl2}}\right)^{q_v} \\ & + \sigma_{vl2}\left(\frac{\underline{\zeta}_{vl}\tilde{\zeta}_{vl}^2}{2\gamma_{vl2}}\right)^{p_v} - \frac{\sigma_{vl1}\underline{\zeta}_{vl}}{\gamma_{vl1}}\tilde{\eta}_{vl}^2 + \frac{\vartheta_{vl1}\sigma_{vl1}\underline{\zeta}_{vl}}{2}\eta_{vl}^2 \\ & - \frac{\sigma_{vl2}\underline{\zeta}_{vl}}{\gamma_{vl2}}\tilde{\zeta}_{vl}^2 + \frac{\vartheta_{vl2}\sigma_{vl2}\underline{\zeta}_{vl}}{2}\zeta_{vl}^2 + 0.557\underline{\zeta}_{vl}\lambda_l\varepsilon_{vl},\end{aligned} \tag{4.36}$$

where $q_v = \frac{r_{v1}+1}{2}$, $p_v = \frac{r_{v2}+1}{2}$, $\gamma_{vl1} = 2\vartheta_{vl1}/(2\vartheta_{vl1} - 1)$, $\gamma_{vl2} = 2\vartheta_{vl2}/(2\vartheta_{vl2} - 1)$ with $\vartheta_{vl1} > \frac{1}{2}$, $\vartheta_{vl2} > \frac{1}{2}$.

Invoking Lemma 4.4 to $\left(\frac{\underline{\zeta}_{vl}\tilde{\eta}_{vl}^2}{2\gamma_{vl1}}\right)^{q_v}$ and $\left(\frac{\underline{\zeta}_{vl}\tilde{\zeta}_{vl}^2}{2\gamma_{vl2}}\right)^{q_v}$, one has

$$\begin{aligned}\left(\frac{\underline{\zeta}_{vl}\tilde{\eta}_{vl}^2}{2\gamma_{vl1}}\right)^{q_v} & \leq \frac{\underline{\zeta}_{vl}\tilde{\eta}_{vl}^2}{2\gamma_{vl1}} + (1 - q_v)\, q_v^{\frac{q_v}{1-q_v}} \\ \left(\frac{\underline{\zeta}_{vl}\tilde{\zeta}_{vl}^2}{2\gamma_{vl2}}\right)^{q_v} & \leq \frac{\underline{\zeta}_{vl}\tilde{\zeta}_{vl}^2}{2\gamma_{vl2}} + (1 - q_v)\, q_v^{\frac{q_v}{1-q_v}}.\end{aligned} \tag{4.37}$$

Substituting (4.37) into (4.36) gives

$$
\begin{aligned}
\dot{L}_{vi} \leq & -c_{i1}|s_{vi}|^{2q_v} - c_{i2}|s_{vi}|^{2p_v} - \sigma_{vi1}\left(\frac{\underline{\zeta}_{vi}\tilde{\eta}_{vi}^2}{2\gamma_{vi1}}\right)^{q_v} \\
& - \sigma_{vi1}\left(\frac{\underline{\zeta}_{vi}\tilde{\eta}_{vi}^2}{2\gamma_{vi1}}\right)^{p_v} - \sigma_{vi2}\left(\frac{\underline{\zeta}_{vi}\tilde{\zeta}_{vi}^2}{2\gamma_{vi2}}\right)^{q_v} - \sigma_{vi2}\left(\frac{\underline{\zeta}_{vi}\tilde{\zeta}_{vi}^2}{2\gamma_{vi2}}\right)^{p_v} \\
& + \sigma_{vi1}\left(\frac{\underline{\zeta}_{vi}\tilde{\eta}_{vi}^2}{2\gamma_{vi1}}\right)^{p_v} + \sigma_{vi2}\left(\frac{\underline{\zeta}_{vi}\tilde{\zeta}_{vi}^2}{2\gamma_{vi2}}\right)^{p_v} - \frac{\sigma_{vi1}\underline{\zeta}_{vi}}{2\gamma_{vi1}}\tilde{\eta}_{vi}^2 \\
& - \frac{\sigma_{vi2}\underline{\zeta}_{vi}}{2\gamma_{vi2}}\tilde{\zeta}_{vi}^2 + \frac{\vartheta_{vi1}\sigma_{vi1}\underline{\zeta}_{vi}}{2}\eta_{vi}^2 + \frac{\vartheta_{vi2}\sigma_{vi2}\underline{\zeta}_{vi}}{2}\zeta_{vi}^2 \\
& + 0.557\underline{\zeta}_{vi}\lambda_i\varepsilon_{vi} + \sum_{j=1}^{2}\left(\sigma_{vij}\left(1-q_v\right)q_v^{\frac{q_v}{1-q_v}}\right).
\end{aligned}
\tag{4.38}
$$

From (4.35), it is reasonable to assume $\sqrt{\underline{\zeta}_{vi}}\,|\tilde{\eta}_{vi}| \leq \Delta_{vi1}$ and $\sqrt{\underline{\zeta}_{vi}}|\tilde{\zeta}_{vi}| \leq \Delta_{vi2}$ with positive constants Δ_{vi1} and Δ_{vi2}. At this point, the following two cases should be considered.

Case 1: $\Delta_{vi1} < \sqrt{2\gamma_{vi1}}$ or $\Delta_{vi2} < \sqrt{2\gamma_{vi2}}$. Then we obtain

$$
\begin{aligned}
& \sigma_{vi1}\left(\frac{\underline{\zeta}_{vi}\tilde{\eta}_{vi}^2}{2\gamma_{vi1}}\right)^{p_v} - \frac{\sigma_{vi1}\underline{\zeta}_{vi}}{2\gamma_{vi1}}\tilde{\eta}_{vi}^2 < 0 \\
& \sigma_{vi2}\left(\frac{\underline{\zeta}_{vi}\tilde{\zeta}_{vi}^2}{2\gamma_{vi2}}\right)^{p_v} - \frac{\sigma_{vi2}\underline{\zeta}_{vi}}{2\gamma_{vi2}}\tilde{\zeta}_{vi}^2 < 0.
\end{aligned}
\tag{4.39}
$$

Case 2: $\Delta_{vi1} \geq \sqrt{2\gamma_{vi1}}$ or $\Delta_{vi2} \geq \sqrt{2\gamma_{vi2}}$. Then one has

$$
\begin{aligned}
& \sigma_{vi1}\left(\frac{\underline{\zeta}_{vi}\tilde{\eta}_{vi}^2}{2\gamma_{vi1}}\right)^{p_v} - \frac{\sigma_{vi1}\underline{\zeta}_{vi}}{2\gamma_{vi1}}\tilde{\eta}_{vi}^2 \leq \sigma_{vi1}\left(\frac{\Delta_{vi1}^2}{2\gamma_{vi1}}\right)^{p_v} - \frac{\sigma_{vi1}}{2\gamma_{vi1}}\Delta_{vi1}^2 \\
& \sigma_{vi2}\left(\frac{\underline{\zeta}_{vi}\tilde{\zeta}_{vi}^2}{2\gamma_{vi2}}\right)^{p_v} - \frac{\sigma_{vi2}\underline{\zeta}_{vi}}{2\gamma_{vi2}}\tilde{\zeta}_{vi}^2 \leq \sigma_{vi2}\left(\frac{\Delta_{vi2}^2}{2\gamma_{vi2}}\right)^{p_v} - \frac{\sigma_{vi2}}{2\gamma_{vi2}}\Delta_{vi2}^2.
\end{aligned}
\tag{4.40}
$$

Cases 1 and 2 can be summarized as

$$
\upsilon_{vi1} = \begin{cases} \sigma_{vi1}\left(\frac{\Delta_{vi1}^2}{2\gamma_{vi1}}\right)^{p_v} - \frac{\sigma_{vi1}}{2\gamma_{vi1}}\Delta_{vi1}^2, & \text{if } \Delta_{vi1} \geq \sqrt{2\gamma_{vi1}} \\ 0, & \text{if } \Delta_{vi1} < \sqrt{2\gamma_{vi1}} \end{cases}
\tag{4.41}
$$

$$\upsilon_{vi2} = \begin{cases} \sigma_{vi2}\left(\frac{\Delta_{vi2}^2}{2\gamma_{vi2}}\right)^{p_v} - \frac{\sigma_{vi2}}{2\gamma_{vi2}}\Delta_{vi2}^2, & \text{if } \Delta_{vi2} \geq \sqrt{2\gamma_{vi2}} \\ 0, & \text{if } \Delta_{vi2} < \sqrt{2\gamma_{vi2}}. \end{cases} \tag{4.42}$$

Substituting (4.41) and (4.42) into (4.38) yields

$$\begin{aligned} \dot{L}_{vi} \leq & -c_{i1}2^{q_v}\left|\frac{s_{vi}^2}{2}\right|^{q_v} - c_{i2}2^{p_v}\left|\frac{s_{vi}^2}{2}\right|^{p_v} - \sigma_{vi1}\left(\frac{\underline{\zeta}_{vi}\tilde{\eta}_{vi}^2}{2\gamma_{vi1}}\right)^{q_v} \\ & - \sigma_{vi1}\left(\frac{\underline{\zeta}_{vi}\tilde{\eta}_{vi}^2}{2\gamma_{vi1}}\right)^{p_v} - \sigma_{vi2}\left(\frac{\underline{\zeta}_{vi}\tilde{\zeta}_{vi}^2}{2\gamma_{vi2}}\right)^{q_v} - \sigma_{vi2}\left(\frac{\underline{\zeta}_{vi}\tilde{\zeta}_{vi}^2}{2\gamma_{vi2}}\right)^{p_v} \\ & + \frac{\vartheta_{vi1}\sigma_{vi1}\underline{\zeta}_{vi}}{2}\eta_{vi}^2 + \frac{\vartheta_{vi2}\sigma_{vi2}\underline{\zeta}_{vi}}{2}\zeta_{vi}^2 + \upsilon_{vi1} + \upsilon_{vi2} \\ & + 0.557\underline{\zeta}_{vi}\lambda_i\varepsilon_{vi} + \sum_{j=1}^{2}\left(\sigma_{vij}\left(1-q_v\right)q_v^{\frac{q_v}{1-q_v}}\right). \end{aligned} \tag{4.43}$$

Then (4.43) can be rewritten as

$$\dot{L}_{vi} \leq -\chi_{vi3}L_{vi}{}^{q_v} - \chi_{vi4}L_{vi}{}^{p_v} + \chi_{vi5}, \tag{4.44}$$

where $\chi_{vi3} = \min\{2^{q_v}c_{i1}, \sigma_{vi1}, \sigma_{vi2}\}$, $\chi_{vi4} = \min\{2^{p_v}c_{i2}, \sigma_{vi1}, \sigma_{vi2}\}$, and $\chi_{vi5} = \frac{\vartheta_{vi1}\sigma_{vi1}\underline{\zeta}_{vi}}{2}\eta_{vi}^2 + \frac{\vartheta_{vi2}\sigma_{vi2}\underline{\zeta}_{vi}}{2}\zeta_{vi}^2 + \upsilon_{vi1} + \upsilon_{vi2} + 0.557\underline{\zeta}_{vi}\lambda_i\varepsilon_{vi} + \sum_{j=1}^{2}\left[\sigma_{vij}\left(1-q_v\right)q_v^{\frac{q_v}{1-q_v}}\right]$.

Recalling Lemma 4.1, it can be concluded that all signals of the translational subsystem are practical fixed-time stable, and we can derive

$$L_{vi} \leq \min\left\{\left(\frac{\chi_{vi5}}{\chi_{vi3}\left(1-\sigma_{vi}\right)}\right)^{\frac{1}{q_v}}, \left(\frac{\chi_{vi5}}{\chi_{vi4}\left(1-\sigma_{vi}\right)}\right)^{\frac{1}{p_v}}\right\}, \tag{4.45}$$

and the setting time denotes

$$T \leq \bar{T}_{vi} := \frac{1}{\chi_{vi3}\sigma_{vi}\left(q_v-1\right)} + \frac{1}{\chi_{vi4}\sigma_{vi}\left(1-p_v\right)}. \tag{4.46}$$

Then, the following inequality holds:

$$|s_{vl}| \leq \bar{\vartheta}_{vl}$$
$$:= \min\left\{\sqrt{2\left(\frac{\chi_{vl5}}{\chi_{vl3}\left(1-\sigma_{vl}\right)}\right)^{\frac{1}{q_v}}}, \sqrt{2\left(\frac{\chi_{vl5}}{\chi_{vl4}\left(1-\sigma_{vi}\right)}\right)^{\frac{1}{p_v}}}\right\}. \quad (4.47)$$

This completes the proof of Theorem 4.1.

Remark 4.7 The convergence speed and precision of the tracking errors can be adjusted by tuning c_{l1}, c_{l2}, σ_{vl1}, and σ_{vl2}. The larger the values, the faster the convergence speed and the better the convergence precision. Furthermore, the convergence speed and precision of tracking errors are affected by r_{v1} and r_{v2}. Thus, by tuning the aforementioned parameters, the convergence speed and the precision of tracking errors can be adjusted as desired.

Theorem 4.2 *Let us consider the attitude dynamics in (4.3) and (4.11). Let the rotational controller be described by (4.52), and let the adaptation tuning laws be given by (4.23)–(4.25). With Assumptions 4.1–4.3, the sliding mode surface $\boldsymbol{s}_{l1}$ is UUB within a fixed time in the presence of the actuator faults and saturation.*

Proof *Step 1*: Taking the time derivative of $\boldsymbol{s}_{l1}$ yields

$$\dot{\boldsymbol{s}}_{l1} = \lambda_l'\left(\boldsymbol{R}_2^{-1}\boldsymbol{\omega}_l - \dot{\boldsymbol{\varphi}}_r\right) - \lambda_{l4}\sum_{j\in\mathcal{N}_l} a_{lj}\dot{\tilde{\boldsymbol{\varphi}}}_j, \quad (4.48)$$

where $\lambda_l' = \lambda_{l3} + \lambda_{l4}\sum_{j\in\mathcal{N}_l} a_{lj}$.

The virtual controller is chosen as

$$\boldsymbol{\omega}_{dl} = -\frac{\boldsymbol{R}_2}{\lambda_l'}\boldsymbol{K}_{l1}\boldsymbol{s}_{l1} + \boldsymbol{R}_2\dot{\boldsymbol{\varphi}}_r - \frac{\lambda_{l4}\boldsymbol{R}_2}{\lambda_l'}\sum_{j\in\mathcal{N}_l} a_{lj}\dot{\tilde{\boldsymbol{\varphi}}}_j, \quad (4.49)$$

where $\boldsymbol{K}_{l1} \in \Re^{3\times3}$ is a symmetric positive definite matrix.

Then, the time derivative of $\frac{1}{2}\boldsymbol{s}_{l1}^{\mathrm{T}}\boldsymbol{s}_{l1}$ along (4.48) and (4.49) is

$$\begin{aligned}
\boldsymbol{s}_{l1}^{\mathrm{T}}\dot{\boldsymbol{s}}_{l1} &= -\boldsymbol{s}_{l1}^{\mathrm{T}}\boldsymbol{K}_{l1}\boldsymbol{s}_{l1} + \lambda_l'\boldsymbol{s}_{l1}^{\mathrm{T}}\boldsymbol{R}_2^{-1}\boldsymbol{s}_{l2} - \boldsymbol{s}_{l1}^{\mathrm{T}}\boldsymbol{\alpha}_l \\
&\leq -\left[\lambda_{\min}\left(\boldsymbol{K}_{l1}\right) - \frac{\lambda_l'\lambda_{\max}\left(\boldsymbol{R}_2^{-1}\right)}{2}\right]\|\boldsymbol{s}_{l1}\|^2 - k_{l1}\|\boldsymbol{s}_{l1}\|^{r_{\varphi1}+1} \\
&\quad + \frac{\lambda_l'\lambda_{\max}\left(\boldsymbol{R}_2^{-1}\right)}{2}\|\boldsymbol{s}_{l2}\|^2 - k_{l2}\|\boldsymbol{s}_{l1}\|^{r_{\varphi2}+1}.
\end{aligned} \quad (4.50)$$

Step 2: Using (4.11) and (4.20) and taking the time derivative of $\boldsymbol{J}_l \boldsymbol{s}_{2l}$, we obtain

$$\boldsymbol{J}_l \dot{\boldsymbol{s}}_{l2} = \boldsymbol{N}_l + \boldsymbol{C}_l \boldsymbol{\rho}_{\delta l} \boldsymbol{\delta}_{lc} + \boldsymbol{C}_l \boldsymbol{\delta}_{lf} + \boldsymbol{d}_{\omega l} + \boldsymbol{\Delta}_{\omega l}, \tag{4.51}$$

where $\boldsymbol{\Delta}_{\omega l} = -\boldsymbol{\omega}_l \times \boldsymbol{J}_l \boldsymbol{\omega}_l - \boldsymbol{J}_l \dot{\boldsymbol{\omega}}_{dl} + \boldsymbol{J}_l (\frac{R_2}{\lambda_l} \dot{\boldsymbol{\alpha}}_l + \frac{\dot{R}_2}{\lambda_l} \boldsymbol{\alpha}_l)$.

Using an RBFNN [33], we attempt to approximate $\boldsymbol{\Delta}_{\omega l}$ in (4.51). Given a prefixed sufficiently large compact set $\boldsymbol{\Omega}_l \in \Re^{n_l}$, there exist a positive integer h_l, a time-varying matrix $\boldsymbol{W}_l^*(t) \in \Re^{h_l \times m_l}$, and a basis function vector $\boldsymbol{\varphi}_l : \Re^{n_l} \rightarrow \Re^{h_l}$ such that $\boldsymbol{\Delta}_{\omega l}$ can be approximated as

$$\boldsymbol{\Delta}_{\omega l} = \boldsymbol{W}_l^{*\mathrm{T}} \boldsymbol{\varphi}_l + \boldsymbol{\varepsilon}_{l0}, \tag{4.52}$$

where $\|\boldsymbol{\varepsilon}_{l0}\| \leq \varepsilon_{\omega l}$ with $\varepsilon_{\omega l}$ being a positive constant.

Remark 4.8 Different from the control methods in [38–41], where the elements of $\boldsymbol{W}_l^*$ were estimated in the controller design, we define $\Phi_{\varphi l} = \sup_{t \geq 0} \|\boldsymbol{W}_l^{*\mathrm{T}}\|^2$. Only $\Phi_{\varphi l}$ needs to be estimated rather than a matrix-valued function, in our proposed control design, and this requirement reduces the computation burden.

From (4.51), define $\eta_{\varphi l} = \sup_{t \geq 0} \left\| \boldsymbol{\varepsilon}_{l0} + \boldsymbol{N}_l + \boldsymbol{C}_l \boldsymbol{\delta}_{lf} + \boldsymbol{d}_{\omega l} \right\|$. Then, let $\hat{\Phi}_{\varphi l}$, $\hat{\eta}_{\varphi l}$ be the estimations of $\Phi_{\varphi l}$, $\eta_{\varphi l}$ respectively, and define $\tilde{\Phi}_{\varphi l} = \hat{\Phi}_{\varphi l} - \Phi_{\varphi l}$, $\tilde{\eta}_{\varphi l} = \hat{\eta}_{\varphi l} - \eta_{\varphi l}$.

The control input $\boldsymbol{\delta}_{lc}$ with saturation is modeled by

$$\boldsymbol{\delta}_{lc} = \mathrm{sat}(\bar{\boldsymbol{\delta}}_{lc}) = \boldsymbol{H}_l \bar{\boldsymbol{\delta}}_{lc}, \tag{4.53}$$

where $\boldsymbol{H}_l = \mathrm{diag}\{h_{l1}, h_{l2}, h_{l3}\}$ is designed as

$$h_{lj} = \begin{cases} \mathrm{sgn}\left(\bar{\delta}_{lcj}\right) \frac{\delta_{lcj,\max}}{\bar{\delta}_{lcj}}, & \text{if } \left|\bar{\delta}_{lcj}\right| > \delta_{lcj,\max} \\ 1, & \text{if } \left|\bar{\delta}_{lcj}\right| \leq \delta_{lcj,\max}. \end{cases} \tag{4.54}$$

From (4.54), there exists $0 < \underline{\zeta}_{\varphi l} \leq 1$ such that $h_{lj} \geq \underline{\zeta}_{\varphi l}$, $j = 1, 2, 3$. Then, we introduce a new adaptation parameter $\hat{\zeta}_{\varphi l}$ as the estimation of $\frac{1}{\underline{\zeta}_{\varphi l}}$.

Consider the following Lyapunov function candidates:

$$\begin{aligned} L'_{\varphi l} &= \frac{1}{2} \boldsymbol{s}_{l2}^{\mathrm{T}} \boldsymbol{J}_l \boldsymbol{s}_{l2} \\ L''_{\varphi l} &= \frac{1}{2 g_l \underline{\zeta}_{\varphi l}} \left(\Phi_{\varphi l} - g_l \underline{\zeta}_{\varphi l} \hat{\Phi}_{\varphi l}\right)^2 + \frac{1}{2 g_l \underline{\zeta}_{\varphi l} \xi_l} \left(\eta_{\varphi l} - g_l \underline{\zeta}_{\varphi l} \xi_l \hat{\eta}_{\varphi l}\right)^2 \\ &\quad + \frac{g_l \underline{\zeta}_{\varphi l}}{2 \beta_l} \left(\hat{\zeta}_{\varphi l}^{-1} - \underline{\zeta}_{\varphi l}\right)^2, \end{aligned} \tag{4.55}$$

where $g_l > 0$ is defined in Assumption 4.3, and ξ_l, β_l are positive design parameters.

Taking the time derivative of $L'_{\varphi i}$ along (4.51)–(4.53) and (4.22) renders

$$\begin{aligned}\dot{L}'_{\varphi i} \leq & \sqrt{\Phi_{\varphi i}} \left\|\boldsymbol{\varphi}_i\right\| \|\boldsymbol{s}_{i2}\| + \eta_{\varphi i} \|\boldsymbol{s}_{i2}\| - g_i\underline{\zeta}_{\varphi i}k_{\phi i}\|\boldsymbol{s}_{i2}\|^4 \\ & - g_i\underline{\zeta}_{\varphi i}k_{si}\|\boldsymbol{s}_{i2}\|^2 - \frac{g_i\underline{\zeta}_{\varphi i}\hat{\Phi}_{\varphi i}\left\|\boldsymbol{\varphi}_i\right\|^2}{k_{\Phi i}}\|\boldsymbol{s}_{i2}\|^2 \\ & - g_i\underline{\zeta}_{\varphi i}\xi_i\hat{\zeta}_{\varphi i}\left(k_{pi}+\hat{\eta}_{\varphi i}\right)\|\boldsymbol{s}_{i2}\|^{2q_\varphi} \\ & - g_i\underline{\zeta}_{\varphi i}\xi_i\hat{\zeta}_{\varphi i}\left(k_{hi}+\hat{\eta}_{\varphi i}\right)\|\boldsymbol{s}_{i2}\|^{2p_\varphi}, \end{aligned} \tag{4.56}$$

where $q_\varphi = \frac{r_{\varphi 1}+1}{2}$ and $p_\varphi = \frac{r_{\varphi 2}+1}{2}$.

Note that $\sqrt{\Phi_{\varphi i}} \left\|\boldsymbol{\varphi}_i\right\| \|\boldsymbol{s}_{i2}\| \leq \frac{\Phi_{\varphi i}}{k_{\Phi i}} \left\|\boldsymbol{\varphi}_i\right\|^2 \|\boldsymbol{s}_{i2}\|^2 + k_{\Phi i}$, and thus it follows that

$$\begin{aligned}\dot{L}'_{\varphi i} \leq & - g_i\underline{\zeta}_{\varphi i}k_{\phi i}\|\boldsymbol{s}_{i2}\|^4 - g_i\underline{\zeta}_{\varphi i}k_{si}\|\boldsymbol{s}_{i2}\|^2 + \eta_{\varphi i}\|\boldsymbol{s}_{i2}\| \\ & + \left(\Phi_{\varphi i} - g_i\underline{\zeta}_{\varphi i}\hat{\Phi}_{\varphi i}\right)\frac{\left\|\boldsymbol{\varphi}_i\right\|^2}{k_{\Phi i}}\|\boldsymbol{s}_{i2}\|^2 + k_{\Phi i} \\ & - g_i\underline{\zeta}_{\varphi i}\xi_i\hat{\zeta}_{\varphi i}\left(k_{pi}+\hat{\eta}_{\varphi i}\right)\|\boldsymbol{s}_{i2}\|^{2q_\varphi} \\ & - g_i\underline{\zeta}_{\varphi i}\xi_i\hat{\zeta}_{\varphi i}\left(k_{hi}+\hat{\eta}_{\varphi i}\right)\|\boldsymbol{s}_{i2}\|^{2p_\varphi}, \end{aligned} \tag{4.57}$$

Then, taking the time derivative of $L''_{\varphi i}$ along (4.56)–(4.57), we obtain

$$\begin{aligned}\dot{L}''_{\varphi i} = & \left(g_i\underline{\zeta}_{\varphi i}\hat{\Phi}_{\varphi i} - \Phi_{\varphi i}\right)\frac{\left\|\boldsymbol{\varphi}_i\right\|^2}{k_{\Phi i}}\|\boldsymbol{s}_{i2}\|^2 \\ & + \gamma_{\Phi i}\left(\Phi_{\varphi i} - g_i\underline{\zeta}_{\varphi i}\hat{\Phi}_{\varphi i}\right)\hat{\Phi}_{\varphi i} \\ & - \left(\eta_{\varphi i} - g_i\underline{\zeta}_{\varphi i}\xi_i\hat{\eta}_{\varphi i}\right)\|\boldsymbol{s}_{i2}\|^{2q_\varphi} - \left(\eta_{\varphi i} - g_i\underline{\zeta}_{\varphi i}\xi_i\hat{\eta}_{\varphi i}\right)\|\boldsymbol{s}_{i2}\|^{2p_\varphi} \\ & + \gamma_{\eta i}\left(\eta_{\varphi i} - g_i\underline{\zeta}_{\varphi i}\xi_i\hat{\eta}_{\varphi i}\right)\hat{\eta}_{\varphi i} - \frac{g_i\underline{\zeta}_{\varphi i}}{\beta_i\hat{\zeta}_{\varphi i}^2}\left(\hat{\zeta}_{\varphi i}^{-1} - \underline{\zeta}_{\varphi i}\right)\dot{\hat{\zeta}}_{\varphi i}. \end{aligned} \tag{4.58}$$

At this point, the following two cases based on (4.58) are discussed.

Case 1 $\hat{\zeta}_{\varphi i} > 1$ or $\hat{\zeta}_{\varphi i} = 1$ and $\alpha_{\varphi i} \geq 0$. Combining (4.57) and (4.58), the time derivative of $L'_{\varphi i} + L''_{\varphi i}$ satisfies

$$\begin{aligned}\dot{L}'_{\varphi i} + \dot{L}''_{\varphi i} \leq & -g_i\underline{\zeta}_{\varphi i}k_{si}\|\boldsymbol{s}_{i2}\|^2 - g_i\underline{\zeta}_{\varphi i}\xi_i\hat{\zeta}_{\varphi i}\left(k_{pi}+\hat{\eta}_{\varphi i}\right)\|\boldsymbol{s}_{i2}\|^{2q_\varphi} \\ & - g_i\underline{\zeta}_{\varphi i}\xi_i\hat{\zeta}_{\varphi i}\left(k_{hi}+\hat{\eta}_{\varphi i}\right)\|\boldsymbol{s}_{i2}\|^{2p_\varphi} + \gamma_{\Phi i}\left(\Phi_{\varphi i} - g_i\underline{\zeta}_{\varphi i}\hat{\Phi}_{\varphi i}\right)\hat{\Phi}_{\varphi i}\end{aligned}$$

$$
\begin{aligned}
&-\left(\eta_{\varphi i}-g_i\underline{\zeta}_{\varphi i}\xi_i\hat{\eta}_{\varphi i}\right)\|\boldsymbol{s}_{i2}\|^{2q_\varphi}-\left(\eta_{\varphi i}-g_i\underline{\zeta}_{\varphi i}\xi_i\hat{\eta}_{\varphi i}\right)\|\boldsymbol{s}_{i2}\|^{2p_\varphi}\\
&+\gamma_{\eta i}\left(\eta_{\varphi i}-g_i\underline{\zeta}_{\varphi i}\xi_i\hat{\eta}_{\varphi i}\right)\hat{\eta}_{\varphi i}+\gamma_{\zeta i}g_i\underline{\zeta}_{\varphi i}\xi_i\hat{\zeta}_{\varphi i}^3\left(\hat{\zeta}_{\varphi i}^{-1}-\underline{\zeta}_{\varphi i}\right)\\
&-g_i\underline{\zeta}_{\varphi i}\xi_i\hat{\zeta}_{\varphi i}\left(\hat{\zeta}_{\varphi i}^{-1}-\underline{\zeta}_{\varphi i}\right)\left(k_{pi}+\hat{\eta}_{\varphi i}\right)\|\boldsymbol{s}_{i2}\|^{2q_\varphi}\\
&-g_i\underline{\zeta}_{\varphi i}\xi_i\hat{\zeta}_{\varphi i}\left(\hat{\zeta}_{\varphi i}^{-1}-\underline{\zeta}_{\varphi i}\right)\left(k_{hi}+\hat{\eta}_{\varphi i}\right)\|\boldsymbol{s}_{i2}\|^{2p_\varphi}\\
&+k_{\Phi i}+\eta_{\varphi i}\|\boldsymbol{s}_{i2}\|.
\end{aligned}
\tag{4.59}
$$

From the definition of $\underline{\zeta}_{\varphi i}$, there exists $\underline{\zeta}_{\varphi i}-1\le 0$. Thus, we derive

$$
\begin{aligned}
&g_i\underline{\zeta}_{\varphi i}\xi_i\hat{\zeta}_{\varphi i}\left(k_{pi}+\hat{\eta}_{\varphi i}\right)\|\boldsymbol{s}_{i2}\|^{2q_\varphi}\left(\underline{\zeta}_{\varphi i}-1-\hat{\zeta}_{\varphi i}^{-1}\right)\\
&\quad\le -g_i\underline{\zeta}_{\varphi i}\xi_i\left(k_{pi}+\hat{\eta}_{\varphi i}\right)\|\boldsymbol{s}_{i2}\|^{2q_\varphi}
\end{aligned}
\tag{4.60}
$$

and

$$
\begin{aligned}
&g_i\underline{\zeta}_{\varphi i}\xi_i\hat{\zeta}_{\varphi i}\left(k_{hi}+\hat{\eta}_{i}\right)\|\boldsymbol{s}_{i2}\|^{p_\varphi}\left(\underline{\zeta}_{\varphi i}-1-\hat{\zeta}_{\varphi i}^{-1}\right)\\
&\quad\le -g_i\underline{\zeta}_{\varphi i}\xi_i\left(k_{hi}+\hat{\eta}_{\varphi i}\right)\|\boldsymbol{s}_{i2}\|^{p_\varphi}.
\end{aligned}
\tag{4.61}
$$

Substituting (4.60) and (4.61) into (4.59) yields

$$
\begin{aligned}
\dot{L}_{\varphi i}'+\dot{L}_{\varphi i}''\le& -g_i\underline{\zeta}_{\varphi i}k_{si}\|\boldsymbol{s}_{i2}\|^2-g_i\underline{\zeta}_{\varphi i}\xi_i k_{pi}\|\boldsymbol{s}_{i2}\|^{2q_\varphi}\\
&-g_i\underline{\zeta}_{\varphi i}\xi_i k_{hi}\|\boldsymbol{s}_{i2}\|^{2p_\varphi}+\eta_{\varphi i}\left(\|\boldsymbol{s}_{i2}\|-\|\boldsymbol{s}_{i2}\|^{2q_\varphi}-\|\boldsymbol{s}_{i2}\|^{2p_\varphi}\right)\\
&+\gamma_{\Phi i}\left(\Phi_{\varphi i}-g_i\underline{\zeta}_{\varphi i}\hat{\Phi}_{\varphi i}\right)\hat{\Phi}_{\varphi i}+\gamma_{\eta i}\left(\eta_{\varphi i}-g_i\underline{\zeta}_{\varphi i}\xi_i\hat{\eta}_{\varphi i}\right)\hat{\eta}_{\varphi i}\\
&+\gamma_{\zeta i}g_i\underline{\zeta}_{\varphi i}\xi_i\hat{\zeta}_{\varphi i}^3\left(\hat{\zeta}_{\varphi i}^{-1}-\underline{\zeta}_{\varphi i}\right)+k_{\Phi i}.
\end{aligned}
\tag{4.62}
$$

From (4.62), the following equation holds true:

$$
\eta_{\varphi i}\left(\|\boldsymbol{s}_{i2}\|-\|\boldsymbol{s}_{i2}\|^{2q_\varphi}-\|\boldsymbol{s}_{i2}\|^{2p_\varphi}\right)=\begin{cases}<0, & \text{if } \|\boldsymbol{s}_{i2}\|\ge 1\\ <\eta_{\varphi i}, & \text{if } \|\boldsymbol{s}_{i2}\|<1.\end{cases}
\tag{4.63}
$$

Using Young's inequality, one arrives at

$$
\gamma_{\Phi i}\left(\Phi_{\varphi i}-g_i\underline{\zeta}_{\varphi i}\hat{\Phi}_{\varphi i}\right)\hat{\Phi}_{\varphi i}\le\frac{\gamma_{\Phi i}\Phi_{\varphi i}^2}{2g_i\underline{\zeta}_{\varphi i}}-\frac{\gamma_{\Phi i}\left(\Phi_{\varphi i}-g_i\underline{\zeta}_{\varphi i}\hat{\Phi}_{\varphi i}\right)^2}{2g_i\underline{\zeta}_{\varphi i}}
$$

$$\gamma_{\eta l}\left(\eta_{\varphi l}-g_l\underline{\zeta}_{\varphi l}\xi_l\hat{\eta}_{\varphi l}\right)\hat{\eta}_{\varphi l}\leq\frac{\gamma_{\eta l}\eta_{\varphi l}^2}{g_l\underline{\zeta}_{\varphi l}\xi_l}-\frac{\gamma_{\eta l}\left(\eta_{\varphi l}-g_l\underline{\zeta}_{\varphi l}\xi_l\hat{\eta}_{\varphi l}\right)^2}{2g_l\underline{\zeta}_{\varphi l}\xi_l}. \tag{4.64}$$

Adding and subtracting $\gamma_{\zeta l}g_l\underline{\zeta}_{\varphi l}\xi_l\left(\hat{\zeta}_{\varphi l}^{-1}-\underline{\zeta}_{\varphi l}\right)^2$ into (4.62) and invoking $0<\hat{\zeta}_{\varphi l}^{-1}\leq 1$, it follows that

$$\gamma_{\zeta l}g_l\underline{\zeta}_{\varphi l}\xi_l\left[\left(\hat{\zeta}_{\varphi l}^{-1}-\underline{\zeta}_{\varphi l}\right)^2+\hat{\zeta}_{\varphi l}^3\left(\hat{\zeta}_{\varphi l}^{-1}-\underline{\zeta}_{\varphi l}\right)\right]$$
$$\leq\gamma_{\zeta l}g_l\underline{\zeta}_{\varphi l}\xi_l\left(1+\bar{\zeta}_{\varphi l}^2\right), \tag{4.65}$$

where $\bar{\zeta}_{\varphi l}$ is the upper bound of $\hat{\zeta}_{\varphi l}$.

Substituting (4.63)–(4.65) into (4.62), one has

$$\begin{aligned}\dot{L}'_{\varphi l}+\dot{L}''_{\varphi l}\leq&-g_l\underline{\zeta}_{\varphi l}k_{sl}\|s_{l2}\|^2-\frac{\gamma_{\Phi l}\left(\Phi_{\varphi l}-g_l\underline{\zeta}_{\varphi l}\hat{\Phi}_{\varphi l}\right)^2}{2g_l\underline{\zeta}_{\varphi l}}\\&-\frac{\gamma_{\eta l}\left(\eta_{\varphi l}-g_l\underline{\zeta}_{\varphi l}\xi_l\hat{\eta}_{\varphi l}\right)^2}{2g_l\underline{\zeta}_{\varphi l}\xi_l}-\gamma_{\zeta l}g_l\underline{\zeta}_{\varphi l}\xi_l\left(\hat{\zeta}_{\varphi l}^{-1}-\underline{\zeta}_{\varphi l}\right)^2\\&+\gamma_{\zeta l}g_l\underline{\zeta}_{\varphi l}\xi_l\left(1+\bar{\zeta}_{\varphi l}^2\right)+\frac{\gamma_{\Phi l}\Phi_{\varphi l}^2}{2g_l\underline{\zeta}_{\varphi l}}+\frac{\gamma_{\eta l}\eta_{\varphi l}^2}{g_l\underline{\zeta}_{\varphi l}\xi_l}+k_{\Phi l}+\eta_{\varphi l}.\end{aligned} \tag{4.66}$$

We consider the following Lyapunov function candidate:

$$L_{\varphi l}=\frac{1}{2}s_{l1}^{\mathrm{T}}s_{l1}+L'_{\varphi l}+L''_{\varphi l}. \tag{4.67}$$

Taking the time derivative of $L_{\varphi l}$ along (4.50) yields

$$\begin{aligned}\dot{L}_{\varphi l}\leq&-\left[\lambda_{\min}\left(\boldsymbol{K}_{1l}\right)-\frac{\lambda'_l\lambda_{\max}\left(\boldsymbol{R}_2^{-1}\right)}{2}\right]\|s_{l1}\|^2\\&-\left[g_l\underline{\zeta}_{\varphi l}k_{sl}-\frac{\lambda'_l\lambda_{\max}\left(\boldsymbol{R}_2^{-1}\right)}{2}\right]\|s_{l2}\|^2-\frac{\gamma_{\Phi l}\left(\Phi_{\varphi l}-g_l\underline{\zeta}_{\varphi l}\hat{\Phi}_{\varphi l}\right)^2}{2g_l\underline{\zeta}_{\varphi l}}\\&-\frac{\gamma_{\eta l}\left(\eta_{\varphi l}-g_l\underline{\zeta}_{\varphi l}\xi_l\hat{\eta}_{\varphi l}\right)^2}{2g_l\underline{\zeta}_{\varphi l}\xi_l}-\gamma_{\zeta l}g_l\underline{\zeta}_{\varphi l}\xi_l\left(\hat{\zeta}_{\varphi l}^{-1}-\underline{\zeta}_{\varphi l}\right)^2+\chi_{\varphi l2}.\end{aligned} \tag{4.68}$$

Let us choose

$$\lambda_{\min}\left(\boldsymbol{K}_{l1}\right) \geqq \frac{\lambda_l' \lambda_{\max}\left(\boldsymbol{R}_2^{-1}\right)}{2},$$

$$k_{sl} \geqq \frac{\lambda_l' \lambda_{\max}\left(\boldsymbol{R}_2^{-1}\right)}{2 g_{l\varphi l}},$$

$$\chi_{\varphi l1} = \min\left\{2\lambda_{\min}\left(\boldsymbol{K}_{l1}\right) - \lambda_l' \lambda_{\max}\left(\boldsymbol{R}_2^{-1}\right),\right.$$
$$\left. 2 g_l \underline{\zeta}_{\varphi l} k_{sl} - \lambda_l' \lambda_{\max}\left(\boldsymbol{R}_2^{-1}\right), \gamma_{\Phi l}, 2\beta_l \gamma_{\zeta l} \xi_l, \gamma_{\eta l}\right\},$$

$$\chi_{\varphi l2} = \gamma_{\xi l} g_l \underline{\zeta}_{\varphi l} \xi_l \left(1 + \bar{\zeta}_{\varphi l}^2\right) + \frac{\gamma_{\Phi l} \Phi_{\varphi l}^2}{2 g_l \underline{\zeta}_{\varphi l}} + \frac{\gamma_{\eta l} \eta_{\varphi l}^2}{g_l \underline{\zeta}_{\varphi l} \xi_l}$$
$$+ k_{\Phi l} + \eta_{\varphi l}.$$

Then we have

$$\dot{L}_{\varphi l} \leq -\chi_{\varphi l1} L_{\varphi l} + \chi_{\varphi l2}. \tag{4.69}$$

For $\left(\boldsymbol{s}_{l1}(0), \boldsymbol{s}_{l2}(0), \hat{\Phi}_{\varphi l}(0), \hat{\eta}_{\varphi l}(0), \hat{\zeta}_{\varphi l}(0)\right) \in \Re^3 \times \Re^3 \times \Re \times \Re \times \Re$, we have $|\Phi_{\varphi l} - g_l \underline{\zeta}_{\varphi l} \hat{\Phi}_{\varphi l}| \leq \mu_{\Phi l}$, $|\eta_{\varphi l} - g_l \underline{\zeta}_{\varphi l} \xi_l \hat{\eta}_{\varphi l}| \leq \mu_{\eta l}'$, and $|\hat{\zeta}_{\varphi l}^{-1} - \underline{\zeta}_{\varphi l}| \leq \mu_{\zeta l}'$ with $\mu_{\Phi l}$, $\mu_{\eta l}'$, and $\mu_{\zeta l}'$ being positive constants.

Further, to prove the fixed-time convergence of $\boldsymbol{s}_{l1}$ and $\boldsymbol{s}_{l2}$, we reconsider (4.59). Based on $\hat{\zeta}_{\varphi l} \geq 1$ and $\|\boldsymbol{\varphi}_l\| \leq h_l$, it holds that

$$\begin{aligned}\dot{L}_{\varphi l}' \leq & -g_l \underline{\zeta}_{\varphi l} k_{\phi l} \|\boldsymbol{s}_{l2}\|^4 - g_l \underline{\zeta}_{\varphi l} k_{sl} \|\boldsymbol{s}_{l2}\|^2 - g_l \underline{\zeta}_{\varphi l} \xi_l k_{pl} \|\boldsymbol{s}_{l2}\|^{2q_\varphi} \\ & - g_l \underline{\zeta}_{\varphi l} \xi_l k_{hl} \|\boldsymbol{s}_{l2}\|^{2p_\varphi} + \eta_{\varphi l}\left(\|\boldsymbol{s}_{l2}\| - \|\boldsymbol{s}_{l2}\|^{2q_\varphi} - \|\boldsymbol{s}_{l2}\|^{2p_\varphi}\right) \\ & + \left(\eta_{\varphi l} - g_l \underline{\zeta}_{\varphi l} \xi_l \hat{\eta}_{\varphi l}\right)\left(\|\boldsymbol{s}_{l2}\|^{2q_\varphi} + \|\boldsymbol{s}_{l2}\|^{2p_\varphi}\right) \\ & + \left(\Phi_{\varphi l} - g_l \underline{\zeta}_{\varphi l} \hat{\Phi}_{\varphi l}\right) \frac{\|\boldsymbol{\varphi}_l\|^2}{k_{\Phi l}} \|\boldsymbol{s}_{l2}\|^2 + k_{\Phi l}.\end{aligned} \tag{4.70}$$

To facilitate the controller design, the following inequalities are needed:

$$\left(\Phi_{\varphi l} - g_l \underline{\zeta}_{\varphi l} \hat{\Phi}_{\varphi l}\right) \frac{\|\boldsymbol{\varphi}_l\|^2}{k_{\Phi l}} \|\boldsymbol{s}_{l2}\|^2 \leq \frac{h_l^4 \mu_{\Phi l}^2}{2 k_{\Phi l}} + \frac{\|\boldsymbol{s}_{l2}\|^4}{2 k_{\Phi l}}$$

$$\left(\eta_{\varphi l} - g_l \underline{\zeta}_{\varphi l} \xi_l \hat{\eta}_{\varphi l}\right) \|\boldsymbol{s}_{l2}\|^{2q_\varphi} \leq \left(\frac{\mu_{\eta l}'^2}{2} + \frac{1}{2}\right) \|\boldsymbol{s}_{l2}\|^{2q_\varphi}$$

$$\left(\eta_{\varphi i}-g_i\underline{\zeta}_{\varphi i}\xi_i\hat{\eta}_{\varphi i}\right)\|s_{i2}\|^{2p_\varphi}\leq\left(\frac{\mu_{\eta i}^{\prime 2}}{2}+\frac{1}{2}\right)\|s_{i2}\|^{2p_\varphi}. \tag{4.71}$$

Let us consider the following Lyapunov function candidate:

$$\bar{L}_{\varphi i}=\frac{1}{2}s_{i1}^{\mathrm{T}}s_{i1}+L_{\varphi i}^{\prime}. \tag{4.72}$$

Combining (4.50) and (4.57), we derive

$$\begin{aligned}\dot{\bar{L}}_{\varphi i}&\leq -k_{i1}\|s_{i1}\|^{2q_\varphi}-k_{i2}\|s_{i1}\|^{2p_\varphi}\\&\quad-\left(g_i\underline{\zeta}_i\xi_ik_{pi}-\frac{1+\mu_{\eta i}^{\prime 2}}{2}\right)\|s_{i2}\|^{2q_\varphi}\\&\quad-\left(g_i\underline{\zeta}_i\xi_ik_{hi}-\frac{1+\mu_{\eta i}^{\prime 2}}{2}\right)\|s_{i2}\|^{2p_\varphi}+\chi_{\varphi i5}\\&\leq-\chi_{\varphi i3}\bar{L}_{\varphi i}^{q_\varphi}-\chi_{\varphi i4}\bar{L}_{\varphi i}^{p_\varphi}+\chi_{\varphi i5},\end{aligned} \tag{4.73}$$

where $\chi_{\varphi i3}=\min\{2^{q_\varphi}k_{i1},2^{q_\varphi}(g_i\underline{\zeta}_i\xi_ik_{pi}-1/2-\mu_{\eta i}^{\prime 2}/2)/J_{\max}\}$, $\chi_{\varphi i4}=\min\{2^{p_\varphi}2k_{i2},2^{p_\varphi}(g_i\underline{\zeta}_i\xi_ik_{hi}-1/2-\mu_{\eta i}^{\prime 2}/2)/J_{\max}\}$, and $\chi_{\varphi i5}=\eta_{\varphi i}+k_{\Phi i}+\frac{h_i^4\mu_{\Phi i}^2}{2k_{\Phi i}}$. Choose $k_{\phi i}>\frac{1}{2k_{\Phi i}g_i\underline{\zeta}_i}$, $k_{pi}>\frac{1+\mu_{\eta i}^{\prime 2}}{2g_i\underline{\zeta}_i\xi_i}$, and $k_{hi}>\frac{1+\mu_{\eta i}^{\prime 2}}{2g_i\underline{\zeta}_i\xi_i}$.

Using Lemma 4.1, it can be concluded that all signals of the rotational subsystems are practical fixed-time stable, and it can be derived that

$$\bar{L}_{\varphi i}\leq\min\left\{\left(\frac{\chi_{\varphi i5}}{\chi_{\varphi i3}\left(1-\sigma_{\varphi i}\right)}\right)^{\frac{1}{q_\varphi}},\left(\frac{\chi_{\varphi i5}}{\chi_{\varphi i4}\left(1-\sigma_{\varphi i}\right)}\right)^{\frac{1}{p_\varphi}}\right\}, \tag{4.74}$$

and the setting time denotes

$$T\leq\bar{T}_{\varphi i1}:=\frac{1}{\chi_{\varphi i3}\sigma_{\varphi i}\left(\frac{q_\varphi}{2}-1\right)}+\frac{1}{\chi_{\varphi i4}\sigma_{\varphi i}\left(1-\frac{p_\varphi}{2}\right)}. \tag{4.75}$$

Then, we have

$$\begin{aligned}\|s_{i1}\|&\leq\bar{\vartheta}_{\varphi i1}\\&:=\min\left\{\sqrt{2\left(\frac{\chi_{\varphi i5}}{\chi_{\varphi i3}\left(1-\sigma_{\varphi i}\right)}\right)^{\frac{1}{q_\varphi}}},\sqrt{2\left(\frac{\chi_{\varphi i5}}{\chi_{\varphi i4}\left(1-\sigma_{\varphi i}\right)}\right)^{\frac{1}{p_\varphi}}}\right\}.\end{aligned} \tag{4.76}$$

Case 2 $\hat{\zeta}_{\varphi i} = 1$ and $\alpha_{\varphi i} < 0$. From (4.25), $\dot{\hat{\zeta}}_{\varphi i} = 0$ can be obtained. In this case, the actuator saturation is no more considered. Substituting $\hat{\zeta}_{\varphi i} = 1$ into (4.22) gives rise to

$$\begin{aligned}\bar{\boldsymbol{\delta}}_{ic} = &- \boldsymbol{C}_i^{\mathrm{T}} \left(k_{\phi i} \|\boldsymbol{s}_{i2}\|^2 + k_{si} + \frac{\hat{\Phi}_{\varphi i} \left\|\boldsymbol{\varphi}_i\right\|^2}{k_{\Phi i}} \right) \boldsymbol{s}_{i2} \\ &- \boldsymbol{C}_i^{\mathrm{T}} \xi_i \left(k_{pi} + \hat{\eta}_{\varphi i}\right) \|\boldsymbol{s}_{i2}\|^{r_{\varphi 1}-1} \boldsymbol{s}_{i2} \\ &- \boldsymbol{C}_i^{\mathrm{T}} \xi_i \left(k_{hi} + \hat{\eta}_{\varphi i}\right) \|\boldsymbol{s}_{i2}\|^{r_{\varphi 2}-1} \boldsymbol{s}_{i2} \end{aligned} \tag{4.77}$$

with the adaptation tuning law as

$$\dot{\hat{\Phi}}_{\varphi i} = \frac{\left\|\boldsymbol{\varphi}_i\right\|^2}{k_{\Phi i}} \|\boldsymbol{s}_{i2}\|^2 - \gamma_{\Phi i} \hat{\Phi}_{\varphi i} \tag{4.78}$$

$$\dot{\hat{\eta}}_{\varphi i} = \|\boldsymbol{s}_{i2}\|^{r_{\varphi 1}+1} + \|\boldsymbol{s}_{i2}\|^{r_{\varphi 2}+1} - \gamma_{\eta i} \hat{\eta}_{\varphi i}. \tag{4.79}$$

Following the similar proof of Case 1, we choose $k_{\phi i} > \frac{1}{2k_{\Phi i} g_i}$, $k_{pi} > \frac{1+\mu_{\eta i}'^2}{2g_i\xi_i}$, and $k_{hi} > \frac{1+\mu_{\eta i}'^2}{2g_i\xi_i}$. Let $\chi_{\varphi i6} = \min\{2^{q_\varphi} k_{i1}, 2^{q_\varphi}(g_i\xi_i k_{pi} - 1/2 - \mu_{\eta i}'^2/2)/J_{\max}\}$, $\chi_{\varphi i7} = \min\{2^{p_\varphi} k_{i2}, 2^{p_\varphi}(g_i\xi_i k_{hi} - 1/2 - \mu_{\eta i}'^2/2)/J_{\max}\}$, and $\chi_{\varphi i8} = \eta_{\varphi i} + k_{\Phi i} + \frac{h_i^4 \mu_{\Phi i}^2}{2k_{\Phi i}}$. We obtain that

$$\dot{\bar{L}}_{\varphi i} \le -\chi_{\varphi i6} \bar{L}_{\varphi i}^{q_\varphi} - \chi_{\varphi i7} \bar{L}_{\varphi i}^{p_\varphi} + \chi_{\varphi i8}. \tag{4.80}$$

Invoking Lemma 4.1, it could be proved that all signals of rotational subsystems are practically fixed-time stable, and it can be obtained that

$$\bar{L}_{\varphi i} \le \min \left\{ \left(\frac{\chi_{\varphi i8}}{\chi_{\varphi i6}\left(1-\sigma_{\varphi i}\right)} \right)^{\frac{1}{q_\varphi}}, \left(\frac{\chi_{\varphi i8}}{\chi_{\varphi i7}\left(1-\sigma_{\varphi i}\right)} \right)^{\frac{1}{p_\varphi}} \right\}, \tag{4.81}$$

and the setting time is given as follows:

$$T \le \bar{T}_{\varphi i2} := \frac{1}{\chi_{\varphi i6}\sigma_{\varphi i}\left(\frac{q_\varphi}{2} - 1\right)} + \frac{1}{\chi_{\varphi i7}\sigma_{\varphi i}\left(1 - \frac{p_\varphi}{2}\right)}. \tag{4.82}$$

Then, it holds that

$$\begin{aligned}\|s_{l1}\| &\leq \bar{\vartheta}_{\varphi l2}\\ &:= \min\left\{\sqrt{2\left(\frac{\chi_{\varphi l8}}{\chi_{\varphi l6}\left(1-\sigma_{\varphi l}\right)}\right)^{\frac{1}{q_\varphi}}}, \sqrt{2\left(\frac{\chi_{\varphi l8}}{\chi_{\varphi l7}\left(1-\sigma_{\varphi l}\right)}\right)^{\frac{1}{p_\varphi}}}\right\}. \end{aligned} \tag{4.83}$$

Based on the analysis of Case 1 and Case 2, it is concluded that for any $\left(s_{l1}(0), s_{l2}(0), \hat{\Phi}_{\varphi l}(0), \hat{\eta}_{\varphi l}(0), \hat{\zeta}_{\varphi l}(0)\right) \in \Re^3 \times \Re^3 \times \Re \times \Re \times \Re$, there exists a finite $\bar{T}_{\varphi l} := \max\left\{\bar{T}_{\varphi l1}, \bar{T}_{\varphi l2}\right\}$ such that the solution $\|s_{l1}\|$ to the rotational subsystem satisfies $\|s_{l1}\| \leq \bar{\vartheta}_{\varphi l} := \max\left\{\bar{\vartheta}_{\varphi l1}, \bar{\vartheta}_{\varphi l2}\right\}$ for all $t \geq \bar{T}_{\varphi l}$.

Hence, the proof of Theorem 4.2 is completed.

Remark 4.9 The precision and convergence speed of tracking errors can be made arbitrarily small by tuning k_{l1}, k_{l2}, k_{pl}, and k_{hl} sufficiently large, and meanwhile the precision and convergence speed of tracking errors are affected by $r_{\varphi 1}$ and $r_{\varphi 2}$. Therefore, by tuning the above parameters, the convergence speed and precision of tracking errors can be adjusted within desired setting time $\bar{T}_{\varphi i}$.

Remark 4.10 The control for velocity dynamics can be achieved if velocity V_l and adaptive parameters $\hat{\Phi}_{\varphi l}$ and $\hat{\eta}_{vl}$ are obtained. Similarly, the control for attitude dynamics can be realized if error s_{l2}, angle ϕ_l, and adaptive parameters $\hat{\eta}_{\varphi l}$ are acquired. As a matter of fact, velocity V_l and angle ϕ_l can be derived by the sensors transmitting data in real time, and $\hat{\Phi}_{\varphi l}$, $\hat{\eta}_{vl}$, and $\hat{\eta}_{\varphi l}$ can be updated by using adaption laws (4.16) and (4.23)–(4.24), so that our proposed control method has a clearer physical meaning.

Remark 4.11 This chapter aims at designing control methods from the perspective of 6-DoF UAVs in the sense of solving practical problems. The spacecraft model focuses more on the attitude control model, while the UAV model is a model of speed and attitude coupling. Therefore, in terms of the complexity of controller design, the networked control of 6-DoF UAVs has more factors to consider and thus is more difficult for control design.

4.4 Simulations and Analyses

Numerical examples of three UAVs are carried out in this section to validate the advantages of our proposed control scheme. The communication topology of the multi-UAV system is described in Fig. 4.2. The structural and aerodynamic parameters are taken from [30]. The upper and lower limits of the actuators are chosen as $T_{xlc,\max} = 250\text{N}$, $T_{xlc,\min} = 0\text{N}$, $\delta_{lcj,\max} = 60^\circ$ $(l = 1, 2, 3)$. And the initial states are listed as $V_1 = V_2 = V_3 = 40\,\text{m/s}$, $\varphi_1 = [0, 1, 0]^{\text{T}}$ deg,

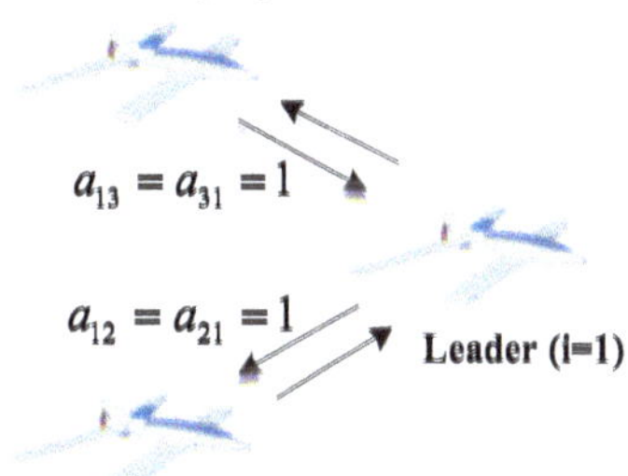

Fig. 4.2 The communication topology of networked UAVs

$\boldsymbol{\varphi}_2 = \boldsymbol{\varphi}_3 = [0.5, 1, 0.5]^{\mathrm{T}}$ deg, $\boldsymbol{\omega}_1 = \boldsymbol{\omega}_2 = \boldsymbol{\omega}_3 = [0, 0, 0]^{\mathrm{T}}$ deg/s. It is required that the velocity and attitude track the trajectories $V_r = 50$ m/s and $[5, 10, 5]^{\mathrm{T}}$deg.

It is supposed that external disturbances are chosen as $\boldsymbol{d}_{v1} = \boldsymbol{d}_{v2} = \boldsymbol{d}_{v3} = [0.3\sin(3t), 0.5\cos(2t), 0.2\cos(t)]^{\mathrm{T}}$ and $\boldsymbol{d}_{\omega 1} = \boldsymbol{d}_{\omega 2} = \boldsymbol{d}_{\omega 3} = [0.3\cos(2t), 0.2\sin(3t), 0.5\sin(t)]^{\mathrm{T}}$. At the same time, by injecting the actuator faults in velocity and attitude channels, the actuator fault signals are assumed as $\rho_{T2} = 0.9$ during $t \in (3, 7]$ and $\boldsymbol{\rho}_{\delta 3} = \mathrm{diag}\{1, 1, 0.9\}$ during $t \in (5, 8]$.

The control parameters are selected as $\lambda_{i1} = \lambda_{i2} = \lambda_{i3} = \lambda_{i4} = 1$, $k_{11} = k_{21} = k_{31} = 2$, and $k_{12} = k_{22} = k_{32} = 2$, which are used to calculate s_{vi}, $\boldsymbol{s}_{i1}$, and $\boldsymbol{s}_{i2}$. And the control parameters are chosen as $r_{v1} = r_{\varphi 1} = 0.8$, $r_{v2} = r_{\varphi 2} = 1.2$, $c_{11} = c_{12} = c_{13} = 10$ $(t \in (0, 3.5])$, $c_{11} = c_{12} = c_{13} = 70$ $(t \in (3.5, 8])$, $c_{21} = c_{22} = c_{23} = 0.1$ $(t \in (0, 3.5])$, $c_{21} = c_{22} = c_{23} = 35000$ $(t \in (3.5, 8])$, $c_{31} = c_{32} = c_{33} = 2$ $(t \in (0, 3.5])$, $c_{31} = c_{32} = c_{33} = 70$ $(t \in (3.5, 8])$, $\varepsilon_{v1} = \varepsilon_{v3} = 1$, $\varepsilon_{v2} = 1000000$ $(t \in (0, 3.5])$, $\varepsilon_{v2} = 1000$ $(t \in (3.5, 8])$, $\boldsymbol{K}_{i1} = \mathrm{diag}\{0.001, 0.001, 0.001\}$, $k_{\phi 1} = k_{s1} = k_{p1} = k_{h1} = 0.001$, $k_{\phi 2} = k_{s2} = k_{p2} = k_{h2} = 0.001$, $k_{\phi 3} = k_{s3} = k_{p3} = k_{h3} = 0.001$, $\xi_1 = \xi_2 = \xi_3 = 0.0001$ to compute $\bar{T}_{xic}$ and $\bar{\boldsymbol{\delta}}_{ic}$.

Then, we give the initial value of $\hat{\zeta}_{vi}(0) = 0$, $\hat{\eta}_{vi}(0) = 0$, $\hat{\Phi}_{\varphi i}(0) = 0$, $\hat{\eta}_{\varphi i}(0) = 0$, and $\hat{\zeta}_{\varphi i}(0) = 1.01$ and select the design parameter $\gamma_{v11} = \gamma_{v12} = 0.001$, $\gamma_{v21} = \gamma_{v22} = 0.0001$, $\gamma_{v31} = \gamma_{v32} = 0.001$, $\sigma_{v11} = \sigma_{v31} = 0.001$, $\sigma_{v21} = 0.0001$, $\sigma_{v12} = \sigma_{v32} = 0.0001$, $\sigma_{v22} = 0.00001$, $k_{\Phi 1} = k_{\Phi 2} = k_{\Phi 3} = 1000$, $\gamma_{\Phi 1} = \gamma_{\Phi 2} = \gamma_{\Phi 3} = 0.001$, $\gamma_{\eta 1} = \gamma_{\eta 2} = \gamma_{\eta 3} = 0.001$, $\gamma_{\zeta 1} = \gamma_{\zeta 2} = \gamma_{\zeta 3} = 1$, $\beta_1 = \beta_2 = \beta_3 = 0.0001$ for adaptation tuning law $\hat{\zeta}_{vi}$, $\hat{\eta}_{vi}$, $\hat{\Phi}_{\varphi i}$, $\hat{\eta}_{\varphi i}$, and $\hat{\zeta}_{\varphi i}$.

To highlight the superiority of the proposed control scheme, our proposed adaptive fixed-time control scheme is compared with two other control methods. The first comparative method is designed by removing the fixed-time terms $\mathrm{sig}(s_{vi})^{r_{v1}}$, $\mathrm{sig}(s_{vi})^{r_{v2}}$, $\|\boldsymbol{s}_{i1}\|^{r_{\varphi 1}-1}$, $\|\boldsymbol{s}_{i1}\|^{r_{\varphi 2}-1}$, $\|\boldsymbol{s}_{i2}\|^{r_{\varphi 1}-1}$, and $\|\boldsymbol{s}_{i2}\|^{r_{\varphi 2}-1}$ from our proposed control scheme, which is marked as *Compared controller 1*. The second comparative approach refers to [27] and is marked as *Compared controller 2*. In the next subsection, the comparative results are analyzed.

Figure 4.3 presents the trajectories of state responses and tracking errors. It can be observed from Fig. 4.3 that our proposed control scheme achieves good velocity

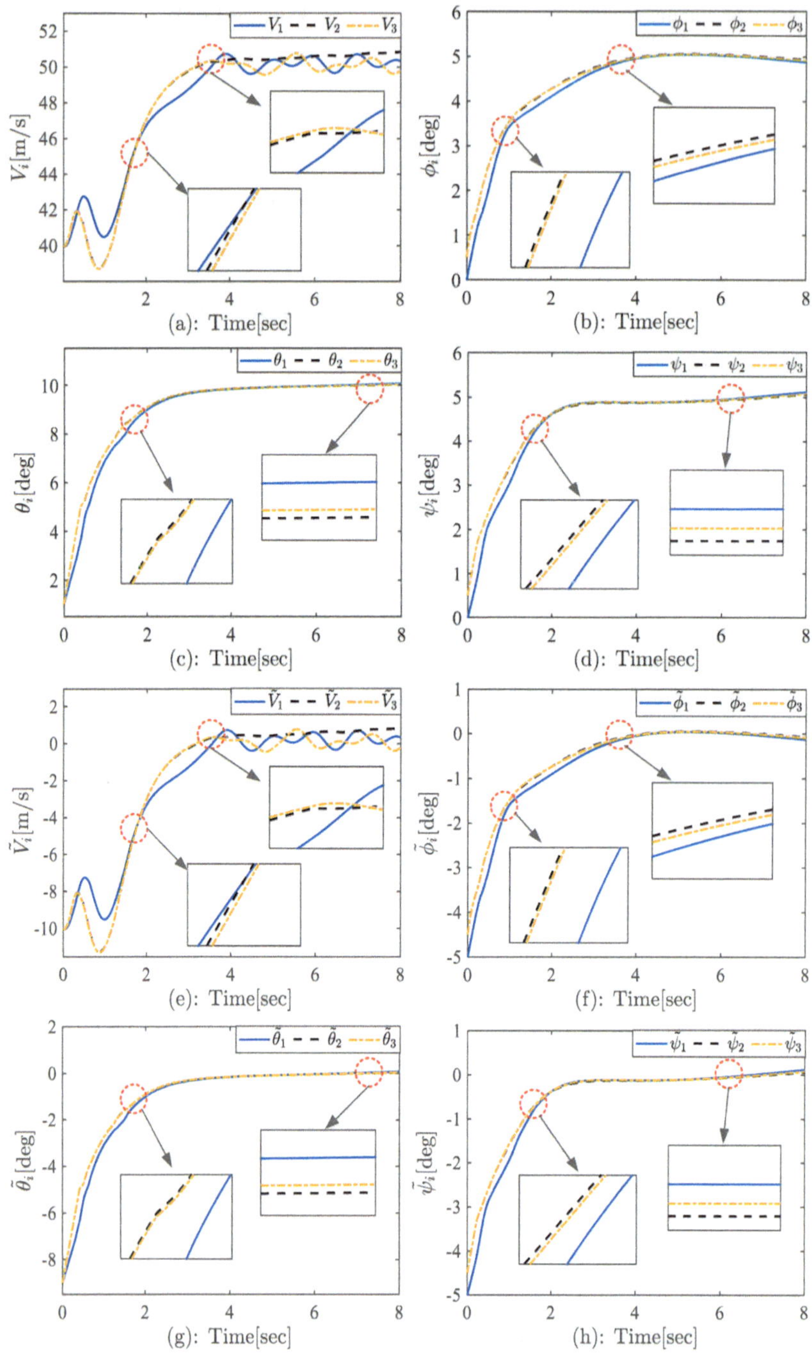

Fig. 4.3 Velocity and attitude tracking with our developed controller. (**a**)–(**h**) show the trajectory curves of state responses and errors. (**a**) V_i response. (**b**) ϕ_i response. (**c**) θ_i response. (**d**) ψ_i response. (**e**) Tracking error of V_i. (**f**) Tracking error of ϕ_i. (**g**) Tracking error of θ_i. (**h**) Tracking error of ψ_i

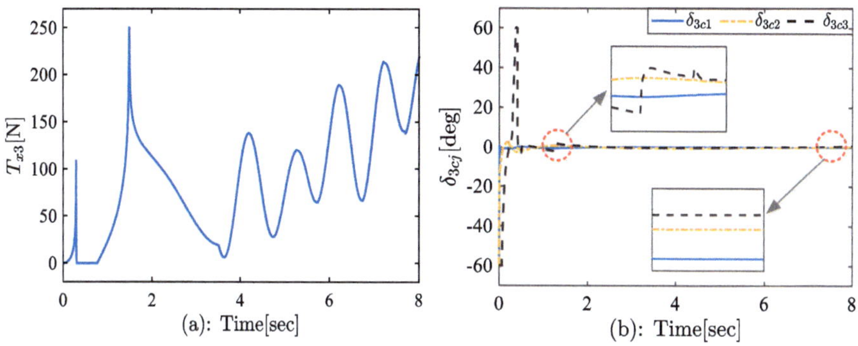

Fig. 4.4 Time response of control signals. (**a**) and (**b**) depict the trajectories of T_{x3} and δ_{3c}, respectively. (**a**) T_{x3}. (**b**) δ_{3c}

and attitude tracking for networked UAVs. Especially when the thrust and control surface failed during $t \in (3, 7]$ and $t \in (5, 8]$, no obvious performance degradations occur. The time responses of designed control actions $\bar{T}_{xic}$ and $\bar{\delta}_{ic}$ (taking follower 1 for example) are presented in Fig. 4.4, and meanwhile the adaptive parameters $\hat{\zeta}_{vi}$, $\hat{\eta}_{vi}$, $\hat{\Phi}_{\varphi i}$, $\hat{\eta}_{\varphi i}$, and $\hat{\zeta}_{\varphi i}$ are described in Fig. 4.5. It can be seen from Figs. 4.4 and 4.5 that $\bar{T}_{xic}$ and $\bar{\delta}_{ic}$ satisfy their saturation constraints, and the involved adaptive parameters are bounded.

Then, we move on to demonstrate the superiority of our proposed controller over *Compared controller 1* and *Compared controller 2*. With respect to follower 1, the comparative simulation results of tracking performances are shown in Fig. 4.6. It can be observed from Fig. 4.6 that the velocity tracking error is non-convergent with *Compared controller 1* and occurs large oscillations when reacting to the actuator faults with *Compared controller 2*. While our proposed method exhibits a better tracking performance, the tracking errors tend to be convergent in a more rapid manner.

Then, it can be concluded that the proposed controller has superiorities over *Compared controller 1* and *Compared controller 2*, such as fixed-time convergence and better fault-tolerant capability, and robustness to external disturbances. To visually evaluate the tracking performances of different controllers, the indices of the offset errors, integral of absolute errors, and convergence time are listed in Table 4.1, which validates the superiority of our proposed control scheme.

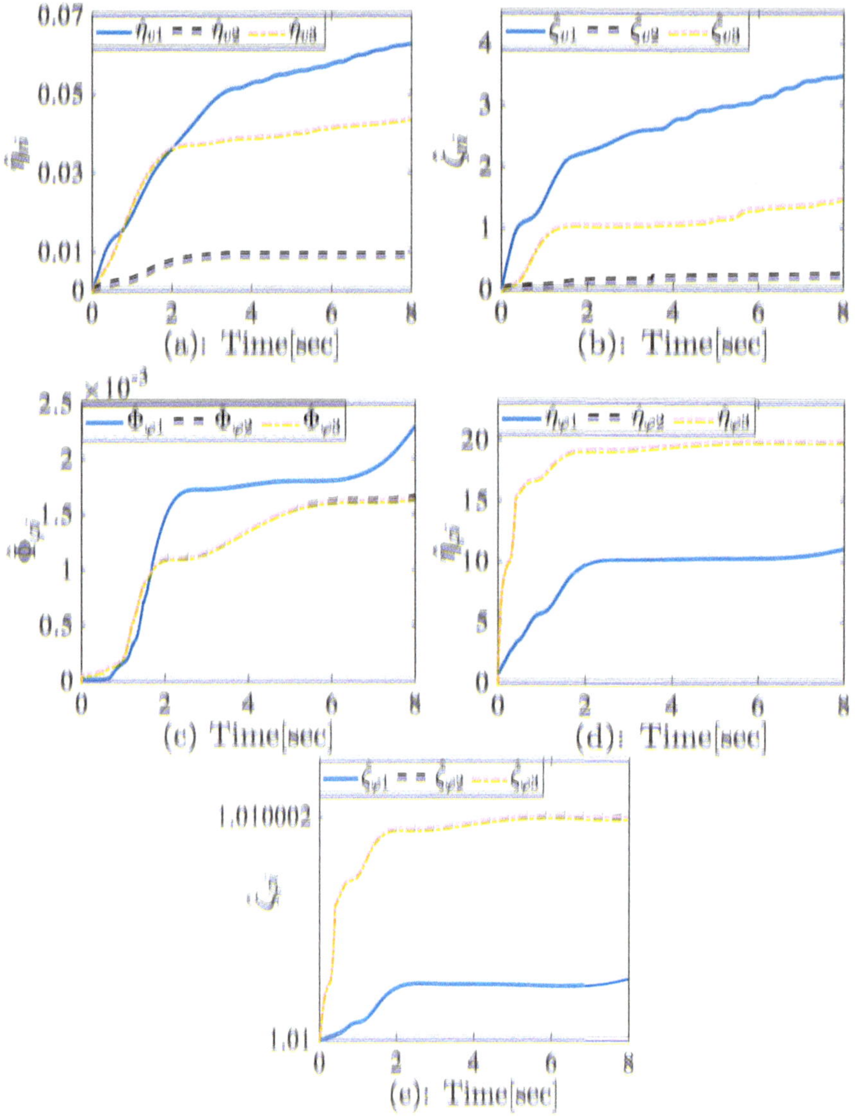

Fig. 4.5 Time response of adaptive parameters in (4.17)–(4.18) and (4.23)–(4.25). (**a**)–(**e**) show the curves of parameters $\hat{\eta}_{vi}$, $\hat{\xi}_{vi}$, $\hat{\Phi}_{\psi i}$, $\hat{\eta}_{\psi i}$, and $\hat{\xi}_{\psi i}$, respectively. (**a**) $\hat{\eta}_{vi}$; (**b**) $\hat{\xi}_{vi}$; (**c**) $\hat{\Phi}_{\psi i}$; (**d**) $\hat{\eta}_{\psi i}$; (**e**) $\hat{\xi}_{\psi i}$

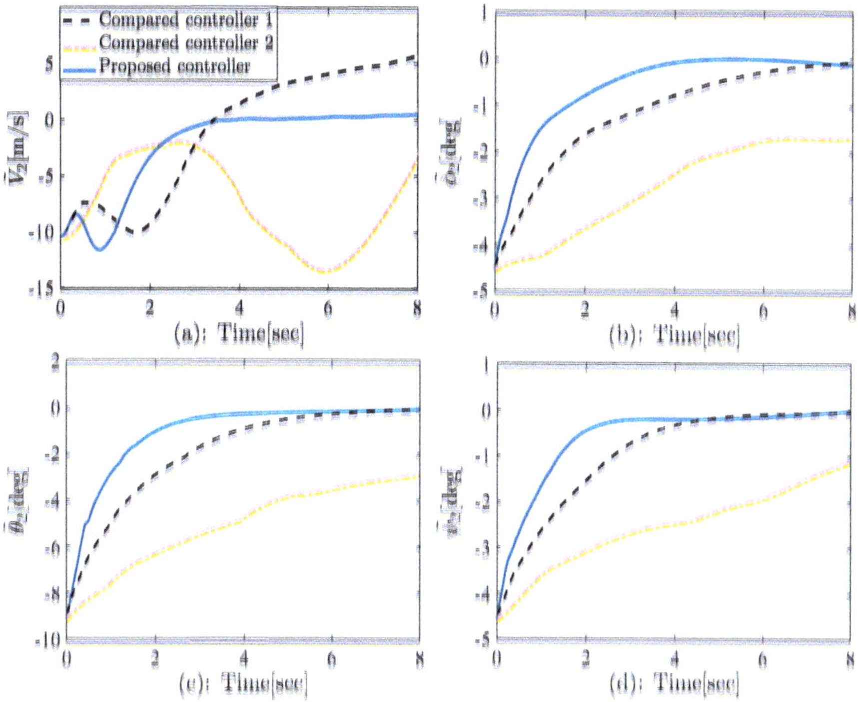

Fig. 4.6 Velocity and attitude tracking under the proposed controller and compared controllers (taking follower 1 as an example). (**a**)–(**d**) depict the trajectories of state tracking errors. (**a**) Tracking error $\tilde{V}_2$. (**b**) Tracking error $\tilde{\phi}_2$. (**c**) Tracking error $\tilde{\theta}_2$. (**d**) Tracking error $\tilde{\psi}_2$

Table 4.1 Transient and steady-state performances under different controllers

	Controller	Offset error	Integral of absolute error	Convergence time
$\tilde{V}_2$	**Proposed Controller**	**0.8373m/s**	**20.4360m**	**3.5s**
	Compared controller 1	5.6874m/s	40.2164m	Unconvergence
	Compared controller 2	− 2.8410 m/s	55.4265m	Unconvergence
$\tilde{\phi}_2$	**Proposed Controller**	**−0.0633deg**	**4.4462deg.s**	**4.2s**
	Compared controller 1	− 0.0905deg	9.3512deg.s	8s
	Compared controller 2	− 1.6104deg	21.2558deg.s	Unconvergence
$\tilde{\theta}_2$	**Proposed Controller**	**0.0417deg**	**7.6916deg.s**	**3s**
	Compared controller 1	− 0.1063deg	15.5718deg.s	6s
	Compared controller 2	− 2.6279deg	38.8895deg.s	Unconvergence
$\tilde{\psi}_2$	**Proposed Controller**	**0.0610deg**	**4.2155deg.s**	**2.4s**
	Compared controller 1	− 0.0475deg	7.6265deg.s	4.3s
	Compared controller 2	− 1.0211deg	19.5014deg.s	Unconvergence

4.5 Conclusions

This chapter has developed a fixed-time fault-tolerant anti-saturation control strategy for velocity and attitude tracking for networked UAVs. Fixed-time-based SMC has been embedded into the proposed control framework to realize the fixed-time tracking property. To tackle actuator faults and network induced errors, the adaptation parameters and well-defined smooth functions have been designed in the controller design. Numerical simulations are given to validate the effectiveness of the developed method. Despite that the proposed strategy is capable of tolerating actuator faults and actuator saturation, issues of sensor fault diagnosis have not yet been considered in the design. Investigation of this factor, which affects the performance of multi-UAV systems, can be our future works. Besides, this work lays down the groundwork for other coordinated control problems of UAVs including motion re-coordination and trajectory planning. This extension of coordinated control to a guidance system for UAVs deserves further investigation.

References

1. G. Gu, P.R. Chandler, C.J. Schumacher, A. Sparks, M. Pachter, Optimal cooperative sensing using a team of UAVs. IEEE Trans. Aerosp. Electron. Syst. **42**(4), 1446–1458 (2006)
2. A. Das, S. Shirazipourazad, D. Hay, A. Sen, Tracking of multiple targets using optimal number of UAVs. IEEE Trans. Aerosp. Electron. Syst. **55**(4), 1769–1784 (2019)
3. N. Wang, C.K. Ahn, Coordinated trajectory tracking control of a marine aerial-surface heterogeneous system. IEEE/ASME Trans. Mechatron. **26**(6), 3198–3210 (2021)
4. S. Kim, C.K. Ahn, P. Shi, Performance recovery tracking-controller for quadcopters via invariant dynamic surface approach. IEEE Trans. Industr. Inform. **15**(9), 5235–5243 (2019)
5. S. Kim, C.K. Ahn, Adaptive nonlinear tracking control algorithm for quadcopter applications. IEEE Trans. Aerosp. Electron. Syst. **56**(1), 84–94 (2020)
6. Y. Yu, J. Guo, C.K. Ahn, Z. Xiang, Neural adaptive distributed formation control of nonlinear multi-UAVs with unmodeled dynamics. IEEE Trans. Neural Networks Learn. Syst. **34**(11), 9555–9561 (2022)
7. C. Deng, W.W. Che, P. Shi, Cooperative fault-tolerant output regulation for multiagent systems by distributed learning control approach. IEEE Trans. Neural Networks Learn. Syst. **31**(11), 4831–4841 (2020)
8. T. Xu, G. Lv, Z. Duan, Z. Sun, J. Yu, Distributed fixed-time triggering-based containment control for networked nonlinear agents under directed graphs. IEEE Trans. Circuits Syst. I Regul. Pap. **67**(10), 3541–3552 (2020)
9. Z. Yu, Y. Zhang, Z. Liu, Y. Qu, C.Y. Su, Distributed adaptive fractional-order fault tolerant cooperative control of networked unmanned aerial vehicles via fuzzy neural networks. IET Control Theory Appl. **13**(17), 2917–2929 (2019)
10. Z. Yu, Y. Qu, Y. Zhang, Distributed fault-tolerant cooperative control for multi-UAVs under actuator fault and input saturation. IEEE Trans. Control Syst. Technol. **27**(6), 2417–2429 (2019)
11. Y. Wang, T. Zhang, Z. Cai, J. Zhao, K. Wu, Multi-UAV coordination control by chaotic grey wolf optimization based distributed MPC with event-triggered strategy. Chin. J. Aeronaut. **33**(11), 2877–2897 (2020)

12. T.Z. Muslimov, R.A. Munasypov, Adaptive decentralized flocking control of multi-UAV circular formations based on vector fields and backstepping. ISA Trans. **107**, 143–159 (2020)
13. Z. Yu, Y. Zhang, Z. Liu, Y. Qu, C.Y. Su, B. Jiang, Decentralized finite-time adaptive fault-tolerant synchronization tracking control for multiple UAVs with prescribed performance. J. Frankl. Inst. **357**(16), 11830–11862 (2020)
14. G. Lin, H. Li, C.K. Ahn, D. Yao, Event-based finite-time neural control for human-in-the-loop UAV attitude systems. IEEE Trans. Neural Networks Learn. Syst. **34**(12), 10387–10397 (2022)
15. J. Zhang, P. Zhang, J. Yan, Distributed adaptive finite-time compensation control for UAV swarm with uncertain disturbances. IEEE Trans. Circuits Syst. I Regul. Pap. **68**(2), 829–841 (2021)
16. F. Wang, H. Gao, K. Wang, C. Zhou, Q. Zong, C. Hua, Disturbance observer-based finite-time control design for a quadrotor UAV with external disturbance. IEEE Trans. Aerosp. Electron. Syst. **57**(2), 834–847 (20210)
17. A. Polyakov, Nonlinear feedback design for fixed-time stabilization of linear control systems. IEEE Trans. Autom. Control **57**(8), 2106–2110 (2012)
18. A.M. Zou, K.D. Kumar, A.H. Ruiter, Fixed-time attitude tracking control for rigid spacecraft. Automatica **113**, 108792 (2020)
19. Y. Huang, Y. Jia, Adaptive fixed-time six-DOF tracking control for noncooperative spacecraft fly-around mission. IEEE Trans. Control Syst. Technol. **27**(4), 1796–1804 (2019)
20. J. Zhang, S. Yu, Y. Yan, D. Wu, Fixed-time output feedback sliding mode tracking control of marine surface vessels under actuator faults with disturbance cancellation. Appl. Ocean Res. **104**, 102378 (2020)
21. N. Wang, H. Li, Leader-follower formation control of surface vehicles: A fixed-time control approach. ISA Trans. **124**, 356–364 (2022)
22. S. Song, J.H. Park, B. Zhang, X. Song, Event-triggered adaptive practical fixed-time trajectory tracking control for unmanned surface vehicle. IEEE Trans. Circuits Syst. II Express Briefs **68**(1), 436–440 (2021)
23. M. Van, D. Ceglarek, Robust fault tolerant control of robot manipulators with global fixed-time convergence. J. Frankl. Inst. **358**(1), 699–722 (2021)
24. H. Yang, B. Jiang, H. Yang, H.H. Liu, Synchronization of multiple 3-DOF helicopters under actuator faults and saturations with prescribed performance. ISA Trans. **75**, 118–126 (2018)
25. Y. Liu, X. Dong, Z. Ren, J. Cooper, Fault-tolerant control for commercial aircraft with actuator faults and constraints. J. Frankl. Inst. **356**(7), 3849–3868 (2019)
26. Z. Zheng, M. Qian, P. Li, H. Yi, Distributed adaptive control for UAV formation with input saturation and actuator fault. IEEE Access **7**, 144638–144647 (2019)
27. Z. Yu, Y. Qu, Y. Zhang Distributed fault-tolerant cooperative control for multi-UAVs under actuator fault and input saturation. IEEE Trans. Control Syst. Technol. **27**(6), 2417–2429 (2019)
28. Z, Yu, Y. Zhang, B. Jiang, C.Y. Su, J. Fu, Y. Jin, T. Chai, Nussbaum-based finite-time fractional-order backstepping fault-tolerant flight control of fixed-wing UAV against input saturation with hardware-in-the-loop validation. Mech. Syst. Signal Process. **153**(15), 107406 (2021)
29. P. Chen, C. Cai, B. Jiang, Y. Zou, Finite time fault tolerant control design for UAV attitude control systems with actuator fault and actuator saturation. IFAC-PapersOnLine **52**(24), 53–58 (2019)
30. O. Espen, S. Rune, K. Raymond, Underactuated waypoint tracking of a fixed-wing UAV. IFAC Proc. **46**(30), 126–133 (2015)
31. B. Zhang, X. Sun, M. Lv, S. Liu, Distributed coordinated control for fixed-wing UAVs with dynamic event-triggered communication. IEEE Trans. Veh. Technol. **71**(5), 4665–4676 (2022)
32. H. Wang, P.X. Liu, X. Zhao, X. Liu, Adaptive fuzzy finite-time control of nonlinear systems with actuator faults. IEEE Trans. Cybern. **50**(5), 1786–1797 (2020)
33. N. Zhou, Y. Kawano, M. Cao, Neural network-based adaptive control for spacecraft under actuator failures and input saturations. IEEE Trans. Neural Networks Learn. Syst. **31**(9), 3696–3710 (2020)

34. Q. Shen, D. Wang, S. Zhu, K. Poh, Finite-time fault-tolerant attitude stabilization for spacecraft with actuator saturation. IEEE Trans. Aerosp. Electron. Syst. **51**(3), 2390–2405 (2015)
35. D. Ba, Y.X. Li, S. Tong, Fixed-time adaptive neural tracking control for a class of uncertain nonstrict nonlinear systems. Neurocomputing **363**, 273–280 (2019)
36. C. Wang, Y. Lin, Decentralized adaptive tracking control for a class of interconnected nonlinear time-varying systems. Automatica **54**, 16–24 (2015)
37. S.S. Ge, C. Wang, Adaptive NN control of uncertain nonlinear pure-feedback systems. Automatica **38**(4), 671–682 (2002)
38. Y. Zou, Z. Zheng, A robust adaptive RBFNN augmenting backstepping control approach for a model-scaled helicopter. IEEE Trans. Control Syst. Technol. **23**(6), 2344–2352 (2015)
39. Q. Shen, D. Wang, S. Zhu, K. Poh, Artificial potential-based adaptive H_∞ synchronized tracking control for accommodation vessel. IEEE Trans. Aerosp. Electron. Syst. **64**(7), 5640–5647 (2017)
40. H. Liang, L. Chen, Y. Pan, H. Lam, Fuzzy-based robust precision consensus tracking for uncertain networked systems with cooperative-antagonistic interactions. IEEE Trans. Fuzzy Syst. **31**(4), 1362–1376 (2022)
41. H. Liang, Z. Du, T. Huang, Y. Pan, Neuroadaptive performance guaranteed control for multiagent systems with power integrators and unknown measurement sensitivity. IEEE Trans. Neural Networks Learn. Syst. **34**(12), 9771–9782 (2022)

Chapter 5
Specified-Time Fault-Tolerant Tracking Control

Chapters 3 and 4 present the fixed-time cooperative control algorithm for the velocity and attitude tracking of multiple UAVs, which effectively enhances the convergence speed. However, as for the fixed-time control, the upper bound of convergence time is influenced by the control parameters, making it difficult to set in advance. To address this issue, researchers have proposed the specified-time control method, with the convergence time being arbitrarily set and independent of initial state and control parameters [1–3]. This characteristic better meets the explicit requirements of convergence time for multi-UAV system. Therefore, Chap. 5 intends to solve the specified-time control problem for the velocity and attitude tracking of multiple UAVs.

5.1 Introduction

Cooperative flying of a team of fixed-wing unmanned aerial vehicles (UAVs) has been extensively researched recently, considering its vital role in possessing the cost effectiveness and flexibility of flight cooperation [4]. Such an issue has been widely discussed in terms of formation keeping and formation reconfiguration [5] and velocity/attitude synchronization tracking for multiple UAVs [6]. Distributed control for coordination is now a research spotlight due to its undeniable benefits that each agent in multi-agent system does not need to obtain the complete system information to determine the control input updates [7]. To date, various advanced distributed control designs for UAVs are currently available in many investigations [8–11]. Typically, with the help of neural network approximation and sliding mode estimators, Yu et al. [8] solved the attitude synchronization tracking problem of networked UAVs with modeling uncertainties and actuator faults in a distributed cooperative control framework. Based on dynamic surface control and disturbance observer, Yu et al. [9] explored the distributed cooperative fault-tolerant control to

M. Lv et al., *Unmanned Aerial Vehicles Formation Control and Safety Guarantee*, Cognitive Intelligence and Robotics, https://doi.org/10.1007/978-981-96-5829-9_5

achieve the longitudinal synchronization tracking for UAVs under input saturation and actuator faults. Further, in [10], a distributed model predictive approach with swarm intelligence was designed for coordinated control and trajectory tracking of multiple UAVs. In [11], a vector-field-based distributed control was proposed to achieve the path following for networked UAVs and coordinate their group to a specified circular motion. Although effective, aforementioned results simplify the multi-UAV system using a two-degree model [10, 11] or just focus on analyzing attitude/longitudinal dynamics while ignoring the fact that the practical UAV system is a complicated nonlinear system with six-degree-of-freedom (6-DoF) dynamics. Actually, it is a challenge to guarantee the satisfaction of control performances for 6-DoF UAV model on account of the fact that the considered 6-DoF UAV model is nonlinear and highly coupled. The complexity of the model will increase the difficulty of controller design.

Besides, above investigations mainly concentrate on the analysis of the resulting closed-loop stability, while ignoring the transient and steady-state performances, such as the minimum convergence rate, maximum overshoot, and maximum steady error. Actually, guaranteeing the prescribed configurations is vital to obtain the reliable tracking and robust performances for UAV flight even in the presence of external disturbances, modeling uncertainties, and actuator faults. Recently, the prescribed performance control (PPC) method [12–15] was proposed as a potential way to handle the aforementioned problem. To date, such an approach has been extended to UAV dynamic systems to gain the guaranteed performance bounds. Typically, an adaptive neural fractional-order tracking control scheme was designed to guarantee the prescribed performance for UAV system [1]. Zhang et al. [2] took the input constraints into consideration and proposed an adaptive constrained control with prescribed performance methodology for the tracking problem of UAVs. Chen et al. [3] incorporated the prescribed performance function into disturbance observer design to achieve effective steady-state and transient performance for UAVs. Backstepping control combined with a prescribed performance function was employed for trajectory tracking with guaranteed transient tracking performances [16].

However, it is noted that using the exponentially decaying functions in the PPC design to build the performance envelops [17–20], the tracking errors are driven to the prespecified residual sets as time goes to infinity. For this, the specified-time control has been evolved as an effective way to fill this gap [21–23]. By specified time, the convergence settling time can be preassigned offline according to the mission-oriented demands. Recently, some works on specified-time controller have been advanced for the second-order multi-agent system [24], longitudinal motion of hypersonic vehicle [25], and attitude tracking of spacecraft [26, 28], by which the tracking errors can be regulated to a prescribed residual set within the specified time, whereas other complicated practical systems are expected to be studied, for example, the 6-DoF UAV system in this work.

Motivated by the foregoing analyses, this work proposes an adaptive specified-time prescribed performance control (SPPC) method for synchronization tracking of multiple UAVs. The contribution of this work is threefold:

- Different from the finite-/fixed-time control methodologies on longitudinal/attitude synchronization [28–30] or 2D motion of UAVs [31–33], this work first considers the SPPC for 6-DoF networked UAVs, where both the synchronization tracking errors of velocity and attitude can evolve along the performance bounds within the specified time, whose settling time can be user-defined.
- In contrast with the exponentially decaying functions in traditional PPC design [17, 18, 20], where the tracking errors are driven to the prespecified residual sets as time goes to infinity, the novel specified-time performance function (STPF) is presented, whose reach time can be arbitrarily selected by users. Further, as opposed to existing STPFs with hyperbolic cosecant function designed by Shao et al. [25] and high-order polynomial proposed by Liu et al. [27], our proposed STPF owns a simpler form and is easier to implement in practice.
- Unlike some nonadaptive controls for multi-UAV system without the consideration of actuator faults and model uncertainties [10, 32, 34], through introducing adaptive robust terms and well-defined smooth functions into the proposed specified-time control scheme, the actuator faults together with the modeling uncertainties can be effectively compensated for.

The remainder of this chapter is arranged as follows: Sect. 5.2 includes the problem formulation and preliminaries. The main results of specified-time control design and corresponding stability analysis are discussed in Sect. 5.3. The experiment and simulation examples are given in Sect. 5.4 to show the effectiveness of theoretical findings. Conclusions and future research directions are detailed in Sect. 5.5.

Remark The main control schemes and contents of this chapter are from the published journal paper "B. Y. Zhang, X. X. Sun, M. L. Lv, Distributed adaptive specified-time synchronization tracking of multiple 6-DOF fixed-wing UAVs with guaranteed performances. ISA transactions. 2022, 129: 260–272." The authors appreciate the permission from Elsevier to rcuse the results published in the relevant journal.

5.2 Problem Formulation

5.2.1 UAV Dynamic Model

Consider the multi-UAV system consisting of N identical fixed wing UAVs. The dynamics of the ith UAV is defined in [35], whose mathematical expressions are given below:

$$\dot{\boldsymbol{p}}_i = \boldsymbol{R}_1\left(\boldsymbol{\varphi}_i\right)\boldsymbol{v}_i \tag{5.1}$$

$$\dot{\boldsymbol{v}}_i = -\boldsymbol{S}\left(\boldsymbol{\omega}_i\right)\boldsymbol{v}_i + \frac{\boldsymbol{T}_i}{m_i} + \boldsymbol{R}_1^{\mathrm{T}}\left(\boldsymbol{\varphi}_i\right)\boldsymbol{g} + \frac{\boldsymbol{F}_i}{m_i} + \boldsymbol{d}_{vi} \tag{5.2}$$

$$\dot{\varphi}_l = R_2^{-1}(\varphi_l)\,\omega_l \tag{5.3}$$

$$J_l\dot{\omega}_l = -\,\omega_l \times J_l\omega_l + N_l + C(\delta_l)\,\delta_l + J_l d_{\omega l} \tag{5.4}$$

with $p_l = [x_l, y_l, z_l]^{\mathrm{T}}$ being the inertial position, $v_l = [u_l, v_l, w_l]^{\mathrm{T}}$ being the linear velocity, $\varphi_l = [\phi_l, \theta_l, \psi_l]^{\mathrm{T}}$ being the attitude angle, $\omega_i = [p_l, q_l, r_l]^{\mathrm{T}}$ being the attitude angular velocity, $T_l = [T_{xl}, 0, 0]^{\mathrm{T}}$ being the engine thrust along x body axis, $\delta_l = [\delta_{al}, \delta_{el}, \delta_{rl}]^{\mathrm{T}}$ being the vector consisting of elevator, ailerons, and rudder, $g = [0, 0, g_z]^{\mathrm{T}}$ being the gravity vector, $d_{vl} \in \Re^3$ and $d_{\omega l} \in \Re^3$ being the bounded disturbance vectors, and J_l being the inertia matrix defined by

$$J_l = \begin{bmatrix} J_{xl} & 0 & -J_{xzl} \\ 0 & J_{yl} & 0 \\ -J_{zxl} & 0 & J_{zl} \end{bmatrix}. \tag{5.5}$$

Besides, the rotation matrixes $R_1(\varphi_l) \in \mathrm{SO}(3)$ and $R_2(\varphi_i) \in \Re^{3\times 3}$ are defined as

$$R_1 = \begin{bmatrix} \mathrm{c}_{\psi l}\mathrm{c}_{\theta l} & -\mathrm{s}_{\psi l}\mathrm{s}_{\phi l} + \mathrm{c}_{\psi l}\mathrm{s}_{\theta l}\mathrm{s}_{\phi l} & \mathrm{s}_{\psi l}\mathrm{s}_{\phi l} + \mathrm{c}_{\psi i}\mathrm{s}_{\theta i}\mathrm{c}_{\phi l} \\ \mathrm{s}_{\psi l}\mathrm{c}_{\theta l} & \mathrm{c}_{\psi l}\mathrm{c}_{\phi l} + \mathrm{s}_{\psi l}\mathrm{s}_{\theta l}\mathrm{s}_{\phi l} & -\mathrm{c}_{\psi l}\mathrm{s}_{\phi l} + \mathrm{s}_{\psi i}\mathrm{s}_{\theta i}\mathrm{c}_{\phi l} \\ -\mathrm{s}_{\theta l} & \mathrm{c}_{\theta l}\mathrm{s}_{\phi l} & \mathrm{c}_{\theta i}\mathrm{c}_{\phi i} \end{bmatrix},$$

$$R_2 = \begin{bmatrix} 1 & 0 & -s_{\psi l} \\ 0 & c_{\psi l} & s_{\psi l}c_{\theta l} \\ 0 & -s_{\theta l} & c_{\psi l}c_{\theta l} \end{bmatrix}, \tag{5.6}$$

where c., s. represent $\cos(\cdot), \sin(\cdot)$, respectively. Note that the force vector $F_l = \left[F_{X_l}, F_{Y_l}, F_{Z_l}\right]^{\mathrm{T}}$ and the moment vector $N_l = \left[F_{L_l}, F_{M_l}, F_{N_l}\right]^{\mathrm{T}}$ are described in terms of aerodynamic coefficients. Let us now define the control gain matrix in the following form:

$$C(\delta_l) = \begin{bmatrix} \bar{q}_l S_l b_l c_{l\delta_{al}} & 0 & \bar{q}_l S_l b_l c_{l\delta_{rl}} \\ 0 & \bar{q}_l S_l \bar{c}_l c_{m\delta_{el}} & 0 \\ \bar{q}_l S_l b_l c_{n\delta_{al}} & 0 & \bar{q}_l S_l b_l c_{n\delta_{rl}} \end{bmatrix}, \tag{5.7}$$

where the detailed calculations of F_l, N_l, and $C(\delta_l)$ can be found in [34] and our previous work [35].

5.2.2 Control-Oriented Model

In line with the discussions [34, 35], the dynamics of UAV system can be decoupled into a translational subsystem (5.2) and a rotational (5.3)–(5.4) subsystem. Considering that the time constants of translational systems are huger than that of the rotational systems, the translational dynamics equals the slow subsystem, while the rotational dynamics is regarded as the fast subsystem. Based on the proposed slow and fast subsets, the synchronization tracking control for the multi-UAV system can be decoupled into the inner- and outer-loop control, which, respectively, realizes the attitude and velocity synchronization tracking.

Let us solve the time derivative of V_l along (5.2) which arrives at

$$\dot{V}_l = \frac{\boldsymbol{v}_l^{\mathrm{T}}\dot{\boldsymbol{v}}_l}{V_l} = \frac{\boldsymbol{v}_l^{\mathrm{T}}}{V_l}\left(-\boldsymbol{S}(\boldsymbol{w}_l)\,\boldsymbol{v}_l + \frac{\boldsymbol{T}_l}{m_l} + \boldsymbol{R}_1^{\mathrm{T}}(\boldsymbol{\varphi}_l)\,\boldsymbol{g} + \frac{\boldsymbol{F}_l}{m_l} + \boldsymbol{d}_{vl}\right). \tag{5.8}$$

With the property $\boldsymbol{v}_l^{\mathrm{T}}\boldsymbol{S}(\boldsymbol{w}_l)\,\boldsymbol{v}_l = 0$, the thrust can be extracted from (5.8)

$$\dot{V}_l = \frac{u_l T_{xl}}{m_l V_l} + \frac{\boldsymbol{v}_l^{\mathrm{T}}}{V_l}\left(\boldsymbol{R}_1^{\mathrm{T}}(\boldsymbol{\varphi}_l)\,\boldsymbol{g} + \frac{\boldsymbol{F}_l}{m_l} + \boldsymbol{d}_{vl}\right). \tag{5.9}$$

Remark 5.1 Since the UAV model is clearly underactuated, T_{xl} simply acts on velocity component u_l, and no direct controls are imposed on velocity components w_l and v_l. Whereas with the relationship $w_l = u_l \tan\alpha_l$ and $v_l = u_l/\sin\beta_l$, it is still controllable by regulating the total velocity V_l and pointing the velocity vector $\boldsymbol{v}_l$ in an expected direction.

In practical flight application, actuator faults are possibly encountered by engine thrust. To describe the actuator faults, the thrust is rewritten as $T_{xl} = \rho_{Tl} T_{xl0} + T_{xlf}$, where $0 \le \rho_{Tl} \le 1$ denotes the actuation effectiveness of thrust, T_{xl0} represents the command thrust control, and T_{xlf} is the additive fault. Moreover, considering the coefficient uncertainties, the accurate value of $\boldsymbol{F}_l$ cannot be known a priori. As a result, $\boldsymbol{F}_l$ can be decomposed into known $\boldsymbol{F}_{l0}$ and unknown $\Delta\boldsymbol{F}_l$. Then, the velocity dynamics can be given by

$$\dot{V}_l = \frac{u_l \rho_{Tl} T_{xl0}}{m_l V_l} + \frac{u_l T_{xlf}}{m_l V_l} + \frac{\boldsymbol{v}_l^{\mathrm{T}}}{V_l}\left(\boldsymbol{R}_1^{\mathrm{T}}(\boldsymbol{\varphi}_l)\,\boldsymbol{g} + \frac{\boldsymbol{F}_{l0}}{m_l} + \frac{\Delta\boldsymbol{F}_l}{m_l} + \boldsymbol{d}_{vl}\right). \tag{5.10}$$

Based on (5.3) and (5.4), the attitude motion has the following dynamics:

$$\dot{\boldsymbol{\varphi}}_l = \boldsymbol{R}_2^{-1}(\boldsymbol{\varphi}_l)\,\boldsymbol{\omega}_l \tag{5.11}$$

$$\dot{\boldsymbol{\omega}}_l = \boldsymbol{J}_l^{-1}\boldsymbol{S}(\boldsymbol{J}_l\boldsymbol{\omega}_l)\,\boldsymbol{\omega}_l + \boldsymbol{J}_l^{-1}\boldsymbol{N}_l + \boldsymbol{J}_l^{-1}\boldsymbol{C}(\boldsymbol{\delta}_l)\,\boldsymbol{\delta}_l + \boldsymbol{d}_{\omega l}. \tag{5.12}$$

During operation, the control surfaces may encounter general faults, modeled by $\boldsymbol{\delta}_l = \boldsymbol{\rho}_{\delta l}\boldsymbol{\delta}_{l0} + \boldsymbol{\delta}_{lf}$, where $\boldsymbol{\delta}_{l0} = [\delta_{al0}, \delta_{el0}, \delta_{rl0}]^{\mathrm{T}}$ denotes the designed control action, $\boldsymbol{\rho}_{\delta l} = \mathrm{diag}\{\rho_{\delta al}, \rho_{\delta el}, \rho_{\delta rl}\}$ is the fault gain, and $\boldsymbol{\delta}_{lf} = \left[\delta_{alf}, \delta_{elf}, \delta_{rlf}\right]^{\mathrm{T}}$ represents the bias fault [8]. While considering the coefficient uncertainties, $\boldsymbol{J}_l^{-1}\boldsymbol{S}(\boldsymbol{J}_l\boldsymbol{\omega}_l)\,\boldsymbol{\omega}_l$ is decomposed into a known component $\boldsymbol{J}_{l0}^{-1}\boldsymbol{S}(\boldsymbol{J}_{l0}\boldsymbol{\omega}_l)\,\boldsymbol{\omega}_l$ and an uncertain one $\Delta\boldsymbol{J}_l^{-1}\boldsymbol{S}(\boldsymbol{J}_l\boldsymbol{\omega}_l)\,\boldsymbol{\omega}_l$. Similarly, $\boldsymbol{J}_l^{-1}\boldsymbol{N}_l$ is composed of a known part $J_{i0}^{-1}N_{i0}$ and an uncertain one $\Delta J_i^{-1}N_i$. Also, $J_i^{-1}C(\delta_i)$ consists of a known component $J_{i0}^{-1}C_0(\delta_i)$ and an uncertain one $\Delta J_i^{-1}C(\delta_i)$. Accordingly, the angular velocity dynamics is found as

$$\dot{\omega}_i = J_{i0}^{-1}S(J_{i0}\omega_i)\omega_i + J_{i0}^{-1}N_{i0} + J_{i0}^{-1}C_0(\delta_i)\rho_{\delta i}\delta_{i0} + \Delta_{\omega i} + d_{\omega i}, \tag{5.13}$$

where $\Delta_{\omega i} = \Delta J_i^{-1}S(J_i\omega_i)\omega_i + \Delta J_i^{-1}N_i + \Delta J_i^{-1}C(\delta_i)\delta_i + J_{i0}^{-1}C_0(\delta_i)\delta_{if}$ denotes the integrated uncertainties comprising modeling uncertainties and actuator faults.

Assumption 5.1 ([35]) Suppose that $\|\Delta J_i^{-1}C(\delta_i)\rho_{\delta i}\left(J_{i0}^{-1}C_0(\delta_i)\right)^{-1}\|_\infty < 1$ and $\partial\Delta_{\omega i}/\partial\delta_{i0} + J_{i0}^{-1}C_0(\delta_i)\rho_{\delta i} \neq 0$, $\forall t > 0$.

Assumption 5.2 ([36]) It holds that $\lambda_{\min}(J_{i0}^{-1}C_0(\delta_i)\rho_{\delta i}\left(J_{i0}^{-1}C_0(\delta_i)\right)^{\mathrm{T}}) \geq g_{i,\min}$ with the positive constant $g_{i,\min}$.

Remark 5.2 Assumptions 5.1–5.2 provide the controllability condition for (5.13), which guarantees the existence of sufficient control action δ_{i0} for each UAV in the presence of modeling uncertainties and actuator faults.

Remark 5.3 From (5.13), the control signal δ_{i0} is involved in $\Delta_{\omega i}$. When using neural networks to directly handle $\Delta_{\omega i}$, the algebraic loops will be introduced into the control scheme. To solve the algebraic loops, Butterworth low-pass filter is introduced to filter the control signal δ_{i0} [29]. Then one has $\Delta_{\omega i} = \Delta'_{\omega i} + l_{\omega i}$, where $\Delta'_{\omega i} = \Delta J_i^{-1}S(J_i\omega_i)\omega_i + \Delta J_i^{-1}N_i + \Delta J_i^{-1}C(\delta_i)\delta'_{i0} + \Delta J_i^{-1}C(\delta_i)\delta_{if} + J_{i0}^{-1}C_0(\delta_i)\delta_{if}$, δ'_{i0} is the filtered signal and $l_{\omega i} \in \Re^3$ is the bounded filter error.

5.2.3 RBF Neural Networks

With the universal approximation feature, the radial basis function neural networks (RBFNNs) are involved to estimate the modeling uncertainties [37]. For $\Delta_{vi}(\Theta_{vi}) = [\Delta_{vi1}, \Delta_{vi2}, \Delta_{vi3}]^{\mathrm{T}}$ and $\Delta'_{\omega i}(\Theta_{\omega i}) = \left[\Delta'_{\omega i1}, \Delta'_{\omega i2}, \Delta'_{\omega i3}\right]^{\mathrm{T}}$, using RBFNNs to approximate yields

$$\begin{aligned}\Delta_{via}(\Theta_{vi}) &= W_{via}^{*\mathrm{T}}\varsigma_{via}(\Theta_{vi}) + \tau_{via}(\Theta_{vi}) \\ \Delta'_{\omega ia}(\Theta_{\omega i}) &= W_{\omega ia}^{*\mathrm{T}}\varsigma_{\omega ia}(\Theta_{\omega i}) + \tau_{\omega ia}(\Theta_{\omega i}),\end{aligned} \tag{5.14}$$

where $\Theta_{vi} = \left[v_i^{\mathrm{T}}, \varphi_i^{\mathrm{T}}, \omega_i^{\mathrm{T}}, \delta_{i0}^{\mathrm{T}}\right]^{\mathrm{T}} \in \Re^{12}$ and $\Theta_{\omega i} = \left[v_i^{\mathrm{T}}, \varphi_i^{\mathrm{T}}, \omega_i^{\mathrm{T}}, \delta_{i0}^{\mathrm{T}}\right]^{\mathrm{T}} \in \Re^{12}$, $\tau_{via}(\Theta_{vi}) \in \Re$ and $\tau_{\omega ia}(\Theta_{\omega i}) \in \Re$ are the functional approximation errors under optimal weight vectors $W_{via}^* \in \Re^{M\times 1}$ and $W_{\omega ia}^* \in \Re^{M\times 1}$ with $M > 1$ being the node number, $\varsigma_{via}(\Theta_{vi}) = \left[\varsigma_{via}^1(\Theta_{vi}), \ldots, \varsigma_{via}^M(\Theta_{vi})\right]^{\mathrm{T}} \in \Re^M$ and

$\varsigma_{\omega ia}(\boldsymbol{\Theta}_{\omega i}) = \left[\varsigma^1_{\omega ia}(\boldsymbol{\Theta}_{\omega i}), \ldots, \varsigma^M_{\omega ij}(\boldsymbol{\Theta}_{\omega i})\right]^{\mathrm{T}} \in \Re^M$ are the basic function vectors with $\varsigma^n_{via}(\boldsymbol{\Theta}_{vi})$ and $\varsigma^n_{\omega ia}(\boldsymbol{\Theta}_{\omega i})$, $n = 1, \ldots, M$, $a = 1, 2, 3$, commonly chosen as Gaussian basis functions:

$$\varsigma^n_{via}(\boldsymbol{\Theta}_{vi}) = \exp\left[-\frac{\left(\boldsymbol{\Theta}_{vi} - \boldsymbol{\Upsilon}^n_{vi}\right)^{\mathrm{T}}\left(\boldsymbol{\Theta}_{vi} - \boldsymbol{\Upsilon}^n_{vi}\right)}{\kappa^2_{vi}}\right]$$
$$\varsigma^n_{\omega ia}(\boldsymbol{\Theta}_{\omega i}) = \exp\left[-\frac{\left(\boldsymbol{\Theta}_{\omega i} - \boldsymbol{\Upsilon}^n_{\omega i}\right)^{\mathrm{T}}\left(\boldsymbol{\Theta}_{\omega i} - \boldsymbol{\Upsilon}^n_{\omega i}\right)}{\kappa^2_{\omega i}}\right], \tag{5.15}$$

where $\boldsymbol{\Upsilon}^n_{vi} \in \Re^{12}$ and $\boldsymbol{\Upsilon}^n_{\omega i} \in \Re^{12}$ stand for the centers of the nth neural cell, and κ_{vi} and $\kappa_{\omega i}$ denote the variances.

In this section, an undirected graph $\mathcal{G} = (\mathcal{V}, \mathcal{E}, \mathcal{A})$ is given to describe the communication topology among N UAVs, where $\mathcal{V} = \{v_1, \cdots, v_N\}$ denotes the node set, $\mathcal{E} = \{(i, j) : i, j \in \mathcal{V}, \text{and } i \neq j\}$ is the edge set, and $\mathcal{A} = [a_{ij}] \in \Re^{n \times n}$ is the weighted adjacency matrix. If $a_{ij} = a_{ji} \neq 0$, it implies that there exists an edge between the ith UAV and jth UAV; otherwise $a_{ij} = a_{ji} = 0$. Moreover, $\mathcal{N}_i = \{v_j : (v_j, v_i) \in \mathcal{E}\}$ represents the neighbors of ith UAV. The Laplacian matrix of $\mathcal{G}$ is defined as $\mathcal{L} = \mathcal{D} - \mathcal{A}$, where $\mathcal{D} = \mathrm{diag}\{d_1, \cdots, d_N\}$ with $d_i = \sum_{j=1}^N a_{ij}$. If there is a path between each pair of UAVs, the undirected communication topology $\mathcal{G}$ is connected.

Assumption 5.3 Based on the leader-following structure, the communication topology $\mathcal{G}$ of considered multi-UAV system is connected with each follower having access to the leader.

For a given leader signal of velocity V_r and attitude $\boldsymbol{\varphi}_r$, the main control goal is to derive the distributed adaptive specified-time controller for the leader-following multi-UAV system in terms of maneuver completion time, permitted response overshoot, and allowed tracking accuracy such that:

- All signals in the resulting closed-loop system are bounded.
- The synchronous velocity and attitude tracking errors for multi-UAVs are guaranteed to converge within the prescribed tolerances in a specified time T_f while exhibiting the guaranteed performances.

To realize the control objective, the assumption regarding the references of velocity and attitude and some useful lemmas are given below.

Assumption 5.4 The leader signal V_r, $\boldsymbol{\varphi}_r$, their derivative $\dot{V}_r$, $\dot{\boldsymbol{\varphi}}_r$, and their second-order derivative $\ddot{V}_r$, $\ddot{\boldsymbol{\varphi}}_r$ are bounded and available for each follower.

Lemma 5.1 ([38]) *The following inequality can be maintained:* $0 \leq |z| - \frac{z^2}{\sqrt{z^2+\varepsilon}} \leq \sqrt{\varepsilon}$ *with scalars* $\varepsilon > 0$, $z \in \Re$.

Lemma 5.2 ([39]) *There exist scalars* $\varepsilon > 0$, $\vartheta \in \Re$ *such that* $0 \le |\vartheta| - \vartheta \tanh\left(\frac{\vartheta}{\varepsilon}\right) \le 0.2785\varepsilon$.

5.2.4 Specified-Time Performance Function

Definition 5.1 An STPF is defined as a C^n (for integer $n \ge 1$) continuous function $k_l(t) : \Re_{\ge 0} \rightarrow \Re_+$ if there exists a specified time $T_f > 0$ such that:

(i) $k_l(t)$ is positive and decreasing for $t \in [0, T_f)$.
(ii) $k_l(t) \equiv k_{l\infty} > 0$ for all $t \ge T_f$, where $k_{l\infty}$ is a positive design parameter.

According to Definition 5.1, a continuously differentiable STPF for velocity tracking is designed as

$$\begin{aligned} k_{vl}(t) =& \frac{b_{vl}(k_{vl0} - k_{vl\infty})}{T_{vf}^2} t^2 - \frac{2b_{vl}(k_{vl0} - k_{vl\infty})}{T_{vf}} t + b_{vi}k_{vi0} + (1 - b_{vl})k_{vl\infty}, \\ & t \in [0, T_{vf}) \\ k_{vl}(t) =& k_{vl\infty}, \quad t \in [T_{vf}, +\infty), \end{aligned} \tag{5.16}$$

where b_{vl}, k_{vl0}, and $k_{vl\infty}$ are positive design parameters. T_{vf} and $k_{vl\infty}$ are the prescribed settling time and maximum steady bound of velocity tracking error, respectively.

Let us define the velocity tracking errors for the ith UAV and jth UAV as $\tilde{V}_i = V_i - V_r$ and $\tilde{V}_j = V_j - V_r$, respectively. The synchronization tracking error of velocity is defined by

$$e_{vl} = \lambda_{l1}\tilde{V}_l + \lambda_{l2}\sum_{j\in\mathcal{N}_i} a_{lj}\left(\tilde{V}_l - \tilde{V}_j\right) = \lambda_l\tilde{V}_l - \lambda_{l2}\sum_{j\in\mathcal{N}_i} a_{lj}\tilde{V}_j, \tag{5.17}$$

where λ_{l1} and λ_{l2} are positive design parameters, which regulate the tracking and synchronization performances of velocity, respectively. And $\lambda_i = \lambda_{l1} + \lambda_{l2}\sum_{j\in\mathcal{N}_i} a_{lj}$.

Combined with $k_{vl}(t)$, the following prescribed bound is imposed on e_{vl}:

$$-\underline{\varepsilon}_{vl}k_{vl}(t) < e_{vl} < \bar{\varepsilon}_{vl}k_{vl}(t) \tag{5.18}$$

where $\underline{\varepsilon}_{vl}$ and $\bar{\varepsilon}_{vl}$ are positive parameters chosen in a manner that

$$\begin{cases} \underline{\varepsilon}_{vl} = 1 \text{ and } 0 < \bar{\varepsilon}_{vl} \le 1, & \text{if } e_{vl}(0) < 0 \\ \bar{\varepsilon}_{vl} = 1 \text{ and } 0 < \underline{\varepsilon}_{vl} \le 1, & \text{if } e_{vl}(0) \ge 0. \end{cases} \tag{5.19}$$

Similar with (5.16), continuously differentiable STPFs for attitude (including ϕ, θ, and ψ) tracking are designed as

$$k_{ll}(t) = \frac{b_{ll}(k_{ll0} - k_{ll\infty})}{T_{lf}^2}t^2 - \frac{2b_{ll}(k_{ll0} - k_{ll\infty})}{T_{lf}}t + b_{ll}k_{ll0} + (1 - b_{ll})k_{ll\infty},$$
$$t \in [0, T_{lf})$$
$$k_{ll}(t) = k_{ll\infty},\ t \in [T_{lf}, +\infty), \tag{5.20}$$

where b_{ll}, k_{ll0}, and $k_{ll\infty}$ are positive design parameters. T_{lf} and $k_{ll\infty}$ denote the preset settling time and maximum tracking errors of attitude during steady-state phase, respectively, $l = \{\phi, \theta, \psi\}$.

From $\tilde{\varphi}_l = \varphi_l - \varphi_r$ and $\tilde{\varphi}_j = \varphi_j - \varphi_r$, define the attitude synchronization tracking error as

$$e_{\varphi l} = \lambda_{l3}\tilde{\varphi}_l + \lambda_{l4}\sum_{j\in\mathcal{N}_l} a_{lj}(\tilde{\varphi}_l - \tilde{\varphi}_j) = \lambda_l'\tilde{\varphi}_l - \lambda_{l4}\sum_{j\in\mathcal{N}_l} a_{lj}\tilde{\varphi}_j, \tag{5.21}$$

where $e_{\varphi l} = [e_{\phi l}, e_{\theta l}, e_{\psi l}]^{\mathrm{T}}$, λ_{l3} and λ_{l4} are positive design parameters, which regulate the tracking and synchronization performances of attitude, respectively, and $\lambda_l' = \lambda_{l3} + \lambda_{l4}\sum_{j\in\mathcal{N}_l} a_{lj}$.

In accordance with $k_{ll}(t)$, the following prescribed bounds are imposed on e_{ll}, $l = \{\phi, \theta, \psi\}$:

$$-\underline{\varepsilon}_{ll}k_{ll}(t) < e_{ll} < \bar{\varepsilon}_{ll}k_{ll}(t), \tag{5.22}$$

where $\underline{\varepsilon}_{ll}$ and $\bar{\varepsilon}_{ll}$ are positive parameters chosen in the same manner as (5.19).

Remark 5.4 The behavioral boundary in traditional PPC [13–15] is always designed as exponential-type function: $k_{sl}(t) = (k_{sl0} - k_{sl\infty})\exp(-l_{sl}t) + k_{sl\infty}$, which implies that e_{sl} enters into $(-\underline{\varepsilon}_{sl}k_{sl\infty}, \bar{\varepsilon}_{sl}k_{sl\infty})$ at the infinite time, $s = \{v, \phi, \theta, \psi\}$. Differently, the proposed STPFs guarantee that the tracking error e_{sl} can reach to prescribed interval $(-\underline{\varepsilon}_{sl}k_{sl\infty}, \bar{\varepsilon}_{sl}k_{sl\infty})$ after specified time T_{sf}, where T_{sf} can be arbitrarily determined by users, which is engineering and mission-oriented. More importantly, as opposed to existing STPFs with hyperbolic cosecant function designed by Shao et al. [25] and high-order polynomial proposed by Liu et al. [27], the proposed STPF owns a simple form and is easy to implement.

5.3 Main Results

5.3.1 Error Transformation

Definition 5.2 (Practically Leader-Following Specified-time Synchronization) For multi-UAV system described as (5.2)–(5.4), the practical leader-following specified-time synchronization can be achieved if there exist velocity and attitude tracking errors such that

$$\lim_{t \geq T_{sf}} -\underline{\varepsilon}_{si} k_{si\infty}(t) < e_{si} < \bar{\varepsilon}_{si} k_{si\infty}(t) \tag{5.23}$$

for $s = \{v, \phi, \theta, \psi\}, i = 1, \ldots, N$.

According to Definition 5.2, error transformations for e_{vi} and $\boldsymbol{e}_{\varphi i}$ are developed, respectively, whose stability can guarantee the performance bounds in (5.23).

We define the error transformation for e_{vi} as

$$\xi_{vi} = k_{vi}^{-1}(t)\, e_{vi}, \tag{5.24}$$

where ξ_{vi} is the transformed error. Introducing (5.24) guarantees that e_{vi} will evolve along its performance bound such that $-\underline{\varepsilon}_{vi} < \xi_{vi} < \bar{\varepsilon}_{vi}$ holds over all time.

Taking the time derivative of ξ_{vi} yields

$$\dot{\xi}_{vi} = k_{vi}^{-1}(t)\left(\dot{e}_{vi} - \dot{k}_{vi}(t)\,\xi_{vi}\right). \tag{5.25}$$

The error transformation for $\boldsymbol{e}_{\varphi i}$ is introduced as

$$\xi_{\phi i} = k_{\phi i}^{-1}(t)\, e_{\phi i}, \quad \xi_{\theta i} = k_{\theta i}^{-1}(t)\, e_{\theta i}, \quad \xi_{\psi i} = k_{\psi i}^{-1}(t)\, e_{\psi i}, \tag{5.26}$$

where ξ_{li}, $l = \{\phi, \theta, \psi\}$ are transformed errors. Inspecting (5.26) indicates that $e_{\phi i}$, $e_{\theta i}$, and $e_{\psi i}$ satisfy their respective performance envelop (5.23), provided that $-\underline{\varepsilon}_{li} < \xi_{li} < \bar{\varepsilon}_{li}$ hold for all time.

Taking the time derivative of ξ_{li} arrives at

$$\dot{\xi}_{li} = k_{li}^{-1}(t)\left(\dot{e}_{li} - \dot{k}_{li}(t)\,\xi_{li}\right). \tag{5.27}$$

Define $\boldsymbol{\xi}_{\varphi i} = \left[\xi_{\phi i}, \xi_{\theta i}, \xi_{\psi i}\right]^{\mathrm{T}}$, $\boldsymbol{k}_{\varphi i} = \operatorname{diag}\left\{k_{\phi i}, k_{\theta i}, k_{\psi i}\right\}$, $\boldsymbol{dk}_{\varphi i} = \operatorname{diag}\{\dot{k}_{\phi i}, \dot{k}_{\theta i}, \dot{k}_{\psi i}\}$. Taking the time derivative of $\boldsymbol{\xi}_{\varphi i}$ can be compactly expressed as

$$\dot{\boldsymbol{\xi}}_{\varphi i} = \boldsymbol{k}_{\varphi i}^{-1}\left(\dot{\boldsymbol{e}}_{\varphi i} - \boldsymbol{dk}_{\varphi i}\boldsymbol{\xi}_{\varphi i}\right). \tag{5.28}$$

Remark 5.5 The dynamic transformed subsystems (5.24) and (5.26) are obtained by incorporating the performance bounds into the synchronization errors e_{vi} and $\boldsymbol{e}_{\varphi i}$.

Hereafter, the control objective equals stabilizing these transformed subsystems, while ensuring that all signals in the resulting closed-loop system are bounded and the constraints $-\underline{\varepsilon}_{si} < \xi_{si} < \bar{\varepsilon}_{si}$, $s = \{v, \phi, \theta, \psi\}$ are never violated.

As shown in Fig. 5.1, the control framework of our proposed design scheme is depicted. Specifically, the detailed designs of T_{xi0} and $\boldsymbol{\delta}_{i0}$ are given as follows.

5.3.2 Adaptive Specified-Time Controller Design for Velocity Subsystem

The adaptive controller T_{xi0} is designed as

$$\begin{aligned} T_{xi0} = & \frac{-\alpha_{vi}\hat{\beta}_{vi}}{\zeta_{vi}} \tanh\left(\frac{\xi_{vi} N_{vi} k_{vi}^{-1} \alpha_{vi}\hat{\beta}_{vi}}{\varepsilon_{vi}}\right) \\ & - \hat{g}_{vi} \tanh\left(\frac{\xi_{vi} N_{vi} k_{vi}^{-1} \hat{g}_{vi}\zeta_{vi}}{\varepsilon_{vi}}\right), \end{aligned} \tag{5.29}$$

and the adaptive tuning laws are proposed as

$$\dot{\hat{\Phi}}_{vi} = \frac{\gamma_{vi1}\xi_{vi} N_{vi} k_{vi}^{-1} \lambda_i \boldsymbol{v}_i^{\mathrm{T}} \boldsymbol{\Xi}_{vi}}{2h_{vi1}^2 V_i^2} - \gamma_{vi1}\sigma_{vi1}\hat{\Phi}_{vi} \tag{5.30}$$

$$\dot{\hat{g}}_{vi} = \gamma_{vi2} \left|\xi_{vi} N_{vi} k_{vi}^{-1}\right| \zeta_{vi} - \gamma_{vi2}\sigma_{vi2}\hat{g}_{vi} \tag{5.31}$$

$$\dot{\hat{\beta}}_{vi} = \gamma_{vi3}\xi_{vi} N_{vi} k_{vi}^{-1} \alpha_{vi} - \gamma_{vi3}\sigma_{vi3}\hat{\beta}_{vi}, \tag{5.32}$$

where σ_{vi1}, σ_{vi2}, σ_{vi3}, and ε_{vi} are selected positive parameters.

5.3.3 Adaptive Specified-Time Controller Design for Attitude Subsystem

With the positive design parameters $\sigma_{\omega i1}$, $\sigma_{\omega i2}$, and $\varepsilon_{\omega i}$, the control input $\boldsymbol{\delta}_{i0}$ and adaptive tuning laws are designed as

$$\boldsymbol{\delta}_{i0} = -\left(\boldsymbol{J}_{i0}^{-1}\boldsymbol{C}_0\right)^{\mathrm{T}} \frac{\boldsymbol{e}_{\omega i}\hat{\beta}_{\omega i}^2 \boldsymbol{\zeta}_{\omega i}^{\mathrm{T}}\boldsymbol{\zeta}_{\omega i}}{\sqrt{\hat{\beta}_{\omega i}^2 \boldsymbol{e}_{\omega i}^{\mathrm{T}}\boldsymbol{e}_{\omega i}\boldsymbol{\zeta}_{\omega i}^{\mathrm{T}}\boldsymbol{\zeta}_{\omega i} + \varepsilon_{\omega i}}} \tag{5.33}$$

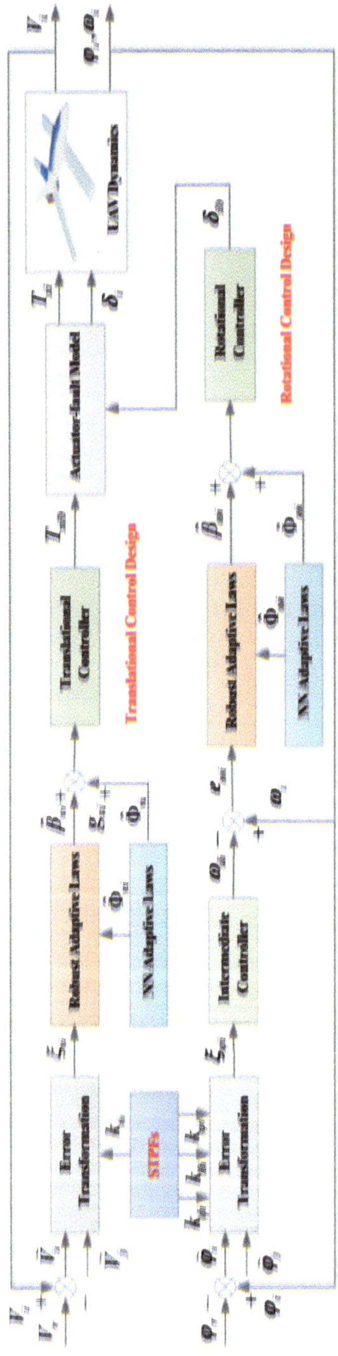

Fig. 5.1 The control framework of our proposed design method

$$\dot{\hat{\Phi}}_{\omega i} = \gamma_{\omega i1} \frac{e_{\omega i}^{\mathrm{T}} \boldsymbol{\Xi}_{\omega i}}{2h_{\omega i1}^2} - \sigma_{\omega i1} \gamma_{\omega i1} \hat{\Phi}_{\omega i} \tag{5.34}$$

$$\dot{\hat{\beta}}_{\omega i} = \gamma_{\omega i2} e_{\omega i}^{\mathrm{T}} \boldsymbol{\zeta}_{\omega i} - \sigma_{\omega i2} \gamma_{\omega i2} \hat{\beta}_{\omega i}. \tag{5.35}$$

5.3.4 Stability Analysis

Theorem 5.1 *Considering the translational dynamics (5.10), design the thrust controller as (5.29) and adaptive tuning laws as (5.30)–(5.32). Under Assumption 5.4, given the STPF k_{vi} and performance bounds $\underline{\varepsilon}_{vi}$, $\bar{\varepsilon}_{vi}$, there exist adjustable parameters c_{vi}, ε_{vi}, h_{vi1}, h_{vi2}, γ_{vi1}, γ_{vi2}, γ_{vi3}, and σ_{vi1}, σ_{vi2}, σ_{vi3} $(i = 1, 2, \ldots, N)$ such that:*

(1) The translational subsystem signals are semi-globally uniformly ultimately bounded (SGUUB) in the presence of external disturbances, modeling uncertainties, and actuator faults.
(2) The velocity synchronization tracking error can evolve along the performance bounds within the specified time T_{vf}.

Proof Using (5.10) and (5.17), (5.25) further becomes

$$\dot{\xi}_{vi} = k_{vi}^{-1} \left(\zeta_{vi} \rho_{Ti} T_{xi0} + \zeta_{vi} T_{xif} + f_{vi} + \lambda_i \frac{\boldsymbol{v}_i^{\mathrm{T}} (\boldsymbol{\Delta}_{vi} + \boldsymbol{d}_{vi})}{V_i} \right), \tag{5.36}$$

where $\zeta_{vi} = \frac{u_i \lambda_i}{m_i V_i}$, $f_{vi} = \lambda_i \frac{\boldsymbol{v}_i^{\mathrm{T}}}{V_i} \left(\boldsymbol{R}_1^{\mathrm{T}} (\boldsymbol{\varphi}_i) \boldsymbol{g} + \frac{\boldsymbol{F}_{i0}}{m_i} \right) - \lambda_i \dot{V}_r - \lambda_{i2} \sum_{j \in \mathcal{N}_i} a_{ij} \dot{\hat{V}}_j - \dot{k}_{vi} \xi_{vi}$, and $\boldsymbol{\Delta}_{vi} = \frac{\Delta \boldsymbol{F}_i}{m_i}$.

Considering RBFNN to approximate the uncertain term $\boldsymbol{\Delta}_{vi}$, it follows from (5.14) that

$$\boldsymbol{v}_i^{\mathrm{T}} (\boldsymbol{\Delta}_{vi} + \boldsymbol{d}_{vi}) = \boldsymbol{v}_i^{\mathrm{T}} \begin{bmatrix} \boldsymbol{W}_{vi1}^{*\mathrm{T}} \boldsymbol{S}_{vi1} + \tau_{vi1} + d_{vi1} \\ \boldsymbol{W}_{vi2}^{*\mathrm{T}} \boldsymbol{S}_{vi2} + \tau_{vi2} + d_{vi2} \\ \boldsymbol{W}_{vi3}^{*\mathrm{T}} \boldsymbol{S}_{vi3} + \tau_{vi3} + d_{vi3} \end{bmatrix} = \sum_{a=1}^{3} v_{ia} \boldsymbol{W}_{via}^{*\mathrm{T}} \boldsymbol{S}_{via} + \sum_{a=1}^{3} v_{ia} (\tau_{via} + d_{via}), \tag{5.37}$$

where v_{ia} is the element of $\boldsymbol{v}_i$.

Based on Young's inequality, one has

$$\frac{\xi_{vi}}{V_i} v_{ia} \boldsymbol{W}_{via}^{*\mathrm{T}} \boldsymbol{S}_{via} \leq \frac{\xi_{vi}^2 v_{ia}^2 \Phi_{vi} \boldsymbol{S}_{via}^{\mathrm{T}} \boldsymbol{S}_{via}}{2h_{vi1}^2 V_i^2} + \frac{h_{vi1}^2}{2}$$

$$\frac{\xi_{vi}}{V_i}v_{ia}\left(\tau_{via}+d_{via}\right)\leq\frac{\xi_{vi}^2v_{ia}^2}{2h_{vi2}^2V_i^2}+\frac{h_{vi2}^2\bar{\tau}_{via}^2}{2},\tag{5.38}$$

where $\Phi_{vi} = \max\left\{\boldsymbol{W}_{vi1}^{*\mathrm{T}}\boldsymbol{W}_{vi1}^{*}, \boldsymbol{W}_{vi2}^{*\mathrm{T}}\boldsymbol{W}_{vi2}^{*}, \boldsymbol{W}_{vi3}^{*\mathrm{T}}\boldsymbol{W}_{vi3}^{*}\right\}$ and $\tau_{via}+d_{via}$ satisfies $|\tau_{via}+d_{via}|\leq\bar{\tau}_{via}$.

Substituting (5.38) into (5.37) yields

$$\frac{\xi_{vi}}{V_i}\boldsymbol{v}_i^{\mathrm{T}}\left(\boldsymbol{\Delta}_{vi}+\boldsymbol{d}_{vi}\right)\leq\sum_{a=1}^{3}\left(\frac{\xi_{vi}^2v_{ia}^2\Phi_{vi}\boldsymbol{\varsigma}_{via}^{\mathrm{T}}\boldsymbol{\varsigma}_{via}}{2h_{vi1}^2V_i^2}+\frac{\xi_{vi}^2v_{ia}^2}{2h_{vi2}^2V_i^2}+\frac{h_{vi1}^2}{2}+\frac{h_{vi2}^2\bar{\tau}_{via}^2}{2}\right).\tag{5.39}$$

Define $\boldsymbol{\Xi}_{vi} = \left[\xi_{vi}v_{i1}\boldsymbol{\varsigma}_{vi1}^{\mathrm{T}}\boldsymbol{\varsigma}_{vi1}, \xi_{vi}v_{i2}\boldsymbol{\varsigma}_{vi2}^{\mathrm{T}}\boldsymbol{\varsigma}_{vi2}, \xi_{vi}v_{i3}\boldsymbol{\varsigma}_{vi3}^{\mathrm{T}}\boldsymbol{\varsigma}_{vi3}\right]^{\mathrm{T}}$. Substituting (5.39) into (5.36), one has

$$\begin{aligned}\xi_{vi}\dot{\xi}_{vi}\leq&\xi_{vi}k_{vi}^{-1}\left(\zeta_{vi}\rho_{Ti}T_{xi0}+\zeta_{vi}T_{xif}+f_{vi}\right)+\xi_{vi}k_{vi}^{-1}\lambda_i\left(\frac{\Phi_{vi}\boldsymbol{v}_i^{\mathrm{T}}\boldsymbol{\Xi}_{vi}}{2h_{vi1}^2V_i^2}+\frac{\boldsymbol{v}_i^{\mathrm{T}}\boldsymbol{v}_i\xi_{vi}}{2h_{vi2}^2V_i^2}\right)\\&+k_{vi}^{-1}\lambda_i\sum_{a=1}^{3}\left(\frac{h_{vi1}^2}{2}+\frac{h_{vi2}^2\bar{\tau}_{via}^2}{2}\right).\end{aligned}\tag{5.40}$$

Define $\beta_{vi} = \frac{1}{\rho_{Ti}}$, $g_{vi} = \beta_{vi}\sup_{t\geq0}\left|T_{xif}\right|$. Let $\tilde{\Phi}_{vi} = \hat{\Phi}_{vi} - \Phi_{vi}$, $\tilde{\beta}_{vi} = \hat{\beta}_{vi} - \beta_{vi}$, $\tilde{g}_{vi} = \hat{g}_{vi} - g_{vi}$, where $\hat{\Phi}_{vi}$, $\hat{\beta}_{vi}$, and $\hat{g}_{vi}$ are the estimations of Φ_{vi}, β_{vi}, and g_{vi}, respectively. To ensure $-\underline{\varepsilon}_{vi} < \xi_{vi} < \bar{\varepsilon}_{vi}$, consider the Lyapunov candidate L_1 as

$$\begin{aligned}L_1=&\frac{1}{2}\left(q\left(\xi_{vi}\right)\log\frac{\underline{\varepsilon}_{vi}^2}{\underline{\varepsilon}_{vi}^2-\xi_{vi}^2}+\left(1-q\left(\xi_{vi}\right)\right)\log\frac{\bar{\varepsilon}_{vi}^2}{\bar{\varepsilon}_{vi}^2-\xi_{vi}^2}\right)\\&+\frac{1}{2\gamma_{vi1}}\tilde{\Phi}_{vi}^2+\frac{\rho_{Ti}}{2\gamma_{vi2}}\tilde{g}_{vi}^2+\frac{\rho_{Ti}}{2\gamma_{vi3}}\tilde{\beta}_{vi}^2\end{aligned}\tag{5.41}$$

with

$$q\left(\xi_{vi}\right)=\begin{cases}0, & \text{if } \xi_{vi}\geq0\\1, & \text{if } \xi_{vi}<0.\end{cases}\tag{5.42}$$

Remark 5.6 Note that if $\underline{\varepsilon}_{vi} = \bar{\varepsilon}_{vi}$, L_1 changes to a symmetric one

$$L_1=\frac{1}{2}\log\frac{\bar{\varepsilon}_{vi}^2}{\bar{\varepsilon}_{vi}^2-\xi_{vi}^2}.\tag{5.43}$$

Both (5.41) and (5.43) follow the similar controller design. In this chapter, the asymmetric form (5.41) is adopted in the following control design.

From (5.40), taking the time derivative of (5.41) arrives at

$$\begin{aligned}\dot{L}_1 =&\xi_{vi} N_{vi} k_{vi}^{-1} \left(\zeta_{vi}\rho_{Ti} T_{xi0} + \zeta_{vi} T_{xif} + f_{vi}\right) + \frac{1}{\gamma_{vi1}} \tilde{\Phi}_{vi} \dot{\hat{\Phi}}_{vi} \\ &+ \xi_{vi} N_{vi} k_{vi}^{-1} \lambda_i \left(\frac{\Phi_{vi} \boldsymbol{v}_i^{\mathrm{T}} \boldsymbol{\Xi}_{vi}}{2h_{vi1}^2 V_i^2} + \frac{\boldsymbol{v}_i^{\mathrm{T}} \boldsymbol{v}_i \xi_{vi}}{2h_{vi2}^2 V_i^2} \right) + \frac{\rho_{Ti}}{\gamma_{vi2}} \tilde{g}_{vi} \dot{\hat{g}}_{vi} \\ &+ N_{vi} k_{vi}^{-1} \lambda_i \sum_{a=1}^{3} \left(\frac{h_{vi1}^2}{2} + \frac{h_{vi2}^2 \bar{\tau}_{via}^2}{2} \right) + \frac{\rho_{Ti}}{\gamma_{vi3}} \tilde{\beta}_{vi} \dot{\hat{\beta}}_{vi},\end{aligned} \tag{5.44}$$

where $N_{vi} = \frac{q(\xi_{vi})}{\underline{\varepsilon}_{vi}^2 - \xi_{vi}^2} + \frac{1-q(\xi_{vi})}{\bar{\varepsilon}_{vi}^2 - \xi_{vi}^2}$.

Recalling the definition of g_{vi} yields

$$\xi_{vi} N_{vi} k_{vi}^{-1} \zeta_{vi} T_{xif} \le \left|\xi_{vi} N_{vi} k_{vi}^{-1}\right| \rho_{Ti} \zeta_{vi} \left(\hat{g}_{vi} - \tilde{g}_{vi}\right). \tag{5.45}$$

Invoking Lemma 5.2, the term $\left|\xi_{vi} N_{vi} k_{vi}^{-1}\right| \rho_{Ti} \zeta_{vi} \hat{g}_{vi}$ follows

$$\begin{aligned}\left|\xi_{vi} N_{vi} k_{vi}^{-1}\right| \rho_{Ti} \zeta_{vi} \hat{g}_{vi} \le& 0.2785 \rho_{Ti} \varepsilon_{vi} + \xi_{vi} N_{vi} k_{vi}^{-1} \rho_{Ti} \zeta_{vi} \hat{g}_{vi} \\ &\cdot \tanh\left(\frac{\xi_{vi} N_{vi} k_{vi}^{-1} \zeta_{vi} \hat{g}_{vi}}{\varepsilon_{vl}} \right).\end{aligned} \tag{5.46}$$

Combined with (5.45) and (5.46), (5.44) becomes

$$\begin{aligned}\dot{L}_1 =&\xi_{vi} N_{vi} k_{vi}^{-1} \zeta_{vi} \rho_{Ti} T_{xi0} + \xi_{vi} N_{vi} k_{vi}^{-1} \left(-c_{vi} k_{vi} \xi_{vi} + \alpha_{vi}\right) \\ &+ \xi_{vi} N_{vi} k_{vi}^{-1} \rho_{Ti} \hat{g}_{vi} \zeta_{vi} \tanh\left(\frac{\hat{g}_{vi} \xi_{vi} N_{vi} k_{vi}^{-1} \zeta_{vi}}{\varepsilon_{vi}} \right) \\ &+ \frac{\tilde{\Phi}_{vi}}{\gamma_{vi1}} \left(\dot{\hat{\Phi}}_{vi} - \frac{\gamma_{vi1} \xi_{vi} N_{vi} k_{vi}^{-1} \lambda_i \boldsymbol{v}_i^{\mathrm{T}} \boldsymbol{\Xi}_{vi}}{2h_{vi1}^2 V_i^2} \right) \\ &+ \frac{\rho_{Ti}}{\gamma_{vi2}} \tilde{g}_{vi} \left(\dot{\hat{g}}_{vi} - \gamma_{vi2} \left|\xi_{vi} N_{vi} k_{vi}^{-1}\right| \zeta_{vi} \right) + \frac{\rho_{Ti}}{\gamma_{vi3}} \tilde{\beta}_{vi} \dot{\hat{\beta}}_{vi} \\ &+ 0.2785 \rho_{Ti} \varepsilon_{vi} + N_{vi} k_{vi}^{-1} \lambda_i \sum_{a=1}^{3} \left(\frac{h_{vi1}^2}{2} + \frac{h_{vi2}^2 \bar{\tau}_{via}^2}{2} \right),\end{aligned} \tag{5.47}$$

where $\alpha_{vi} = c_{vi} k_{vi} \xi_{vi} + f_{vi} + \lambda_i \frac{\hat{\Phi}_{vi} \boldsymbol{v}_i^{\mathrm{T}} \boldsymbol{\Xi}_{vi}}{2h_{vi1}^2 V_i^2} + \lambda_i \frac{\boldsymbol{v}_i^{\mathrm{T}} \boldsymbol{v}_i \xi_{vi}}{2h_{vi2}^2 V_i^2}$.

Invoking Lemma 5.2 and (5.29), one has

$$\xi_{vl} N_{vl} k_{vl}^{-1} \zeta_{vl} \rho_{Tl} T_{xl0} \leq -\rho_{Tl} \xi_{vl} N_{vl} k_{vl}^{-1} \alpha_{vl} \hat{\beta}_{vl} + 0.2785 \rho_{Tl} \varepsilon_{vl} - \rho_{Tl} \xi_{vl} N_{vl} k_{vl}^{-1} \hat{g}_{vl} \zeta_{vl} \tanh \left(\frac{\xi_{vl} N_{vl} k_{vl}^{-1} \hat{g}_{vl} \zeta_{vl}}{\varepsilon_{vl}} \right). \tag{5.48}$$

Recalling the definition of β_{vl} yields

$$-\rho_{Tl} \xi_{vl} N_{vl} k_{vl}^{-1} \alpha_{vl} \hat{\beta}_{vl} = -\xi_{vl} N_{vl} k_{vl}^{-1} \alpha_{vl} - \rho_{Tl} \xi_{vl} N_{vl} k_{vl}^{-1} \alpha_{vl} \tilde{\beta}_{vl}. \tag{5.49}$$

Substituting (5.30)–(5.49) into (5.47) gives

$$\dot{L}_1 \leq -c_{vl} N_{vl} \xi_{vl}^2 - \sigma_{vl1} \tilde{\Phi}_{vl} \hat{\Phi}_{vl} - \rho_{Tl} \sigma_{vl2} \tilde{g}_{vl} \hat{g}_{vl} - \rho_{Tl} \sigma_{vl3} \tilde{\beta}_{vl} \hat{\beta}_{vl} + 0.557 \rho_{Tl} \varepsilon_{vl} + N_{vl} k_{vl}^{-1} \lambda_l \sum_{a=1}^{3} \left(\frac{h_{vl1}^2}{2} + \frac{h_{vl2}^2 \bar{\tau}_{vla}^2}{2} \right). \tag{5.50}$$

Due to the fact that $-\tilde{\Phi}_{vl} \hat{\Phi}_{vl} \leq -\frac{1}{2} \tilde{\Phi}_{vl}^2 + \frac{1}{2} \Phi_{vl}^2$, $-\tilde{g}_{vl} \hat{g}_{vl} \leq -\frac{1}{2} \tilde{g}_{vl}^2 + \frac{1}{2} g_{vl}^2$, $-\tilde{\beta}_{vl} \hat{\beta}_{vl} \leq -\frac{1}{2} \tilde{\beta}_{vl}^2 + \frac{1}{2} \beta_{vl}^2$, $\log \frac{\underline{\varepsilon}_{vl}^2}{\underline{\varepsilon}_{vl}^2 - \xi_{vl}^2} \leq \frac{\xi_{vl}^2}{\underline{\varepsilon}_{vl}^2 - \xi_{vl}^2}$, and $\log \frac{\bar{\varepsilon}_{vl}^2}{\bar{\varepsilon}_{vl}^2 - \xi_{vl}^2} \leq \frac{\xi_{vl}^2}{\bar{\varepsilon}_{vl}^2 - \xi_{vl}^2}$ with $\xi_{vl} \in \left(-\underline{\varepsilon}_{vl}, \bar{\varepsilon}_{vl} \right)$, it follows that

$$\begin{aligned} \dot{L}_1 \leq & -c_{vl} N_{vl} \xi_{vl}^2 - \frac{\sigma_{vl1}}{2} \tilde{\Phi}_{vl}^2 - \frac{\rho_{Tl} \sigma_{vl2}}{2} \tilde{g}_{vl}^2 - \frac{\rho_{Tl} \sigma_{vl3}}{2} \tilde{\beta}_{vl}^2 + \frac{\sigma_{vl1}}{2} \Phi_{vl}^2 + \frac{\rho_{Tl} \sigma_{vl2}}{2} g_{vl}^2 \\ & + \frac{\rho_{Tl} \sigma_{vl3}}{2} \beta_{vl}^2 + 0.557 \varepsilon_{vl} \rho_{Tl} + N_{vl} k_{vl}^{-1} \lambda_l \sum_{a=1}^{3} \left(\frac{h_{vl1}^2}{2} + \frac{h_{vl2}^2 \bar{\tau}_{vla}^2}{2} \right) \\ \leq & -\kappa_{1l} L_1 + \chi_{1l}, \end{aligned} \tag{5.51}$$

where $\kappa_{1l} = \min \{ c_{vl}, \gamma_{vl1} \sigma_{vl1}, \gamma_{vl2} \sigma_{vl2}, \gamma_{vl3} \sigma_{vl3} \}$ and $\chi_{1l} = \frac{\sigma_{vl1}}{2} \Phi_{vl}^2 + \frac{\rho_{Tl} \sigma_{vl2}}{2} g_{vl}^2 + \frac{\rho_{Tl} \sigma_{vl3}}{2} \beta_{vl}^2 + 0.557 \rho_{Tl} \varepsilon_{vl} + N_{vl} k_{vl}^{-1} \lambda_l \sum_{a=1}^{3} \left(\frac{h_{vl1}^2}{2} + \frac{h_{vl2}^2 \bar{\tau}_{vla}^2}{2} \right)$.

From (5.51), it can be obtained that

$$0 \leq L_1 \leq \left(\frac{\chi_{1l}}{\kappa_{1l}} + L_1(0) \right) e^{-\kappa_{1l} t} + \frac{\chi_{1l}}{\kappa_{1l}}, \quad t \geq 0. \tag{5.52}$$

As a result, it can be concluded that all signals of translational dynamics are bounded, which implies that the error signal e_{vl} remains in the interval $-\underline{\varepsilon}_{vl} k_{vl} < e_{vl} < \bar{\varepsilon}_{vl} k_{vl}$, $\forall t \geq 0$, provided that $-\underline{\varepsilon}_{vl} k_{vl}(0) < e_{vl}(0) < \bar{\varepsilon}_{vl} k_{vl}(0)$.

This completes the proof of Theorem 5.1.

Theorem 5.2 *Considering the rotational dynamics (5.11), (5.13), design the surface deflection controller (5.33) and adaptive tuning laws (5.34)–(5.35). Under*

Assumptions 5.1–5.4, given the STPFs k_{li} and performance bounds $\underline{\varepsilon}_{li}$, $\bar{\varepsilon}_{li}$, $l = \{\phi, \theta, \psi\}$, there exist adjustable parameters $\boldsymbol{K}_{1i}$, $\boldsymbol{K}_{2i}$, $\varepsilon_{\omega i}$, $h_{\omega i1}$, $h_{\omega i2}$, $\gamma_{\omega i1}$, $\gamma_{\omega i2}$, $\sigma_{\omega i1}$, and $\sigma_{\omega i2}$ ($i = 1, 2, \ldots, N$) such that:

(1) All signals in the resulting rotational subsystem are SGUUB in spite of external disturbances, modeling uncertainties, and actuator faults.
(2) The synchronization tracking error of attitude can evolve along the performance bounds within the appointed times T_{lf}, $l = \{\phi, \theta, \psi\}$.

Proof The rotational control design comprises two steps:
Step 1: Using (5.11) and (5.28), take the time derivative of $\boldsymbol{\xi}_{\varphi i}$ as

$$\dot{\boldsymbol{\xi}}_{\varphi i} = \boldsymbol{k}_{\varphi i}^{-1}\left(\lambda_i'\left(\boldsymbol{R}_2^{-1}\boldsymbol{\omega}_i - \dot{\boldsymbol{\varphi}}_r\right) - \lambda_{i4}\sum_{j\in\mathcal{N}_i} a_{ij}\dot{\tilde{\boldsymbol{\varphi}}}_j - \boldsymbol{d}\boldsymbol{k}_{\varphi i}\boldsymbol{\xi}_{\varphi i}\right). \tag{5.53}$$

To guarantee $-\underline{\varepsilon}_{li} < \xi_{li} < \bar{\varepsilon}_{li}$, $l = \{\phi, \theta, \psi\}$, consider the Lyapunov candidate asymmetric BLF as

$$L_2 = L_{2\phi} + L_{2\theta} + L_{2\psi} \tag{5.54}$$

with $l = \{\phi, \theta, \psi\}$

$$L_{2l} = \frac{1}{2}\left(q\left(\xi_{li}\right)\log\frac{\underline{\varepsilon}_{li}^2}{\underline{\varepsilon}_{li}^2 - \xi_{li}^2} + \left(1 - q\left(\xi_{li}\right)\right)\log\frac{\bar{\varepsilon}_{li}^2}{\bar{\varepsilon}_{li}^2 - \xi_{li}^2}\right), \tag{5.55}$$

where $q\left(\xi_{li}\right)$ is set as same as (5.42).

Remark 5.7 Note that if $\underline{\varepsilon}_{li} = \bar{\varepsilon}_{li}$, $l = \{\phi, \theta, \psi\}$, L_{2l} changes to a symmetric one

$$L_{2l} = \frac{1}{2}\log\frac{\bar{\varepsilon}_{li}^2}{\bar{\varepsilon}_{li}^2 - \xi_{li}^2}, \tag{5.56}$$

where (5.55) and (5.56) follow the similar control design processes. In this work, the asymmetric form (5.55) is taken in the following control design.

Take the time derivative of L_2 in (5.54) as

$$\dot{L}_2 = \boldsymbol{\xi}_{\varphi i}^{\mathrm{T}}\boldsymbol{N}_{\varphi i}\boldsymbol{k}_{\varphi i}^{-1}\left(\lambda_i'\left(\boldsymbol{R}_2^{-1}\boldsymbol{\omega}_i - \dot{\boldsymbol{\varphi}}_r\right) - \lambda_{i4}\sum_{j\in\mathcal{N}_i} a_{ij}\dot{\tilde{\boldsymbol{\varphi}}}_j - \boldsymbol{d}\boldsymbol{k}_{\varphi i}\boldsymbol{\xi}_{\varphi i}\right), \tag{5.57}$$

where $\boldsymbol{N}_{\varphi i} = \operatorname{diag}\left\{N_{\phi i}, N_{\theta i}, N_{\psi i}\right\}$, defined by $l = \{\phi, \theta, \psi\}$

$$N_{li} = q\left(\xi_{li}\right)\frac{1}{\underline{\varepsilon}_{li}^2 - \xi_{li}^2} + \left(1 - q\left(\xi_{li}\right)\right)\frac{1}{\bar{\varepsilon}_{li}^2 - \xi_{li}^2}. \tag{5.58}$$

Then the virtual controller is designed as

$$\boldsymbol{\omega}_{di}=\frac{\boldsymbol{R}_2}{\lambda_i'}\left(-\boldsymbol{k}_{\varphi i}\boldsymbol{K}_{1i}\boldsymbol{\xi}_{\varphi i}+\lambda_i'\dot{\boldsymbol{\varphi}}_r+\lambda_{i4}\sum_{j\in\mathcal{N}_i}a_{ij}\dot{\tilde{\boldsymbol{\varphi}}}_j+\boldsymbol{dk}_{\varphi i}\boldsymbol{\xi}_{\varphi i}\right) \tag{5.59}$$

with $\boldsymbol{K}_{1i}\in\Re^{3\times3}$ being a selected positive diagonal matrix.

Define $\boldsymbol{e}_{\omega i}=\boldsymbol{\omega}_i-\boldsymbol{\omega}_{di}$. It follows from (5.57) and (5.59) that

$$\dot{L}_2=-\boldsymbol{\xi}_{\varphi i}^{\mathrm{T}}\boldsymbol{N}_{\varphi i}\boldsymbol{K}_{1i}\boldsymbol{\xi}_{\varphi i}+\lambda_i'\boldsymbol{\xi}_{\varphi i}^{\mathrm{T}}\boldsymbol{N}_{\varphi i}\boldsymbol{k}_{\varphi i}^{-1}\boldsymbol{R}_2^{-1}\boldsymbol{e}_{\omega i}. \tag{5.60}$$

Step 2: Taking the time derivative of $\frac{1}{2}\boldsymbol{e}_{\omega i}^{\mathrm{T}}\boldsymbol{e}_{\omega i}$ along (5.13) denotes

$$\begin{aligned}\boldsymbol{e}_{\omega i}^{\mathrm{T}}\dot{\boldsymbol{e}}_{\omega i}=&\,\boldsymbol{e}_{\omega i}^{\mathrm{T}}\left(\boldsymbol{J}_{i0}^{-1}\boldsymbol{C}_0\boldsymbol{\rho}_{\delta i}\boldsymbol{\delta}_{i0}+\boldsymbol{J}_{i0}^{-1}\boldsymbol{S}\left(\boldsymbol{J}_{i0}\boldsymbol{\omega}_i\right)\boldsymbol{\omega}_i+\boldsymbol{J}_{i0}^{-1}\boldsymbol{N}_{i0}\right)\\&+\boldsymbol{e}_{\omega i}^{\mathrm{T}}\left(\boldsymbol{\Delta}_{\omega i}'+\boldsymbol{l}_{\omega i}+\boldsymbol{d}_{\omega i}-\dot{\boldsymbol{\omega}}_{di}\right).\end{aligned} \tag{5.61}$$

From (5.14), RBFNN is utilized to approximate $\boldsymbol{\Delta}_{\omega i}'$ as

$$\begin{aligned}\boldsymbol{e}_{\omega i}^{\mathrm{T}}\left(\boldsymbol{\Delta}_{\omega i}'+\boldsymbol{l}_{\omega i}+\boldsymbol{d}_{\omega i}\right)=&\sum_{a=1}^{3}e_{\omega ia}\boldsymbol{W}_{\omega ia}^{*\mathrm{T}}\boldsymbol{\varsigma}_{\omega ia}\\&+\sum_{a=1}^{3}e_{\omega ia}\left(\tau_{\omega ia}+l_{\omega ia}+d_{\omega ia}\right).\end{aligned} \tag{5.62}$$

Using Young's inequality, one has

$$\begin{aligned}e_{\omega ia}\boldsymbol{W}_{\omega ia}^{*\mathrm{T}}\boldsymbol{\varsigma}_{\omega ia}&\le\frac{e_{\omega ia}^2\Phi_{\omega i}\boldsymbol{\varsigma}_{\omega ia}^{\mathrm{T}}\boldsymbol{\varsigma}_{\omega ia}}{2h_{\omega i1}^2}+\frac{h_{\omega i1}^2}{2}\\e_{\omega ia}\left(\tau_{\omega ia}+l_{\omega ia}+d_{\omega ia}\right)&\le\frac{e_{\omega ia}^2}{2h_{\omega i2}^2}+\frac{h_{\omega i2}^2\bar{\tau}_{\omega ia}^2}{2},\end{aligned} \tag{5.63}$$

where $\Phi_{\omega i}=\max\left\{\boldsymbol{W}_{\omega i1}^{*\mathrm{T}}\boldsymbol{W}_{\omega i1}^{*},\boldsymbol{W}_{\omega i2}^{*\mathrm{T}}\boldsymbol{W}_{\omega i2}^{*},\boldsymbol{W}_{\omega i3}^{*\mathrm{T}}\boldsymbol{W}_{\omega i3}^{*}\right\}$ and $|\tau_{\omega ia}+l_{\omega ia}+d_{\omega ia}|\le\bar{\tau}_{\omega ia}$.

Substituting (5.63) into (5.62) yields

$$\begin{aligned}\boldsymbol{e}_{\omega i}^{\mathrm{T}}\left(\boldsymbol{\Delta}_{\omega i}'+\boldsymbol{l}_{\omega i}+\boldsymbol{d}_{\omega i}\right)\le&\,\boldsymbol{e}_{\omega i}^{\mathrm{T}}\left(\frac{\Phi_{\omega i}\boldsymbol{\Xi}_{\omega i}}{2h_{\omega i1}^2}+\frac{\boldsymbol{e}_{\omega i}}{2h_{\omega i2}^2}\right)\\&+\sum_{a=1}^{3}\left(\frac{h_{\omega i1}^2}{2}+\frac{h_{\omega i2}^2\bar{\tau}_{\omega ia}^2}{2}\right),\end{aligned} \tag{5.64}$$

where $\boldsymbol{\Xi}_{\omega i} = \left[e_{\omega i1}\varsigma_{\omega i1}^{\mathrm{T}}\varsigma_{\omega i1}, e_{\omega i2}\varsigma_{\omega i2}^{\mathrm{T}}\varsigma_{\omega i2}, e_{\omega i3}\varsigma_{\omega i3}^{\mathrm{T}}\varsigma_{\omega i3}\right]^{\mathrm{T}}$.

Define $g_{\omega i} = \inf_{t\geq 0}\lambda_{\min}(\boldsymbol{J}_{i0}^{-1}\boldsymbol{C}_0(\boldsymbol{\delta}_i)\boldsymbol{\rho}_{\delta i}(\boldsymbol{J}_{i0}^{-1}\boldsymbol{C}_0(\boldsymbol{\delta}_i))^{\mathrm{T}})$ and $\beta_{\omega i} = \frac{1}{g_{\omega i}}$ by Assumption 5.2. Let $\hat{\Phi}_{\omega i}$ and $\hat{\beta}_{\omega i}$ be the estimations of $\Phi_{\omega i}$, $\beta_{\omega i}$, respectively, with $\tilde{\Phi}_{\omega i} = \hat{\Phi}_{\omega i} - \Phi_{\omega i}$ and $\tilde{\beta}_{\omega i} = \hat{\beta}_{\omega i} - \beta_{\omega i}$.

Consider the following Lyapunov function:

$$L_3 = L_2 + \frac{1}{2}\boldsymbol{e}_{\omega i}^{\mathrm{T}}\boldsymbol{e}_{\omega i} + \frac{1}{2\gamma_{\omega i1}}\tilde{\Phi}_{\omega i}^2 + \frac{g_{\omega i}}{2\gamma_{\omega i2}}\tilde{\beta}_{\omega i}^2, \tag{5.65}$$

where $\gamma_{\omega i1}$ and $\gamma_{\omega i2}$ are positive design parameters.

By checking (5.60), (5.61), and (5.64), the time derivative of (5.65) can be deduced by

$$\begin{aligned}\dot{L}_3 =& -\boldsymbol{\xi}_{\varphi i}^{\mathrm{T}}\boldsymbol{N}_{\varphi i}\boldsymbol{K}_{1i}\boldsymbol{\xi}_{\varphi i} + \boldsymbol{e}_{\omega i}^{\mathrm{T}}\boldsymbol{J}_{i0}^{-1}\left(\boldsymbol{C}_0\boldsymbol{\rho}_{\delta i}\boldsymbol{\delta}_{i0} + \boldsymbol{S}(\boldsymbol{J}_{i0}\boldsymbol{\omega}_i) + \boldsymbol{N}_{i0}\right)\\ &+ \boldsymbol{e}_{\omega i}^{\mathrm{T}}\left(\frac{\Phi_{\omega i}\boldsymbol{\Xi}_{\omega i}}{2h_{\omega i1}^2} + \frac{\boldsymbol{e}_{\omega i}}{2h_{\omega i2}^2} + \lambda_i'\left(\boldsymbol{N}_{\varphi i}\boldsymbol{k}_{\varphi i}^{-1}\boldsymbol{R}_2^{-1}\right)^{\mathrm{T}}\boldsymbol{\xi}_{\varphi i} - \dot{\boldsymbol{\omega}}_{di}\right)\\ &+ \frac{1}{\gamma_{\omega i1}}\tilde{\Phi}_{\omega i}\dot{\hat{\Phi}}_{\omega i} + \frac{g_{\omega i}}{\gamma_{\omega i2}}\tilde{\beta}_{\omega i}\dot{\hat{\beta}}_{\omega i} + \sum_{a=1}^{3}\left(\frac{h_{\omega i1}^2}{2} + \frac{h_{\omega i2}^2\bar{\tau}_{\omega ia}^2}{2}\right).\end{aligned} \tag{5.66}$$

Introducing $\boldsymbol{\zeta}_{\omega i} = \boldsymbol{K}_{2i}\boldsymbol{e}_{\omega i} + \boldsymbol{J}_{i0}^{-1}\boldsymbol{S}(\boldsymbol{J}_{i0}\boldsymbol{\omega}_i)\boldsymbol{\omega}_i + \boldsymbol{J}_{i0}^{-1}\boldsymbol{N}_{i0} + \frac{\hat{\Phi}_{\omega i}\boldsymbol{\Xi}_{\omega i}}{2h_{\omega i1}^2} + \frac{\boldsymbol{e}_{\omega i}}{2h_{\omega i2}^2} + \lambda_i'\left(\boldsymbol{N}_{\varphi i}\boldsymbol{k}_{\varphi i}^{-1}\boldsymbol{R}_2^{-1}\right)^{\mathrm{T}}\boldsymbol{\xi}_{\varphi i} - \dot{\boldsymbol{\omega}}_{di}$, (5.63) becomes

$$\begin{aligned}\dot{L}_3 =&\ \boldsymbol{\xi}_{\varphi i}^{\mathrm{T}}\boldsymbol{N}_{\varphi i}\boldsymbol{K}_{1i}\boldsymbol{\xi}_{\varphi i} - \boldsymbol{e}_{\omega i}^{\mathrm{T}}\boldsymbol{K}_{2i}\boldsymbol{e}_{\omega i} + \boldsymbol{e}_{\omega i}^{\mathrm{T}}\boldsymbol{J}_{i0}^{-1}\boldsymbol{C}_0\boldsymbol{\rho}_{\delta i}\boldsymbol{\delta}_{i0} + \boldsymbol{e}_{\omega i}^{\mathrm{T}}\boldsymbol{\zeta}_{\omega i}\\ &+ \frac{1}{\gamma_{\omega i1}}\tilde{\Phi}_{\omega i}\left(\dot{\hat{\Phi}}_{\omega i} - \gamma_{\omega i1}\frac{\boldsymbol{e}_{\omega i}^{\mathrm{T}}\boldsymbol{\Xi}_{\omega i}}{2h_{\omega i1}^2}\right) + \frac{g_{\omega i}}{\gamma_{\omega i2}}\tilde{\beta}_{\omega i}\dot{\hat{\beta}}_{\omega i}\\ &+ \sum_{a=1}^{3}\left(\frac{h_{\omega i1}^2}{2} + \frac{h_{\omega i2}^2\bar{\tau}_{\omega ia}^2}{2}\right).\end{aligned} \tag{5.67}$$

Noting Lemma 5.1 and (5.33), the term of $\boldsymbol{e}_{\omega i}^{\mathrm{T}}\boldsymbol{J}_{i0}^{-1}\boldsymbol{C}_0\boldsymbol{\rho}_{\delta i}\boldsymbol{\delta}_{i0}$ renders

$$\begin{aligned}\boldsymbol{e}_{\omega i}^{\mathrm{T}}\boldsymbol{J}_{i0}^{-1}\boldsymbol{C}_0\boldsymbol{\rho}_{\delta i}\boldsymbol{\delta}_{i0} =& -\boldsymbol{e}_{\omega i}^{\mathrm{T}}\boldsymbol{J}_{i0}^{-1}\boldsymbol{C}_0\boldsymbol{\rho}_{\delta i}\left(\boldsymbol{J}_{i0}^{-1}\boldsymbol{C}_0\right)^{\mathrm{T}}\frac{\boldsymbol{e}_{\omega i}\hat{\beta}_{\omega i}^2\boldsymbol{\zeta}_{\omega i}^{\mathrm{T}}\boldsymbol{\zeta}_{\omega i}}{\sqrt{\hat{\beta}_{\omega i}^2\boldsymbol{e}_{\omega i}^{\mathrm{T}}\boldsymbol{e}_{\omega i}\boldsymbol{\zeta}_{\omega i}^{\mathrm{T}}\boldsymbol{\zeta}_{\omega i} + \varepsilon_{\omega i}}}\\ \leq& \frac{-g_{\omega i}\hat{\beta}_{\omega i}^2\boldsymbol{e}_{\omega i}^{\mathrm{T}}\boldsymbol{e}_{\omega i}\boldsymbol{\zeta}_{\omega i}^{\mathrm{T}}\boldsymbol{\zeta}_{\omega i}}{\sqrt{\hat{\beta}_{\omega i}^2\boldsymbol{e}_{\omega i}^{\mathrm{T}}\boldsymbol{e}_{\omega i}\boldsymbol{\zeta}_{\omega i}^{\mathrm{T}}\boldsymbol{\zeta}_{\omega i} + \varepsilon_{\omega i}}} \leq g_{\omega i}\sqrt{\varepsilon_{\omega i}} - g_{\omega i}|\hat{\beta}_{\omega i}|\,\|\boldsymbol{e}_{\omega i}\|\,\|\boldsymbol{\zeta}_{\omega i}\|\\ \leq& g_{\omega i}\sqrt{\varepsilon_{\omega i}} - g_{\omega i}\hat{\beta}_{\omega i}\boldsymbol{e}_{\omega i}^{\mathrm{T}}\boldsymbol{\zeta}_{\omega i}.\end{aligned} \tag{5.68}$$

Due to the fact $g_{\omega l}(\hat{\beta}_{\omega l} - \tilde{\beta}_{\omega l}) = g_{\omega l}\beta_{\omega l} = 1$, substituting (5.34)–(5.68) into (5.67) yields

$$\dot{L}_3 \leq -\boldsymbol{\xi}_{\varphi l}^{\mathrm{T}} \boldsymbol{N}_{\varphi l} \boldsymbol{K}_{1l} \boldsymbol{\xi}_{\varphi l} - \boldsymbol{e}_{\omega l}^{\mathrm{T}} \boldsymbol{K}_{2l} \boldsymbol{e}_{\omega l} - \sigma_{\omega l 1} \tilde{\Phi}_{\omega l} \hat{\Phi}_{\omega l} - g_{\omega i} \sigma_{\omega i 2} \tilde{\beta}_{\omega l} \hat{\beta}_{\omega l} + g_{\omega l} \sqrt{\varepsilon_{\omega l}} + \sum_{a=1}^{3} \left(\frac{h_{\omega l 1}^2}{2} + \frac{h_{\omega l 2}^2 \bar{\tau}_{\omega l a}^2}{2} \right). \tag{5.69}$$

Recalling $-\tilde{\Phi}_{\omega l} \hat{\Phi}_{\omega l} \leq -\frac{1}{2}\tilde{\Phi}_{\omega l}^2 + \frac{1}{2}\Phi_{\omega l}^2$, $-\tilde{\beta}_{\omega l} \hat{\beta}_{\omega l} \leq -\frac{1}{2}\tilde{\beta}_{\omega i}^2 + \frac{1}{2}\beta_{\omega l}^2$, $\log \frac{\underline{\varepsilon}_{ll}^2}{\underline{\varepsilon}_{ll}^2 - \xi_{ll}^2} \leq \frac{\xi_{ll}^2}{\underline{\varepsilon}_{ll}^2 - \xi_{ll}^2}$, and $\log \frac{\bar{\varepsilon}_{ll}^2}{\bar{\varepsilon}_{ll}^2 - \xi_{ll}^2} \leq \frac{\xi_{ll}^2}{\bar{\varepsilon}_{ll}^2 - \xi_{ll}^2}$ with $\xi_{ll} \in (-\underline{\varepsilon}_{ll}, \bar{\varepsilon}_{ll})$, $l = \{\phi, \theta, \psi\}$, it follows that

$$\begin{aligned} \dot{L}_3 \leq & -\boldsymbol{\xi}_{\varphi l}^{\mathrm{T}} \boldsymbol{N}_{\varphi l} \boldsymbol{K}_{1l} \boldsymbol{\xi}_{\varphi l} - \boldsymbol{e}_{\omega l}^{\mathrm{T}} \boldsymbol{K}_{2l} \boldsymbol{e}_{\omega l} - \frac{\sigma_{\omega l 1}}{2} \tilde{\Phi}_{\omega l}^2 - \frac{g_{\omega i} \sigma_{\omega i 2}}{2} \tilde{\beta}_{\omega l}^2 + \frac{\sigma_{\omega l 1}}{2} \Phi_{\omega l}^2 \\ & + \frac{g_{\omega l} \sigma_{\omega l 2}}{2} \beta_{\omega l}^2 + g_{\omega l} \sqrt{\varepsilon_{\omega l}} + \sum_{a=1}^{3} \left(\frac{h_{\omega l 1}^2}{2} + \frac{h_{\omega l 2}^2 \bar{\tau}_{\omega i a}^2}{2} \right) \\ \leq & -\kappa_{3l} L_3 + \chi_{3l}, \end{aligned} \tag{5.70}$$

where $\kappa_{3l} = \min\{2\lambda_{\min}(\boldsymbol{K}_1), 2\lambda_{\min}(\boldsymbol{K}_2), \sigma_{\omega l 1}\gamma_{\omega l 1}, \sigma_{\omega l 2}\gamma_{\omega i 2}\}$ and $\chi_{3l} = \sum_{a=1}^{3} \left(\frac{h_{\omega l 1}^2}{2} + \frac{h_{\omega l 2}^2 \bar{\tau}_{\omega l a}^2}{2} \right) + \frac{\sigma_{\omega l 1}}{2} \Phi_{\omega l}^2 + \frac{g_{\omega l} \sigma_{\omega l 2}}{2} \beta_{\omega l}^2 + g_{\omega l} \sqrt{\varepsilon_{\omega l}}$.

Integrating (5.70) over $[0, t]$, one concludes

$$0 \leq L_3 \leq \left(\frac{\chi_{3l}}{\kappa_{3l}} + L_3(0) \right) e^{-\kappa_{3l} t} + \frac{\chi_{3l}}{\kappa_{3l}}. \tag{5.71}$$

Consequently, it can be concluded that all signals of rotational dynamics are bounded, which indicates that the error signals $e_{\phi l}$, $e_{\theta l}$, and $e_{\psi l}$ evolve along the interval $-\underline{\varepsilon}_{ll} k_{ll} < e_{ll} < \bar{\varepsilon}_{ll} k_{ll}$, $\forall t \geq 0$, provided that $-\underline{\varepsilon}_{ll} k_{ll}(0) < e_{ll}(0) < \bar{\varepsilon}_{ll} k_{ll}(0)$, $l = \{\phi, \theta, \psi\}$.

This ends the proof of Theorem 5.2.

As a series of design parameters are involved in the controller design, a trial and error procedure for parameters selection is summarized as follows:

Step 1: Calculate $e_{sl}(0)$, $s = \{v, \phi, \theta, \psi\}$. If $e_{sl}(0) \geq 0$, set $\bar{\varepsilon}_{sl} = 1$; else set $\underline{\varepsilon}_{sl} = 1$.

Step 2: Select the values of $\underline{\varepsilon}_{sl}$ or $\bar{\varepsilon}_{sl}$, $s = \{v, \phi, \theta, \psi\}$ according to the level of overshoot permitted by the mission.

Step 3: Choose appropriate b_{sl} and k_{sl0}, while giving T_{sf} and $k_{sl\infty}$, $s = \{v, \phi, \theta, \psi\}$, based on the mission-oriented requirements on the velocity and attitude synchronization tracking errors. Then, solving (5.19) and (5.23) arrives at k_{sl}, $s = \{v, \phi, \theta, \psi\}$.

Step 4: Determine the control parameters c_{vi}, h_{vi1}, h_{vi2}, and ε_{vi} by trial and error to render reasonable control efforts of T_{xi0}. Meanwhile, choose the control parameters $\boldsymbol{K}_{1i}$, $\boldsymbol{K}_{2i}$, $h_{\omega i1}$, $h_{\omega i2}$, and $\varepsilon_{\omega i}$ by trial and error to render proper virtual control $\boldsymbol{\omega}_{di}$ and control efforts of $\boldsymbol{\delta}_{i0}$.
Step 5: Choose γ_{vi1}, γ_{vi2}, γ_{vi3}, σ_{vi1}, σ_{vi2}, σ_{vi3}. Larger γ_{vi1}, γ_{vi2}, and γ_{vi3} would lead to faster convergence of $\hat{\Phi}_{vi}$, $\hat{g}_{vi}$, and $\hat{\beta}_{vi}$. Meanwhile, select $\gamma_{\omega i1}$, $\gamma_{\omega i2}$, $\sigma_{\omega i1}$, and $\sigma_{\omega i2}$. Larger $\gamma_{\omega i1}$ and $\gamma_{\omega i2}$ would result in faster convergence of $\hat{\Phi}_{\omega i}$ and $\hat{\beta}_{\omega i}$.

Remark 5.8 It is worth remarking that the modeling uncertainties and actuator faults are effectively compensated through combining the bound estimation approach and smooth function $\frac{z}{\sqrt{z^2+\varepsilon}}$ in Lemma 5.2. Actually, the obstacle caused by $\boldsymbol{J}_{i0}^{-1}\boldsymbol{C}_0\boldsymbol{\rho}_{\delta i}$ is difficult to deal with since this term multiplies the control action $\boldsymbol{\delta}_i$. In this work, with the help of Assumption 5.2 and the introduction of $g_{\omega i}$, the unknown time-varying term of $\boldsymbol{J}_{i0}^{-1}\boldsymbol{C}_0\boldsymbol{\rho}_{\delta i}$ is successfully circumvented by adaptive tuning law for $\hat{g}_{\omega i}$ in (5.35).

Remark 5.9 As opposed to traditional PPC [1–3, 15], where the exponentially decaying functions are used to drive the tracking errors to the prespecified residual sets as time goes to infinity, the structural error transformations with STPFs as defined by (5.16) and (5.20) are presented in proposed control design, whereby combined with BLFs in (5.41) and (5.55), the specified-time controllers are developed to not only guarantee the user-specified-time but also specify the velocity and attitude errors within prescribed bounds.

Remark 5.10 In comparison with most available works on finite/fixed-time control for UAV dynamics [28–30], where the estimation approaches for convergence time were used with conservations, our scheme is able to fulfill the more challenging work that the settling time can be offline preassigned according to mission requirements without estimations and conservations. Such an achievement is attributed to our proposed STPFs, which guarantee that tracking error e_{si} can reach to prescribed interval $\left(-\underline{\varepsilon}_{si}k_{si\infty}, \bar{\varepsilon}_{si}k_{si\infty}\right)$ after specified time T_{sf}.

5.4 Simulations and Analyses

To verify the applicability of the proposed control scheme, the experiments have been performed on semi-physical UAV Platform (conducted by Beijing Links Co., Ltd.). The detailed experimental setup can refer to our previous work [40]. The experiments and simulations shown in Sect. 5.4.1 cover the proposed specified-time tracking for velocity and attitude in terms of different specified times and three cases with different actuator faults and inertia uncertainties. Comparative simulation results against state-of-the-art approaches are discussed in Sect. 5.4.2 in order to demonstrate the superiority of the proposed control scheme in terms of convergence time and convergence error.

5.4.1 Support of Specified-Time Tracking for Velocity and Attitude

To demonstrate the effectiveness of the proposed control scheme, experiment and simulation validations for three UAV systems are carried. The structural parameters and coefficients for each UAV are adopted from Table I in [35], and the corresponding communication graph is illustrated in Fig. 5.2. Choose the initial flight states as $V_1 = V_2 = V_3 = 40\,\mathrm{m/s}$, $\boldsymbol{\varphi}_1 = [0, 1, 0]^{\mathrm{T}}\,\mathrm{deg}$, $\boldsymbol{\varphi}_2 = \boldsymbol{\varphi}_3 = [0.5, 1, 0.5]^{\mathrm{T}}\,\mathrm{deg}$, $\boldsymbol{\omega}_1 = \boldsymbol{\omega}_2 = \boldsymbol{\omega}_3 = [0, 0, 0]^{\mathrm{T}}\,\mathrm{deg/s}$. Suppose that the velocity is kept as $40\,\mathrm{m/s}$, and the attitude tracks the reference $\phi_r = 5°$, $\theta_r = 10°$, and $\psi_r = 5°$. The external disturbances in attitude motion are described as $\boldsymbol{d}_{\omega 2} = \boldsymbol{d}_{\omega 3} = [0.3\cos(2t), 0.2\sin(3t), 0.5\sin(t)]^{\mathrm{T}}$. Meanwhile, it is assumed that follower 2 suffers from the actuator fault during $t \in (2, 5]\,\mathrm{s}$, given by $\boldsymbol{\rho}_{\delta 3} = \mathrm{diag}\{1, 1, 0.9\}$.

The prescribed parameters of novel STPFs are chosen as $T_{jf} = 8s$, $b_{ji} = 1$, $k_{ji\infty} = 1$, $k_{\phi i0} = k_{\psi i0} = 15$, $k_{\theta i0} = 20$, $\underline{\varepsilon}_{ji} = \bar{\varepsilon}_{ji} = 0.5$ ($i = 1, 2, 3$, $j = \{\phi, \theta, \psi\}$). And the controller parameters are selected as $\boldsymbol{K}_{1i} = \mathrm{diag}\{2, 2, 4\}$, $\boldsymbol{K}_{21} = \boldsymbol{K}_{22} = \mathrm{diag}\{10^{-4}, 10^{-4}, 10^{-4}\}$, $\boldsymbol{K}_{23} = \mathrm{diag}\{10^{-5}, 10^{-5}, 10^{-5}\}$, $h_{\omega i2} = 100$, $\varepsilon_{\omega i} = 1000$, $\gamma_{\omega l2} - 10^{-5}$, $\sigma_{\omega i2} = 10^{-2}$ ($i = 1, 2, 3$).

The specified-time responses for the attitude tracking are sketched in Fig. 5.3, from which it is evident that the attitude synchronization errors evolve strictly inside their respective performance envelops in terms of settling time ($T_f = 8s$). The steady-state accuracy of attitude synchronization errors is guaranteed as 0.5. This in turn indicates that the proposed control scheme enhances the fault-tolerant property. Since the designed STPFs are judiciously specified to meet the mission demands, the attitude synchronization tracking can be achieved with required tracking accuracy.

Further, according to the requirement of flight mission, different specified times should be considered, wherein the specified times are set to $T_f = 6s$ and $T_f = 7s$, respectively. The corresponding attitude responses are shown in Fig. 5.4a–f, from which it notes that the synchronization tracking errors effectively converge within the boundary constraints under different specified times. From Figs. 5.3 and 5.4,

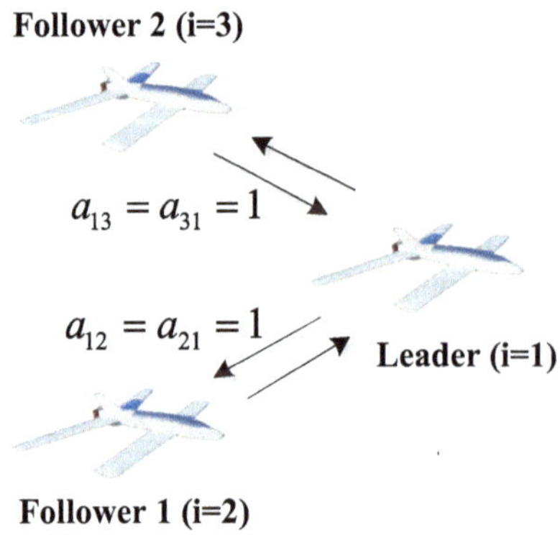

Fig. 5.2 Communication graph for UAVs

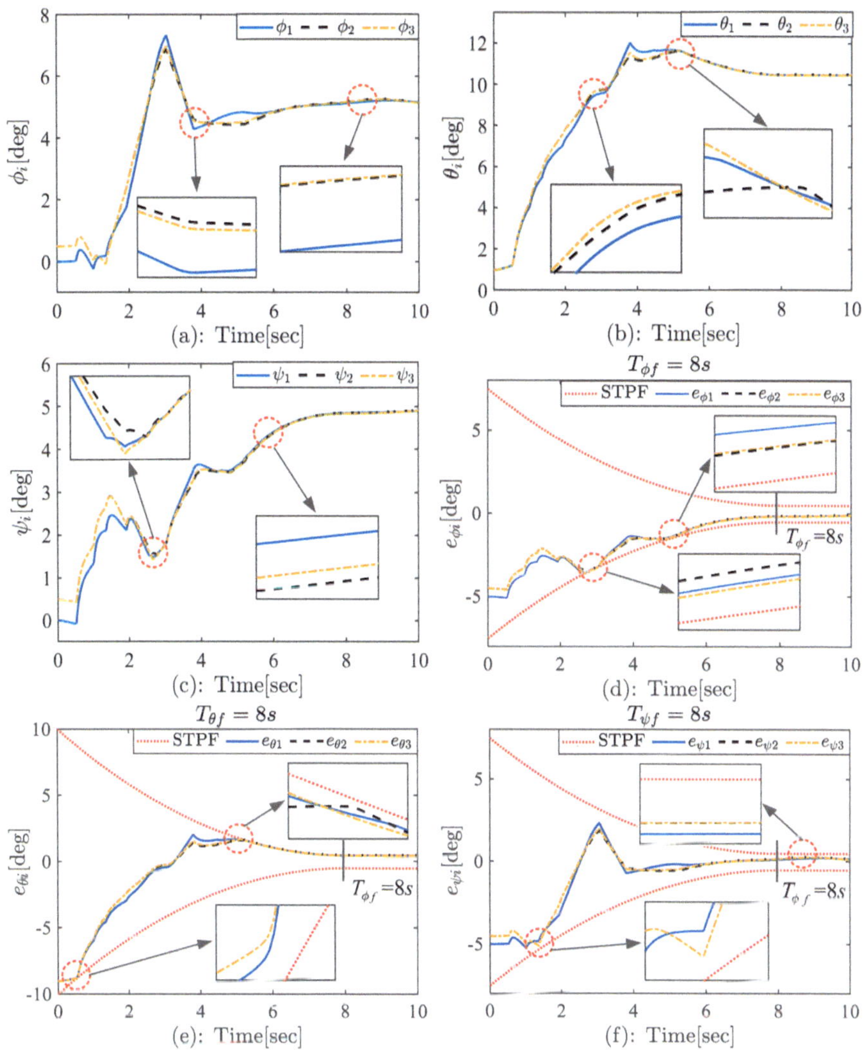

Fig. 5.3 Attitude synchronization tracking within $T_f = 8$ s. (**a**)–(**f**) show the responses of attitude angles and their corresponding synchronization tracking errors. (**a**) ϕ_i. (**b**) θ_i. (**c**) ψ_i. (**d**) e_{ϕ_i}. (**e**) e_{θ_i}. (**f**) e_{ψ_i}

it can be seen that the proposed specified-time controller can adapt to the need of different user-defined settling times.

To give a comprehensive insight into the robustness of proposed specified-time control scheme, different cases with actuator faults and inertia uncertainties are considered. To this end, we provide the following three cases:

Case 1: All UAVs are healthy during flight, with no considerations of actuator faults and inertia uncertainties.

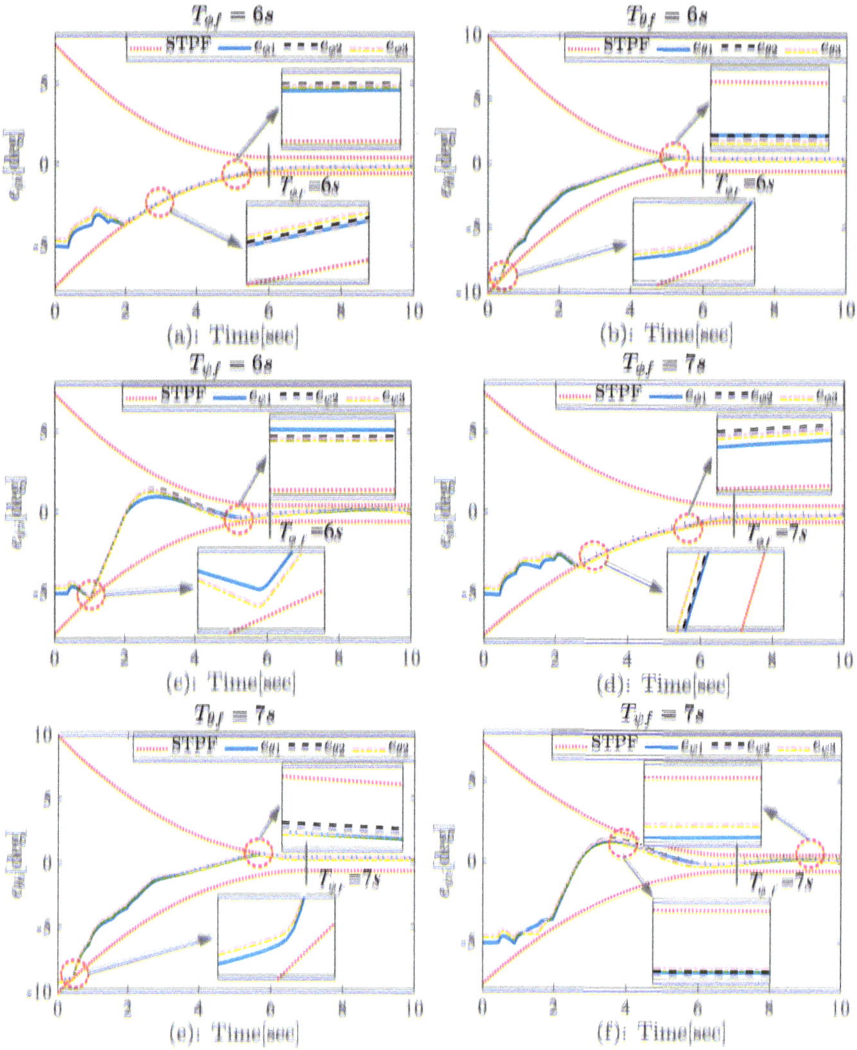

Fig. 5.4 Attitude synchronization tracking within different specified time. (**a**)–(**f**) show the synchronization error of attitude within $T_f = 6$ s and $T_f = 7$ s, respectively. (**a**) e_{ϕ_i} for $T_f = 6$ s. (**b**) e_{θ_i} for $T_f = 6$ s. (**c**) e_{ψ_i} for $T_f = 6$ s. (**d**) e_{ϕ_i} for $T_f = 7$s. (**e**) e_{θ_i} for $T_f = 7$ s (**f**) e_{ψ_i} for $T_f = 7$ s

Case 2: follower 2 suffers from actuator fault during $t \in (2, 5]\, s$, denoted by $\boldsymbol{\rho}_{\delta 3} = \text{diag}\{1, 1, 0.9\}$.

Case 3: Based on Case 2, the inertia uncertainties are further taken into account. Given this, the actual inertia matrix is given by $\boldsymbol{J}_l = \boldsymbol{J}_{l0} + \Delta \boldsymbol{J}_l$, of which the ranges are tabulated by $\pm$ 10%.

Taking follower 2 (the user-oriented time $T_f = 8$ s) for example, the robustness with proposed controller is verified in terms of aforementioned three cases in Fig. 5.5. It can be seen from Fig. 5.5 that the developed control algorithm is not only robust against actuator faults but also able to accommodate the modeling uncertainties and external disturbances. In response to the actuator faults and inertia uncertainties, the synchronization tracking errors under proposed control scheme have small deviations from the normal responses. These results are presented to

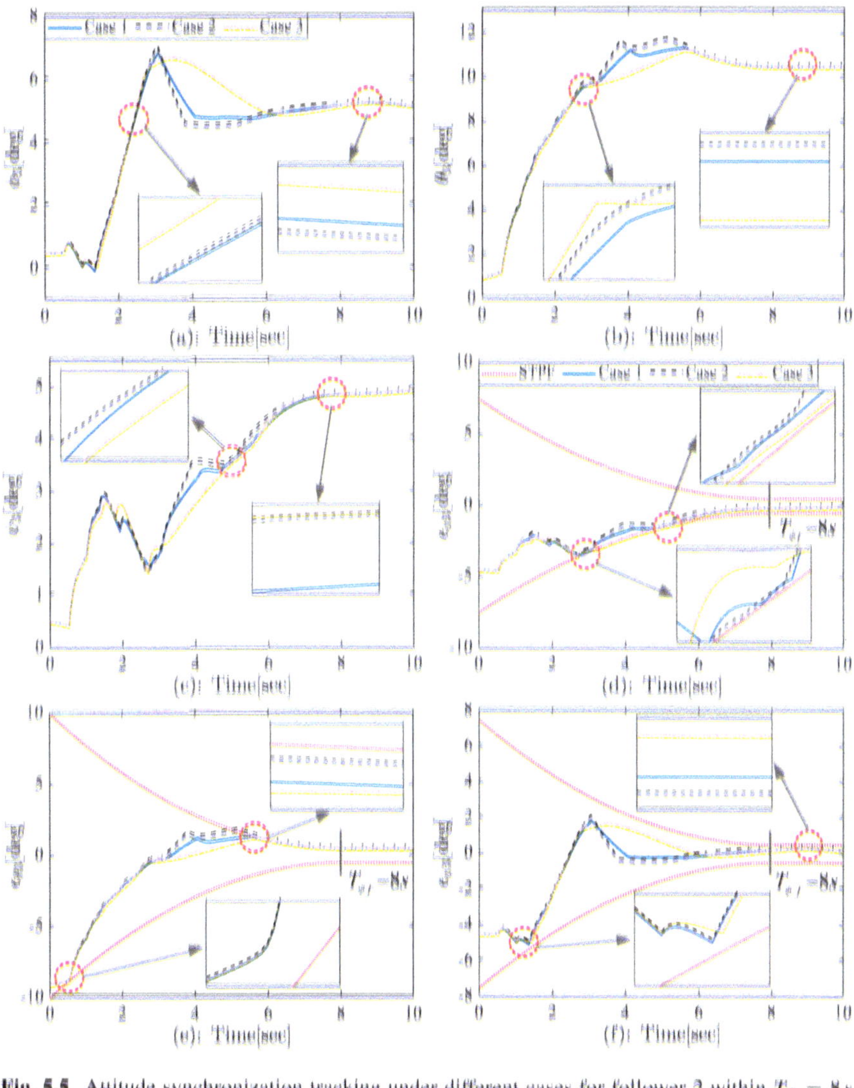

Fig. 5.5 Attitude synchronization tracking under different cases for follower 2 within $T_f = 8$ s. **(a)**–**(f)** show the responses of attitude angles and their corresponding synchronization tracking errors. **(a)** ϕ_3, **(b)** θ_3, **(c)** ψ_3, **(d)** e_{ϕ_3}, **(e)** e_{θ_3}, **(f)** e_{ψ_3}

illustrate the effectiveness of the proposed fault-tolerant property and disturbance compensation feature.

5.4.2 Comparative Simulation with State-of-the-Art

To illustrate the superiority of proposed control scheme, comparative simulation results against state-of-the-art approaches are performed. As a representative example, a finite-time controller designed by Yu et al. [28] is considered. To provide a fair comparison, the same flight test in one sortie is conducted. Taking above Case 2 within $T_f = 8$ s for example, Fig. 5.6 shows the attitude tracking under compared finite-time controller. From Fig. 5.6a–c, there are two obvious disadvantages of compared finite-time controller: Firstly, the synchronization attitude errors under finite-time controller violate the boundary constraints, and secondly, the steady-state errors under finite-time controller are larger than the proposed specified-time controller. Thus, it is concluded that the proposed controller outperforms the compared finite-time controller especially in terms of the steady-state errors.

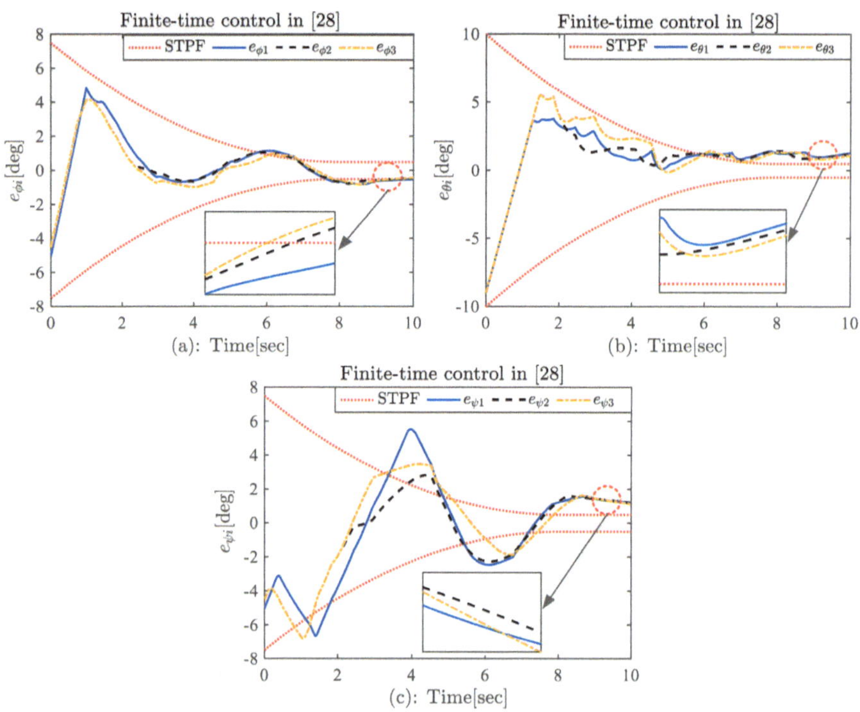

Fig. 5.6 Comparison results under finite-time controller in [28]. **(a)**–**(c)** show the curves of attitude synchronization errors. **(a)** e_{ϕ_i}. **(b)** e_{θ_i}. **(c)** e_{ψ_i}

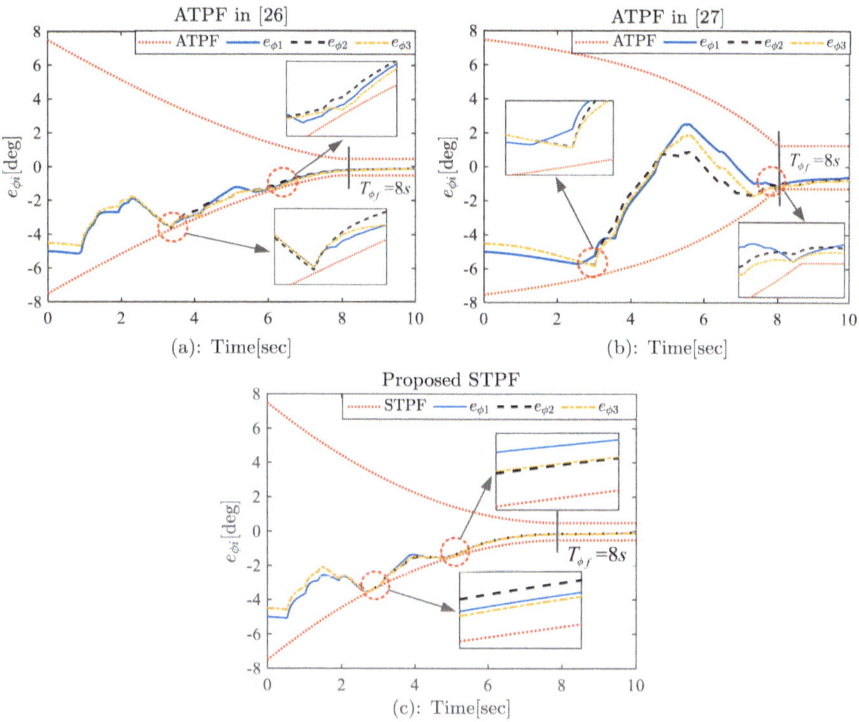

Fig. 5.7 Comparison results under different specified-time performance functions (taking ϕ_i tracking for example). (**a**)–(**c**) show the curves of ϕ_i synchronization errors. (**a**) Synchronization error of ϕ_i under ATPF [26]. (**b**) Synchronization error of ϕ_i under ATPF [27]. (**c**) Synchronization error of ϕ_i under proposed STPF

Further, to show the advantages of our proposed STPF, two other appointed-time performance functions (ATPFs) are evaluated in this comparative simulation: (i) ATPF of $\alpha(t)$ designed by Yin et al. [26], where choose $\alpha_0 = 15$, $\alpha_T = 1$, $\gamma = 0.3$, and $T = 8$. (ii) ATPF of $\eta_{\sigma i}(t)$ devised by Liu et al. [27], where choose $a_{\sigma i,0} = 15$, $a_{\sigma i,1} = -0.5$, $a_{\sigma i,2} = -0.06$, $a_{\sigma i,3} = -0.001$, $a_{\sigma i,4} = -0.001$, $\eta_{\sigma i,s} = 1$, and $t_f = 8$. To provide a fair result for the compared [26, 27] and proposed STPF, select the same control parameters, and conduct the flight tests in one sortie. Taking the ϕ_i tracking within $T_{\phi f} = 8$ s for example, Fig. 5.7 shows the comprehensive synchronization errors under the compared ATPFs and proposed STPF. It is observed that in contrast with the compared ATPFs [26, 27], the synchronization error responses with proposed STPF are faster and smoother and can converge to a smaller region around origin. To visually evaluate the tracking performance specification under compared ATPFs and proposed STPF, the performance indices of the maximum error fluctuations, offset errors, and integral of absolute errors are listed in Table 5.1, which validates the effectiveness of our theoretical findings.

Table 5.1 Transient and steady-state performance comparisons under ATPFs and our proposed STPF

	Controller	Maximum error fluctuations	Offset error	Integral of absolute error
e_{ϕ_1}	**Proposed STPF**	**5.0731**	**−0.0836**	**15.9140**
	Compared ATPF in [26]	5.1368	−0.0928	18.4815
	Compared ATPF in [27]	5.7149	−0.6007	26.1281
e_{ϕ_2}	**Proposed STPF**	**4.5671**	**−0.0797**	**15.2617**
	Compared ATPF in [26]	4.6450	−0.0886	17.6886
	Compared ATPF in [27]	5.7806	−0.7081	23.8607
e_{ϕ_3}	**Proposed STPF**	**4.5671**	**−0.0796**	**15.2327**
	Compared ATPF in [26]	4.6450	−0.0896	18.0372
	Compared ATPF in [27]	5.8161	−0.7689	24.9152

5.5 Conclusions

This work investigated the specified-time fault-tolerant coordinated control problem for multiple 6-DoF UAVs subject to external disturbances, modeling uncertainties and actuator faults. Novel STPFs were developed, which is capable of guaranteeing the tracking of velocity and attitude along the prespecified performance bounds in a user-specified time, as well as the boundedness of all closed-loop signals. Moreover, the proposed control scheme was not only robust against external disturbances and modeling uncertainties but also able to accommodate the actuator faults. These results are obtained with the help of the adaptive estimation methods and appropriate smooth functions. Experiment and simulation results have demonstrated the validation and superiority of proposed control scheme. Currently, it is supposed that the communication among UAVs is ideal with no consideration of packet dropouts and time delay. However, in practical flight, the complex interferences may be encountered by the communication network which leads to the packet dropouts and time delay of data transmission. Future works should include the extended cases involving packet dropouts [41] and time delay [42] in the proposed control design.

References

1. S. Shao, M. Chen, Adaptive neural discrete-time fractional-order control for a UAV system with prescribed performance using disturbance observer. IEEE Trans. Syst. Man Cybern. **51**(2), 742–754 (2021)

2. Y. Zhang, S. Wang, B. Chang, W. Wu, Adaptive constrained backstepping controller with prescribed performance methodology for carrier-based UAV. Aerosp. Sci. Technol. **92**, 55–65 (2019)
3. J. Chen, R. Sun, B. Zhu, Disturbance observer-based control for small nonlinear UAV systems with transient performance constraint. Aerosp. Sci. Technol. **105**, 106028 (2020)
4. A. Das, S. Shirazipourazad, D. Hay, A. Sen, Tracking of multiple targets using optimal number of UAVs. IEEE Trans. Aerosp. Electron. Syst. **55**(4), 1769–1784 (2019)
5. Y. Chen, D. Chang, C. Zhang, Autonomous tracking using a swarm of UAVs: a constrained multi-agent reinforcement learning approach. IEEE Trans. Veh. Technol. **69**(11), 13702–13717 (2020)
6. N. Farmani, L. Sun, D.J. Pack, A scalable multitarget tracking system for cooperative unmanned aerial vehicles. IEEE Trans. Aerosp. Electron. Syst. **53**(4), 1947–1961 (2017)
7. M. Lv, W. Yu, J. Cao, S. Baldi, Consensus in high-power multi-agent systems with mixed unknown control directions via hybrid Nussbaum-based control. IEEE Trans. Cybern. **52**(6), 5184–5196 (2020)
8. Z. Yu, Y. Zhang, Z. Liu, Y. Qu, C.Y. Su, Distributed adaptive fractional-order fault tolerant cooperative control of networked unmanned aerial vehicles via fuzzy neural networks. IET Control Theory **13**(17), 2917–2929 (2019)
9. Z. Yu, Y. Qu, Y. Zhang, Distributed fault-tolerant cooperative control for multi-UAVs under actuator fault and input saturation. IEEE Trans. Control Syst. Technol. **27**(6), 2417–2429 (2019)
10. Y. Wang, T. Zhang, Z. Cai, J. Zhao, K. Wu, Multi-UAV coordination control by chaotic grey wolf optimization based distributed MPC with event-triggered strategy. Chin. J. Aeronaut. **33**(11), 2877–2897 (2020)
11. T.Z. Muslimov, R.A. Munasypov, Adaptive decentralized flocking control of multi-UAV circular formations based on vector fields and backstepping. ISA Trans. **107**, 143–159 (2020)
12. Z. Yu, Y. Zhang, Z. Liu, Y. Qu, C.Y. Su, B. Jiang, Decentralized finite-time adaptive fault-tolerant synchronization tracking control for multiple UAVs with prescribed performance. J. Frankl. Inst. **357**(16), 11830–11862 (2020)
13. C.P. Bechlioulis, G.A. Rovithakis, Robust adaptive control of feedback linearizable MIMO nonlinear systems with prescribed performance. IEEE Trans. Autom. Control **53**(9), 2090–2099 (2008)
14. C.P. Bechlioulis, G.A. Rovithakis, Prescribed performance adaptive control for multi-input multi-output affine in the control nonlinear systems. IEEE Trans. Autom. Control **55**(5), 1220–1226 (2010)
15. I.S. Dimanidis, C.P. Bechlioulis, G.A. Rovithakis, Output feedback approximation-free prescribed performance tracking control for uncertain MIMO nonlinear systems. IEEE Trans. Autom. Control **65**(12), 5058–5069 (2020)
16. N. Koksal, H. An, B. Fidan, Backstepping-based adaptive control of a quadrotor UAV with guaranteed tracking performance. ISA Trans. **105**, 98–110 (2020)
17. W. Sun, Y. Wu, Z. Sun, Command filter-based finite-time adaptive fuzzy control for uncertain nonlinear systems with prescribed performance. IEEE Trans. Fuzzy Syst. **28**(12), 3161–3170 (2020)
18. Z. Li, H. Ji, Finite-time consensus and tracking control of a class of nonlinear multiagent systems. IEEE Trans. Autom. Control **63**(12), 4413–4420 (2018)
19. S. Dai, K. Lu, X. Jin, Fixed-Time formation control of unicycle-type mobile robots with visibility and performance constraints. IEEE Trans. Ind. Electron. **68**(12), 12615–12625 (2021)
20. J. Ke, W. Huang, J. Wang, J. Zeng, Fixed-time consensus control for multi-agent systems with prescribed performance under matched and mismatched disturbances. ISA Trans. **119**, 135–151 (2022)
21. Y. Liu, Y. Zhao, W. Ren, G. Chen, Appointed-time consensus: Accurate and practical designs. Automatica **89**, 425–429 (2018)
22. Y. Zhao, Y. Zhou, Y. Liu, G. Wen, P. Huang, Fixed-time bipartite synchronization with a pre-appointed settling time over directed cooperative-antagonistic networks. Automatica **123**, 109301 (2021)

23. Y. Zhao, Y. Liu, G. Wen, W. Ren, G. Chen, Designing distributed specified-time consensus protocols for linear multiagent systems over directed graph. IEEE Trans. Autom. Control **64**(7), 2945–2952 (2019)
24. C. Wei, J. Luo, Z. Yin, J. Yuan, Leader-following consensus of second-order multi-agent systems with arbitrarily appointed-time prescribed performance. IET Control Theory **104**, 105939 (2020)
25. X. Shao, Y. Shi, W. Zhang, Fault-tolerant quantized control for flexible air-breathing hypersonic vehicles with appointed-time tracking performances. IEEE Trans. Aerosp. Electron. Syst. **57**(2), 1261–1273 (2021)
26. Z. Yin, A. Suleman, J. Luo, C. Wei, Appointed-time prescribed performance attitude tracking control via double performance functions. Aerosp. Sci. Technol. **93**, 105337 (2019)
27. M. Liu, X. Shao, G. Ma, Appointed-time fault-tolerant attitude tracking control of spacecraft with double-level guaranteed performance bounds. Aerosp. Sci. Technol. **92**, 337–346 (2019)
28. Z. Yu, Z. Liu, Y. Zhang, Y. Qu, C. Su, Distributed finite-time fault-tolerant containment control for multiple unmanned aerial vehicle. IEEE Trans. Neural Networks Learn. Syst. **31**(6), 2077–2091 (2020)
29. Z. Yu, Y. Zhang, Z. Liu, Y. Qu, C. Su, B. Jiang, Decentralized finite-time adaptive fault-tolerant synchronization tracking control for multiple UAVs with prescribed performance. J. Frankl. Inst. **357**(16), 11830–11862 (2020)
30. Z. Yu, Y. Zhang, B. Jiang, C.Y. Su, J. Fu, Y. Jin, T. Chai, Nussbaum-based finite-time fractional-order backstepping fault-tolerant flight control of fixed-wing UAV against input saturation with hardware-in-the-loop validation. Mech. Syst. Signal Process. **153**, 107406 (2021)
31. T.Z. Muslimov, R.A. Munasypov, Consensus-based cooperative control of parallel fixed-wing UAV formations via adaptive backstepping. Aerosp. Sci. Technol. **109**, 106416 (2021)
32. H. Chen, X. Wang, L. Shen, R.A. Munasypov, Formation flight of fixed-wing UAV swarms: A group-based hierarchical approach. Chin. J. Aeronaut. **134**(2), 504–515 (2021)
33. Z. Tang, R. Cunha, T. Hamel, C. Silvestre, Aircraft landing using dynamic two-dimensional image-based guidance control. IEEE Trans. Aerosp. Electron. Syst. **55**(5), 2104–2117 (2019)
34. O. Espen, S. Rune, K. Raymond, Underactuated waypoint tracking of a fixed-wing UAV. IFAC Proceedings Volumes **46**(30), 126–133 (2015)
35. B. Zhang, X. Sun, S. Liu, M. Lv, X. Deng, Event-triggered adaptive fault-tolerant synchronization tracking control for multiple 6-DOF fixed-wing UAVs. IEEE Trans. Veh. Technol. **71**(1), 148–161 (2021)
36. H. Yang, B. Jiang, H. Yang, H.H. Liu, Synchronization of multiple 3-DoF helicopters under actuator faults and saturations with prescribed performance. ISA Trans. **75**, 118–126 (2018)
37. M. Lv, W. Yu, J. Cao, S. Baldi, A separation-based methodology to consensus tracking of switched high-order nonlinear multi-agent systems. IEEE Trans. Neural Networks Learn. Syst. **33**(10), 5467–5479 (2021)
38. C. Wang, Y. Lin, Decentralized adaptive tracking control for a class of interconnected nonlinear time-varying systems. Automatica **54**, 16–24 (2015)
39. S.S. Ge, C. Wang, Adaptive NN control of uncertain nonlinear pure-feedback systems. Automatica **38**(4), 671–682 (2002)
40. B. Zhang, X. Sun, M. Lv, S. Liu, L, Li, Distributed adaptive fixed-time fault-tolerant control for multiple 6-DOF UAVs with full-state constraints guarantee. IEEE Syst. J. **16**(3), 4792–4803 (2021)
41. D. Zhang, Q.G. Wang, L. Yu, Q.K. Shao, H_∞ filtering for networked systems with multiple time-varying transmissions and random packet dropouts. IEEE Trans. Industr. Inform. **9**(3), 1705–1716 (2013)
42. C. Lin, Q.G. Wang, T.H. Lee, Y. He, B. Chen, Observer-based H_∞ control for T-S fuzzy systems with time-delay: delay-dependent design method. IEEE Trans. Syst. Man Cybern. **37**(4), 1030–1038 (2007)

Part III
Event-Triggered and Communication-Efficient Control

Chapter 6
Event-Triggered Fault-Tolerant Control

Chapters 3, 4, and 5 study the cooperative control problems of multiple UAVs with fixed-time and preset-time convergence, meeting the requirements of rapid convergence for velocity and attitude tracking. In addition to convergence time, factors affecting UAV control performance also include the constraints of actual UAV systems, among which the limitation on actuator response frequency is one of the challenges in studying UAV system constraints. Typically, UAV control systems adopt a time-driven strategy, calculating and implementing control based on fixed-period updates. However, when the state of the UAV system approaches stability, high-frequency controller updates can lead to frequent actuator responses, resulting in unnecessary actuator wear. Therefore, this chapter focuses on how to reduce the frequency of controller updates while ensuring the control performance of multiple UAVs.

6.1 Introduction

Research on cooperative flight of multiple fixed-wing unmanned aerial vehicles (UAVs) has attracted increasing attention recently in lots of applications due to the advantages such as mission cooperation, flexibility, and cost-effectiveness [1, 2]. In general, the coordinated control problem has been investigated in several aspects, including the task assignment and motion re-coordination for UAVs in a group [3] and synchronous tracking of velocity and attitude for UAV formation [4] (that is to achieve $V_l \rightarrow V_r$, $V_l \rightarrow V_j$, $(\phi_l, \theta_l, \psi_l) \rightarrow (\phi_r, \theta_r, \psi_r)$, and $(\phi_l, \theta_l, \psi_l) \rightarrow (\phi_j, \theta_j, \psi_j)$). In particular, the distributed control strategy has particularly been attached tremendous attentions due to the fact that global information is no longer required in control design while preserving closed-loop stability [5, 6]. Driven by the emerging distributed cooperative control research on multiple UAVs, some results have appeared in [7–10]. To list a few, Yu et al. [7] presented a distributed

M. Lv et al., *Unmanned Aerial Vehicles Formation Control and Safety Guarantee*, Cognitive Intelligence and Robotics, https://doi.org/10.1007/978-981-96-5829-9_6

cooperative control design for networked UAVs against actuator faults and model uncertainties while maintaining the synchronization of flight attitude. In [8], for longitudinal synchronization tracking of multiple UAVs, a distributed cooperative fault-tolerant controller was proposed in the presence of input saturation. Based on swarm intelligence, Wang et al. [9] proposed a distributed model predictive approach for coordination control of multiple UAVs. In [10], with path-following vector fields, the UAV group achieves a circular motion around the target while rendering the course angles and velocities to converge to the vector field-specified values.

However, it is worth mentioning that all aforementioned methods [7–10] are performed based on the assumption that either the UAV dynamics can be simplified to be two-degree models [9, 10], or only attitude/longitudinal models are taken into account [7, 8]. Such a model simplification unavoidably lowers the system description precision. In other words, the dynamics in [7–10] cannot accurately describe the actual UAV system due to the fact that the fixed-wing UAV system is actually a nonlinear dynamic model with six-degree-of-freedom (6-DoF) with 12 state variables, which can be decomposed into rotational and translational dynamics, respectively. Therefore, the motivation of investigation on 6-DoF UAV dynamics naturally arises.

Moreover, note that the cooperative flight in aforementioned studies is achieved within continuous-time framework. For a multi-UAV system, which is actually a networked control system consisting of various components such as sensors, controllers, and actuators, the limited bandwidth of signal transmission is a major concern that affects system performance [5, 6]. Especially in cooperative flight of UAVs, lots of signal transmissions are required. Therefore, the traditional periodical updating of control signals is likely to cause transmission channel congestion. In addition, when implemented within the framework of continuous-time control, the control updating is executed at every fixed sampling period no matter whether it is necessary or not, which would cause overloading of transmission burden in some cases. Above phenomena will be more apparent as the number of UAVs grows. As a result, the periodic sampled controls are inappropriate for multi-UAV system. The main concern is the control design with event-triggered mechanism, which functions as an important solution to reducing the signal transmission frequency.

The idea of event-triggered control is firstly presented in [11]. Different from traditional periodic or time-triggered control, the signal transmission between different modules relies on a state-, input-, or output-dependent triggering condition, which significantly reduces the transmission burden. This technique has been successfully applied to UAV systems. Based on a distributed model predictive control scheme, the event-triggered mechanism was developed in [9, 12, 13] for cooperative path following of multi-UAVs. In [14], an event-triggered output feedback control method was presented for UAV system over cognitive radio networks. Further, Tian et al. [15] proposed an event-trigger-based attitude control for UAV system against external disturbances. Although effective performances are obtained in aforementioned results [9, 12–15] with event-triggered mechanism, two problems are worth mentioning: The first is that the event-triggered controls were proposed only for single UAV in [14, 15]. The second problem is that, even if the multi-UAVs were considered in [9, 12, 13], only a two-degree model was discussed, being far from the actual UAV system. Considering the 6-DoF UAVs, the complex dynamics makes the event-triggered control design more challenging. Therefore, the novel event-triggered control scheme for 6-DoF multi-UAV system should be further exploited.

In formation flight of multiple UAVs, actuator faults not only have adverse impacts on the safety of the individual UAV but also induce the instability of the overall networked UAVs [16, 17]. To avoid the catastrophes induced by the actuator faults, the fault-tolerant cooperative control (FTCC) is progressively investigated for multiple UAVs in recent years, and the individual FTCC is most investigated to compensate the fault effects for the faulty UAV. In studies [8, 18], the FTCC schemes were developed for multi-UAVs in the longitudinal plane within a distributed control design. Studies [7, 19–22] further investigated the FTCC for attitude dynamics of multi-UAVs in the presence of actuator faults. However, it should be noted that the FTCC methods presented in [7, 8, 18–22] were developed to only regulate the longitudinal motions and attitude dynamics of multi-UAVs, with no consideration of 6-DOF dynamics. Regarding the 6-DoF multi-UAV system, the strongly nonlinear dynamics of UAV makes the distributed FTCC design more complex, and the aforementioned results cannot be directly extended to the synchronization tracking of velocity and attitude for multiple 6-DoF UAVs. Therefore, more new and effective FTCC scheme should be further developed.

Motivated by above discussions, the main contribution of our chapter is threefold below:

- In comparison with the existing results that are concentrated on two-degree model [9, 10], longitudinal motion [8, 18], and attitude dynamics [7, 19–21], this study considers the distributed cooperative control for multi-UAVs where 6-DoF UAV dynamics is considered subject to actuator faults, modeling uncertainties, and external disturbances.
- Compared with the event-triggered control schemes that regulate on the single UAV [14, 15] or the two-degree model of multi-UAVs [9, 12, 13], this chapter develops the distributed event-triggered control for 6-DoF multi-UAVs, where the event-triggering mechanisms are introduced in translational and rotational subsystems, respectively, for each UAV, which considerably reduces the signal transmission load.
- Unlike most of the studies dealing with the control of UAV flight [3, 9–12, 25, 26], adaptive backstepping method is used together with the neural network approximation, bound estimation approach, and well-defined smooth functions to accurately estimate and compensate the actuator faults, external disturbances, and model uncertainties in the same time with the obtaining of global stability.

The rest of chapter is organized as follows. In Sect. 6.2, the UAV dynamics, actuator fault model, graph theory, and control objective are presented. Section 6.3 gives the details of event-triggered adaptive control design and stability analysis. Section 6.4 shows the simulation results, and Sect. 6.5 draws the conclusions.

Remark The main control schemes and contents of this chapter are from the published journal paper "B. Y. Zhang, X. X. Sun, S. G. Liu, M. L. Lv, Event-triggered adaptive fault-tolerant synchronization tracking control for multiple 6-DOF fixed-wing UAVs. IEEE Transactions on Vehicular Technology. 2021, 71(1): 148–161." The authors appreciate the permission from Institute of Electrical and Electronics Engineers to reuse the results published in the relevant journal.

6.2 Problem Formulation

6.2.1 UAV Dynamic Model

Consider a team of N identical fixed-wing UAVs. Using Newton-Euler convention, the full 6-DoF dynamics of the ith UAV is given by Espen et al. [25], Castaneda et al. [26]

$$\dot{\boldsymbol{p}}_i = \boldsymbol{R}_1\left(\boldsymbol{\varphi}_i\right)\boldsymbol{v}_i, \tag{6.1}$$

$$\dot{\boldsymbol{v}}_i = -\boldsymbol{S}\left(\boldsymbol{\omega}_i\right)\boldsymbol{v}_i + \frac{\boldsymbol{T}_i}{m_i} + \boldsymbol{R}_1^{\mathrm{T}}\left(\boldsymbol{\varphi}_i\right)\boldsymbol{g} + \frac{\boldsymbol{F}_i}{m_i} + \boldsymbol{d}_{vi}, \tag{6.2}$$

$$\dot{\boldsymbol{\varphi}}_i = \boldsymbol{R}_2^{-1}\left(\boldsymbol{\varphi}_i\right)\boldsymbol{\omega}_i, \tag{6.3}$$

$$\boldsymbol{J}_i\dot{\boldsymbol{\omega}}_i = -\boldsymbol{\omega}_i \times \boldsymbol{J}_i\boldsymbol{\omega}_i + \boldsymbol{N}_i + \boldsymbol{C}_i\boldsymbol{\delta}_i + \boldsymbol{J}_i\boldsymbol{d}_{\omega i}, \tag{6.4}$$

where $\boldsymbol{p}_i = [x_i, y_i, z_i]^{\mathrm{T}}$ denotes the inertial position, $\boldsymbol{v}_i = [u_i, v_i, w_i]^{\mathrm{T}}$ is the non-inertial (body fixed frame coordinates) expression of the linear velocity, $\boldsymbol{\varphi}_i = [\phi_i, \theta_i, \psi_i]^{\mathrm{T}}$ represents the attitude described by Euler angles, and $\boldsymbol{\omega}_i = [p_i, q_i, r_i]^{\mathrm{T}}$ is the non-inertial expression of the angular velocity. Moreover, $\boldsymbol{T}_i = [T_{xi}, 0, 0]^{\mathrm{T}}$ denotes the thrust vector along x body axis, $\boldsymbol{\delta}_i = [\delta_{ai}, \delta_{ei}, \delta_{ri}]^{\mathrm{T}}$ is the control surface, where $\delta_{ei}, \delta_{ai}, \delta_{ri}$ represent the deflection of elevator, aileron, and rudder, respectively, m_i is the vehicle mass, $\boldsymbol{g} = [0, 0, g_z]^{\mathrm{T}}$ is the gravity acceleration in inertia frame, $\boldsymbol{d}_{vi} \in \Re^3$ and $\boldsymbol{d}_{\omega i} \in \Re^3$ are the unknown bounded external disturbances, and $\boldsymbol{J}_i$ is the inertia tensor with the symmetric $x - z$ plane which is expressed as

$$\boldsymbol{J}_i = \begin{bmatrix} J_{xxi} & 0 & -J_{xzi} \\ 0 & J_{yyi} & 0 \\ -J_{zxi} & 0 & J_{zzi} \end{bmatrix}. \tag{6.5}$$

Besides, $\boldsymbol{R}_1\left(\boldsymbol{\varphi}_i\right) \in \mathrm{SO}(3)$ is the rotation matrix transforming the body frame coordinates to inertial axis coordinates, and the matrix $\boldsymbol{R}_2\left(\boldsymbol{\varphi}_i\right) \in \Re^{3\times 3}$ maps the time derivative of the Euler angles to the non-inertial expression of the angular velocity. Both matrices are given as follows:

$$R_1(\varphi_i) = \begin{bmatrix} c_{\psi i}c_{\theta i} & -s_{\psi i}s_{\phi i}+c_{\psi i}s_{\theta i}s_{\phi i} & s_{\psi i}s_{\phi i}+c_{\psi i}s_{\theta i}c_{\phi i} \\ s_{\psi i}c_{\theta i} & c_{\psi i}c_{\phi i}+s_{\psi i}s_{\theta i}s_{\phi i} & -c_{\psi i}s_{\phi i}+s_{\psi i}s_{\theta i}c_{\phi i} \\ -s_{\theta i} & c_{\theta i}s_{\phi i} & c_{\theta i}c_{\phi i} \end{bmatrix}$$

$$R_2(\varphi_i) = \begin{bmatrix} 1 & 0 & -s_{\psi i} \\ 0 & c_{\psi i} & s_{\psi i}c_{\theta i} \\ 0 & -s_{\theta i} & c_{\psi i}c_{\theta i} \end{bmatrix}, \tag{6.6}$$

where s_a and c_a stand for $\sin(a)$ and $\cos(a)$ functions, respectively. The aerodynamic force $F_i = [F_{Xi}, F_{Yi}, F_{Zi}]^{\mathrm{T}}$ and moment $N_i = [F_{Li}, F_{Mi}, F_{Ni}]^{\mathrm{T}}$ are calculated by means of aerodynamic coefficients:

$$F_i = \bar{q}_i S_i R_3^{-1}(\alpha_i, \beta_i)[-C_{Di}, C_{Yi}, -C_{Li}]^{\mathrm{T}}, \tag{6.7}$$

$$N_i = \bar{q}_i S_i \left[b_i C'_{li}, \bar{c}_i C'_{Mi}, b_i C'_{ni}\right]^{\mathrm{T}}, \tag{6.8}$$

where $\alpha_i = \arctan(w_i/u_i)$ and $\beta_i = \arcsin(u_i/v_i)$ denote the attack angle and sideslip angle, respectively. The dynamic pressure $\bar{q}_i = \frac{1}{2}\rho V_i^2$ is a function of V_i, which is the total velocity in body fixed frame. Air density ρ , wingspan b_i, wing surface area S_i, and mean aerodynamic chord $\bar{c}_i$ are constant parameters. The transformation matrix $R_3(\alpha_i, \beta_i)$ is

$$R_3(\alpha_i, \beta_i) = \begin{bmatrix} c_{\alpha i}c_{\beta i} & s_{\beta i} & s_{\alpha i}c_{\beta i} \\ -c_{\alpha i}s_{\beta i} & c_{\beta i} & -s_{\alpha i}s_{\beta i} \\ -s_{\alpha i} & 0 & c_{\alpha i} \end{bmatrix}. \tag{6.9}$$

And C_{Di}, C_{Yi}, C_{Li}, C'_{li}, C'_{Mi}, and C'_{ni} are the dimensionless coefficients in force/moment expressions, of which the corresponding detailed descriptions are shown in [25]. The control effectiveness matrix is defined as

$$C_i = \begin{bmatrix} \bar{q}_i S_i b_i c_{l\delta ai} & 0 & \bar{q}_i S_i b_i c_{l\delta ri} \\ 0 & \bar{q}_i S_i \bar{c}_i c_{m\delta ei} & 0 \\ \bar{q}_i S_i b_i c_{n\delta ai} & 0 & \bar{q}_i S_i b_i c_{n\delta ri} \end{bmatrix}. \tag{6.10}$$

6.2.2 Control-Oriented Model

According to studies [25, 26], the whole system can be decomposed into two interconnected subsystems, i.e., translational kinematics (6.2) and rotational kinematics

(6.3)–(6.4). Then the controller for each subsystem is designed independently for each UAV.

From (6.2), the time derivative of V_i is

$$\dot{V}_i = \frac{\boldsymbol{v}_i^{\mathrm{T}} \dot{\boldsymbol{v}}_i}{V_i} = \frac{\boldsymbol{v}_i^{\mathrm{T}}}{V_i} \left(-\boldsymbol{S}\left(\boldsymbol{w}_i\right) \boldsymbol{v}_i + \frac{\boldsymbol{T}_i}{m_i} + \boldsymbol{R}_1^{\mathrm{T}}\left(\boldsymbol{\varphi}_i\right) \boldsymbol{g} + \frac{\boldsymbol{F}_i}{m_i} + \boldsymbol{d}_{vi} \right). \tag{6.11}$$

Based on (6.11), the thrust is extracted, and using the property $\boldsymbol{v}_i^{\mathrm{T}} \boldsymbol{S}\left(\boldsymbol{w}_i\right) \boldsymbol{v}_i = 0$, it holds that

$$\dot{V}_i = \frac{u_i T_{xi}}{m_i V_i} + \frac{\boldsymbol{v}_i^{\mathrm{T}}}{V_i} \left(\boldsymbol{R}_1^{\mathrm{T}}\left(\boldsymbol{\varphi}_i\right) \boldsymbol{g} + \frac{\boldsymbol{F}_i}{m_i} + \boldsymbol{d}_{vi} \right). \tag{6.12}$$

Remark 6.1 From (6.12), even if there are no direct controls for velocity components v_i and w_i, it is still controllable to point $\boldsymbol{v}_i$ in a desired direction by controlling V_i, α_i, and β_i.

During operation, the thrust possibly suffers from actuator faults, modeled by [22]

$$T_{xi} = \rho_{Ti} T_{xi0} + T_{xif}, \tag{6.13}$$

where T_{xi0} is the designed control input, ρ_{Ti} is the unknown actuator efficiency factor satisfying $0 \leq \rho_{Ti} \leq 1$, and T_{xif} is the bounded unknown stuck fault or bias fault. Note that (6.13) implies the following four cases:

(i) $\rho_{Ti} = 1$ and $T_{xif} = 0$. This means the fault-free case.
(ii) $0 < \rho_{Ti} < 1$ and $T_{xif} = 0$. This indicates the partial loss of effectiveness.
(iii) $\rho_{Ti} = 1$ and $T_{xif} \neq 0$. This indicates the bias fault.
(iv) $\rho_{Ti} = 0$ and $T_{xif} \neq 0$. This means that T_{xi} is stuck at bounded time-varying function T_{xif}.

Then with actuator faults (6.13), the dynamics of (6.12) is described as

$$\dot{V}_i = \frac{u_i \rho_{Ti} T_{xi0}}{m_i V_i} + \frac{u_i T_{xif}}{m_i V_i} + \frac{\boldsymbol{v}_i^{\mathrm{T}}}{V_i} \left(\boldsymbol{R}_1^{\mathrm{T}}\left(\boldsymbol{\varphi}_i\right) \boldsymbol{g} + \frac{\boldsymbol{F}_i}{m_i} + \boldsymbol{d}_{vi} \right). \tag{6.14}$$

Note that the accurate information of $\boldsymbol{F}_i$ cannot be known a prior due to the coefficient uncertainties [27, 28]. In this sense, $\boldsymbol{F}_i$ is decomposed into a known component $\boldsymbol{F}_{i0}$ and an uncertain one $\Delta \boldsymbol{F}_i$. Therefore, the velocity kinematics is reformulated as

$$\dot{V}_i = \frac{u_i \rho_{Ti} T_{xi0}}{m_i V_i} + \frac{u_i T_{xif}}{m_i V_i} + \frac{\boldsymbol{v}_i^{\mathrm{T}}}{V_i} \left(\boldsymbol{R}_1^{\mathrm{T}}\left(\boldsymbol{\varphi}_i\right) \boldsymbol{g} + \frac{\boldsymbol{F}_{i0}}{m_i} + \boldsymbol{\Delta}_{vi} + \boldsymbol{d}_{vi} \right), \tag{6.15}$$

where $\boldsymbol{\Delta}_{vi} = \Delta \boldsymbol{F}_i / m_i$.

From (6.3)–(6.4), the dynamical attitude motion in inertial reference frame is rewritten as

$$\dot{\varphi}_i = R_2^{-1}(\varphi_i)\,\omega_i, \tag{6.16}$$

$$\dot{\omega}_i = J_i^{-1}S(J_i\omega_i) + J_i^{-1}N_i + J_i^{-1}C_i\delta_i + d_{\omega i}. \tag{6.17}$$

Considering the coefficient uncertainties [29], $J_i^{-1}S(J_i\omega_i)$ is decomposed into a known component $J_{i0}^{-1}S(J_{i0}\omega_i)$ and an uncertain one $\Delta J_i^{-1}S(J_i\omega_i)$. Similarly, $J_i^{-1}N_i$ is composed of a known part $J_{i0}^{-1}N_{i0}$ and an uncertain one $\Delta J_i^{-1}N_i$. Also, $J_i^{-1}C_i$ consists of a known component $J_{i0}^{-1}C_{i0}$ and an uncertain one $\Delta J_i^{-1}C_i$. Then the angular velocity model (6.17) is reformulated as

$$\begin{aligned}\dot{\omega}_i = {} & J_{i0}^{-1}S(J_{i0}\omega_i)+\Delta J_i^{-1}S(J_i\omega_i)+J_{i0}^{-1}N_{i0}+\Delta J_i^{-1}N_i \\ & +\left(J_{i0}^{-1}C_{i0}+\Delta J_i^{-1}C_i\right)\delta_i + d_{\omega i}.\end{aligned} \tag{6.18}$$

During flight, the flight control surfaces inevitably suffer from actuator faults, which are considered as [7]

$$\delta_{ji} = \rho_{\delta ji}\delta_{ji0} + \delta_{jif}, \tag{6.19}$$

where δ_{ji0} is the jth applied control input with $j = \{a, e, r\}$, $\rho_{\delta ji}$ is the unknown actuator efficiency factor satisfying $0 \le \rho_{\delta ji} \le 1$, and δ_{jif} is the bounded unknown stuck fault or bias fault. Consider that (6.19) satisfies the same cases as (6.13) in the following:

(i) $\rho_{\delta ji} = 1$ and $\delta_{jif} = 0$. This refers to the fault-free case.
(ii) $0 < \rho_{\delta ji} < 1$ and $\delta_{jif} = 0$. This means the partial loss of effectiveness.
(iii) $\rho_{\delta ji} = 1$ and $\delta_{jif} \neq 0$. This indicates the bias fault.
(iv) $\rho_{\delta ji} = 0$ and $\delta_{jif} \neq 0$. This means that δ_{ji} is stuck at bounded time-varying function δ_{jif}.

For the simplicity of presentation, the actuator fault model (6.19) is formulated by

$$\delta_i = \rho_{\delta i}\delta_{i0} + \delta_{if}, \tag{6.20}$$

where $\delta_{i0} = [\delta_{ai0}, \delta_{ei0}, \delta_{ri0}]^{\mathrm{T}}$, $\rho_{\delta i} = \mathrm{diag}\{\rho_{\delta ai}, \rho_{\delta ei}, \rho_{\delta ri}\}$, $\delta_{if} = \left[\delta_{aif}, \delta_{eif}, \delta_{rif}\right]^{\mathrm{T}}$.

In the presence of actuator faults, the dynamic model of angular velocity can be formulated by

$$\dot{\omega}_i = J_{i0}^{-1}S(J_{i0}\omega_i)+J_{i0}^{-1}N_{i0}+J_{i0}^{-1}C_{i0}\rho_{\delta i}\delta_{i0}+\Delta_{\omega i}+d_{\omega i}, \tag{6.21}$$

where $\Delta_{\omega i} = \Delta J_i^{-1}S(J_i\omega_i) + \Delta J_i^{-1}N_i + \Delta J_i^{-1}C_i\delta_i + J_{i0}^{-1}C_0(\delta_i)\,\delta_{if}$ is the lumped uncertainties induced by the actuator faults and modeling uncertainties.

Assumption 6.1 ([30]) For all $t > 0$, it is supposed that $\|\Delta \boldsymbol{J}_l^{-1} \boldsymbol{C}_l \boldsymbol{\rho}_{\delta l} \left(\boldsymbol{J}_{l0}^{-1} \boldsymbol{C}_{l0}\right)^{-1} \|_\infty < 1$ and $\partial \boldsymbol{\Delta}_{\omega l} / \partial \boldsymbol{\delta}_{l0} + \boldsymbol{J}_{l0}^{-1} \boldsymbol{C}_{l0} \boldsymbol{\rho}_{\delta l} \neq 0$.

Assumption 6.2 ([31]) There exists a positive constant $g_{l,\min}$ such that

$$\lambda_{\min}\left(\boldsymbol{J}_{l0}^{-1} \boldsymbol{C}_{l0} \boldsymbol{\rho}_{\delta l} \left(\boldsymbol{J}_{l0}^{-1} \boldsymbol{C}_{l0}\right)^{\mathrm{T}}\right) \geq g_{l,\min}.$$

Remark 6.2 Assumptions 6.1–6.2 substantially guarantee the controllability of angular velocity dynamics (6.21), which implies that there still exists sufficient control action $\boldsymbol{\delta}_{l0}$ even in the presence of actuator faults for each UAV. Note that $g_{l,\min}$ is only used for analysis and not required a priori.

Remark 6.3 Considering that $\boldsymbol{\Delta}_{\omega l}$ involves the control signal $\boldsymbol{\delta}_{l0}$, if neural networks are used to approximate $\boldsymbol{\Delta}_{\omega l}$, the algebraic loops inevitably exist. To break the algebraic loops, the low-pass filter technique can be employed to filter $\boldsymbol{\delta}_{l0}$ [32]. Then $\boldsymbol{\Delta}_{\omega l} = \boldsymbol{\Delta}'_{\omega l} + \boldsymbol{l}_{\omega l}$, where $\boldsymbol{l}_{\omega l} \in \Re^{3\times 1}$ is the filtering error. At this point, the angular velocity kinematics can be described by

$$\begin{aligned}\dot{\boldsymbol{\omega}}_l = &\boldsymbol{J}_{l0}^{-1} \boldsymbol{S}\left(\boldsymbol{J}_{l0} \boldsymbol{\omega}_l\right) + \boldsymbol{J}_{l0}^{-1} \boldsymbol{N}_{l0} + \boldsymbol{J}_{l0}^{-1} \boldsymbol{C}_{i0} \boldsymbol{\rho}_{\delta i} \boldsymbol{\delta}_{l0} \\ &+ \boldsymbol{\Delta}'_{\omega l} + \boldsymbol{l}_{\omega l} + \boldsymbol{d}_{\omega l}.\end{aligned} \tag{6.22}$$

6.2.3 *Neural Network Approximation*

As the universal approximation property, the RBFNNs are employed in this chapter to compensate the modeling uncertainties [19, 21, 33]. Denote $\boldsymbol{\Theta}_{vl} = \left[\boldsymbol{v}_l^{\mathrm{T}}, \boldsymbol{\varphi}_l^{\mathrm{T}}, \boldsymbol{\omega}_l^{\mathrm{T}}, \boldsymbol{\delta}_{l0}^{\mathrm{T}}\right]^{\mathrm{T}} \in \Re^{12}$ and $\boldsymbol{\Theta}_{\omega l} = \left[\boldsymbol{v}_l^{\mathrm{T}}, \boldsymbol{\varphi}_l^{\mathrm{T}}, \boldsymbol{\omega}_l^{\mathrm{T}}, \boldsymbol{\delta}_{l0}^{\mathrm{T}}\right]^{\mathrm{T}} \in \Re^{12}$. For unknown continuous terms $\boldsymbol{\Delta}_{vl}\left(\boldsymbol{\Theta}_{vl}\right) = \left[\Delta_{vl1}, \Delta_{vl2}, \Delta_{vi3}\right]^{\mathrm{T}}$ and $\boldsymbol{\Delta}'_{\omega l}\left(\boldsymbol{\Theta}_{\omega l}\right) = \left[\Delta'_{\omega l1}, \Delta'_{\omega l2}, \Delta'_{\omega l3}\right]^{\mathrm{T}}$, there exist RBFNNs such that

$$\begin{aligned}\Delta_{vlj}\left(\boldsymbol{\Theta}_{vl}\right) &= \boldsymbol{W}_{vlj}^{*\mathrm{T}} \boldsymbol{\varsigma}_{vlj}\left(\boldsymbol{\Theta}_{vl}\right) + \tau_{vlj}\left(\boldsymbol{\Theta}_{vi}\right), \\ \Delta'_{\omega lj}\left(\boldsymbol{\Theta}_{\omega l}\right) &= \boldsymbol{W}_{\omega lj}^{*\mathrm{T}} \boldsymbol{\varsigma}_{\omega lj}\left(\boldsymbol{\Theta}_{\omega l}\right) + \tau_{\omega lj}\left(\boldsymbol{\Theta}_{\omega i}\right),\end{aligned} \tag{6.23}$$

where $j = 1, 2, 3$, and $\tau_{vlj}\left(\boldsymbol{\Theta}_{vl}\right) \in \Re$ and $\tau_{\omega lj}\left(\boldsymbol{\Theta}_{\omega i}\right) \in \Re$ are the bounded approximation errors. $\boldsymbol{W}_{vlj}^{*} \in \Re^{M\times 1}$ and $\boldsymbol{W}_{\omega lj}^{*} \in \Re^{M\times 1}$ are idealized weight matrixes with M nodes. $\boldsymbol{\varsigma}_{vlj}\left(\boldsymbol{\Theta}_{vl}\right) = \left[\varsigma_{vlj}^{1}\left(\boldsymbol{\Theta}_{vl}\right), \ldots, \varsigma_{vlj}^{M}\left(\boldsymbol{\Theta}_{vl}\right)\right]^{\mathrm{T}} \in \Re^{M}$ and $\boldsymbol{\varsigma}_{\omega lj}\left(\boldsymbol{\Theta}_{\omega l}\right) = \left[\varsigma_{\omega lj}^{1}\left(\boldsymbol{\Theta}_{\omega l}\right), \ldots, \varsigma_{\omega lj}^{M}\left(\boldsymbol{\Theta}_{\omega l}\right)\right]^{\mathrm{T}} \in \Re^{M}$ are known vectors consisted of Gaussian basis functions $\varsigma_{vlj}^{l}\left(\boldsymbol{\Theta}_{vl}\right)$ and $\varsigma_{\omega lj}^{l}\left(\boldsymbol{\Theta}_{\omega l}\right)$, $l = 1, \ldots, M$, commonly selected as the following exponential form:

$$\varsigma_{vij}^{l}(\boldsymbol{\Theta}_{vi}) = \exp\left[-\frac{(\boldsymbol{\Theta}_{vi}-\boldsymbol{\Upsilon}_{vi})^{\mathrm{T}}(\boldsymbol{\Theta}_{vi}-\boldsymbol{\Upsilon}_{vi})}{\kappa_{vi}^{2}}\right],$$

$$\varsigma_{\omega ij}^{l}(\boldsymbol{\Theta}_{\omega i}) = \exp\left[-\frac{(\boldsymbol{\Theta}_{\omega i}-\boldsymbol{\Upsilon}_{\omega i})^{\mathrm{T}}(\boldsymbol{\Theta}_{\omega i}-\boldsymbol{\Upsilon}_{\omega i})}{\kappa_{\omega i}^{2}}\right], \tag{6.24}$$

where $\boldsymbol{\Upsilon}_{vi} \in \Re^{12}$ and $\boldsymbol{\Upsilon}_{\omega i} \in \Re^{12}$ are the centers of the receptive fields, and κ_{vi} and $\kappa_{\omega i}$ are the width of the Gaussian basis functions.

The undirected graph always describes the connections among the multi-UAV system. Given an undirected graph $\mathcal{G} = (\mathcal{V}, \mathcal{E}, \mathcal{A})$, it consists of nodes $\mathcal{V} = \{v_1, \cdots, v_n\}$ and the sets of edges $\mathcal{E} = \{(i, j), \ i, j \in \mathcal{V}, \text{and } i \neq j\}$, and $\mathcal{A} = [a_{ij}] \in \Re^{n\times n}$ denotes the weighted adjacency matrix of $\mathcal{G}$. If there exists an edge between nodes i and j, then $a_{ij} = a_{ji} \neq 0$, and otherwise $a_{ij} = a_{ji} = 0$. Moreover, a_{ii} equals 0 for all $i \in \mathcal{V}$. The neighbors of node i is denoted by $\mathcal{N}_i = \{v_j : (v_j, v_i) \in \mathcal{E}\}$. The Laplacian matrix is defined by $\mathcal{L} = \mathcal{D} - \mathcal{A}$, where $\mathcal{D} = \mathrm{diag}\{d_1, \cdots, d_n\}$ with $d_i = \sum_{j=1}^{n} a_{ij}$. $\mathcal{G}$ is connected if there is a path from every node to each other node.

For a leader-following multi-UAV system, the augmented graph $\bar{\mathcal{G}} \in (\bar{\mathcal{V}}, \bar{\mathcal{E}}, \bar{\mathcal{A}})$ for n UAVs including the leader is concerned. Let $\mathcal{H} = \mathrm{diag}\{h_1, \cdots, h_n\}$ denote the leader adjacency matrix of the graph $\bar{\mathcal{G}}$. If the ith follower receives the information from the leader UAV, then $h_i > 0$ and otherwise $h_i = 0$.

Assumption 6.3 The topology of considered multi-UAV system is an undirected graph with at least one node having access to the leading node.

Lemma 6.1 ([34]) *For an undirected connected graph with the Laplacian matrix $\mathcal{L}$, if at least one $h_i > 0$, then symmetric matrix $\mathcal{L} + \mathcal{H}$ associated with $\bar{\mathcal{G}}$ is positive definite.*

Remark 6.4 Assumption 6.3 and Lemma 6.1 preserve that the followers can receive the necessary information from the leader.

6.2.4 Control Objective

Given the reference commands V_r and φ_r of V_i and φ_i, respectively, the goal is to design the cooperative distributed event-triggered tracking controller such that:

- All the closed-loop signals remain bounded, and the synchronous velocity and attitude tracking errors of each UAV are SGUUB in the presence of actuator faults, modeling uncertainties, and external disturbances.
- The signal transmission burden is considerably reduced for multi-UAV system. Also, the Zeno phenomenon is precluded such that the time between two consecutive switching is lower bounded by a positive constant.

To achieve the control objective, the following assumption and lemmas are instrumental to stability analysis.

Assumption 6.4 The desired references V_r, φ_r and their time derivatives up to second order $\ddot{V}_r$, $\ddot{\varphi}_r$ are continuous and bounded.

Lemma 6.2 ([35]) *For any scalars* $\varepsilon > 0$, $z \in \Re$, *the following equation holds:* $0 \le |z| - \frac{z^2}{\sqrt{z^2+\varepsilon}} \le \sqrt{\varepsilon}$.

Lemma 6.3 ([36]) *For any scalars* $\varepsilon > 0$, $\vartheta \in \Re$, $0 \le |\vartheta| - \vartheta \tanh\left(\frac{\vartheta}{\varepsilon}\right) \le 0.2785\varepsilon$ *holds.*

6.3 Main Results

6.3.1 Event-Triggered Control Mechanism

We are now at the position to provide the following triggering condition for translational and rotational kinematics, respectively.

The event-triggered control mechanism for velocity tracking is presented as

$$\begin{aligned} T_{xi0}(t) &= \hat{T}_{xi0}\left(t_k^v\right), \quad \forall t \in \left[t_k^v, t_{k+1}^v\right) \\ t_{k+1} &= \inf\{t > t_k^v \,|\, |\hat{T}_{xi0}(t) - \hat{T}_{xi0}\left(t_k^v\right)| \ge a_1 |\hat{T}_{xi0}\left(t_k^v\right)| + a_2\}, \end{aligned} \tag{6.25}$$

where $k = 0, 1, 2, \ldots$, $t_0 := 0$, $\hat{T}_{xi0}(t)$ is calculated by designed controller and will be presented in the following section, and $0 < a_1 < 1$, $a_2 > 0$ are design parameters.

In view of (6.25), it follows that $|\hat{T}_{xi0}(t) - \hat{T}_{xi0}\left(t_k^v\right)| \le a_1 |\hat{T}_{xi0}\left(t_k^v\right)| + a_2$ and $T_{xi0}(t) = \hat{T}_{xi0}\left(t_k^v\right)$ for all $t \in \left[t_k^v, t_{k+1}^v\right)$. Since $t_0 := 0$, these results hold for each $k = 1, 2, 3 \ldots$, and there exists

$$|\hat{T}_{xi0}(t) - T_{xi0}(t)| \le a_1 |T_{xi0}(t)| + a_2, \quad \forall t \ge 0. \tag{6.26}$$

Define $p_1(t) = \frac{\hat{T}_{xi0}(t) - T_{xi0}(t)}{a_1 |T_{xi0}(t)| + a_2}$. It follows that $|p_1(t)| \le 1$. This fact implies the following inequality for all time:

$$\hat{T}_{xi0}(t) - T_{xi0}(t) = a_1 p_1(t) T_{xi0}(t) + a_2 p_1(t). \tag{6.27}$$

Using (6.27) yields

$$T_{xi0} = \frac{\hat{T}_{xi0}}{1 + a_1 p_1} - \frac{a_2 p_1}{1 + a_1 p_1}. \tag{6.28}$$

Then the event-triggered control mechanism for attitude tracking gives

$$\begin{aligned}\boldsymbol{\delta}_{i0}(t) =& \hat{\boldsymbol{\delta}}_{i0}\left(t_k^{\varphi}\right), \quad \forall t \in \left[t_k^{\varphi}, t_{k+1}^{\varphi}\right), \\ t_{k+1}^{\varphi} =& \inf\{t > t_k^{\varphi} \mid |\hat{\delta}_{ai0}(t) - \hat{\delta}_{ai0}\left(t_k^{\varphi}\right)| \geq b_1|\hat{\delta}_{ai0}\left(t_k^{\varphi}\right)| + b_2 \\ & \text{or}|\hat{\delta}_{ei0}(t) - \hat{\delta}_{ei0}\left(t_k^{\varphi}\right)| \geq b_1|\hat{\delta}_{ei0}\left(t_k^{\varphi}\right)| + b_2 \\ & \text{or}|\hat{\delta}_{ri0}(t) - \hat{\delta}_{ri0}\left(t_k^{\varphi}\right)| \geq b_1|\hat{\delta}_{ri0}\left(t_k^{\varphi}\right)| + b_2\},\end{aligned} \tag{6.29}$$

where $\hat{\boldsymbol{\delta}}_{i0}(t) = [\hat{\delta}_{ai0}(t), \hat{\delta}_{ei0}(t), \hat{\delta}_{ri0}(t)]^{\mathrm{T}}$ is produced by designed controller and will be described in the following section, and $0 < b_1 < 1$, $b_2 > 0$ are design parameters.

Similarly, we define $p_2(t) = \frac{\hat{\delta}_{ai0}(t)-\delta_{ai0}(t)}{b_1|\delta_{ai0}(t)|+b_2}$, $p_3(t) = \frac{\hat{\delta}_{ei0}(t)-\delta_{ei0}(t)}{b_1|\delta_{ei0}(t)|+b_2}$, $p_4(t) = \frac{\hat{\delta}_{ri0}(t)-\delta_{ri0}(t)}{b_1|\delta_{ri0}(t)|+b_2}$. Then we arrive at

$$\boldsymbol{\delta}_{i0} = \boldsymbol{D}_i\hat{\boldsymbol{\delta}}_{i0} + \boldsymbol{\sigma}_i, \tag{6.30}$$

where $\boldsymbol{D}_i = \operatorname{diag}\{D_{i1}, D_{i2}, D_{i3}\}$ and $\boldsymbol{\sigma}_i = [\sigma_{i1}, \sigma_{i2}, \sigma_{i3}]^{\mathrm{T}}$ with

$$D_{ij} = \frac{1}{1 + b_1 p_{j+1}\operatorname{sign}\left(\delta_{i0j}\right)}, \sigma_{ij} = \frac{-b_2 p_{j+1}}{1 + b_1 p_{j+1}\operatorname{sign}\left(\delta_{i0j}\right)}.$$

Remark 6.5 Differently from existing works [25, 26], the event-triggered mechanisms are first introduced in thrust T_{xi0} and control surface $\boldsymbol{\delta}_{i0}$ to obtain $\hat{T}_{xi0}$ and $\hat{\boldsymbol{\delta}}_{i0}$. This operation ensures that the time instants of updating control signals are determined by event-triggering mechanism instead of traditional periodical updating, which significantly reduces the signal transmission frequency, while saving on system resources.

Following the double-layer control structure, the translational and rotational control designs are proposed independently. Specifically, $\hat{T}_{xi0}$ and $\hat{\boldsymbol{\delta}}_{i0}$ designs are given. To facilitate readers' comprehension, the overall block diagram of the proposed control scheme is presented in Fig. 6.1

6.3.2 *Event-Triggered Adaptive Controller Design for Velocity Subsystem*

Based on velocity tracking errors $\tilde{V}_i = V_i - V_r$ and $\tilde{V}_j = V_j - V_r$, the objective of guaranteeing velocity synchronization tracking is equivalent to make $\tilde{V}_i' \rightarrow 0$.

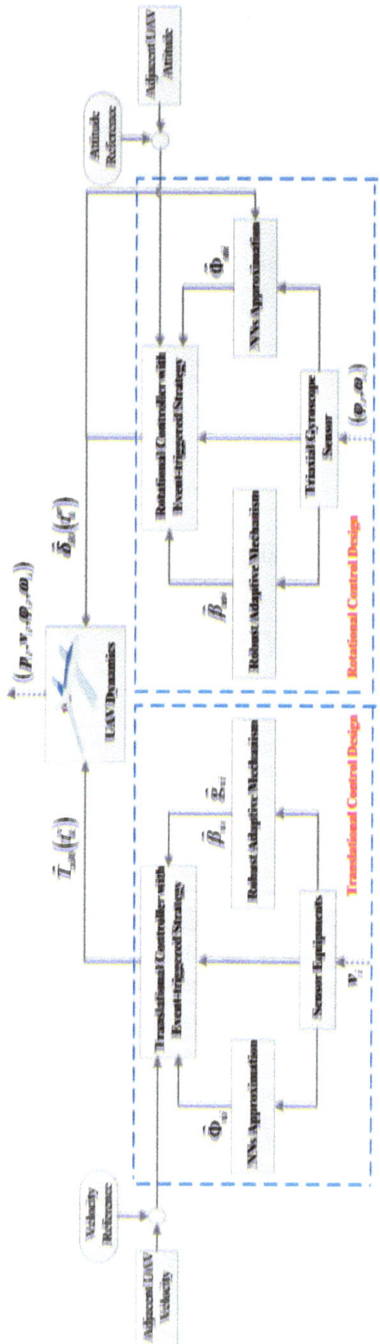

Fig. 6.1 Overall framework of the proposed control strategy

$$\tilde{V}_l' = \lambda_{l1}\tilde{V}_l + \lambda_{l2}\sum_{j\in\mathcal{N}_l} a_{lj}(\tilde{V}_l - \tilde{V}_j), \tag{6.31}$$

where λ_{l1} and λ_{l2} are positive design parameters.

We employ the following dynamic controller $\hat{T}_{xl0}$:

$$\hat{T}_{xl0} = \frac{-(1+a_1)\,B_{vl}}{\zeta_{vl}}, \tag{6.32}$$

where $\zeta_{vl} = \frac{u_l\lambda_l}{m_l V_l}$ with $\lambda_l = \lambda_{l1} + \lambda_{l2}\sum_{j\in\mathcal{N}_l} a_{lj}$, and B_{vl} is defined as

$$\begin{aligned} B_{vl} =& (\alpha_{vl} - r_{vl})\hat{\beta}_{vl}\tanh\left(\frac{\tilde{V}_l'(\alpha_{vl} - r_{vl})\hat{\beta}_{vl}}{\varepsilon_{vl}}\right) \\ &+ \hat{g}_{vl}\zeta_{vl}\tanh\left(\frac{\tilde{V}_l'\hat{g}_{vl}\zeta_{vl}}{\varepsilon_{vl}}\right), \end{aligned}$$

where $\alpha_{vl} = c_{vl}\tilde{V}_l' + \lambda_l(\frac{\boldsymbol{v}_l^{\mathrm{T}}}{V_l}(\boldsymbol{R}_1^{\mathrm{T}}\boldsymbol{g} + \frac{\boldsymbol{F}_{l0}}{m_l}) + \frac{\hat{\Phi}_{vl}\boldsymbol{v}_l^{\mathrm{T}}\boldsymbol{\Xi}_{vl}}{2h_{vl1}^2 V_l} + \frac{\boldsymbol{v}_l^{\mathrm{T}}\boldsymbol{v}_l\tilde{V}_l'}{2h_{vl2}^2 V_l^2})$ with $\boldsymbol{\Xi}_{vl} = [\frac{\tilde{V}_l' v_{l1}\boldsymbol{S}_{vl1}^{\mathrm{T}}\boldsymbol{S}_{vl1}}{V_l}, \frac{\tilde{V}_l' v_{l2}\boldsymbol{S}_{vl2}^{\mathrm{T}}\boldsymbol{S}_{vl2}}{V_l}, \frac{\tilde{V}_l' v_{l3}\boldsymbol{S}_{vl3}^{\mathrm{T}}\boldsymbol{S}_{vl3}}{V_l}]^{\mathrm{T}}$ and $r_{vl} = \lambda_l\dot{V}_r + \lambda_{l2}\sum_{j\in\mathcal{N}_l} a_{lj}\dot{\tilde{V}}_j$. The adaptive parameters $\hat{\Phi}_{vl}$, $\hat{g}_{vl}$, and $\hat{\beta}_{vl}$ are updated by

$$\dot{\hat{\Phi}}_{vl} = \lambda_l\frac{\gamma_{vl1}\tilde{V}_l'\boldsymbol{v}_l^{\mathrm{T}}\boldsymbol{\Xi}_{vl}}{2h_{vl1}^2 V_l} - \sigma_{vl1}\hat{\Phi}_{vl}, \tag{6.33}$$

$$\dot{\hat{g}}_{vl} = \gamma_{vl2}|\tilde{V}_l'|\zeta_{vl} - \sigma_{vl2}\hat{g}_{vl}, \tag{6.34}$$

$$\dot{\hat{\beta}}_{vl} = \gamma_{vl3}\tilde{V}_l'(\alpha_{vl} - r_{vl}) - \sigma_{vl3}\hat{\beta}_{vl}, \tag{6.35}$$

where $h_{vl1}, h_{vl2}, \gamma_{vl1}, \gamma_{vl2}, \gamma_{vl3}, \sigma_{vl1}, \sigma_{vl2}, \sigma_{vl3}, \varepsilon_{vl}$ are positive design parameters.

6.3.3 Event-Triggered Adaptive Controller Design for Attitude Subsystem

Similarly, the attitude synchronization tracking error is expressed as

$$\boldsymbol{e}_{\varphi l}' = \lambda_{l3}\boldsymbol{e}_{\varphi l} + \lambda_{l4}\sum_{j\in\mathcal{N}_l} a_{lj}\left(\boldsymbol{e}_{\varphi l} - \boldsymbol{e}_{\varphi j}\right), \tag{6.36}$$

where $\boldsymbol{e}_{\varphi l} = \boldsymbol{\varphi}_l - \boldsymbol{\varphi}_r$, $\boldsymbol{e}_{\varphi j} = \boldsymbol{\varphi}_j - \boldsymbol{\varphi}_r$, and λ_{l3}, λ_{l4} are positive design parameters.

We propose the dynamic controller $\hat{\boldsymbol{\delta}}_{i0}$ as follows:

$$\hat{\boldsymbol{\delta}}_{i0} = -\left(\boldsymbol{J}_{i0}^{-1}\boldsymbol{C}_{i0}\right)^{\mathrm{T}} \frac{\boldsymbol{e}_{\omega i}\hat{\beta}_{\omega i}^{2}\boldsymbol{\zeta}_{\omega i}^{\mathrm{T}}\boldsymbol{\zeta}_{\omega i}}{\sqrt{\hat{\beta}_{\omega i}^{2}\boldsymbol{e}_{\omega i}^{\mathrm{T}}\boldsymbol{e}_{\omega i}\boldsymbol{\zeta}_{\omega i}^{\mathrm{T}}\boldsymbol{\zeta}_{\omega i} + \varepsilon_{\omega i}}}, \tag{6.37}$$

where $\boldsymbol{\zeta}_{\omega i} = \boldsymbol{K}_{2i}\boldsymbol{e}_{\omega i} + \boldsymbol{J}_{i0}^{-1}\boldsymbol{S}(\boldsymbol{J}_{i0}\boldsymbol{\omega}_i) + \boldsymbol{J}_{i0}^{-1}\boldsymbol{N}_{i0} + \frac{\hat{\Phi}_{\omega i}\boldsymbol{\Xi}_{\omega i}}{2h_{\omega i1}^2} + \frac{\boldsymbol{e}_{\omega i}}{2h_{\omega i2}^2} + \lambda_i'(\boldsymbol{R}_2^{-1})^{\mathrm{T}}\boldsymbol{e}_{\varphi i}' - \dot{\boldsymbol{\omega}}_{di}$ with $\lambda_i' = \lambda_{i3} + \lambda_{i4}\sum_{j\in\mathcal{N}_i} a_{ij}$ and $\boldsymbol{\Xi}_{\omega i} = \left[e_{\omega i1}\boldsymbol{\varsigma}_{\omega i1}^{\mathrm{T}}\boldsymbol{\varsigma}_{\omega i1}, e_{\omega i2}\boldsymbol{\varsigma}_{\omega i2}^{\mathrm{T}}\boldsymbol{\varsigma}_{\omega i2}, e_{\omega i3}\boldsymbol{\varsigma}_{\omega i3}^{\mathrm{T}}\boldsymbol{\varsigma}_{\omega i3}\right]^{\mathrm{T}}$. The adaptive parameters $\hat{\Phi}_{\omega i}$ and $\hat{\beta}_{\omega i}$ are proposed as

$$\dot{\hat{\Phi}}_{\omega i} = \gamma_{\omega i1}\frac{\boldsymbol{e}_{\omega i}^{\mathrm{T}}\boldsymbol{\Xi}_{\omega i}}{2h_{\omega i1}^2} - \sigma_{\omega i1}\hat{\Phi}_{\omega i}, \tag{6.38}$$

$$\dot{\hat{\beta}}_{\omega i} = \gamma_{\omega i2}\boldsymbol{e}_{\omega i}^{\mathrm{T}}\boldsymbol{\zeta}_{\omega i} - \sigma_{\omega i2}\hat{\beta}_{\omega i}, \tag{6.39}$$

where $h_{\omega i1}, h_{\omega i2}, \gamma_{\omega i1}, \gamma_{\omega i2}, \sigma_{\omega i1}, \sigma_{\omega i2}, \varepsilon_{\omega i}$ are positive design parameters.

6.3.4 Stability Analysis

Theorem 6.1 *Consider the translational kinematics (6.15) composed of the adaptive tracking controller (6.32) and parameter adaptation laws (6.33)–(6.35). Let Assumptions 6.3–6.4 hold. There exist adjustable parameters $\lambda_{i1}, \lambda_{i2}, c_{vi}, \varepsilon_{vi}, h_{vi1}, h_{vi2}, \gamma_{vij}$, and σ_{vij} $(i = 1, 2, \ldots, N,\ j = 1, 2, 3)$ such that:*

(1) All signals of the translational subsystem are SGUUB in the presence of actuator faults, modeling uncertainties, and external disturbances.
(2) Velocity synchronization tracking error $\tilde{V}_i'$ is such that $\lim_{t\to\infty}|\tilde{V}_i'| \le \mu_{Vi}$ with $\mu_{Vi} > 0$ a constant.
(3) The signal transmission burden for translational subsystem is considerably reduced, and meanwhile the Zeno phenomenon is precluded.

Proof Noting (6.15) and (6.31), we obtain the derivative of $\tilde{V}_i'$ as

$$\begin{aligned}\dot{\tilde{V}}_i' = {} & \lambda_i\left[\frac{\boldsymbol{v}_i^{\mathrm{T}}}{V_i}\left(\boldsymbol{R}_1^{\mathrm{T}}\boldsymbol{g} + \frac{\boldsymbol{F}_{i0}}{m_i}\right) + \frac{\boldsymbol{v}_i^{\mathrm{T}}}{V_i}(\boldsymbol{\Delta}_{vi} + \boldsymbol{d}_{vi})\right] + \lambda_i\frac{u_i\rho_{Ti}T_{xi0}}{m_iV_i} \\ & + \lambda_i\frac{u_iT_{xif}}{m_iV_i} - \lambda_i\dot{V}_r - \lambda_{i2}\sum_{j\in\mathcal{N}_i} a_{ij}\dot{\tilde{V}}_j.\end{aligned} \tag{6.40}$$

Substituting (6.28) into (6.40) gives

$$\dot{\tilde{V}}_i' = \lambda_i \left[\frac{\boldsymbol{v}_i^{\mathrm{T}}}{V_i} \left(\boldsymbol{R}_1^{\mathrm{T}} \boldsymbol{g} + \frac{\boldsymbol{F}_{i0}}{m_i} \right) + \frac{\boldsymbol{v}_i^{\mathrm{T}}}{V_i} (\boldsymbol{\Delta}_{vi} + \boldsymbol{d}_{vi}) \right] + \frac{\zeta_{vi} \rho_{Ti} \hat{T}_{xi0}}{1 + a_1 p_1}$$
$$+ \zeta_{vi} \varpi_{vi} - r_{vi}, \tag{6.41}$$

where $\varpi_{vi} = T_{xif} - \frac{a_2 p_1 \rho_{Ti}}{1+a_1 p_1}$.

Integrating (6.23) leads to

$$\boldsymbol{v}_i^{\mathrm{T}} (\boldsymbol{\Delta}_{vi} + \boldsymbol{d}_{vi}) = \boldsymbol{v}_i^{\mathrm{T}} \begin{bmatrix} \boldsymbol{W}_{vi1}^{*\mathrm{T}} \boldsymbol{\varsigma}_{vi1} + \tau_{vi1} + d_{vi1} \\ \boldsymbol{W}_{vi2}^{*\mathrm{T}} \boldsymbol{\varsigma}_{vi2} + \tau_{vi2} + d_{vi2} \\ \boldsymbol{W}_{vi3}^{*\mathrm{T}} \boldsymbol{\varsigma}_{vi3} + \tau_{vi3} + d_{vi3} \end{bmatrix}$$
$$= \sum_{j=1}^{3} v_{ij} \boldsymbol{W}_{vij}^{*\mathrm{T}} \boldsymbol{\varsigma}_{vij} + \sum_{j=1}^{3} v_{ij} \left(\tau_{vij} + d_{vij} \right), \tag{6.42}$$

where $\boldsymbol{v}_i = [v_{i1}, v_{i2}, v_{i3}]^{\mathrm{T}} = [u_i, v_i, w_i]^{\mathrm{T}}$.

Using Young's inequality, we arrive at

$$\frac{\tilde{V}_i'}{V_i} v_{ij} \boldsymbol{W}_{vij}^{*\mathrm{T}} \boldsymbol{\varsigma}_{vij} \le \frac{\tilde{V}_i'^2 v_{ij}^2 \varPhi_{vi} \boldsymbol{\varsigma}_{vij}^{\mathrm{T}} \boldsymbol{\varsigma}_{vij}}{2 h_{vi1}^2 V_i^2} + \frac{h_{vi1}^2}{2},$$
$$\frac{\tilde{V}_i'}{V_i} v_{ij} \left(\tau_{vij} + d_{vij} \right) \le \frac{\tilde{V}_i'^2 v_{ij}^2}{2 h_{vi2}^2 V_i^2} + \frac{h_{vi2}^2 \bar{\tau}_{vij}^2}{2}, \tag{6.43}$$

where $\varPhi_{vi} = \max \left\{ \boldsymbol{W}_{vi1}^{*\mathrm{T}} \boldsymbol{W}_{vi1}^*, \boldsymbol{W}_{vi2}^{*\mathrm{T}} \boldsymbol{W}_{vi2}^*, \boldsymbol{W}_{vi3}^{*\mathrm{T}} \boldsymbol{W}_{vi3}^* \right\}$ and $\tau_{vij} + d_{vij}$ satisfies $\left| \tau_{vij} + d_{vij} \right| \le \bar{\tau}_{vij}$.

Substituting (6.43) into (6.42) yields

$$\frac{\tilde{V}_i'}{V_i} \boldsymbol{v}_i^{\mathrm{T}} (\boldsymbol{\Delta}_{vi} + \boldsymbol{d}_{vi}) \le \sum_{j=1}^{3} \left(\frac{\tilde{V}_i'^2 v_{ij}^2 \varPhi_{vi} \boldsymbol{\varsigma}_{vij}^{\mathrm{T}} \boldsymbol{\varsigma}_{vij}}{2 h_{vi1}^2 V_i^2} + \frac{\tilde{V}_i'^2 v_{ij}^2}{2 h_{vi2}^2 V_i^2} \right.$$
$$\left. + \frac{h_{vi1}^2}{2} + \frac{h_{vi2}^2 \bar{\tau}_{vij}^2}{2} \right). \tag{6.44}$$

Combining (6.41) and (6.44), it holds that

$$\tilde{V}_i' \dot{\tilde{V}}_i' \le \lambda_i \tilde{V}_i' \left[\frac{\boldsymbol{v}_i^{\mathrm{T}}}{V_i} \left(\boldsymbol{R}_1^{\mathrm{T}} \boldsymbol{g} + \frac{\boldsymbol{F}_{i0}}{m_i} \right) + \frac{\varPhi_{vi} \boldsymbol{v}_i^{\mathrm{T}} \boldsymbol{\Xi}_{vi}}{2 h_{vi1}^2 V_i} + \frac{\boldsymbol{v}_i^{\mathrm{T}} \boldsymbol{v}_i \tilde{V}_i'}{2 h_{vi2}^2 V_i^2} \right]$$
$$+ \tilde{V}_i' \left(\frac{\zeta_{vi} \rho_{Ti} \hat{T}_{xi0}}{1 + a_1 p_1} + \zeta_{vi} \varpi_{vi} \right) - \tilde{V}_i' r_{vi}$$

$$+\lambda_l \sum_{j=1}^{3} \left(\frac{h_{vl1}^2}{2} + \frac{h_{vl2}^2 \bar{\tau}_{vlj}^2}{2} \right). \tag{6.45}$$

Define $\beta_{vl} = \frac{1}{\rho_{Tl}}$, $g_{vl} = \beta_{vl} \sup_{t \geq 0} \varpi_{vl}$. Let $\tilde{\Phi}_{vl} = \hat{\Phi}_{vl} - \Phi_{vl}$, $\tilde{\beta}_{vl} = \hat{\beta}_{vl} - \beta_{vl}$, $\tilde{g}_{vl} = \hat{g}_{vl} - g_{vl}$, where $\hat{\Phi}_{vl}$, $\hat{\beta}_{vl}$, and $\hat{g}_{vl}$ are the estimations of Φ_{vl}, β_{vl}, and g_{vl}, respectively. Invoking above definitions yields

$$\tilde{V}_l' \zeta_{vl} \varpi_{vl} \leq \rho_{Tl} g_{vl} |\tilde{V}_l'| \zeta_{vl} = \rho_{Tl} \hat{g}_{vl} |\tilde{V}_l'| \zeta_{vl} - \rho_{Tl} \tilde{g}_{vl} |\tilde{V}_l'| \zeta_{vl}. \tag{6.46}$$

According to Lemma 6.3, $\rho_{Tl} \hat{g}_{vl} |\tilde{V}_l'| \zeta_{vl}$ satisfies

$$\rho_{Tl} \hat{g}_{vl} |\tilde{V}_l'| \zeta_{vl} \leq \rho_{Tl} \left(\tilde{V}_l' \hat{g}_{vl} \zeta_{vl} \tanh \left(\frac{\tilde{V}_l' \hat{g}_{vl} \zeta_{vl}}{\varepsilon_{vl}} \right) + 0.2785 \varepsilon_{vl} \right). \tag{6.47}$$

Take the following Lyapunov function candidate:

$$L_1 = \frac{1}{2} \tilde{V}_l'^2 + \frac{1}{2\gamma_{vl1}} \tilde{\Phi}_{vl}^2 + \frac{\rho_{Tl}}{2\gamma_{vl2}} \tilde{g}_{vl}^2 + \frac{\rho_{Tl}}{2\gamma_{vl3}} \tilde{\beta}_{vl}^2. \tag{6.48}$$

Using (6.45)–(6.47) and taking the derivative of (6.48), we have

$$\begin{aligned}
\dot{L}_1 \leq\ & \tilde{V}_l' \left[\rho_{Tl} \hat{g}_{vl} \zeta_{vl} \tanh \left(\frac{\tilde{V}_l' \hat{g}_{vl} \zeta_{vl}}{\varepsilon_{vl}} \right) + \alpha_{vl} - r_{vl} \right] \\
& + \frac{\tilde{V}_l' \zeta_{vl} \rho_{Tl} \hat{T}_{xl0}}{1 + a_1 p_1} - \tilde{V}_l' \left(c_{vl} \tilde{V}_l' + \lambda_l \frac{\tilde{\Phi}_{vl} \boldsymbol{v}_l^{\mathrm{T}} \boldsymbol{\Xi}_{vl}}{2h_{vl1}^2 V_l} \right) \\
& + \frac{\rho_{Tl}}{\gamma_{vl2}} \tilde{g}_{vl} (\dot{\hat{g}}_{vl} - \gamma_{vl2} |\tilde{V}_l'| \zeta_{vl}) + \frac{1}{\gamma_{vl1}} \tilde{\Phi}_{vl} \dot{\hat{\Phi}}_{vl} \\
& + \frac{\rho_{Tl}}{\gamma_{vl3}} \tilde{\beta}_{vl} \dot{\hat{\beta}}_{vl} + 0.2785 \varepsilon_{vl} \rho_{Tl} \\
& + \lambda_l \sum_{j=1}^{3} \left(\frac{h_{vl1}^2}{2} + \frac{h_{vl2}^2 \bar{\tau}_{vlj}^2}{2} \right).
\end{aligned} \tag{6.49}$$

It follows from (6.32) that

$$\frac{\tilde{V}_l' \zeta_{vl} \rho_{Tl} \hat{T}_{xl0}}{1 + a_1 p_1} = \frac{-\rho_{Tl} (1 + a_1)}{1 + a_1 p_1} \tilde{V}_l' B_{vl}. \tag{6.50}$$

Bearing in mind that $-a\tanh(a/\vartheta) \leq 0$ with $\forall a \in \Re$ and $\vartheta > 0$, we obtain $-\tilde{V}_l' B_{vl} \leq 0$. Since $|p_1(t)| \leq 1$, one has $1 - a_1 \leq 1 + a_1 p_1 \leq 1 + a_1$. Invoking Lemma 6.3 yields

$$\begin{aligned}\frac{-\rho_{Tl}(1+a_1)}{1+a_1 p_1}\tilde{V}_l' B_{vl} &\leq -\rho_{Tl}\tilde{V}_l'\hat{g}_{vl}\zeta_{vl}\tanh\left(\frac{\tilde{V}_l'\hat{g}_{vl}\zeta_{vl}}{\varepsilon_{vl}}\right) \\ &= \rho_{Tl}\tilde{V}_l'(\alpha_{vl} - r_{vl})\hat{\beta}_{vl} + 0.2785\varepsilon_{vl}\rho_{Tl}. \end{aligned} \tag{6.51}$$

Noting $-\rho_{Tl}\tilde{V}_l'(\alpha_{vl} - r_{vl})\hat{\beta}_{vl} = -\tilde{V}_l'(\alpha_{vl} - r_{vl}) - \rho_{Tl}\tilde{V}_l'(\alpha_{vl} - r_{vl})\tilde{\beta}_{vl}$ and substituting (6.33)–(6.35) and (6.51) into (6.49) result in

$$\begin{aligned}\dot{L}_1 \leq &-c_{vl}\tilde{V}_l'^2 - \frac{\sigma_{vl1}}{\gamma_{vl1}}\tilde{\Phi}_{vl}\hat{\Phi}_{vl} - \frac{\rho_{Tl}\sigma_{vl2}}{\gamma_{vl2}}\tilde{g}_{vl}\hat{g}_{vl} - \frac{\rho_{Tl}\sigma_{vl3}}{\gamma_{vl3}}\tilde{\beta}_{vl}\hat{\beta}_{vl} \\ &+ 0.557\varepsilon_{vl}\rho_{Tl} + \lambda_l\sum_{j=1}^{3}\left(\frac{h_{vl1}^2}{2} + \frac{h_{vl2}^2\bar{\tau}_{vlj}^2}{2}\right). \end{aligned} \tag{6.52}$$

Using Young's inequality gives rise to $-\tilde{\Phi}_{vl}\hat{\Phi}_{vl} \leq -\frac{1}{2}\tilde{\Phi}_{vl}^2 + \frac{1}{2}\Phi_{vl}^2$, $-\tilde{g}_{vl}\hat{g}_{vl} \leq -\frac{1}{2}\tilde{g}_{vl}^2 + \frac{1}{2}g_{vl}^2$, $-\tilde{\beta}_{vl}\hat{\beta}_{vl} \leq -\frac{1}{2}\tilde{\beta}_{vl}^2 + \beta_{vl}^2$. Then (6.52) is formulated as

$$\begin{aligned}\dot{L}_1 \leq &-c_{vl}\tilde{V}_l'^2 - \frac{\sigma_{vl1}}{2\gamma_{vl1}}\tilde{\Phi}_{vl}^2 - \frac{\rho_{Tl}\sigma_{vl2}}{2\gamma_{vl2}}\tilde{g}_{vl}^2 - \frac{\rho_{Tl}\sigma_{vl3}}{2\gamma_{vl3}}\tilde{\beta}_{vl}^2 \\ &+ \frac{\sigma_{vl1}}{2\gamma_{vl1}}\Phi_{vl}^2 + \frac{\rho_{Tl}\sigma_{vl2}}{2\gamma_{vl2}}g_{vl}^2 + \frac{\rho_{Tl}\sigma_{vl3}}{2\gamma_{vl3}}\beta_{vl}^2 \\ &+ 0.557\varepsilon_{vl}\rho_{Tl} + \lambda_l\sum_{j=1}^{3}\left(\frac{h_{vl1}^2}{2} + \frac{h_{vl2}^2\bar{\tau}_{vlj}^2}{2}\right) \\ \leq &-\kappa_{vl}L_2 + \chi_{vl}, \end{aligned} \tag{6.53}$$

where $\kappa_{vl} = \min\{2c_{vl}, \sigma_{vl1}, \sigma_{vl2}, \sigma_{vl3}\}$ and $\chi_{vl} = \frac{\sigma_{vl1}}{2\gamma_{vl1}}\Phi_{vl}^2 + \frac{\rho_{Tl}\sigma_{vl2}}{2\gamma_{vl2}}g_{vl}^2 + \frac{\rho_{Tl}\sigma_{vl3}}{2\gamma_{vl3}}\beta_{vl}^2 + \lambda_l\sum_{j=1}^{3}(\frac{h_{vl1}^2}{2} + \frac{h_{vl2}^2\bar{\tau}_{vlj}^2}{2}) + 0.557\varepsilon_{vl}\rho_{Tl}$.

From (6.53), it follows that

$$0 \leq L_1 \leq \left[\frac{\chi_{vl}}{\kappa_{vl}} + L_1(0)\right]e^{-\kappa_{vl}t} + \frac{\chi_{vl}}{\kappa_{vl}}, \quad t \geq 0. \tag{6.54}$$

As a result, all the closed-loop signals in translational subsystem are SGUUB. It should be noticed that $\tilde{V}_l'$ can be made arbitrarily small by increasing c_{vl}, γ_{vl1}, γ_{vl2}, and γ_{vl3} and meanwhile decreasing h_{vl1}, h_{vl2}, σ_{vl1}, σ_{vl2}, σ_{vl3}, and ε_{vl}. Therefore, the velocity synchronization tracking error can be made arbitrarily small by appropriate choice of the design parameters. Moreover, the choice of a_1 and a_2 should balance the system performance and the number of triggering events. This will be further shown in the following numerical example.

Next, we prove that our proposed control scheme can avoid Zeno behavior. That is, it should ensure that there exists a $t^* > 0$ satisfying $\left\{t_{k+1}^v - t_k^v\right\} \geq t^*$. Since $e_{Ti} = \hat{T}_{xi0}(t) - T_{xi0}(t)$, $\forall t \in \left\{t_{k+1}^v - t_k^v\right\}$, there exists $\frac{d}{dt}|e_{Ti}| = \operatorname{sign}(e_{Ti})\, \dot{e}_{Ti} \leq |\dot{\hat{T}}_{xi0}|$. From (6.32), $\hat{T}_{xi0}$ is differentiable, and $\dot{\hat{T}}_{xi0}$ is a function of all closed-loop signals. Therefore, there exists a constant μ_{Ti} such that $|\dot{\hat{T}}_{xi0}(t)| \leq \mu_{Ti}$. Since $e_{Ti}\left(t_k^v\right) = 0$ and $\lim\limits_{t \to t_{k+1}^v} e_{Ti}(t) = a_2$, it can be seen that the lower bound of t^* satisfies $t^* \leq \mu_{Ti} / a_2$, which means that the Zeno behavior is successfully excluded. This completes the proof of Theorem 6.1.

Theorem 6.2 *Consider the rotational kinematics (6.16), (6.22) composed of the adaptive tracking controller (6.37) and parameter adaptation laws (6.38)–(6.39). Let Assumptions 6.1–6.4 hold. There exist adjustable parameters* λ_{i3}, λ_{i4}, $\boldsymbol{K}_{ji}$, $\varepsilon_{\omega i}$, $h_{\omega ij}$, $\gamma_{\omega ij}$, *and* $\sigma_{\omega ij}$ $(i = 1, 2, \ldots, N,\ j = 1, 2)$ *such that:*

(1) All signals of the rotational subsystem are SGUUB in the presence of actuator faults, modeling uncertainties, and external disturbances.

(2) Attitude synchronization tracking error $\boldsymbol{e}'_{\varphi i}$ *is such that* $\lim_{t \to \infty} \left\|\boldsymbol{e}'_{\varphi i}\right\| \leq \mu_{\varphi i}$ *with* $\mu_{\varphi i} > 0$ *a constant.*

(3) The signal transmission burden for rotational subsystem is considerably reduced, and meanwhile the Zeno phenomenon is avoided.

Proof *Step 1:* Choose the following Lyapunov function:

$$L_2 = \frac{1}{2} \boldsymbol{e}'^{\mathrm{T}}_{\varphi i} \boldsymbol{e}'_{\varphi i}. \tag{6.55}$$

Define $\boldsymbol{e}_{\omega i} = \boldsymbol{\omega}_i - \boldsymbol{\omega}_{di}$. Taking the derivative of (6.55) along the trajectories of (6.16) and (6.36) arrives at

$$\dot{L}_2 = \boldsymbol{e}'^{\mathrm{T}}_{\varphi i} \left[\lambda'_i \left(\boldsymbol{R}_2^{-1} \boldsymbol{\omega}_{di} + \boldsymbol{R}_2^{-1} \boldsymbol{e}_{\omega i} - \dot{\boldsymbol{\varphi}}_r \right) - \lambda_{i4} \sum_{j \in \mathcal{N}_i} a_{ij} \dot{\boldsymbol{e}}_{\varphi j} \right]. \tag{6.56}$$

We design the intermediate controller $\boldsymbol{\omega}_{di}$ as

$$\boldsymbol{\omega}_{di} = \frac{\boldsymbol{R}_2}{\lambda'_i} \left(-\boldsymbol{K}_{1i} \boldsymbol{e}'_{\varphi i} + \lambda'_i \dot{\boldsymbol{\varphi}}_r + \lambda_{i4} \sum_{j \in \mathcal{N}_i} a_{ij} \dot{\boldsymbol{e}}_{\varphi j} \right), \tag{6.57}$$

where $\boldsymbol{K}_{1i} \in \Re^{3 \times 3}$ is a symmetric positive definite matrix.

Substituting (6.57) into (6.56) leads to

$$\dot{L}_2 = -\boldsymbol{e}_{\varphi i}^{\prime \mathrm{T}} \boldsymbol{K}_{1i} \boldsymbol{e}_{\varphi i}^{\prime} + \lambda_i^{\prime} \boldsymbol{e}_{\omega i}^{\mathrm{T}} (\boldsymbol{R}_2^{-1})^{\mathrm{T}} \boldsymbol{e}_{\varphi i}^{\prime}. \tag{6.58}$$

Step 2: For the purpose of forcing $\boldsymbol{\omega}_i$ to track $\boldsymbol{\omega}_{di}$, from (6.22), it holds that

$$\begin{aligned}\boldsymbol{e}_{\omega i}^{\mathrm{T}} \dot{\boldsymbol{e}}_{\omega i} &= \boldsymbol{e}_{\omega i}^{\mathrm{T}} (\boldsymbol{J}_{i0}^{-1} \boldsymbol{S}(\boldsymbol{J}_{i0} \boldsymbol{\omega}_i) + \boldsymbol{J}_{i0}^{-1} \boldsymbol{N}_{i0} + \boldsymbol{J}_{i0}^{-1} \boldsymbol{C}_{i0} \boldsymbol{\rho}_{\delta i} \boldsymbol{D}_i \hat{\boldsymbol{\delta}}_{i0}) \\ &\quad + \boldsymbol{e}_{\omega i}^{\mathrm{T}} \left(\boldsymbol{\sigma}_i^{\prime} + \boldsymbol{\Delta}_{\omega i}^{\prime} + \boldsymbol{l}_{\omega i} + \boldsymbol{d}_{\omega i} - \dot{\boldsymbol{\omega}}_{di} \right),\end{aligned} \tag{6.59}$$

where $\boldsymbol{\sigma}_i^{\prime} = \boldsymbol{J}_{i0}^{-1} \boldsymbol{C}_0 \boldsymbol{\rho}_{\delta i} \boldsymbol{\sigma}_i$ is bounded.

Using (6.23) yields

$$\begin{aligned}\boldsymbol{e}_{\omega i}^{\mathrm{T}} \left(\boldsymbol{\Delta}_{\omega i}^{\prime} + \boldsymbol{\sigma}_i^{\prime} + \boldsymbol{l}_{\omega i} + \boldsymbol{d}_{\omega i} \right) &= \sum_{j=1}^{3} e_{\omega ij} \boldsymbol{W}_{\omega ij}^{*\mathrm{T}} \boldsymbol{\varsigma}_{\omega ij} + \sum_{j=1}^{3} e_{\omega ij} \cdot \\ &\quad (\tau_{\omega ij} + \sigma_{ij}^{\prime} + l_{\omega ij} + d_{\omega ij}).\end{aligned} \tag{6.60}$$

Using Young's inequality, one has

$$\begin{aligned} e_{\omega ij} \boldsymbol{W}_{\omega ij}^{*\mathrm{T}} \boldsymbol{\varsigma}_{\omega ij} &\le \frac{e_{\omega ij}^2 \Phi_{\omega i} \boldsymbol{\varsigma}_{\omega ij}^{\mathrm{T}} \boldsymbol{\varsigma}_{\omega ij}}{2h_{\omega i1}^2} + \frac{h_{\omega i1}^2}{2}, \\ e_{\omega ij} \left(\tau_{\omega ij} + \sigma_{ij}^{\prime} + l_{\omega ij} + d_{\omega ij} \right) &\le \frac{e_{\omega ij}^2}{2h_{\omega i2}^2} + \frac{h_{\omega i2}^2 \bar{\tau}_{\omega ij}^2}{2}, \end{aligned} \tag{6.61}$$

where $\Phi_{\omega i} = \max\left\{ \boldsymbol{W}_{\omega i1}^{*\mathrm{T}} \boldsymbol{W}_{\omega i1}^{*}, \boldsymbol{W}_{\omega i2}^{*\mathrm{T}} \boldsymbol{W}_{\omega i2}^{*}, \boldsymbol{W}_{\omega i3}^{*\mathrm{T}} \boldsymbol{W}_{\omega i3}^{*} \right\}$, and the term $\left(\tau_{\omega ij} + \sigma_{ij}^{\prime} + l_{\omega ij} + d_{\omega ij} \right)$ satisfies $\left| \tau_{\omega ij} + \sigma_{ij}^{\prime} + l_{\omega ij} + d_{\omega ij} \right| \le \bar{\tau}_{\omega ij}$.

Substituting (6.61) into (6.60) gives

$$\begin{aligned}\boldsymbol{e}_{\omega i}^{\mathrm{T}} \left(\boldsymbol{\Delta}_{\omega i}^{\prime} + \boldsymbol{\sigma}_i^{\prime} + \boldsymbol{l}_{\omega i} + \boldsymbol{d}_{\omega i} \right) &\le \boldsymbol{e}_{\omega i}^{\mathrm{T}} \left(\frac{\Phi_{\omega i} \boldsymbol{\Xi}_{\omega i}}{2h_{\omega i1}^2} + \frac{\boldsymbol{e}_{\omega i}}{2h_{\omega i2}^2} \right) \\ &\quad + \sum_{j=1}^{3} \left(\frac{h_{\omega i1}^2}{2} + \frac{h_{\omega i2}^2 \bar{\tau}_{\omega ij}^2}{2} \right).\end{aligned} \tag{6.62}$$

Since $\lambda_{\min}(\boldsymbol{J}_{i0}^{-1} \boldsymbol{C}_{i0} \boldsymbol{\rho}_{\delta i} (\boldsymbol{J}_{i0}^{-1} \boldsymbol{C}_{i0})^{\mathrm{T}}) \ge g_{i,\min}$ by Assumption 6.2, we arrive at

$$\lambda_{\min}(\boldsymbol{J}_{i0}^{-1} \boldsymbol{C}_{i0} \boldsymbol{\rho}_{\delta i} \boldsymbol{D}_i (\boldsymbol{J}_{i0}^{-1} \boldsymbol{C}_{i0})^{\mathrm{T}}) \ge \frac{g_{i,\min}}{1 + b_1} > 0. \tag{6.63}$$

Define $g_{\omega l} = \inf_{t\geq 0} \lambda_{\min}(\boldsymbol{J}_{l0}^{-1}\boldsymbol{C}_{l0}\boldsymbol{\rho}_{\delta l}\boldsymbol{D}_l(\boldsymbol{J}_{l0}^{-1}\boldsymbol{C}_{l0})^{\mathrm{T}})$, $\beta_{\omega i} = \frac{1}{g_{\omega i}}$. Let $\tilde{\Phi}_{\omega l} = \hat{\Phi}_{\omega l} - \Phi_{\omega l}$ and $\tilde{\beta}_{\omega l} = \hat{\beta}_{\omega l} - \beta_{\omega l}$ with $\hat{\Phi}_{\omega l}$ and $\hat{\beta}_{\omega l}$ being the estimations of $\Phi_{\omega l}$ and $\beta_{\omega l}$, respectively.

Consider the following quadratic function candidate:

$$L_3 = L_2 + \frac{1}{2}\boldsymbol{e}_{\omega l}^{\mathrm{T}}\boldsymbol{e}_{\omega l} + \frac{1}{2\gamma_{\omega l1}}\tilde{\Phi}_{\omega l}^2 + \frac{g_{\omega i}}{2\gamma_{\omega i2}}\tilde{\beta}_{\omega i}^2. \tag{6.64}$$

It follows from (6.58)–(6.59) and (6.62) that the time derivative of (6.64) denotes

$$\begin{aligned}
\dot{L}_3 = & -\boldsymbol{e}_{\varphi l}^{\prime\mathrm{T}}\boldsymbol{K}_{1l}\boldsymbol{e}_{\varphi l}^{\prime} - \boldsymbol{e}_{\omega l}^{\mathrm{T}}\boldsymbol{K}_{2l}\boldsymbol{e}_{\omega l} + \boldsymbol{e}_{\omega l}^{\mathrm{T}}\boldsymbol{\zeta}_{\omega l} - \boldsymbol{e}_{\omega i}^{\mathrm{T}}\frac{\tilde{\Phi}_{\omega l}\boldsymbol{\Xi}_{\omega l}}{2h_{\omega l1}^2} \\
& + \boldsymbol{e}_{\omega l}^{\mathrm{T}}\boldsymbol{J}_{l0}^{-1}\boldsymbol{C}_{l0}\boldsymbol{\rho}_{\delta l}\boldsymbol{D}_l\hat{\boldsymbol{\delta}}_{l0} + \sum_{j=1}^{3}\left(\frac{h_{\omega l1}^2}{2} + \frac{h_{\omega i2}^2\bar{\tau}_{\omega lj}^2}{2}\right) \\
& + \frac{1}{\gamma_{\omega l1}}\tilde{\Phi}_{\omega l}\dot{\hat{\Phi}}_{\omega l} + \frac{g_{\omega l}}{\gamma_{\omega l2}}\tilde{\beta}_{\omega l}\dot{\hat{\beta}}_{\omega l}.
\end{aligned} \tag{6.65}$$

Noting (6.40) and invoking Lemma 6.2, it follows that

$$\begin{aligned}
& \boldsymbol{e}_{\omega l}^{\mathrm{T}}\boldsymbol{J}_{l0}^{-1}\boldsymbol{C}_{l0}\boldsymbol{\rho}_{\delta l}\boldsymbol{D}_l\hat{\boldsymbol{\delta}}_{l0} = -\boldsymbol{e}_{\omega l}^{\mathrm{T}}\boldsymbol{J}_{l0}^{-1}\boldsymbol{C}_{l0}\boldsymbol{\rho}_{\delta l}\boldsymbol{D}_l\left(\boldsymbol{J}_{l0}^{-1}\boldsymbol{C}_{l0}\right)^{\mathrm{T}} \cdot \\
& \frac{\boldsymbol{e}_{\omega l}\hat{\beta}_{\omega l}^2\boldsymbol{\zeta}_{\omega l}^{\mathrm{T}}\boldsymbol{\zeta}_{\omega l}}{\sqrt{\hat{\beta}_{\omega l}^2\boldsymbol{e}_{\omega l}^{\mathrm{T}}\boldsymbol{e}_{\omega l}\boldsymbol{\zeta}_{\omega l}^{\mathrm{T}}\boldsymbol{\zeta}_{\omega l} + \varepsilon_{\omega l}}} \leq \frac{-g_{\omega l}\hat{\beta}_{\omega l}^2\boldsymbol{e}_{\omega l}^{\mathrm{T}}\boldsymbol{e}_{\omega l}\boldsymbol{\zeta}_{\omega l}^{\mathrm{T}}\boldsymbol{\zeta}_{\omega l}}{\sqrt{\hat{\beta}_{\omega l}^2\boldsymbol{e}_{\omega l}^{\mathrm{T}}\boldsymbol{e}_{\omega l}\boldsymbol{\zeta}_{\omega l}^{\mathrm{T}}\boldsymbol{\zeta}_{\omega l} + \varepsilon_{\omega l}}} \\
& \leq g_{\omega l}\sqrt{\varepsilon_{\omega l}} - g_{\omega l}|\hat{\beta}_{\omega l}|\,\|\boldsymbol{e}_{\omega l}\|\,\|\boldsymbol{\zeta}_{\omega l}\| \\
& \leq g_{\omega l}\sqrt{\varepsilon_{\omega l}} - g_{\omega l}\hat{\beta}_{\omega l}\boldsymbol{e}_{\omega l}^{\mathrm{T}}\boldsymbol{\zeta}_{\omega l}.
\end{aligned} \tag{6.66}$$

Substituting (6.38)–(6.39) and (6.66) into (6.65), it holds that

$$\begin{aligned}
\dot{L}_3 \leq & -\boldsymbol{e}_{\varphi l}^{\prime\mathrm{T}}\boldsymbol{K}_{1l}\boldsymbol{e}_{\varphi l}^{\prime} - \boldsymbol{e}_{\omega l}^{\mathrm{T}}\boldsymbol{K}_{2l}\boldsymbol{e}_{\omega l} + \boldsymbol{e}_{\omega l}^{\mathrm{T}}\boldsymbol{\zeta}_{\omega l} + g_{\omega l}\sqrt{\varepsilon_{\omega l}} \\
& - g_{\omega l}(\hat{\beta}_{\omega l} - \tilde{\beta}_{\omega l})\boldsymbol{e}_{\omega l}^{\mathrm{T}}\boldsymbol{\zeta}_{\omega l} + \sum_{j=1}^{3}\left(\frac{h_{\omega l1}^2}{2} + \frac{h_{\omega l2}^2\bar{\tau}_{\omega lj}^2}{2}\right) \\
& - \frac{\sigma_{\omega l1}}{\gamma_{\omega l1}}\tilde{\Phi}_{\omega l}\hat{\Phi}_{\omega l} - \frac{g_{\omega l}\sigma_{\omega l2}}{\gamma_{\omega l2}}\tilde{\beta}_{\omega l}\hat{\beta}_{\omega l}.
\end{aligned} \tag{6.67}$$

Following $g_{\omega l}(\hat{\beta}_{\omega l} - \tilde{\beta}_{\omega l}) = g_{\omega l}\beta_{\omega l} = 1$, one arrives at

$$\dot{L}_3 \leq -e_{\varphi i}^{\prime \mathrm{T}} K_{1i} e_{\varphi i}^{\prime} - e_{\omega i}^{\mathrm{T}} K_{2i} e_{\omega i} + \sum_{j=1}^{3} \left(\frac{h_{\omega i1}^2}{2} + \frac{h_{\omega i2}^2 \bar{\tau}_{\omega ij}^2}{2} \right)$$
$$+ g_{\omega i} \sqrt{\varepsilon_{\omega i}} - \frac{\sigma_{\omega i1}}{\gamma_{\omega i1}} \tilde{\Phi}_{\omega i} \hat{\Phi}_{\omega i} - \frac{g_{\omega i} \sigma_{\omega i2}}{\gamma_{\omega i2}} \tilde{\beta}_{\omega i} \hat{\beta}_{\omega i}. \tag{6.68}$$

Using Young's inequality for the terms of $\tilde{\Phi}_{\omega i} \hat{\Phi}_{\omega i}$ and $\tilde{\beta}_{\omega i} \hat{\beta}_{\omega i}$, it can be checked that

$$\dot{L}_3 \leq -\kappa_{\omega i} L_3 + \chi_{\omega i}, \tag{6.69}$$

where $\kappa_{\omega i} = \min \{2\lambda_{\min}(K_{1i}), 2\lambda_{\min}(K_{2i}), \sigma_{\omega i1}, \sigma_{\omega i2}\}$ and $\chi_{\omega i} = g_{\omega i} \sqrt{\varepsilon_{\omega i}} + \sum_{j=1}^{3} \left(\frac{h_{\omega i1}^2}{2} + \frac{h_{\omega i2}^2 \bar{\tau}_{\omega ij}^2}{2} \right) + \frac{\sigma_{\omega i1}}{2\gamma_{\omega i1}} \Phi_{\omega i}^2 + \frac{g_{\omega i} \sigma_{\omega i2}}{2\gamma_{\omega i2}} \beta_{\omega i}^2$.

Integrating (6.69) over $[0, t]$ leads to

$$0 \leq L_3 \leq \left[\frac{\chi_{\omega i}}{\kappa_{\omega i}} + L_3(0) \right] e^{-\kappa_{\omega i} t} + \frac{\chi_{\omega i}}{\kappa_{\omega i}}. \tag{6.70}$$

Therefore, it can be seen from (6.70) that all closed-loop signals are SGUUB. Note that the convergence region of $e_{\varphi i}^{\prime}$ can be made smaller by increasing K_{1i}, K_{2i}, $\gamma_{\omega i1}$, and $\gamma_{\omega i2}$ and meanwhile decreasing $h_{\omega i1}$, $h_{\omega i2}$, $\sigma_{\omega i1}$, $\sigma_{\omega i2}$, and $\varepsilon_{\omega i}$. Thus attitude synchronization tracking error can be made as small as desired by appropriate choice of the design parameters. Moreover, the choice of b_1 and b_2 requires to balance the control performance and the number of triggering events. This will be further shown in the following simulations.

Further, it is proved that the Zeno behavior can be avoided. That is to guarantee that there exists a $t^* > 0$ such that $\left\{ t_{k+1}^{\varphi} - t_k^{\varphi} \right\} \geq t^*$. Since $e_{\delta i} = \hat{\delta}_{i0}(t) - \hat{\delta}_{i0}(t_k)$, $\forall t \in \left\{ t_{k+1}^{\varphi} - t_k^{\varphi} \right\}$, we obtain $\dot{e}_{\delta i}(t) = \dot{\hat{\delta}}_{i0}(t)$. From (6.37), $\hat{\delta}_{i0}(t)$ is a function of bounded closed-loop signals. It is clear that $\dot{e}_{\delta i}(t)$ is bounded and $|\dot{e}_{\delta ij}(t)| = |\dot{\hat{\delta}}_{i0j}(t)| \leq \mu_{\delta ij}$ with positive constant $\mu_{\delta ij}$, $j = 1, 2, 3$, where $e_{\delta ij}$ and $\hat{\delta}_{i0j}$ are the jth elements of $e_{\delta i}$ and $\hat{\delta}_{i0}$, respectively. Considering $e_{\delta ij}\left(t_k^{\varphi}\right) = 0$ and $\lim_{t \to t_{k+1}^{\varphi}} e_{\delta ij}(t) = b_2$, it can be judged that $t^* \geq \mu_{\delta ij} / b_2$; hence the Zeno behavior is successfully precluded. This completes the proof of Theorem 6.2.

In Theorems 6.1 and 6.2, we have designed the controllers that guarantee event-triggered convergence and fault tolerance for velocity and attitude synchronization tracking simultaneously under model uncertainties, external disturbances, and actuator faults. The proposed controllers have the following features in comparison with the related existing controllers.

- Differently from control schemes for single 6-DoF fixed-wing UAV of [25, 26] with no consideration of fault-tolerant and event-triggered strategy, our scheme extends design to event-triggered adaptive fault-tolerant control scheme for multiple 6-DoF fixed-wing UAVs.

- The modeling uncertainties are effectively handled with the help of neural network approximations for $\boldsymbol{\Delta}_{vi}$ and $\boldsymbol{\Delta}'_{\omega i}$. Meanwhile, to deal with the actuator faults, the bound estimations for parameters $\beta_{vi} = 1/\rho_{Ti}$, $g_{vi} = \beta_{vi} \sup_{t\geq 0} T_{xif}$ in translational control protocol and $\beta_{\omega i} = 1/g_{\omega i}$, $g_{\omega i} = \inf_{t\geq 0} \lambda_{\min}\left(\boldsymbol{J}_{i0}^{-1}\boldsymbol{C}_0(\boldsymbol{\delta}_i)\,\boldsymbol{\rho}_{\delta i}\boldsymbol{D}_i\left(\boldsymbol{J}_{i0}^{-1}\boldsymbol{C}_0(\boldsymbol{\delta}_i)\right)^{\mathrm{T}}\right)$ in rotational control design are introduced in (6.34), (6.35), and (6.39). Further, we employ the features of the smooth function $\vartheta \tanh\left(\frac{\vartheta}{\varepsilon}\right)$ in design of $\hat{T}_{xi0}$ in (6.32) and function $\frac{z^2}{\sqrt{z^2+\varepsilon}}$ when designing $\hat{\boldsymbol{\delta}}_{i0}$ in (6.38).

6.4 Simulations and Analyses

In order to illustrate the effectiveness and superiority of the proposed event-triggered adaptive control scheme, several representative simulations are conducted in this section. The structure parameters and coefficient values are referred to [25] and listed in Table 6.1. Three networked fixed-wing UAVs are considered to complete the mission of coordinated turn and climb. The initial conditions are given in Table 6.2. The corresponding communication topology is illustrated in Fig. 6.2. The

Table 6.1 Structural and aerodynamic parameter values of UAV

Coefficient	Value	Unit	Coefficient	Value	Unit
m_i	20.64	kg	b_i	1.96	m
S_i	1.37	m^2	$\bar{c}_i$	0.76	m
ρ	1.29	$\mathrm{kg}\cdot\mathrm{m}^{-3}$	g	9.8	$\mathrm{m}\cdot\mathrm{s}^{-2}$
J_{xxi}	1.607	$\mathrm{kg}\cdot\mathrm{m}^2$	J_{yyi}	7.51	$\mathrm{kg}\cdot\mathrm{m}^2$
J_{zzi}	7.18	$\mathrm{kg}\cdot\mathrm{m}^2$	J_{xzi}	0.59	$\mathrm{kg}\cdot\mathrm{m}^2$
c_{L0i}	0.1	/	$c_{L\alpha i}$	0.25	rad^{-1}
c_{D0i}	0.5	/	$c_{Y\beta i}$	−0.1	rad^{-1}
c_{l0i}	−0.001	/	$c_{l\beta i}$	−0.038	rad^{-1}
c_{lpi}	−0.213	$\mathrm{rad}^{-1}\cdot\mathrm{s}$	c_{lri}	0.114	$\mathrm{rad}^{-1}\cdot\mathrm{s}$
$c_{l\delta ai}$	−0.056	rad^{-1}	$c_{l\delta ri}$	0.014	rad^{-1}
c_{m0i}	0.022	/	$c_{m\alpha i}$	−0.473	rad^{-1}
c_{mqi}	−3.449	$\mathrm{rad}^{-1}\cdot\mathrm{s}$	$c_{m\delta ei}$	−0.364	rad^{-1}
c_{n0i}	0.022	/	$c_{n\beta i}$	0.036	rad^{-1}
c_{npi}	−0.151	$\mathrm{rad}^{-1}\cdot\mathrm{s}$	c_{nri}	−0.195	$\mathrm{rad}^{-1}\cdot\mathrm{s}$
$c_{n\delta ai}$	−0.036	rad^{-1}	$c_{n\delta ri}$	−0.055	rad^{-1}

Table 6.2 Initial states of UAVs

	$x_i(0)$	$y_i(0)$	$z_i(0)$	$V_i(0)$	$\phi_i(0)$	$\theta_i(0)$	$\psi_i(0)$	$p_i(0)$	$q_i(0)$	$r_i(0)$
Leader	−20 m	20 m	100 m	40 m/s	0°	1°	0°	0 rad/s	0 rad/s	0 rad/s
Follower 1	0 m	0 m	100 m	40.5 m/s	0°	1°	0°	0 rad/s	0 rad/s	0 rad/s
Follower 2	−40 m	40 m	100 m	40 m/s	0°	1°	0°	0 rad/s	0 rad/s	0 rad/s

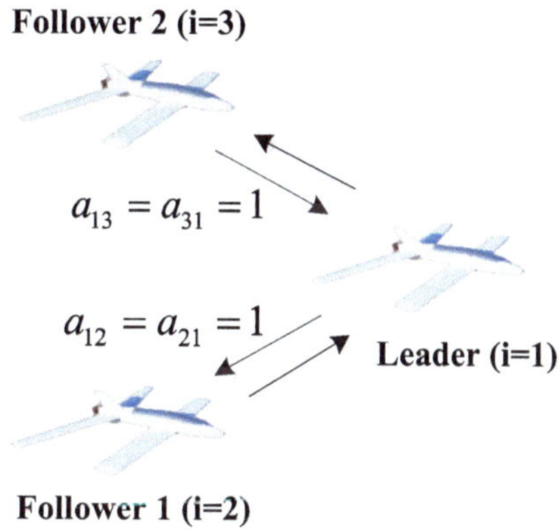

Fig. 6.2 Communication topology of UAVs

reference commands are set as

$$V_r = \begin{cases} 40\,\mathrm{m/s}, & 0\mathrm{s} \le t \le 4\mathrm{s} \\ 45\,\mathrm{m/s}, & 4\mathrm{s} < t \le 15\mathrm{s} \\ 50\,\mathrm{m/s}, & 15\mathrm{s} < t \le 24\mathrm{s}, \end{cases}$$

$$[\phi_r, \theta_r, \psi_r]^{\mathrm{T}} = \begin{cases} [0\,\mathrm{deg}, 0\,\mathrm{deg}, 0\,\mathrm{deg}]^{\mathrm{T}}, & 0\mathrm{s} \le t \le 4\,\mathrm{s} \\ [5\,\mathrm{deg}, 5\,\mathrm{deg}, 10\,\mathrm{deg}]^{\mathrm{T}}, & 4\mathrm{s} < t \le 15\,\mathrm{s} \\ [0\,\mathrm{deg}, 0\,\mathrm{deg}, 10\,\mathrm{deg}]^{\mathrm{T}}, & 15\mathrm{s} < t \le 24\,\mathrm{s}. \end{cases}$$

During the flight, suppose that the external disturbances in velocity and attitude motion are described as $\boldsymbol{d}_{vi} = [0.1\sin(\pi t), 0.1\sin(4t), 0.1\cos(\pi t)]^{\mathrm{T}}$ and $\boldsymbol{d}_{\omega i} = [0.2\cos(\pi t), 0.1\sin(\pi t), 0.2\sin(\pi t)]^{\mathrm{T}}$ $(i = 1, 2, 3)$; meanwhile, the rudder of follower 2 loses its effectiveness described by $\boldsymbol{\rho}_{\delta 3} = \mathrm{diag}\{1, 1, 0.9\}$ during $t \in (19, 22]$ s, while the leader and follower 1 are healthy during flight. The main parameters of the proposed control scheme are set in Table 6.3.

Scenario 1 In the first scenario, the simulation is conducted by applying the proposed event-triggered adaptive control scheme to the multi-UAV dynamic models in the above simulation scenario. Figure 6.3 depicts the flight trajectories of UAV formation, from which we can observe that the coordinated turn and climb with satisfactory velocity and attitude synchronization tracking is achieved. The detailed velocity and attitude tracking performances are shown in Fig. 6.4. From Fig. 6.4a–d, it can be observed that velocity and attitude both converge availably to the desired trajectories, which validates the effectiveness of proposed schemes.

Table 6.3 Controller parameters

Coefficient	Value	Coefficient	Value
c_{v1}, c_{v2}, c_{v3}	10	a_2	$0.5\ (t \le 14)$
$\varepsilon_{v1}, \varepsilon_{v2}, \varepsilon_{v3}$	0.1		$1\ (t > 14)$
$\varepsilon_{\omega 1}, \varepsilon_{\omega 2}, \varepsilon_{\omega 3}$	1	b_1	0.01
a_1	0.01	b_2	1
$\gamma_{\omega l1}, \gamma_{\omega l2}$	0.01	$\sigma_{\omega l1}, \sigma_{\omega l2}$	25
$\boldsymbol{K}_{11}, \boldsymbol{K}_{12}, \boldsymbol{K}_{13}$	$\mathrm{diag}\{50, 50, 51\}\ (t \le 5)$	$\boldsymbol{K}_{21}, \boldsymbol{K}_{22}, \boldsymbol{K}_{23}$	$\mathrm{diag}\{45, 45, 46\}\ (5 < t \le 15)$

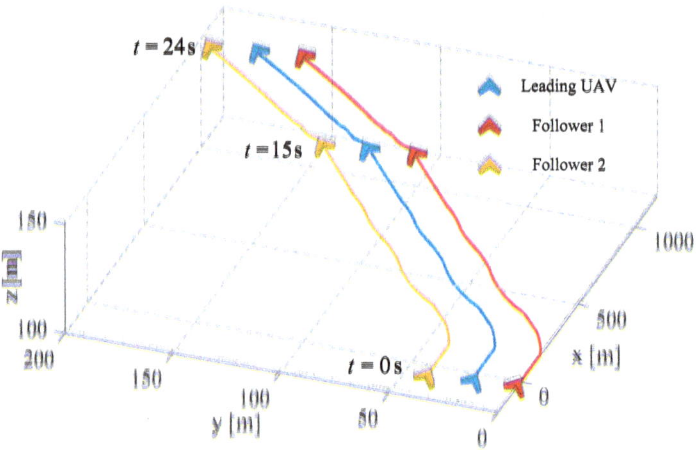

Fig. 6.3 Trajectories of UAV formation

Additionally, the tracking errors of velocity and attitude are depicted in Fig. 6.4e–h. From Fig. 6.4i–l, we see that the proposed scheme can guarantee the synchronization tracking of velocity and attitude for UAVs. With the proposed control scheme, the time intervals of each event for UAVs are presented in Fig. 6.5a–l, and meanwhile Fig. 6.6a–c lists the time at which the events are triggered. Then, the numbers of events triggered of control inputs for each UAV during 24 s are calculated, as shown in Fig. 6.7.

Scenario 2 To illustrate the superiority of proposed control scheme, two comparative control methods are conducted. The first comparative method is obtained by removing the event-triggered strategy from our proposed control scheme, which is used to show that the event-triggered feature reduces the control transmission burden and meanwhile maintains the control performances (denoted as Compared controller 1 hereafter). Moreover, we find the control approach designed in [22] (regarded as Compared controller 2 hereafter) worth for comparison because it deals with the flight control for 6-DoF fixed-wing UAV as well, however without fault-tolerant and event-triggered control. As a consequence, the comparative simulation results are depicted in Fig. 6.8 (taking follower 2 for example), which shows the tracking curves of velocity and attitude under the three control schemes. Obviously,

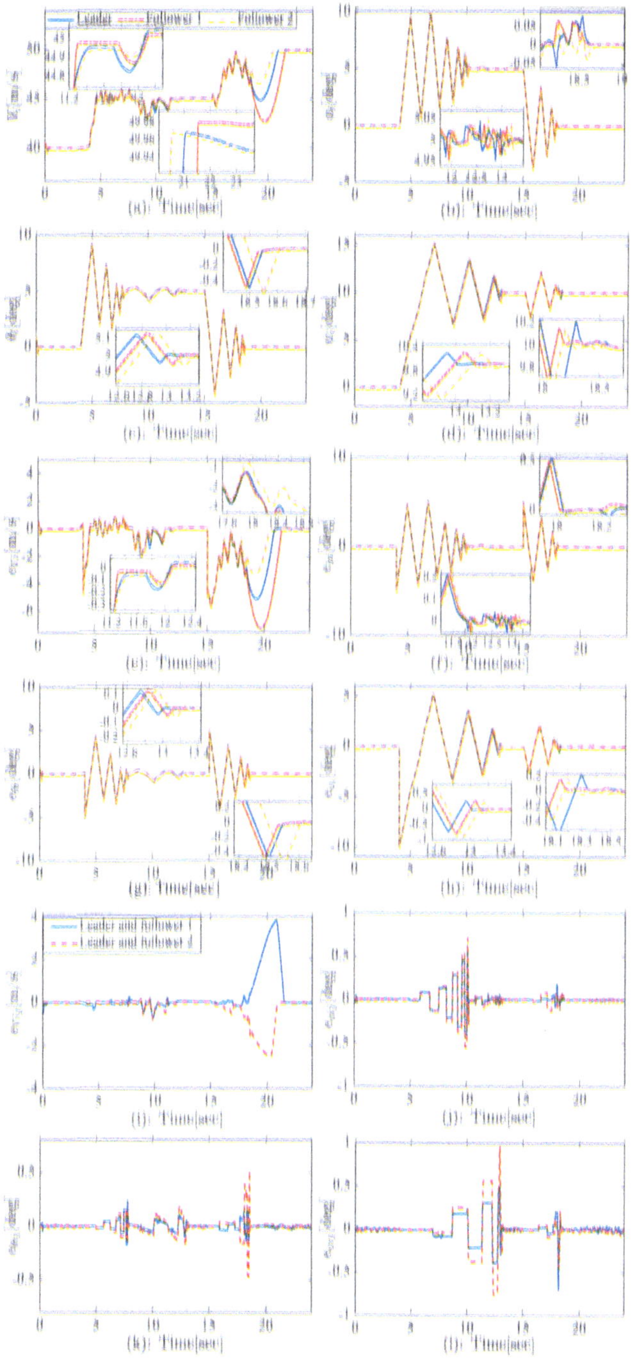

Fig. 6.4 Velocity and attitude angle tracking results and errors by using proposed controller. (a)–(l) show the curves of state responses and errors

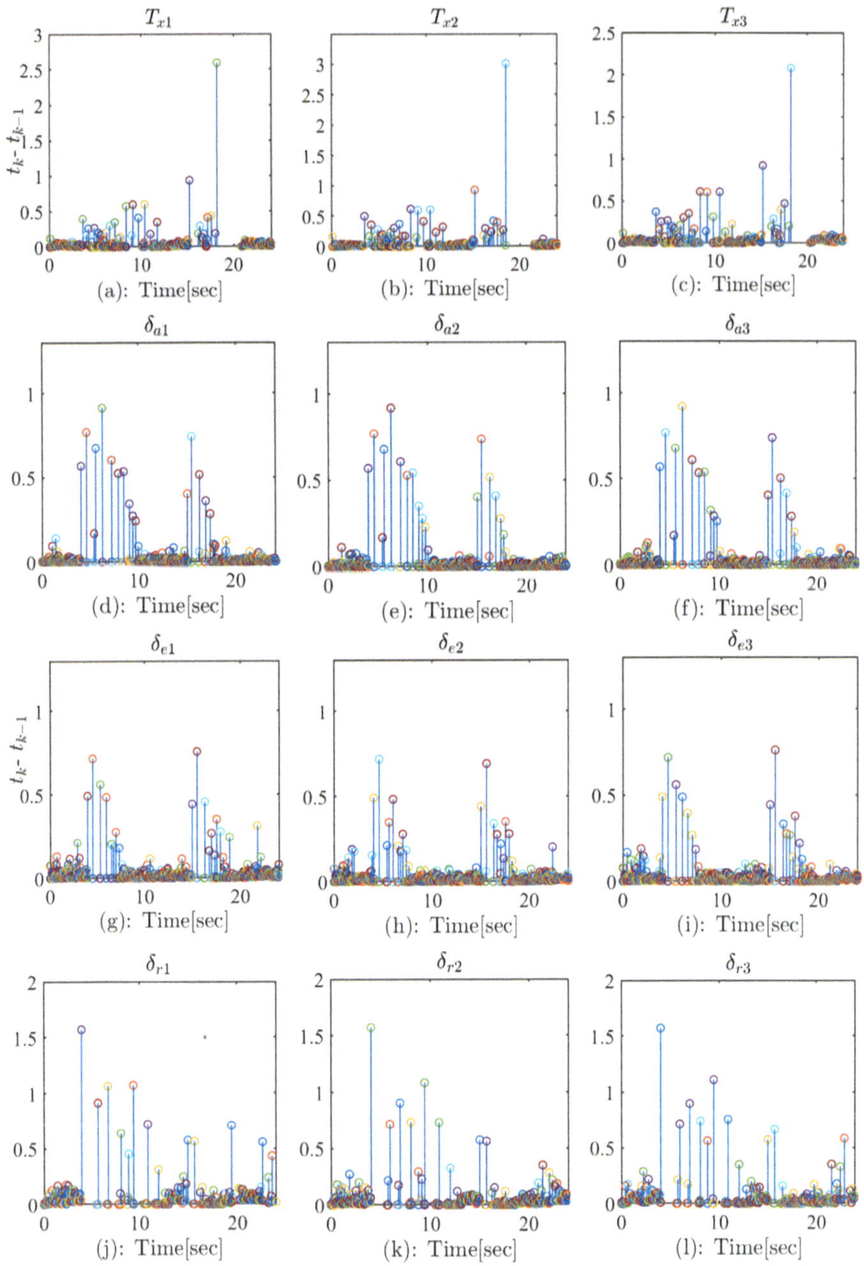

Fig. 6.5 Time interval of triggering events. (**a**)–(**l**) show the curves of triggering time intervals. (**a**) Control signal T_{x1}. (**b**) Control signal T_{x2}. (**c**) Control signal T_{x3}. (**d**) Control signal δ_{a1}. (**e**) Control signal δ_{a2}. (**f**) Control signal δ_{a3}. (**g**) Control signal δ_{e1}. (**h**) Control signal δ_{e2}. (**i**) Control signal δ_{e3}. (**j**) Control signal δ_{r1}. (**k**) Control signal δ_{r2}. (**l**) Control signal δ_{r3}

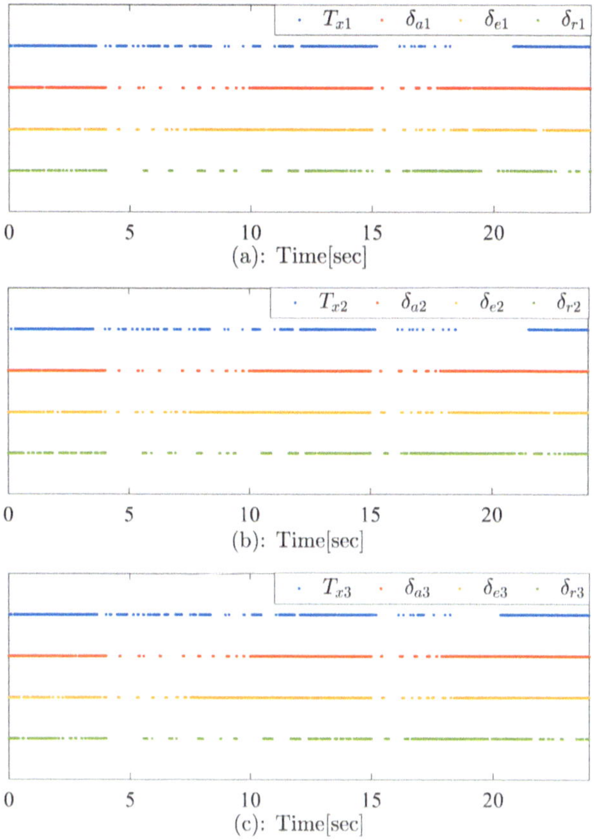

Fig. 6.6 Triggering time of control signals. (**a**)–(**c**) show the curves of triggering event. (**a**) Leading UAV. (**b**) Follower 1. (**c**) Follower 2

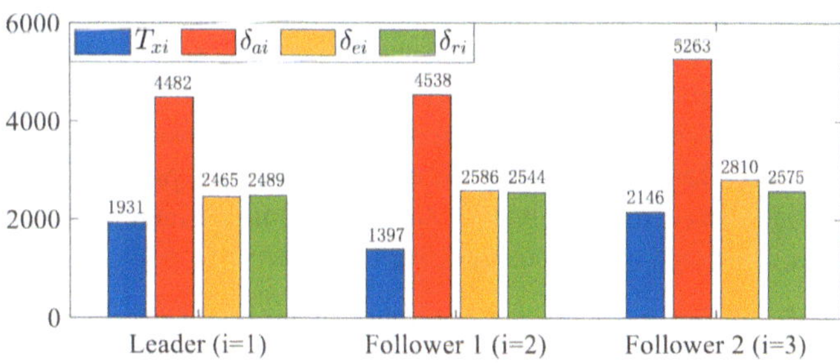

Fig. 6.7 Number of times for control transmission

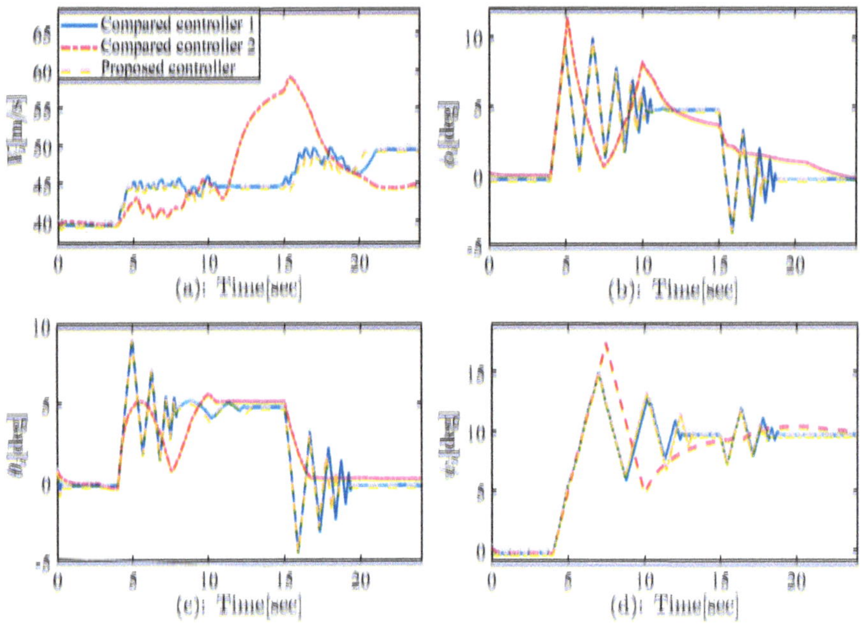

Fig. 6.8 Control performance comparison of velocity and attitude tracking under two comparative controllers and proposed controller (taking the follower 2 for example). (**a**)–(**d**) show the curves of state responses. (**a**) Velocity response. (**b**) Roll angle response. (**c**) Pitch angle response. (**d**) Yaw angle response

Table 6.4 IAE assessment of tracking performance for scenario 2

Control schemes	V_3	ϕ_3	θ_3	ψ_3
Compared controller 1	15.8	18	14.3	25
Compared controller 2	88.5	33.1	19.2	37.4
Proposed controller	17.8	16.4	13.4	25.6

Compared controller 2 breaks down in velocity and attitude tracking when it reacts to the external disturbances and actuator faults, which obtains the worst performance of three control schemes. In addition, the integral absolute error (IAE) and control transmission times are utilized here as performance indices. The results are summarized in Tables 6.4 and 6.5. It is explicit that the proposed scheme saves 86.67% of control transmission burden and meanwhile maintains the good control performances compared to two comparative controllers.

Scenario 3 To further evaluate the sensitivity of proposed event-triggered conditions, we consider the following four different fault modes:

Fault Mode 1 The rudder of follower 2 loses its effectiveness described by $\boldsymbol{\rho}_{\delta_3} = \text{diag}\{1, 1, 0.9\}$ during $t \in (19, 22]$ s.

Table 6.5 Control transmission times for scenario 2

Control schemes	T_{V3}	δ_{a3}	δ_{e3}	δ_{r3}
Compared controller 1	24000	24000	24000	24000
Compared controller 2	24000	24000	24000	24000
Proposed controller	2146	5263	2810	2575

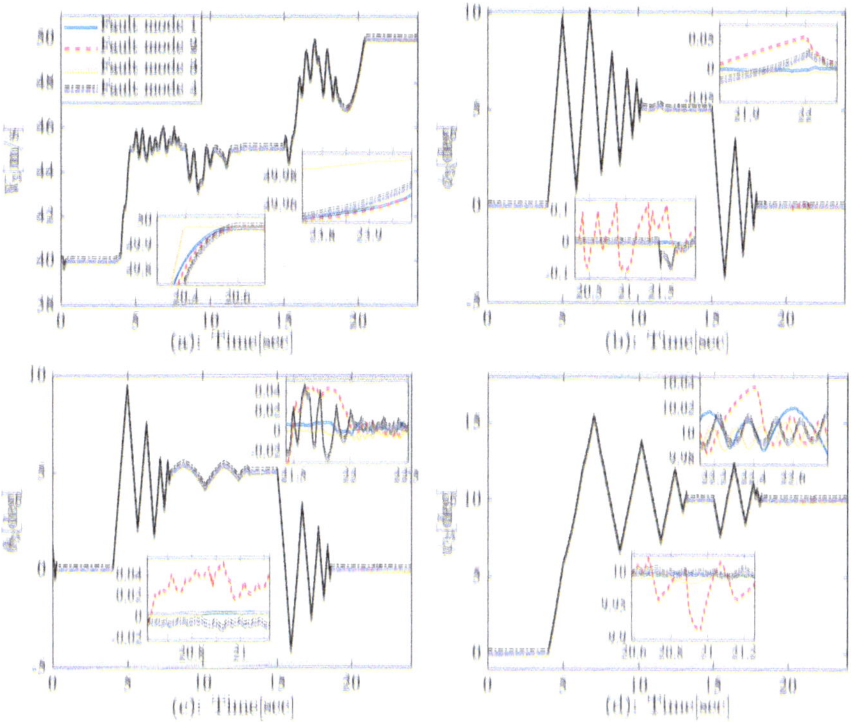

Fig. 6.9 Velocity and attitude tracking of follower 2 under different fault modes. **(a)**–**(d)** show the curves of state responses. **(a)** Velocity response. **(b)** Roll angle response. **(c)** Pitch angle response. **(d)** Yaw angle response

Fault Mode 2 Based on fault mode 1, the rudder of follower 2 also suffers from the bias fault introduced as $\boldsymbol{\delta}_{3f} = [0, 0, 1]^T$ during $t \in (19, 22]$ s.

Fault Mode 3 The elevator of follower 2 loses its effectiveness also with the bias fault described by $\boldsymbol{\rho}_{\delta_3} = \text{diag}\{0.9, 1, 1\}$ and $\boldsymbol{\delta}_{3f} = [0.5, 0, 0]^T$ during $t \in (19, 22]$ s.

Fault Mode 4 The aileron of follower 2 loses its effectiveness described by $\boldsymbol{\rho}_{\delta_3} = \text{diag}\{1, 0.9, 1\}$ and meanwhile suffers from the bias fault as $\boldsymbol{\delta}_{3f} = [0, 1, 0]^T$ during $t \in (19, 22]$ s.

Based on above fault modes, the fault-tolerant performances of follower 2 for different fault modes are shown in Fig. 6.9. Desired robustness with proposed controller is demonstrated in case of the above four fault modes in Fig. 6.9. From Fig. 6.9, we can notice that the fault-tolerant property guarantees the stable tracking of velocity and attitude for follower 2 in the presence of different faults on three control surfaces.

6.5 Conclusions

An event-triggered cooperative synchronization fault-tolerant control methodology has been designed for multiple UAVs subject to actuator faults, modeling uncertainties, and external disturbances. As compared with most existing works concerning two-degree or only attitude/longitudinal dynamics, 6-DoF dynamics with 12 state variables has been considered. Specifically, event-triggering mechanisms are introduced into translational and rotational subsystems independently, which reduces the signal transmission burden, while saving on system resources. The actuator faults and the network induced errors are dealt with by the bound estimation approach and some well defined smooth functions. All closed-loop signals are SGUUB, and meanwhile the synchronization tracking of velocity and attitude for UAVs is achieved well. We note that the extension of event-triggering mechanism to communication between UAVs is open. Also, involving output/state constrains and prescribed performances into our proposed scheme remains a good point for future research.

References

1. T. Kim, D. Qiao, Energy-Efficient Data Collection for IoT Networks via Cooperative Multi-Hop UAV Networks. IEEE Trans. Veh. Technol. **70**, 471–486 (2017)
2. Y. Chen, D. Chang, C. Zhang, Autonomous tracking using a swarm of UAVs: a constrained multi-agent reinforcement learning approach. IEEE Trans. Veh. Technol. **69**(11), 13702–13717 (2020)
3. Y. Wang, M. Shan, D. Wang, Motion capability analysis for multiple fixed-wing UAV formations with speed and heading rate constraints. IEEE Trans. Control Network Syst. **7**(2), 977–989 (2020)
4. Z. Yu, Y. Zhang, B. Jiang, C.Y. Su, J. Fu, Y. Jin, T. Chai, Fractional-order adaptive fault-tolerant synchronization tracking control of networked fixed-wing UAVs against actuator-sensor faults via intelligent learning mechanism. IEEE Trans. Neural Networks Learn. Syst.. **32**(12), 5539–5553 (2021)
5. M. Lv, B.D. Schutter, C. Shi, S. Baldi, Logic-based distributed switching control for agents in power chained form with multiple unknown control directions. Automatica **137**, 110143 (2022)
6. S. Li, X. Wang, Finite-time consensus and collision avoidance control algorithms for multiple AUVs. Automatica **49**(11), 3359–3367 (2013)
7. Z. Yu, Y. Zhang, Z. Liu, Y. Qu, C.Y. Su, Distributed adaptive fractional-order fault tolerant cooperative control of networked unmanned aerial vehicles via fuzzy neural networks. IET Control Theory Appl. **13**(17), 2917–2929 (2019)

8. Z. Yu, Y. Qu, Y. Zhang, Distributed fault-tolerant cooperative control for multi-UAVs under actuator fault and input saturation. IEEE Trans. Control Syst. Technol. **27**(6), 2417–2429 (2019)
9. Y. Wang, T. Zhang, Z. Cai, J. Zhao, K. Wu, Multi-UAV coordination control by chaotic grey wolf optimization based distributed MPC with event-triggered strategy. Chin. J. Aeronaut. **33**(11), 2877–2897 (2020)
10. T.Z. Muslimov, R.A. Munasypov, Adaptive decentralized flocking control of multi-UAV circular formations based on vector fields and backstepping. ISA Trans. **107**, 143–159 (2020)
11. L. Xing, C. Wen, Z. Liu, H. Su, J. Cai, Event-triggered adaptive control for a class of uncertain nonlinear systems. IEEE Trans. Autom. Control **62**(4), 2071–2076 (2017)
12. Z. Cai, H. Zhou, J. Zhao, K. Wu, Y. Wang, Formation control of multiple unmanned aerial vehicles by event-triggered distributed model predictive control. IEEE Access **6**, 55614–55627 (2018)
13. N. Hung, A. Pascoal, T. Johansen, Cooperative path following of constrained autonomous vehicles with model predictive control and event-triggered communications. Int. J. Robust Nonlinear Control **30**(7), 2644–2670 (2020)
14. Z. Wu, J. Xiong, M. Xie, A switching method to event-triggered output feedback control for unmanned aerial vehicles over cognitive radio networks. IEEE Trans. Syst. Man Cybern.: Syst. **51**(12), 7530–7541 (2020)
15. B. Tian, J. Cui, H. Lu, L. Liu, Q. Zong, Attitude control of UAVs based on event-triggered super-twisting algorithm. IEEE Trans. Industr. Inform. **17**(2), 1029–1038 (2020)
16. H. Yang, Q.L. Han, X. Ge, L. Ding, Y. Xu, B. Jiang, D. Zhou, Fault-tolerant cooperative control of multi-agent systems: a survey of trends and methodologies. IEEE Trans. Industr. Inform. **16**(1), 4–17 (2020)
17. M.A. Kamel, X. Yu, Y. Zhang, Formation control and coordination of multiple unmanned ground vehicles in normal and faulty situations: a review. Annu. Rev. Control. **49**, 128–144 (2020)
18. Z. Yu, Z. Liu, Y. Zhang, Y. Qu, C.Y. Su, Distributed finite-time fault-tolerant containment control for multiple unmanned aerial vehicles. IEEE Trans. Neural Networks Learn. Syst. **31**(6), 2077–2091 (2020)
19. Z. Yu, Y. Zhang, Z. Liu, Y. Qu, C. Su, B. Jiang, Decentralized finite-time adaptive fault-tolerant synchronization tracking control for multiple UAVs with prescribed performance. J. Franklin Inst-. Eng. Appl. Math. **357**(16), 11830–11862 (2020)
20. Z. Yu, Y. Zhang, B. Jiang, C Y. Su, J. Fu, Y. Jin, T. Chai, Decentralized fractional order backstepping fault-tolerant control of multi-UAVs against actuator faults and wind effects. Aerosp. Sci. Technol. **104**, 105939 (2020)
21. Z. Yu, Y. Zhang, B. Jiang, C.Y. Su, J. Fu, Y. Jin, T. Chai, Fractional order adaptive fault tolerant synchronization tracking control of networked fixed wing UAVs against actuator sensor faults via intelligent learning mechanism. IEEE Trans. Neural Networks Learn. Syst. **32**(12), 5539–5553 (2021)
22. Z. Yu, Y. Zhang, B. Jiang, C.Y. Su, J. Fu, Y. Jin, T. Chai, Nussbaum based finite time fractional order backstepping fault tolerant flight control of fixed wing UAV against input saturation with hardware in the loop validation. Mech. Syst. Signal Process. **153**, 107406 (2021)
23. W. Cai, X. Liao, Y. Song, Indirect robust adaptive fault-tolerant control for attitude tracking of spacecraft. J. Guid. Control. Dyn. **31**(5), 1456–1463 (2008)
24. W. Liu, Y. Geng, B. Wu, D. Wang, Neural-network-based adaptive event-triggered control for spacecraft attitude tracking. IEEE Trans. Neural Networks Learn. Syst. **31**(10), 4015–4024 (2019)
25. O. Espen, S. Rune, K. Raymond, Underactuated waypoint tracking of a fixed-wing UAV. IFAC Proc. **46**(30), 126–133 (2015)
26. H. Castaneda, O.S. Salas-Pena, J. Leon-Morales, Extended observer based on adaptive second order sliding mode control for a fixed wing UAV. ISA Trans. **66**, 226–232 (2017)
27. H. Wang, P.X. Liu, X. Zhao, X. Liu, Adaptive fuzzy finite-time control of nonlinear systems with actuator faults. IEEE Trans. Cybern. **50**(5), 1786–1797 (2020)

28. D. Wang, Q. Zong, B. Tian, S. Shao, X. Zhang, X. Zhao, Neural network disturbance observer-based distributed finite-time formation tracking control for multiple unmanned helicopters. ISA Trans. **73**, 208–226 (2018)
29. S. Shao, M. Chen, Y. Zhang, Adaptive discrete-time flight control using disturbance observer and neural networks. IEEE Trans. Neural Networks Learn. Syst. **30**(12), 3708–3721 (2019)
30. G. Lai, Z. Liu, Y. Zhang, C. Philip, Adaptive position/attitude tracking control of aerial robot with unknown inertial matrix based on a new robust neural identifier. IEEE Trans. Neural Networks Learn. Syst. **27**(1), 18–31 (2016)
31. S.C. Tong, Y.M. Li, P. Shi, Observer-based adaptive fuzzy backstepping output feedback control of uncertain MIMO pure-feedback nonlinear systems. IEEE Trans. Fuzzy Syst. **20**(4), 771–785 (2012)
32. H. Yang, B. Jiang, H. Yang, H.H. Liu, Synchronization of multiple 3-DoF helicopters under actuator faults and saturations with prescribed performance. ISA Trans. **75**, 118–126 (2018)
33. M. Lv, W. Yu, J. Cao, S. Baldi, A separation-based methodology to consensus tracking of switched high-order nonlinear multi-agent systems. IEEE Trans. Neural Networks Learn. Syst. **33**(10), 5467–5479 (2021)
34. M. Lv, W. Yu, J. Cao, S. Baldi, Consensus in high-power multi-agent systems with mixed unknown control directions via hybrid Nussbaum-based control. IEEE Trans. Cybern. **52**(6), 5184–5196 (2020)
35. L. Xing, C. Wen, Y. Zhu, H. Su, Z. Liu, Output feedback control for uncertain nonlinear systems with input quantization. Automatica **65**, 191–202 (2016)
36. E. Cruz-Zavala, J. Moreno, L. Fridman, Uniform robust exact differentiator. IEEE Trans. Autom. Control **56**(11), 2727–2733 (2011)

Chapter 7
Event-Triggered Control with Dynamic Communication

Chapter 6 investigates the event-triggered control method for multiple UAVs, addressing the issue of limited actuator response frequency. Besides, the limited communication resource denotes another significant challenge to affect the control performance of multi-UAV system. Typically, the multi-UAV system employs a periodic communication mechanism. The frequent communication leads to an amount of unnecessary transmission when the multi-UAV system reaches stability, resulting in the network congestion. Thus, this chapter explores how to reduce the frequency of communication while ensuring the control performance of multi-UAV system.

7.1 Introduction

Coordinated flight of multiple fixed-wing unmanned aerial vehicles (UAVs) has been extensively studied over the past decades, due to its important role in achieving flexibility and cost-effectiveness of mission cooperation [1–3]. Generally, coordinated control problem has been investigated in several aspects, including formulating and maintaining a particular formation shape for UAVs in a group [4] and synchronous tracking of velocity and attitude for UAV formation [5]. In particular, distributed control for coordination part is now in research spotlight thanks to its undeniable benefits that global information is no longer required in control design while preserving closed-loop stability [6, 7]. Driven by the distributed coordinated control, various advanced results for multiple UAVs are currently available in [8–11]. Typically, Yu et al. [8] investigated the attitude synchronization tracking problem of networked UAVs in a distributed cooperative control framework. Yu et al. [9] explored the distributed cooperative fault-tolerant control for longitudinal synchronization tracking of multiple UAVs with actuator faults and input saturation. Further, in [10], a distributed model predictive approach

M. Lv et al., *Unmanned Aerial Vehicles Formation Control and Safety Guarantee*, Cognitive Intelligence and Robotics, https://doi.org/10.1007/978-981-96-5829-9_7

with swarm intelligence was designed for the coordinated control of multiple UAVs. In [11], a distributed vector field-based control scheme was proposed to achieve the path following and circular motion around the target for UAV formation.

However, all aforementioned methods [8–11] are performed based on the simplified two-degree models [10, 11] or just taking the attitude/longitudinal dynamics into consideration [12–14]. Such a model simplification unavoidably lowers the system description precision and thus cannot accurately describe the practical UAV system. Actually, the fixed-wing UAV system is a typical nonlinear model with decomposed translational and rotational dynamics. Therefore, designing the velocity and attitude control for translational and rotational dynamics is an open and important problem.

Moreover, note that existing literatures [8–14] focus on the coordinated control problem of UAVs subject to continuous-time information exchange regardless of the consumption of network resources such as bandwidth and energy. Especially for the cooperative UAV flight, this drawback will become more apparent as the number of UAVs increases. Therefore, in order to mitigate the unnecessary information exchanges and reduce the channel load, an energy-efficient option, namely event-triggered communication strategy [15], has been developed, which schedules the interagent communication when the triggering condition is satisfied.

Different from the traditional periodic communication, the event-triggered communication relies on a state-dependent triggering condition. Based on the event-triggered communication mechanism, distributed tracking control protocols are designed for single-integrator agents [16], second-order agents [17], linear agents [18], and nonlinear multi-agent systems [19, 20]. Although effective performances with event-triggered communications are obtained in [16–20], it can be noted that these triggering functions are with fixed threshold values. Nevertheless, for some complex dynamic situations, it is unreasonable to permanently fix the threshold parameter. In this case, the threshold value should dynamically change to reflect the real-time data transmission rates [21]. Therefore, the dynamic triggering rules are proposed and successfully applied in the linearized dynamics of multi-agent system [22, 23] and linear cases of individual system [24, 25]. However, it is noted that studies of [22–25] just consider the linear dynamics of agent systems without regard to nonlinear dynamics, let alone the practical UAV system. Therefore, the crucial question of how to design a novel dynamic triggering mechanism for UAV systems still remains open. Moreover, since practical UAV system inevitably suffers from actuator faults, which would degrade the flight performance and even lead to instability, fault-tolerant control has become another significant research problem, and some effective fault-tolerant control approaches for UAVs have been reported in [8, 9, 26, 27].

Motivated by the foregoing analyses, the main contributions of this work are threefold below:

- As opposed to the existing results [8–11] which are either concentrated on attitude/longitudinal motions or two-degree dynamics, this study proposes an event-based fault-tolerant coordinated control scheme for multi-UAVs such that the consensus tracking of velocity and attitude is achieved subject to actuator faults, external disturbances, and modeling uncertainties.

- In contrast to the static event-triggered communication mechanisms [16–20], the proposed triggering function using state values is determined by a dynamic rule. It can achieve a better trade-off between reducing communication burden and preserving favorable tracking performance. Meanwhile, the Zeno phenomenon is precluded to guarantee that the time between two consecutive communications is lower bounded by a positive constant.
- Unlike the existing cooperative flight of UAVs dependent on the periodic communication of neighbors' states [8–14], the proposed distributed control scheme merely utilizes locally triggered states of underlying neighbors at discretized event-based instants, thus significantly relaxing the pressure on the communication network.

The rest of chapter is organized as follows. The multi-UAV dynamics, communication theory, and control objective are presented in Sect. 7.2. Section 7.3 gives the details of cooperative event-based control design. Section 7.4 shows the simulation validations, and Sect. 7.5 draws the conclusions.

Remark The main control schemes and contents of this chapter are from the published journal paper "B. Y. Zhang, X. X. Sun, M. L. Lv, S. G. Liu, Distributed coordinated control for fixed-wing UAVs with dynamic event-triggered communication. IEEE Transactions on Vehicular Technology. 2022, 71(5): 4665–4676." The authors appreciate the permission from Institute of Electrical and Electronics Engineers to reuse the results published in the relevant journal.

7.2 Problem Formulation

7.2.1 UAV Dynamic Model

Consider a team of N identical fixed-wing UAVs. For the ith UAV, the full dynamics is given by Zhang et al. [28]

$$\dot{\boldsymbol{p}}_i = \boldsymbol{R}_1\left(\boldsymbol{\varphi}_i\right)\boldsymbol{v}_i \tag{7.1}$$

$$\dot{\boldsymbol{v}}_i = -\boldsymbol{S}\left(\boldsymbol{\omega}_i\right)\boldsymbol{v}_i + \frac{\boldsymbol{T}_i}{m_i} + \boldsymbol{R}_1^{\mathrm{T}}\left(\boldsymbol{\varphi}_i\right)\boldsymbol{g} + \frac{\boldsymbol{F}_i}{m_i} + \boldsymbol{d}_{vi} \tag{7.2}$$

$$\dot{\boldsymbol{\varphi}}_i = \boldsymbol{R}_2^{-1}\left(\boldsymbol{\varphi}_i\right)\boldsymbol{\omega}_i \tag{7.3}$$

$$\boldsymbol{J}_i\dot{\boldsymbol{\omega}}_i = -\boldsymbol{\omega}_i \times \boldsymbol{J}_i\boldsymbol{\omega}_i + \boldsymbol{N}_i + \boldsymbol{C}\left(\boldsymbol{\delta}_i\right)\boldsymbol{\delta}_i + \boldsymbol{J}_i\boldsymbol{d}_{\omega i}, \tag{7.4}$$

where $\boldsymbol{p}_i = [x_i, y_i, z_i]^{\mathrm{T}}$ denotes the inertial position, $\boldsymbol{v}_i = [u_i, v_i, w_i]^{\mathrm{T}}$ is the linear velocity, $\boldsymbol{\varphi}_i = [\phi_i, \theta_i, \psi_i]^{\mathrm{T}}$ represents the attitude angle, and $\boldsymbol{\omega}_i = [p_i, q_i, r_i]^{\mathrm{T}}$ denotes the angular velocity. Moreover, $\boldsymbol{T}_i = [T_{xi}, 0, 0]^{\mathrm{T}}$ is the thrust vector along

x body axis, $\boldsymbol{\delta}_l = [\delta_{al}, \delta_{el}, \delta_{rl}]^{\mathrm{T}}$ is the control input, where $\delta_{ai}, \delta_{el}, \delta_{rl}$ represent the deflection of aileron, elevator, and rudder, respectively, m_i is the vehicle mass, $\boldsymbol{g} = [0, 0, g_z]^{\mathrm{T}}$ is the gravity acceleration in inertia frame, $\boldsymbol{d}_{vi} \in \Re^3$ and $\boldsymbol{d}_{\omega l} \in \Re^3$ are the unknown bounded external disturbances, and $\boldsymbol{J}_l$ is the inertia tensor with the symmetric $x - z$ plane:

$$\boldsymbol{J}_l = \begin{bmatrix} J_{xl} & 0 & -J_{xzl} \\ 0 & J_{yl} & 0 \\ -J_{zxl} & 0 & J_{zl} \end{bmatrix}. \tag{7.5}$$

Besides, $\boldsymbol{R}_1(\boldsymbol{\varphi}_l) \in \mathrm{SO}(3)$ is the rotation matrix transforming the body frame coordinates to inertial axis coordinates, and the matrix $\boldsymbol{R}_2(\boldsymbol{\varphi}_i) \in \Re^{3\times3}$ maps the time derivative of attitude angles to the non-inertial expression of angular velocity.

$$\boldsymbol{R}_1(\boldsymbol{\varphi}_l) = \begin{bmatrix} c_{\psi_l} c_{\theta_l} & -s_{\psi_l} s_{\phi_l} + c_{\psi_l} s_{\theta_l} s_{\phi_l} & s_{\psi_l} s_{\phi_i} + c_{\psi_i} s_{\theta_l} c_{\phi_l} \\ s_{\psi_l} c_{\theta_l} & c_{\psi_l} c_{\phi_l} + s_{\psi_l} s_{\theta_l} s_{\phi_l} & -c_{\psi_l} s_{\phi_i} + s_{\psi_i} s_{\theta_l} c_{\phi_l} \\ -s_{\theta_l} & c_{\theta_l} s_{\phi_l} & c_{\theta_i} c_{\phi_l} \end{bmatrix}$$

$$\boldsymbol{R}_2(\boldsymbol{\varphi}_l) = \begin{bmatrix} 1 & 0 & -s_{\psi_l} \\ 0 & c_{\psi_i} & s_{\psi_l} c_{\theta_l} \\ 0 & -s_{\theta_l} & c_{\psi_l} c_{\theta_l} \end{bmatrix}, \tag{7.6}$$

where s_a and c_a stand for $\sin a$ and $\cos a$ functions, respectively. The aerodynamic force $\boldsymbol{F}_l = [F_{Xl}, F_{Yl}, F_{Zl}]^{\mathrm{T}}$ and moment $\boldsymbol{N}_l = [F_{Ll}, F_{Ml}, F_{Nl}]^{\mathrm{T}}$ are calculated in terms of aerodynamic coefficients.

$$\boldsymbol{F}_l = \bar{q}_l S_l \boldsymbol{R}_3^{-1}(\alpha_l, \beta_l)[-C_{Dl}, C_{Yl}, -C_{Ll}]^{\mathrm{T}} \tag{7.7}$$

$$\boldsymbol{N}_l = \bar{q}_l S_l \left[b_l C'_{ll}, \bar{c}_l C'_{Ml}, b_l C'_{nl}\right]^{\mathrm{T}} \tag{7.8}$$

where $\alpha_l = \arctan(w_l/u_l)$ and $\beta_l = \arcsin(u_l/v_l)$ denote the attack angle and sideslip angle, respectively. $\bar{q}_l = \frac{1}{2}\rho V_l^2$ is the dynamic pressure with total velocity $V_l = \sqrt{u_l^2 + v_l^2 + w_l^2}$. Air density ρ, wingspan b_l, wing surface area S_l, and mean aerodynamic chord $\bar{c}_l$ are constants. The transformation matrix $\boldsymbol{R}_3(\alpha_l, \beta_l)$ is

$$\boldsymbol{R}_3(\alpha_l, \beta_l) = \begin{bmatrix} c_{\alpha_l} c_{\beta_l} & s_{\beta_l} & s_{\alpha_l} c_{\beta_l} \\ -c_{\alpha_l} s_{\beta_l} & c_{\beta_l} & -s_{\alpha_l} s_{\beta_l} \\ -s_{\alpha_l} & 0 & c_{\alpha_l} \end{bmatrix}. \tag{7.9}$$

And C_{Dl}, C_{Yl}, C_{Ll}, C'_{ll}, C'_{Ml}, and C'_{nl} are the dimensionless coefficients in force/moment expression, of which the detailed descriptions are shown in [29]. The control effectiveness matrix $\boldsymbol{C}(\boldsymbol{\delta}_l)$ is presented as

$$\boldsymbol{C}(\boldsymbol{\delta}_l)=\begin{bmatrix} \bar{q}_l S_l b_l c_{l\delta_{al}} & 0 & \bar{q}_l S_l b_l c_{l\delta_{rl}} \\ 0 & \bar{q}_l S_l \bar{c}_l c_{m\delta_{el}} & 0 \\ \bar{q}_l S_l b_l c_{n\delta_{al}} & 0 & \bar{q}_l S_l b_l c_{n\delta_{rl}} \end{bmatrix}. \tag{7.10}$$

7.2.2 Control-Oriented Model

Following the descriptions in [28, 29], the fixed-wing UAV system can be decomposed into two interconnected subsystems, i.e., translational dynamics (7.2) and rotational dynamics (7.3)–(7.4). This allows the translational and rotational dynamics to remain in their original forms. In line with [30], considering the time response of translation dynamics with regard to rotational dynamics, it is reasonable to deal with the rotational control problem and translational control problem separately, such that the rotational control action is to be executed, followed by the translational controller.

From (7.2), taking the time derivative of velocity denotes

$$\dot{V}_l=\frac{\boldsymbol{v}_l^{\mathrm{T}}\dot{\boldsymbol{v}}_l}{V_l}=\frac{\boldsymbol{v}_l^{\mathrm{T}}}{V_l}\left(-\boldsymbol{S}(\boldsymbol{\omega}_l)\boldsymbol{v}_l+\frac{\boldsymbol{T}_l}{m_l}+\boldsymbol{R}_1^{\mathrm{T}}(\boldsymbol{\varphi}_l)\boldsymbol{g}+\frac{\boldsymbol{F}_l}{m_l}+\boldsymbol{d}_{vl}\right). \tag{7.11}$$

Based on (7.11), the thrust is extracted. With the property $\boldsymbol{v}_l^{\mathrm{T}}\boldsymbol{S}(\boldsymbol{\omega}_l)\boldsymbol{v}_l = 0$, it follows that

$$\dot{V}_l = \frac{u_l T_{xl}}{m_l V_l} + \frac{\boldsymbol{v}_l^{\mathrm{T}}}{V_l}\left(\boldsymbol{R}_1^{\mathrm{T}}(\boldsymbol{\varphi}_l)\boldsymbol{g} + \frac{\boldsymbol{F}_l}{m_l} + \boldsymbol{d}_{vl}\right). \tag{7.12}$$

Remark 7.1 Since the UAV model is clearly underactuated, T_{xl} simply acts on velocity component u_l, and there are no direct controls for velocity components w_l and v_l, whereas with the relationship $w_l = u_l \tan\alpha_l$ and $v_l = u_l/\sin\beta_l$, it is still controllable to regulate the components w_l, v_l, that is, to point $\boldsymbol{v}_l$ in a desired direction by adding additional controls on α_l and β_l.

During flight, the thrust possibly suffers from actuator faults, modeled by Zhang et al. [28]

$$T_{xl} = \rho_{T_l} T_{xl0} + T_{xlf}, \tag{7.13}$$

where T_{xl0} is the designed control input, ρ_{T_l} is the unknown actuator efficiency factor satisfying $0 \le \rho_{T_l} \le 1$, and T_{xlf} is the bounded unknown bias fault. It follows along (7.13) that:

(i) $\rho_{T_l} = 1$ and $T_{xlf} = 0$. This means the fault-free case.

(ii) $0 < \rho_{T_i} < 1$ and $T_{xif} = 0$. This indicates the partial loss of effectiveness.
(iii) $\rho_{T_i} = 1$ and $T_{xif} \neq 0$. This indicates the bias fault.
(iv) $\rho_{T_i} = 0$ and $T_{xif} \neq 0$. This means that T_{xi} is stuck at bounded time-varying function T_{xif}.

Besides, note that the accurate information of $\boldsymbol{F}_i$ cannot be known a priori due to the coefficient uncertainties. In this sense, $\boldsymbol{F}_i$ is decomposed into a known component $\boldsymbol{F}_{i0}$ and an uncertain one $\Delta \boldsymbol{F}_i$.

Then with actuator faults and modeling uncertainties, the dynamics of (7.12) is rewritten as

$$\dot{V}_i = \frac{u_i T_{xi0}}{m_i V_i} + \frac{\boldsymbol{v}_i^{\mathrm{T}}}{V_i} \left(\boldsymbol{R}_1^{\mathrm{T}} (\boldsymbol{\varphi}_i) \boldsymbol{g} + \frac{\boldsymbol{F}_{i0}}{m_i} \right) + \Delta_{vi}, \tag{7.14}$$

where $\Delta_{vi} = \frac{(\rho_{Ti}-1) u_i T_{xi0}}{m_i V_i} + \frac{u_i T_{xif}}{m_i V_i} + \frac{\boldsymbol{v}_i^{\mathrm{T}}}{V_i} (\frac{\Delta \boldsymbol{F}_i}{m_i} + \boldsymbol{d}_{vi})$ is the lumped disturbances induced by the actuator faults and modeling uncertainties.

From (7.4), considering the coefficient uncertainties, $\boldsymbol{\omega}_i \times \boldsymbol{J}_i \boldsymbol{\omega}_i$ is decomposed into a known component $\boldsymbol{\omega}_i \times \boldsymbol{J}_{i0} \boldsymbol{\omega}_i$ and an uncertain one $\boldsymbol{\omega}_i \times \Delta \boldsymbol{J}_i \boldsymbol{\omega}_i$. Similarly, $\boldsymbol{N}_i$ is composed of a known part $\boldsymbol{N}_{i0}$ and an uncertain one $\Delta \boldsymbol{N}_i$. Also, $\boldsymbol{C}(\boldsymbol{\delta}_i)$ consists of a known component $\boldsymbol{C}_0(\boldsymbol{\delta}_i)$ and an uncertain term $\Delta \boldsymbol{C}(\boldsymbol{\delta}_i)$.

Moreover, during flight the flight control surfaces inevitably suffer from actuator faults, given by Zhang et al. [28]

$$\delta_{li} = \rho_{\delta_{li}} \delta_{li0} + \delta_{lif}, \tag{7.15}$$

where δ_{li0} is the designed control input with $l = \{a, e, r\}$, $\rho_{\delta_{li}}$ is the unknown actuator efficiency factor satisfying $0 \leq \rho_{\delta_{li}} \leq 1$, and δ_{lif} is the bounded unknown bias fault. Based on (7.15), consider the following cases:

(i) $\rho_{\delta_{li}} = 1$ and $\delta_{lif} = 0$. This refers to the fault-free case.
(ii) $0 < \rho_{\delta_{li}} < 1$ and $\delta_{lif} = 0$. This indicates the partial loss of effectiveness.
(iii) $\rho_{\delta_{li}} = 1$ and $\delta_{lif} \neq 0$. This means the bias fault.
(iv) $\rho_{\delta_{li}} = 1$ and $\delta_{lif} \neq 0$. This implies that δ_{li} is stuck at bounded time-varying function δ_{lif}.

Then, the actuator fault model (7.15) is reformulated as

$$\boldsymbol{\delta}_i = \boldsymbol{\rho}_{\delta_i} \boldsymbol{\delta}_{i0} + \boldsymbol{\delta}_{if}, \tag{7.16}$$

where $\boldsymbol{\delta}_{i0} = [\delta_{ai0}, \delta_{ei0}, \delta_{ri0}]^{\mathrm{T}}$, $\boldsymbol{\rho}_{\delta_i} = \mathrm{diag}\left\{\rho_{\delta_{ai}}, \rho_{\delta_{ei}}, \rho_{\delta_{ri}}\right\}$, $\boldsymbol{\delta}_{if} = \left[\delta_{aif}, \delta_{eif}, \delta_{rif}\right]^{\mathrm{T}}$.

Considering above actuator faults and modeling uncertainties, the angular velocity model (7.4) is rewritten as

$$\boldsymbol{J}_{i0} \dot{\boldsymbol{\omega}}_i = -\boldsymbol{\omega}_i \times \boldsymbol{J}_{i0} \boldsymbol{\omega}_i + \boldsymbol{N}_{i0} + \boldsymbol{C}_{i0} \boldsymbol{\delta}_{i0} + \boldsymbol{\Delta}_{\omega i}, \tag{7.17}$$

where $\boldsymbol{\Delta}_{\omega i} = -\Delta \boldsymbol{J}_i \dot{\boldsymbol{\omega}}_i - \boldsymbol{\omega}_i \times \Delta \boldsymbol{J}_i \boldsymbol{\omega}_i + \Delta \boldsymbol{N}_i + \boldsymbol{C}_{i0} \boldsymbol{\delta}_{if} + \boldsymbol{C}_{i0} \left(\boldsymbol{\rho}_i - \boldsymbol{I}_3\right) \boldsymbol{\delta}_{i0} + \Delta \boldsymbol{C}_i \boldsymbol{\delta}_i + \boldsymbol{J}_i \boldsymbol{d}_{\omega i}$ is the lumped uncertainties induced by the actuator faults and modeling uncertainties.

Remark 7.2 In [8], only the attitude model (rotational dynamics) is studied. While in [9], the longitudinal dynamics is just considered. Meanwhile, [10] and [11] regard each UAV as a mass point, neglecting the problem of body posture. In the perspective of fixed-wing UAV, this chapter focuses on the underlying control of both translational and rotational dynamics. With the control law stabilizing the rotational dynamics through δ_{ai}, δ_{ei}, and δ_{ri}, the translational dynamics are controlled by T_{xi}.

7.2.3 *Basic Graph Theory*

The undirected graph always describes the connections among the multiple UAVs. Given an undirected graph $\mathcal{G} = (\mathcal{V}, \mathcal{E}, \mathcal{A})$, it consists of nodes $\mathcal{V} = \{v_1, \cdots, v_n\}$ and the sets of edges $\mathcal{E} = \{(i, j),\ i, j \in \mathcal{V}, \text{and } i \neq j\}$, and $\mathcal{A} = [a_{ij}] \in \Re^{n\times n}$ denotes the weighted adjacency matrix of $\mathcal{G}$. If there exists an edge between nodes i and j, then $a_{ij} = a_{ji} \neq 0$, and otherwise $a_{ij} = a_{ji} = 0$. Moreover, a_{ii} equals 0 for all $i \in \mathcal{V}$. The neighbors of node i are denoted by the set $\mathcal{N}_i = \{v_j : (v_j, v_i) \in \mathcal{E}\}$. The Laplacian matrix is defined by $\mathcal{L} = \mathcal{D} - \mathcal{A}$, where $\mathcal{D} = \text{diag}\{d_1, \cdots, d_n\}$ with $d_i = \sum_{j=1}^{n} a_{ij}$. $\mathcal{G}$ is connected if there is a path from every node to each other node. Moreover, consider the virtual leader governed by the same dynamics as (7.1)–(7.4), where V_r and $\boldsymbol{\varphi}_r$ are velocity and attitude angles of the virtual leader, respectively.

For analysis, some useful lemmas and assumptions are made as follows:

Lemma 7.1 ([31]) *The Laplacian matrix $\mathcal{L}$ is symmetric positive semidefinite under the assumption that $\mathcal{G}$ is an undirected graph.*

Lemma 7.2 ([32]) *For $\boldsymbol{x}_i \in \Re^m$, $\boldsymbol{y}_i \in \Re^m$, $\boldsymbol{y}_j \in \Re^m$, $\boldsymbol{K}_i \in \Re^{m\times m}$, and $\mathcal{A} = [a_{ij}] \in \Re^{n\times n}$, if $\mathcal{A}$ is symmetric, it holds that*

$$\sum_{i=1}^{n}\sum_{j=1}^{n} a_{ij}\boldsymbol{x}_i^{\mathrm{T}} \tanh\left(\boldsymbol{K}_i\left(\boldsymbol{y}_i - \boldsymbol{y}_j\right)\right) = \frac{1}{2}\sum_{i=1}^{n}\sum_{j=1}^{n} a_{ij}\left(\boldsymbol{x}_i - \boldsymbol{x}_j\right)^{\mathrm{T}} \tanh\left(\boldsymbol{K}_i\left(\boldsymbol{y}_i - \boldsymbol{y}_j\right)\right). \tag{7.18}$$

Lemma 7.3 ([33]) *Given $\boldsymbol{x}_i \in \Re^n$, the following inequation holds:*

$$\boldsymbol{x}_i^{\mathrm{T}} \tanh\left(\boldsymbol{x}_i\right) \geq \tanh^{\mathrm{T}}\left(\boldsymbol{x}_i\right) \tanh\left(\boldsymbol{x}_i\right) \geq 0. \tag{7.19}$$

Assumption 7.1 The graph $\mathcal{G}$ is undirected and connected, and V_r, $\boldsymbol{\varphi}_r$ are directly available to all UAVs.

7.2.4 Control Objective

Consider a group of N UAVs with an event-triggered communication graph. The aim is to design the distributed cooperative tracking controller with a dynamic event-triggered communication mechanism such that:

- All the closed-loop signals are uniformly ultimately bounded (UUB).
- The leader-following consensus tracking for velocity and attitude is guaranteed by $\lim_{t\to\infty}|V_i - V_r| = 0$, $\lim_{t\to\infty}\|\boldsymbol{\varphi}_i - \boldsymbol{\varphi}_r\| = 0$, $i = 1, 2, \ldots N$, in the presence of actuator faults, external disturbances, and model uncertainties.
- The communication burden for multi-UAV system is considerably reduced. Meantime, the Zeno behavior is precluded such that the time between two consecutive broadcasts is lower bounded by a positive constant.

7.3 Main Results

7.3.1 Dynamic Event-Triggered Communication Mechanism

We are at the position to provide the triggering communication conditions for translational and rotational kinematics, respectively.

Define $\tilde{V}_i = V_i - V_r$. The triggering function for velocity tracking is presented as

$$f_i\left(t, e_{vi}(t)\right) = \left|e_{vi}(t)\right| - \varsigma_{v2} - \frac{\alpha_v \sum\limits_{j\in\mathcal{N}_i} a_{ij}\left|\tanh\left(\tilde{V}_i\left(t_{k_i}^{vi}\right) - \tilde{V}_j\left(t_{k_j}^{vj}\right)\right)\right|^2}{2\left(\left|\sum\limits_{j\in\mathcal{N}_i} a_{ij}\tanh\left(\tilde{V}_i\left(t_{k_i}^{vi}\right) - \tilde{V}_j\left(t_{k_j}^{vj}\right)\right)\right| + \varsigma_{v1}\right)}, \tag{7.20}$$

where $0 < \alpha_v < 1$, $\varsigma_{v1} > 0$, $\varsigma_{v2} = e^{c_{v1}\tanh(c_{v2}t)}$ is a turning function with $c_{v1} > 0$, $c_{v2} > 0$, $t_{k_i}^{vi}$ $(k_i = 1, 2, \ldots)$ is the latest triggering instant of velocity channel, $e_{vi}(t)$ is the state-based error between the sampled state and continuous state, defined as $e_{vi}(t) = \tilde{V}_i\left(t_{k_i}^{vi}\right) - \tilde{V}_i(t)$, $t \in \left[t_{k_i}^{vi}, t_{k_i+1}^{vi}\right)$, where $t_{k_i+1}^{vi} := \inf\{t > t_{k_i}^{vi} + \tau_{\min}^{vi} | f_i(t, e_{vi}(t)) > 0\}$ with $\tau_{\min}^{vi} \in \Re_{>0}$ being the minimum inter-event time of velocity tracking.

Define $\tilde{\varphi}_i = \varphi_i - \varphi_r$, $\tilde{\omega}_i = \omega_i - \omega_{di}$, where ω_{di} is the designed intermediate control. Then the triggering function for attitude tracking gives

$$f_i(t, e_{\varphi i}(t)) = \left\|e_{\varphi i}(t)\right\| - \varsigma_{\varphi 2} - \frac{\alpha_\varphi \sum_{j\in\mathcal{N}_i} a_{ij} \left\|\tanh\left(\tilde{\omega}_i\left(t_{k_i}^{\varphi i}\right) - \tilde{\omega}_j\left(t_{k_j}^{\varphi j}\right)\right)\right\|^2}{2\left(\left\|\sum_{j\in\mathcal{N}_i} a_{ij}\tanh\left(\tilde{\omega}_i\left(t_{k_i}^{\varphi i}\right) - \tilde{\omega}_j\left(t_{k_j}^{\varphi j}\right)\right)\right\| + \varsigma_{\varphi 1}\right)}, \tag{7.21}$$

where $0 < \alpha_\varphi < 1$, $\varsigma_{\varphi 1} > 0$, $\varsigma_{\varphi 2} = e^{c_{\varphi 1}\tanh(c_{\varphi 2}t)}$ with $c_{\varphi 1} > 0$, $c_{\varphi 2} > 0$, $t_{k_i}^{\varphi i}$ is the latest triggering instant of attitude dynamics, $e_{\varphi i}(t) = \tilde{\omega}_i(t_{k_i}^{\varphi i}) - \tilde{\omega}_i(t)$, $t \in [t_{k_i}^{\varphi i}, t_{k_i+1}^{\varphi i})$ with $t_{k_i+1}^{\varphi i} := \inf\{t > t_{k_i}^{\varphi i} + \tau_{\min}^{\varphi i} | f_i(t, e_{\varphi i}(t)) > 0\}$, where $\tau_{\min}^{\varphi i} \in \Re_{>0}$ is the lower bound on inter-event time of attitude tracking.

Remark 7.3 Under the proposed event-triggered communication strategy, if $|e_{vi}(t)|$ and $\left\|e_{\varphi i}(t)\right\|$ are greater than their corresponding thresholds, i.e., $f_i(t, e_{vi}(t)) > 0$, $f_i\left(t, e_{\varphi i}(t)\right) > 0$, the communication is triggered, and the velocity and attitude information is allowed to broadcast to its neighbors. That is, there is no communication between the ith UAV and its neighbors during $t \in [t_{k_i}^{vi}, t_{k_i+1}^{vi})$ and $t \in [t_{k_i}^{\varphi i}, t_{k_i+1}^{\varphi i})$, which hence provides more efficient use of the communication network and, however, in turn, causes allowable degradation of control performance. More specifically, decreasing $c_{v1}, c_{\varphi 1}, c_{v2}, c_{\varphi 2}, \alpha_v$, and α_φ can lead to more accurate tracking maneuvers, but at the cost of more communication resource. Conversely, increasing c_{v1}, $c_{\varphi 1}$, c_{v2}, $c_{\varphi 2}$, α_v, and α_φ results in less communication consumption but larger tracking errors. Thus a compromise between the control performance and communication load should be made by choosing appropriate design parameters of $c_{v1}, c_{\varphi 1}, c_{v2}, c_{\varphi 2}, \alpha_v$, and α_φ.

Remark 7.4 Different from existing triggering mechanisms [14–18], wherein threshold parameters are permanently fixed, the specified thresholds of proposed triggering functions (7.20)–(7.21) are based on a state-dependent term and a time-dependent term. Specifically, the state-dependent term is constant in interevent interval unless its neighbors broadcast the new states to the ith UAV, while the time-dependent term initially increases and then keeps constant over time, which plays an important role in the elimination of Zeno behavior. Meanwhile, compared with the dynamic triggering rules for linearized dynamics of multi-agent system [22, 23] or linear cases of individual system [24, 25], this chapter focuses on the multi-UAV system, integrates the exchange of velocity and attitude states into the triggering functions, and then proposes the novel dynamic triggering communication mechanisms for translational and rotational dynamics, respectively.

The event-based control protocol combines the communication triggering strategy within the coordinated controller design. Each UAV has three tasks: (i) receiving the neighbors' latest transmitted information $\tilde{V}_j(t_{k_j}^{vj})$, $\tilde{\boldsymbol{\omega}}_j(t_{k_j}^{\varphi j})$ and updating its controller immediately, (ii) monitoring its own state continuously and processing the triggered data with respect to every event-based condition, and (iii) updating the controller and broadcasting the latest triggered data to the neighbors as soon as the event-based condition is satisfied. Based on the above analysis, T_{xi0} and $\boldsymbol{\delta}_{i0}$ are constructed independently. The overall block diagram of the proposed control scheme is depicted in Fig. 7.1.

7.3.2 Event-Triggered Controller Design for Velocity Subsystem

To compensate the unknown disturbance Δ_{vi}, the disturbance observer is designed as

$$\begin{cases} \hat{\Delta}_{vi} = \hat{x}_{vi} + k_{i1}\tilde{V}_i \\ \dot{\hat{x}}_{vi} = -k_{i1}\left(\frac{u_i T_{xi0}}{m_i V_i} + f_{vi} + \hat{x}_{vi} + k_{i1}\tilde{V}_i\right), \end{cases} \tag{7.22}$$

where k_{i1} is a positive design parameter and $\hat{x}_{vi}$ is the state of disturbance observer.

Let us now design the control law T_{xi0} as follows

$$\begin{aligned} T_{xi0} = & -\sigma_{vi}\frac{m_i V_i}{u_i}\sum_{j\in\mathcal{N}_i} a_{ij}\tanh\left(\tilde{V}_i\left(t_{k_i}^{vi}\right) - \tilde{V}_j\left(t_{k_j}^{vj}\right)\right) \\ & -\frac{m_i V_i}{u_i}\left(k_{i2}\tilde{V}_i + f_{vi} + \hat{\Delta}_{vi}\right), \end{aligned} \tag{7.23}$$

where σ_{vi} and k_{i2} are positive design parameters.

7.3.3 Event-Triggered Controller Design for Attitude Subsystem

To estimate the lumped disturbance $\boldsymbol{\Delta}_{\omega i}$, design the disturbance observer as

$$\begin{cases} \hat{\boldsymbol{\Delta}}_{\omega i} = \hat{\boldsymbol{x}}_{\omega i} + \boldsymbol{K}_{i2}\tilde{\boldsymbol{\omega}}_i \\ \dot{\hat{\boldsymbol{x}}}_{\omega i} = -\boldsymbol{K}_{i2}\boldsymbol{J}_{i0}^{-1}\left(\boldsymbol{G}_i + \boldsymbol{C}_{i0}\boldsymbol{\delta}_{i0} + \hat{\boldsymbol{x}}_{\omega i} + \boldsymbol{K}_{i2}\tilde{\boldsymbol{\omega}}_i\right), \end{cases} \tag{7.24}$$

where $\boldsymbol{K}_{i2} \in \Re^{3\times 3}$ is symmetric and positive definite. $\hat{\boldsymbol{x}}_{\omega i} = \left[\hat{x}_{\omega 1}, \hat{x}_{\omega 2}, \hat{x}_{\omega 3}\right]^{\mathrm{T}}$ is the state of the proposed disturbance observer.

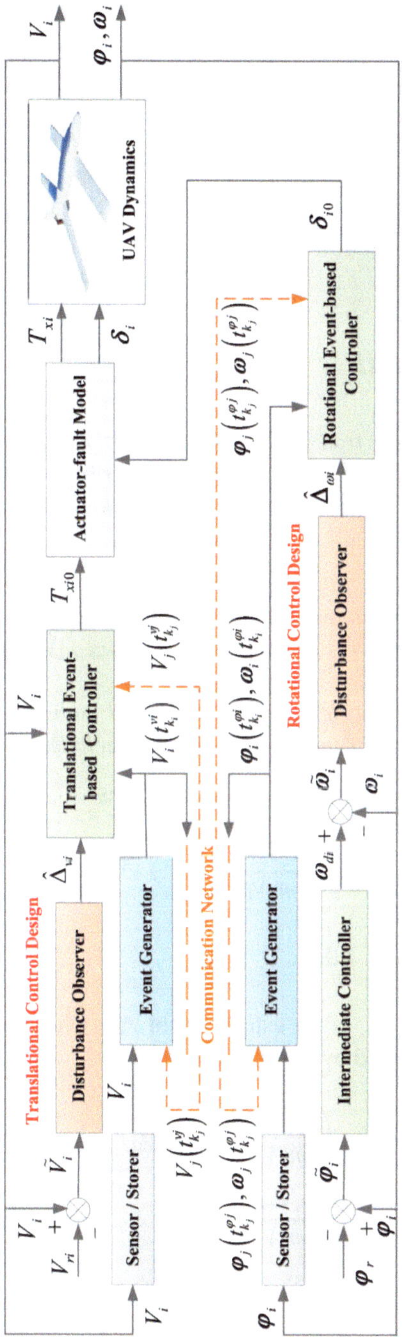

Fig. 7.1 Overall framework of the proposed control strategy

Design the event-based coordinated controller as

$$\begin{aligned}\boldsymbol{\delta}_{i0} =& -\sigma_{\varphi i}\boldsymbol{C}_{i0}^{-1}\sum_{j\in\mathcal{N}_i} a_{ij}\tanh\left(\tilde{\boldsymbol{\omega}}_i\left(t_{k_i}^{\varphi i}\right)-\tilde{\boldsymbol{\omega}}_j\left(t_{k_j}^{\varphi j}\right)\right)\\ &-\boldsymbol{C}_{i0}^{-1}\left(\boldsymbol{K}_{i3}\tilde{\boldsymbol{\omega}}_i+\left(\boldsymbol{R}_2^{-1}\right)^{\mathrm{T}}\tilde{\boldsymbol{\varphi}}_i+\boldsymbol{G}_i+\hat{\boldsymbol{\Delta}}_{\omega i}\right),\end{aligned} \tag{7.25}$$

where $\sigma_{\varphi i}$ is a positive design parameter and $\boldsymbol{K}_{i3} \in \Re^{3\times 3}$ is a symmetric positive definite matrix.

7.3.4 Stability Analysis

Theorem 7.1 *Consider the translational kinematics (7.14) composed of the coordinated tracking controller (7.23) and disturbance observer (7.22). Let Assumption 7.1 hold. There exist adjustable parameters k_{i1}, k_{i2}, and σ_{vi} ($i = 1, 2, \ldots, N$) such that:*

(1) All signals of the translational subsystem are UUB in the presence of actuator faults, modeling uncertainties, and external disturbances.
(2) Velocity tracking error $\tilde{V}_i$ is guaranteed that $\lim_{t\to\infty}|\tilde{V}_i| \le \mu_{Vi}$ with $\mu_{Vi} > 0$ a constant.
(3) The communication burden for translational subsystem is considerably reduced. Meanwhile the Zeno phenomenon is precluded.

Proof Noting (7.14), we obtain the derivative of $\tilde{V}_i$ as

$$\dot{\tilde{V}}_i = \frac{u_i T_{xi0}}{m_i V_i} + f_{vi} + \Delta_{vi}, \tag{7.26}$$

where $f_{vi} = \frac{\boldsymbol{v}_i^{\mathrm{T}}}{V_i}(\boldsymbol{R}_1^{\mathrm{T}}(\boldsymbol{\varphi}_i)\boldsymbol{g} + \frac{\boldsymbol{F}_{i0}}{m_i}) - \dot{V}_r$.

With (7.22), the term $\hat{\Delta}_{vi}$ is the estimation of Δ_{vi}. The evolution of $\dot{\hat{\Delta}}_{vi}$ along (7.22) and (7.26) is described by

$$\begin{aligned}\dot{\hat{\Delta}}_{vi} =& -k_{i1}\left(\frac{u_i T_{xi0}}{m_i V_i} + f_{vi} + \hat{x}_{vi} + k_{i1}\tilde{V}_i\right)\\ &+k_{i1}\left(\frac{u_i T_{xi0}}{m_i V_i} + f_{vi} + \Delta_{vi}\right) = k_{i1}\tilde{\Delta}_{vi}\end{aligned} \tag{7.27}$$

where $\tilde{\Delta}_{vi} = \Delta_{vi} - \hat{\Delta}_{vi}$ denotes the estimation error of Δ_{vi}.

Consider the candidate Lyapunov function

$$L_1 = \frac{1}{2}\sum_{i=1}^{N}\tilde{V}_i^2 + \frac{1}{2}\sum_{i=1}^{N}\tilde{\Delta}_{vi}^2. \tag{7.28}$$

Taking the time derivative of (7.28) along (7.26) and (7.27) gives

$$\begin{aligned}\dot{L}_1 =& \sum_{i=1}^{N}\tilde{V}_i\left(\frac{u_i T_{xi0}}{m_i V_i} + f_{vi} + \Delta_{vi}\right)\\ &+ \sum_{i=1}^{N}\left(-k_{i1}\tilde{\Delta}_{vi}^2 + \tilde{\Delta}_{vi}\dot{\Delta}_{vi}\right).\end{aligned} \tag{7.29}$$

Substituting (7.23) into (7.29) and invoking Lemma 7.2, it follows that

$$\begin{aligned}\dot{L}_1 \leq& -\sum_{i=1}^{N}\sigma_{vi}\tilde{V}_i\sum_{j\in\mathcal{N}_i}a_{ij}\tanh\left(\tilde{V}_i\left(t_{k_i}^{vi}\right) - \tilde{V}_j\left(t_{k_j}^{vj}\right)\right)\\ &- \sum_{i=1}^{N}k_{i2}\tilde{V}_i^2 + \sum_{i=1}^{N}\tilde{V}_i\tilde{\Delta}_{vi} - \sum_{i=1}^{N}\left(k_{i1}\tilde{\Delta}_{vi}^2 - \tilde{\Delta}_{vi}\dot{\Delta}_{vi}\right)\\ =& \sum_{i=1}^{N}\sigma_{vi}e_{vi}\sum_{j\in\mathcal{N}_i}a_{ij}\tanh\left(\tilde{V}_i\left(t_{k_i}^{vi}\right) - \tilde{V}_j\left(t_{k_j}^{vj}\right)\right)\\ &- \frac{1}{2}\sum_{i=1}^{N}\sigma_{vi}\sum_{j\in\mathcal{N}_i}a_{ij}\left(\tilde{V}_i\left(t_{k_i}^{vi}\right) - \tilde{V}_j\left(t_{k_j}^{vj}\right)\right)\tanh\left(\tilde{V}_i\left(t_{k_i}^{vi}\right) - \tilde{V}_j\left(t_{k_j}^{vj}\right)\right)\\ &- \sum_{i=1}^{N}k_{i2}\tilde{V}_i^2 + \sum_{i=1}^{N}\tilde{V}_i\tilde{\Delta}_{vi} - \sum_{i=1}^{N}\left(k_{i1}\tilde{\Delta}_{vi}^2 - \tilde{\Delta}_{vi}\dot{\Delta}_{vi}\right).\end{aligned} \tag{7.30}$$

Taking Lemma 7.3 into account, (7.30) further becomes

$$\begin{aligned}\dot{L}_1 \leq& \sum_{i=1}^{N}\sigma_{vi}\left|e_{vi}\right|\left|\sum_{j\in\mathcal{N}_i}a_{ij}\tanh\left(\tilde{V}_i\left(t_{k_i}^{vi}\right) - \tilde{V}_j\left(t_{k_j}^{vj}\right)\right)\right|\\ &- \frac{1}{2}\sum_{i=1}^{N}\sigma_{vi}\sum_{j\in\mathcal{N}_i}a_{ij}\left|\tanh\left(\tilde{V}_i\left(t_{k_i}^{vi}\right) - \tilde{V}_j\left(t_{k_j}^{vj}\right)\right)\right|^2\\ &- \sum_{i=1}^{N}k_{i2}\tilde{V}_i^2 + \sum_{i=1}^{N}\tilde{V}_i\tilde{\Delta}_{vi} - \sum_{i=1}^{N}\left(k_{i1}\tilde{\Delta}_{vi}^2 - \tilde{\Delta}_{vi}\dot{\Delta}_{vi}\right).\end{aligned} \tag{7.31}$$

Under the proposed event-triggered communication strategy, $f_i(t, e_{vi}(t)) \leq 0$ holds during $t \in \left[t_{k_i}^{vi}, t_{k_i+1}^{vi}\right)$, that is,

$$|e_{vi}| \leq \frac{\alpha_v \sum_{j \in \mathcal{N}_i} a_{ij} \left|\tanh\left(\tilde{V}_i\left(t_{k_i}^{vi}\right) - \tilde{V}_j\left(t_{k_j}^{vj}\right)\right)\right|^2}{2\left(\left|\sum_{j \in \mathcal{N}_i} a_{ij} \tanh\left(\tilde{V}_i\left(t_{k_i}^{vi}\right) - \tilde{V}_j\left(t_{k_j}^{vj}\right)\right)\right| + \varsigma_{v1}\right)} + \varsigma_{v2}. \tag{7.32}$$

Substituting (7.32) into (7.31) yields

$$\begin{aligned}
\dot{L}_1 \leq & \frac{\alpha_v - 1}{2} \sum_{i=1}^{N} \sigma_{vi} \sum_{j \in \mathcal{N}_i} a_{ij} \left(\left|\tanh\left(\tilde{V}_i\left(t_{k_i}^{vi}\right) - \tilde{V}_j\left(t_{k_j}^{vj}\right)\right)\right| - \frac{\varsigma_{v2}}{1 - \alpha_v}\right)^2 \\
& + \frac{\varsigma_{v2}^2}{2(1 - \alpha_v)} \sum_{i=1}^{N} \sigma_{vi} \sum_{j \in \mathcal{N}_i} a_{ij} - \sum_{i=1}^{N} k_{i2} \tilde{V}_i^2 \\
& + \sum_{i=1}^{N} \tilde{V}_i \tilde{\Delta}_{vi} - \sum_{i=1}^{N} \left(k_{i1} \tilde{\Delta}_{vi}^2 - \tilde{\Delta}_{vi} \dot{\Delta}_{vi}\right) \\
\leq & - \sum_{i=1}^{N} k_{i2} \tilde{V}_i^2 + \sum_{i=1}^{N} \tilde{V}_i \tilde{\Delta}_{vi} - \sum_{i=1}^{N} \left(k_{i1} \tilde{\Delta}_{vi}^2 - \tilde{\Delta}_{vi} \dot{\Delta}_{vi}\right) \\
& + \frac{\varsigma_{v2}^2}{2(1 - \alpha_v)} \sum_{i=1}^{N} \sigma_{vi} \sum_{j \in \mathcal{N}_i} a_{ij}.
\end{aligned} \tag{7.33}$$

Using Young's inequality, it can be checked that

$$\begin{aligned}
\dot{L}_1 \leq & - \sum_{i=1}^{N} \left(k_{i2} - \frac{1}{2}\right) \tilde{V}_i^2 - \sum_{i=1}^{N} (k_{i1} - 1) \tilde{\Delta}_{vi}^2 \\
& + \sum_{i=1}^{N} \frac{1}{2} \dot{\Delta}_{vi}^2 + \frac{\varsigma_{v2}^2}{2(1 - \alpha_v)} \sum_{i=1}^{N} \sigma_{vi} \sum_{j \in \mathcal{N}_i} a_{ij} \\
\leq & - \kappa_{vi} L_1 + \chi_{vi},
\end{aligned} \tag{7.34}$$

where from the boundedness of ρ_{T_i}, $\boldsymbol{v}_i$, T_{xi0}, $\underline{T_{xif}}$, $\Delta \boldsymbol{F}_i$, and $\boldsymbol{d}_{vi}$, there exists a positive constant $\overline{\Delta}_{vi}$ such that $|\dot{\Delta}_{vi}| \leq \overline{\Delta}_{vi}$. It can be concluded that $\kappa_{vi} = \min\{2k_{i2} - 1, 2k_{i1} - 2\}$ with $k_{i1} > 1$, $k_{i2} > \frac{1}{2}$, $\chi_{vi} = \frac{1}{2} \sum_{i=1}^{N} \overline{\Delta}_{vi}^2 + \frac{\varsigma_{v2}^2}{2(1 - \alpha_{v2})} \sum_{i=1}^{N} \sigma_{vi} \sum_{j \in \mathcal{N}_i} a_{ij}$.

Integrating (7.34) over $[0, t]$ leads to

$$0 \leq L_1 \leq \left[\frac{\chi_{vi}}{\kappa_{vi}} + L_1(0) \right] e^{-\kappa_{vi} t} + \frac{\chi_{vi}}{\kappa_{vi}}. \tag{7.35}$$

As a result, all closed-loop signals in translational dynamics are UUB. Next, we prove that no Zeno behavior exhibits during maneuvers. That is, there exists a $t_v^* > 0$ satisfying $\left\{ t_{k_i+1}^{vi} - t_{k_i}^{vi} \right\} \geq t_v^*$. Recalling the definition of $e_{vi}(t)$, the evolution of $|e_{vi}(t)|$ over the time interval $t \in [t_{k_i}^{\varphi i}, t_{k_i+1}^{\varphi i})$ satisfies

$$\frac{d}{dt} |e_{vi}| \leq \left| \frac{d}{dt} \left(\tilde{V}_i \left(t_{k_i}^{vi} \right) - \tilde{V}_i(t) \right) \right| = \left| \dot{\tilde{V}}_i(t) \right|. \tag{7.36}$$

$\dot{\tilde{V}}_i(t)$ has been proved bounded in the previous analysis. Therefore, there exists a positive constant μ_{vi} such that $|\dot{\tilde{V}}_i(t)| \leq \mu_{vi}$. Since $e_{vi}\left(t_{k_i}^{vi}\right) = 0$, we see that $|e_{vi}| \leq \mu_{vi} t_v^*$. Further, we have

$$\frac{\alpha_v \sum_{j \in \mathcal{N}_i} a_{ij} \left| \tanh \left(\tilde{V}_i \left(t_{k_i}^{vi} \right) - \tilde{V}_j \left(t_{k_j}^{vj} \right) \right) \right|^2}{2 \left(\left| \sum_{j \in \mathcal{N}_i} a_{ij} \tanh \left(\tilde{V}_i \left(t_{k_i}^{vi} \right) - \tilde{V}_j \left(t_{k_j}^{vj} \right) \right) \right| + \varsigma_{v1} \right)} + \varsigma_{v2} \leq \mu_{vi} t_v^*. \tag{7.37}$$

From (7.37), $\mu_{vi} t_v^* \geq \varsigma_{v2}$ holds. Additionally, considering $1 \leq \varsigma_{v2} \leq e^{c_{v1}}$, it is verified that $t_v^* \geq 1/\mu_{vi} > 0$. Clearly, the Zeno behavior is successfully avoided.

This completes the proof of Theorem 7.1.

Remark 7.5 It should be noticed that $\tilde{V}_i$ can be made arbitrarily small by increasing k_{i1}, k_{i2}, and σ_{vi}. Moreover, the choice of c_{v1}, c_{v2}, and α_v should balance the control performance and triggering number of communication. This will be shown in the following numerical example.

Remark 7.6 Compared with the cooperative controllers of UAVs containing continuous exchanges of state information among neighbors [8–14], the proposed T_{xi0} only uses the neighbors state $\tilde{V}_j(t_{k_j}^{vj})$ in control protocol (7.23) at event-triggered instants, which greatly reduces the communication burden during $t \in [t_{k_j}^{vj}, t_{k_j+1}^{vj})$.

Theorem 7.2 *Consider the rotational kinematics (7.3), (7.17) composed of the coordinated tracking controller (7.25) and disturbance observer (7.24). Let Assumption 7.1 hold. There exist adjustable parameters* $\boldsymbol{K}_{i1}$, $\boldsymbol{K}_{i2}$, $\boldsymbol{K}_{i3}$, $\sigma_{\varphi i}$ *(*$i = 1, 2, \ldots, N$*) such that:*

(1) All signals of the rotational subsystem are UUB in the presence of actuator faults, modeling uncertainties, and external disturbances.

(2) *Attitude tracking error* $\tilde{\boldsymbol{\varphi}}_l$ *is guaranteed that* $\lim_{t\to\infty}\|\tilde{\boldsymbol{\varphi}}_i\| \leq \mu_{\varphi l}$ *with* $\mu_{\varphi l} > 0$ *a constant.*
(3) *The communication burden for rotational subsystem is considerably reduced. Meanwhile the Zeno phenomenon is avoided.*

Proof The rotational control design comprises two steps:

Step 1: Taking the time derivative of $\frac{1}{2}\sum_{l=1}^{N}\tilde{\boldsymbol{\varphi}}_l^{\mathrm{T}}\tilde{\boldsymbol{\varphi}}_l$ along (7.3) yields

$$\sum_{l=1}^{N}\tilde{\boldsymbol{\varphi}}_l^{\mathrm{T}}\dot{\tilde{\boldsymbol{\varphi}}}_l = \sum_{l=1}^{N}\tilde{\boldsymbol{\varphi}}_l^{\mathrm{T}}\left[\boldsymbol{R}_2^{-1}\left(\tilde{\boldsymbol{\omega}}_l + \boldsymbol{\omega}_{dl}\right) - \dot{\boldsymbol{\varphi}}_r\right]. \tag{7.38}$$

We design the intermediate control signal as

$$\boldsymbol{\omega}_{dl} = \boldsymbol{R}_2\left(-\boldsymbol{K}_{l1}\tilde{\boldsymbol{\varphi}}_l + \dot{\boldsymbol{\varphi}}_r\right), \tag{7.39}$$

where $\boldsymbol{K}_{l1} \in \Re^{3\times 3}$ is symmetric and positive definite.

Substituting (7.39) into (7.38) leads to

$$\sum_{l=1}^{N}\tilde{\boldsymbol{\varphi}}_l^{\mathrm{T}}\dot{\tilde{\boldsymbol{\varphi}}}_l = \sum_{l=1}^{N}\left(-\tilde{\boldsymbol{\varphi}}_l^{\mathrm{T}}\boldsymbol{K}_{l1}\tilde{\boldsymbol{\varphi}}_l + \tilde{\boldsymbol{\varphi}}_l^{\mathrm{T}}\boldsymbol{R}_2^{-1}\tilde{\boldsymbol{\omega}}_l\right). \tag{7.40}$$

Step 2: For the purpose of forcing $\boldsymbol{\omega}_l$ to track $\boldsymbol{\omega}_{dl}$, from (7.17), it holds that

$$\begin{aligned}\boldsymbol{J}_{l0}\dot{\tilde{\boldsymbol{\omega}}}_l &= -\boldsymbol{\omega}_l \times \boldsymbol{J}_{l0}\boldsymbol{\omega}_l + \boldsymbol{N}_{l0} + \boldsymbol{C}_{l0}\boldsymbol{\delta}_{l0} + \boldsymbol{\Delta}_{\omega l} - \boldsymbol{J}_{l0}\dot{\boldsymbol{\omega}}_{dl}\\ &= \boldsymbol{G}_l + \boldsymbol{C}_{l0}\boldsymbol{\delta}_{l0} + \boldsymbol{\Delta}_{\omega l},\end{aligned} \tag{7.41}$$

where $\boldsymbol{G}_l = -\boldsymbol{\omega}_l \times \boldsymbol{J}_{l0}\boldsymbol{\omega}_l + \boldsymbol{N}_{l0} - \boldsymbol{J}_{l0}\dot{\boldsymbol{\omega}}_{dl}$.

Taking the time derivative of (7.24), one has

$$\begin{aligned}\dot{\tilde{\boldsymbol{\Delta}}}_{\omega l} &= -\boldsymbol{K}_{l2}\boldsymbol{J}_{l0}^{-1}\left(\boldsymbol{G}_l + \boldsymbol{C}_{l0}\boldsymbol{\delta}_{l0} + \hat{\boldsymbol{x}}_{\omega l} + \boldsymbol{K}_{l2}\tilde{\boldsymbol{\omega}}_l\right)\\ &\quad + \boldsymbol{K}_{l2}\boldsymbol{J}_{l0}^{-1}\left(\boldsymbol{G}_l + \boldsymbol{C}_{l0}\boldsymbol{\delta}_{l0} + \boldsymbol{\Delta}_{\omega l}\right) = \boldsymbol{K}_{l2}\boldsymbol{J}_{l0}^{-1}\tilde{\boldsymbol{\Delta}}_l,\end{aligned} \tag{7.42}$$

where $\tilde{\boldsymbol{\Delta}}_{\omega l} = \boldsymbol{\Delta}_{\omega l} - \hat{\boldsymbol{\Delta}}_{\omega l}$ is the estimation error of $\boldsymbol{\Delta}_{\omega l}$.

Choose the candidate Lyapunov function as

$$L_2 = \frac{1}{2}\sum_{l=1}^{N}\tilde{\boldsymbol{\varphi}}_l^{\mathrm{T}}\tilde{\boldsymbol{\varphi}}_l + \frac{1}{2}\sum_{l=1}^{N}\tilde{\boldsymbol{\omega}}_l^{\mathrm{T}}\boldsymbol{J}_{l0}\tilde{\boldsymbol{\omega}}_l + \frac{1}{2}\sum_{l=1}^{N}\tilde{\boldsymbol{\Delta}}_{\omega i}^{\mathrm{T}}\boldsymbol{J}_{i0}\tilde{\boldsymbol{\Delta}}_{\omega l}. \tag{7.43}$$

From (7.40), (7.41), and (7.42), taking the time derivative of L_2 leads to

$$\dot{L}_2 = \sum_{l=1}^{N}\left(-\tilde{\boldsymbol{\varphi}}_l^{\mathrm{T}}\boldsymbol{K}_{l1}\tilde{\boldsymbol{\varphi}}_l + \tilde{\boldsymbol{\varphi}}_l^{\mathrm{T}}\boldsymbol{R}_2^{-1}\tilde{\boldsymbol{\omega}}_l\right)$$
$$+\sum_{l=1}^{N}\tilde{\boldsymbol{\omega}}_l^{\mathrm{T}}\left(\boldsymbol{G}_l + \boldsymbol{C}_{l0}\boldsymbol{\delta}_{l0} + \boldsymbol{\Delta}_{\omega l}\right)$$
$$-\sum_{l=1}^{N}\left(\tilde{\boldsymbol{\Delta}}_{\omega l}^{\mathrm{T}}\boldsymbol{K}_{l2}\tilde{\boldsymbol{\Delta}}_{\omega l} - \tilde{\boldsymbol{\Delta}}_{\omega l}^{\mathrm{T}}\boldsymbol{J}_{l0}\dot{\boldsymbol{\Delta}}_{\omega l}\right). \tag{7.44}$$

Substituting the controller (7.25) into (7.44) gives

$$\dot{L}_2 = -\sum_{l=1}^{N}\sigma_{\varphi l}\tilde{\boldsymbol{\omega}}_l^{\mathrm{T}}\sum_{j\in\mathcal{N}_l}a_{lj}\tanh\left(\tilde{\boldsymbol{\omega}}_l\left(t_{k_l}^{\varphi l}\right) - \tilde{\boldsymbol{\omega}}_j\left(t_{k_j}^{\varphi j}\right)\right)$$
$$-\sum_{l=1}^{N}\tilde{\boldsymbol{\varphi}}_l^{\mathrm{T}}\boldsymbol{K}_{l1}\tilde{\boldsymbol{\varphi}}_l - \sum_{l=1}^{N}\tilde{\boldsymbol{\omega}}_l^{\mathrm{T}}\boldsymbol{K}_{l3}\tilde{\boldsymbol{\omega}}_l + \sum_{l=1}^{N}\tilde{\boldsymbol{\omega}}_l^{\mathrm{T}}\tilde{\boldsymbol{\Delta}}_{\omega l}$$
$$-\sum_{l=1}^{N}\left(\tilde{\boldsymbol{\Delta}}_{\omega l}^{\mathrm{T}}\boldsymbol{K}_{l2}\tilde{\boldsymbol{\Delta}}_{\omega l} - \tilde{\boldsymbol{\Delta}}_{\omega l}^{\mathrm{T}}\boldsymbol{J}_{l0}\dot{\boldsymbol{\Delta}}_{\omega l}\right). \tag{7.45}$$

By invoking Lemmas 7.2 and 7.3, (7.45) can be rewritten as

$$\dot{L}_2 \leq \sum_{l=1}^{N}\sigma_{\varphi l}\left\|\boldsymbol{e}_{\varphi l}\right\|\left\|\sum_{j\in\mathcal{N}_l}a_{lj}\tanh\left(\tilde{\boldsymbol{\omega}}_l\left(t_{k_l}^{\varphi l}\right) - \tilde{\boldsymbol{\omega}}_j\left(t_{k_j}^{\varphi j}\right)\right)\right\|$$
$$-\frac{1}{2}\sum_{l=1}^{N}\sigma_{\varphi l}\sum_{j\in\mathcal{N}_l}a_{lj}\left\|\tanh\left(\tilde{\boldsymbol{\omega}}_l\left(t_{k_l}^{\varphi l}\right) - \tilde{\boldsymbol{\omega}}_j\left(t_{k_j}^{\varphi j}\right)\right)\right\|^2$$
$$-\sum_{l=1}^{N}\tilde{\boldsymbol{\varphi}}_l^{\mathrm{T}}\boldsymbol{K}_{l1}\tilde{\boldsymbol{\varphi}}_l - \sum_{l=1}^{N}\tilde{\boldsymbol{\omega}}_l^{\mathrm{T}}\boldsymbol{K}_{l3}\tilde{\boldsymbol{\omega}}_l + \sum_{l=1}^{N}\tilde{\boldsymbol{\omega}}_l^{\mathrm{T}}\tilde{\boldsymbol{\Delta}}_{\omega l}$$
$$-\sum_{l=1}^{N}\left(\tilde{\boldsymbol{\Delta}}_{\omega l}^{\mathrm{T}}\boldsymbol{K}_{l2}\tilde{\boldsymbol{\Delta}}_{\omega l} - \tilde{\boldsymbol{\Delta}}_{\omega l}^{\mathrm{T}}\boldsymbol{J}_{l0}\dot{\boldsymbol{\Delta}}_{\omega l}\right). \tag{7.46}$$

With the proposed event-triggered communication mechanism, $f_l\left(t, \boldsymbol{e}_{\varphi l}(t)\right) \leq 0$ satisfies in the interval $t \in [t_{k_l}^{\varphi l}, t_{k_l+1}^{\varphi l})$, that is,

$$\left\|\boldsymbol{e}_{\varphi i}\right\| \leq \frac{\alpha_{\varphi} \sum_{j\in\mathcal{N}_i} a_{ij}\left\|\tanh\left(\tilde{\boldsymbol{\omega}}_i\left(t_{k_i}^{\varphi i}\right)-\tilde{\boldsymbol{\omega}}_j\left(t_{k_j}^{\varphi j}\right)\right)\right\|^2}{2\left(\left\|\sum_{j\in\mathcal{N}_i} a_{ij}\tanh\left(\tilde{\boldsymbol{\omega}}_i\left(t_{k_i}^{\varphi i}\right)-\tilde{\boldsymbol{\omega}}_j\left(t_{k_j}^{\varphi j}\right)\right)\right\|+\varsigma_{\varphi 1}\right)} + \varsigma_{\varphi 2}. \tag{7.47}$$

Inserting (7.47) into (7.46), we have

$$\begin{aligned}
\dot{L}_2 \leq & \frac{\alpha_{\varphi}-1}{2}\sum_{i=1}^{N}\sigma_{\varphi i}\sum_{j\in\mathcal{N}_i} a_{ij}\left(\left\|\tanh\left(\tilde{\boldsymbol{\omega}}_i\left(t_{k_i}^{\varphi i}\right)-\tilde{\boldsymbol{\omega}}_j\left(t_{k_j}^{\varphi j}\right)\right)\right\|\right. \\
& \left.-\frac{\varsigma_{\varphi 2}}{1-\alpha_{\varphi}}\right)^2 + \frac{\varsigma_{\varphi 2}^2}{2\left(1-\alpha_{\varphi}\right)}\sum_{i=1}^{N}\sigma_{\varphi i}\sum_{j\in\mathcal{N}_i} a_{ij} \\
& -\sum_{i=1}^{N}\tilde{\boldsymbol{\varphi}}_i^{\mathrm{T}}\boldsymbol{K}_{i1}\tilde{\boldsymbol{\varphi}}_i - \sum_{i=1}^{N}\tilde{\boldsymbol{\omega}}_i^{\mathrm{T}}\boldsymbol{K}_{i3}\tilde{\boldsymbol{\omega}}_i + \sum_{i=1}^{N}\tilde{\boldsymbol{\omega}}_i^{\mathrm{T}}\tilde{\boldsymbol{\Delta}}_{\omega i} \\
& -\sum_{i=1}^{N}\left(\tilde{\boldsymbol{\Delta}}_{\omega i}^{\mathrm{T}}\boldsymbol{K}_{i2}\tilde{\boldsymbol{\Delta}}_{\omega i} - \tilde{\boldsymbol{\Delta}}_{\omega i}^{\mathrm{T}}\boldsymbol{J}_{i0}\dot{\boldsymbol{\Delta}}_{\omega i}\right) \\
\leq & -\sum_{i=1}^{N}\tilde{\boldsymbol{\varphi}}_i^{\mathrm{T}}\boldsymbol{K}_{i1}\tilde{\boldsymbol{\varphi}}_i - \sum_{i=1}^{N}\tilde{\boldsymbol{\omega}}_i^{\mathrm{T}}\boldsymbol{K}_{i3}\tilde{\boldsymbol{\omega}}_i + \sum_{i=1}^{N}\tilde{\boldsymbol{\omega}}_i^{\mathrm{T}}\tilde{\boldsymbol{\Delta}}_{\omega i} \\
& -\sum_{i=1}^{N}\left(\tilde{\boldsymbol{\Delta}}_{\omega i}^{\mathrm{T}}\boldsymbol{K}_{i2}\tilde{\boldsymbol{\Delta}}_{\omega i} - \tilde{\boldsymbol{\Delta}}_{\omega i}^{\mathrm{T}}\boldsymbol{J}_{i0}\dot{\boldsymbol{\Delta}}_{\omega i}\right) \\
& + \frac{\varsigma_{\varphi 2}^2}{2\left(1-\alpha_{\varphi}\right)}\sum_{i=1}^{N}\sigma_{\varphi i}\sum_{j\in\mathcal{N}_i} a_{ij}.
\end{aligned} \tag{7.48}$$

Rearranging (7.48) gives

$$\begin{aligned}
\dot{L}_2 \leq & -\sum_{i=1}^{N}\lambda_{\min}\left(\boldsymbol{K}_{i1}\right)\tilde{\boldsymbol{\varphi}}_i^{\mathrm{T}}\tilde{\boldsymbol{\varphi}}_i - \sum_{i=1}^{N}\left(\lambda_{\min}\left(\boldsymbol{K}_{i3}\right)-\frac{1}{2}\right)\tilde{\boldsymbol{\omega}}_i^{\mathrm{T}}\tilde{\boldsymbol{\omega}}_i \\
& +\sum_{i=1}^{N}\frac{\lambda_{\max}\left(\boldsymbol{J}_{i0}\right)}{2}\dot{\boldsymbol{\Delta}}_{\omega i}^{\mathrm{T}}\dot{\boldsymbol{\Delta}}_{\omega i} + \frac{\varsigma_{\varphi 2}^2}{2\left(1-\alpha_{\varphi}\right)}\sum_{i=1}^{N}\sigma_{\varphi i}\sum_{j\in\mathcal{N}_i} a_{ij} \\
& -\sum_{i=1}^{N}\left(\lambda_{\min}\left(\boldsymbol{K}_{i2}\right)-\frac{\lambda_{\max}\left(\boldsymbol{J}_{i0}\right)+1}{2}\right)\tilde{\boldsymbol{\Delta}}_{\omega i}^{\mathrm{T}}\tilde{\boldsymbol{\Delta}}_{\omega i} \\
\leq & -\kappa_{\omega i}L_2 + \chi_{\omega i},
\end{aligned} \tag{7.49}$$

where from the boundedness of $\Delta \boldsymbol{J}_i$, $\dot{\boldsymbol{\omega}}_i$, $\boldsymbol{\omega}_i$, $\Delta \boldsymbol{N}_i$, $\boldsymbol{\delta}_{if}$, $\boldsymbol{C}_{i0}$, $\boldsymbol{\rho}_i$, $\boldsymbol{\delta}_{i0}$, $\Delta \boldsymbol{C}_i$, and $\boldsymbol{d}_{\omega i}$, there exists a positive constant $\overline{\Delta}_{\omega i}$ such that $\|\dot{\boldsymbol{\Delta}}_{\omega i}\| \leq \overline{\Delta}_{\omega i}$. It can be obtained that $\kappa_{\omega i} = \min\{2\lambda_{\min}(\boldsymbol{K}_{i1}), \frac{2\lambda_{\min}(\boldsymbol{K}_{i3})-1}{\lambda_{\max}(\boldsymbol{J}_{i0})}, \frac{2\lambda_{\min}(\boldsymbol{K}_{i2})-1}{\lambda_{\max}(\boldsymbol{J}_{i0})} - 1\}$ with $\boldsymbol{K}_{i2}$ and $\boldsymbol{K}_{i3}$ satisfying $\lambda_{\min}(\boldsymbol{K}_{i2}) > \frac{\lambda_{\max}(\boldsymbol{J}_{i0})+1}{2}$, $\lambda_{\min}(\boldsymbol{K}_{i3}) > \frac{1}{2}$, and $\chi_{\omega i} = \sum_{i=1}^{N} \frac{\lambda_{\max}(\boldsymbol{J}_{i0})}{2}\overline{\Delta}_{\omega i}^2 + \frac{\varsigma_{\varphi 2}^2}{2(1-\alpha_\varphi)} \sum_{i=1}^{N} \sigma_{\varphi i} \sum_{j\in\mathcal{N}_i} a_{ij}$.

Integrating (7.49) over $[0, t]$ yields

$$0 \leq L_2 \leq \left[\frac{\chi_{\omega i}}{\kappa_{\omega i}} + L_2(0)\right] e^{-\kappa_{\omega i} t} + \frac{\chi_{\omega i}}{\kappa_{\omega i}}. \tag{7.50}$$

As a result, all closed-loop signals of rotational subsystem are UUB. Further it is proved that the Zeno behavior can be avoided. That is, there exists a $t_\varphi^* > 0$ satisfying $\{t_{k_i+1}^{\varphi i} - t_{k_i}^{\varphi i}\} \geq t_\varphi^*$. Considering the definition of $\boldsymbol{e}_{\varphi i}$, differentiating $\left\|\boldsymbol{e}_{\varphi i}\right\|$ over time interval $t \in [t_{k_i}^{\varphi i}, t_{k_i+1}^{\varphi i})$ yields

$$\frac{d}{dt}\left\|\boldsymbol{e}_{\varphi i}(t)\right\| \leq \left\|\frac{d}{dt}\left(\tilde{\boldsymbol{\omega}}_i\left(t_{k_i}^{\varphi i}\right) - \tilde{\boldsymbol{\omega}}_i(t)\right)\right\| = \left\|\dot{\tilde{\boldsymbol{\omega}}}_i(t)\right\|. \tag{7.51}$$

$\dot{\tilde{\boldsymbol{\omega}}}_i(t)$ has been proved bounded based on the previous analysis. Therefore, there exists a positive constant $\mu_{\varphi i}$ such that $\|\dot{\tilde{\boldsymbol{\omega}}}_i(t)\| \leq \mu_{\varphi i}$. Since $\|\boldsymbol{e}_{\varphi i}(t_{k_i}^{\varphi i})\| = 0$, we obtain that $\left\|\boldsymbol{e}_{\varphi i}(t)\right\| \leq \mu_{\varphi i} t_\varphi^*$. In view of (7.47), there exists

$$\frac{\alpha_\varphi \sum_{j\in\mathcal{N}_i} a_{ij}\left\|\tanh\left(\tilde{\boldsymbol{\omega}}_i\left(t_{k_i}^{\varphi i}\right) - \tilde{\boldsymbol{\omega}}_j\left(t_{k_j}^{\varphi j}\right)\right)\right\|^2}{2\left(\left\|\sum_{j\in\mathcal{N}_i} a_{ij}\tanh\left(\tilde{\boldsymbol{\omega}}_i\left(t_{k_i}^{\varphi i}\right) - \tilde{\boldsymbol{\omega}}_j\left(t_{k_j}^{\varphi j}\right)\right)\right\| + \varsigma_{\varphi 1}\right)} + \varsigma_{\varphi 2} \leq \mu_{\varphi i} t_\varphi^*. \tag{7.52}$$

From (7.52), there exists $\mu_{\varphi i} t_\varphi^* \geq \varsigma_{\varphi 2}$. Therefore, with $1 \leq \varsigma_{\varphi 2} \leq e^{c_{\varphi 1}}$, it is verified that $t_\varphi^* \geq 1/\mu_{\varphi i} > 0$. Clearly, the Zeno behavior is successfully precluded.

This completes the proof of Theorem 7.2.

Remark 7.7 Note that $\left\|\tilde{\boldsymbol{\varphi}}_i\right\|$ can be made arbitrarily small by increasing $\boldsymbol{K}_{i1}$, $\boldsymbol{K}_{i2}$, $\boldsymbol{K}_{i3}$, and $\sigma_{\varphi i}$. Moreover, a good balance of convergence performance and triggering number of communication should be made by choosing reasonable parameters $c_{\varphi 1}$, $c_{\varphi 2}$, and α_φ. This will be further shown in the following numerical simulation.

Remark 7.8 Different from existing results [8–14] with the continuous state information exchanges among neighboring UAVs in controller design, our proposed $\boldsymbol{\delta}_{i0}$ simply uses discrete neighbors' state $\tilde{\boldsymbol{\omega}}_j(t_{k_j}^{\varphi j})$ in control protocol (7.25), which lightens the communication exchange during $t \in [t_{k_j}^{\varphi j}, t_{k_j+1}^{\varphi j})$.

Remark 7.9 Compared with existing literature [27, 34–37] using neural network learning mechanisms to cope with lumped uncertainties caused by actuator faults, external disturbances, and model uncertainties, the proposed disturbance observer compensates for compounded disturbances with less online learning parameters and simpler internal structure, thus reducing the computation burden and obtaining faster disturbance estimation. As a result, the designed disturbance observer is more applicable to disturbance rejection for a practical UAV system.

7.4 Simulations and Analyses

To verify the effectiveness of the proposed scheme, a simulation scenario is considered to address a coordinated tracking problem for three networked fixed-wing UAVs. The structure parameters and coefficient values of each UAV are listed in Table 7.1 [28]. The corresponding communication topology is illustrated in Fig. 7.2.

The initial conditions are summarized as $V_{10} = V_{20} = V_{30} = 20\,\text{m/s}$, $\boldsymbol{\varphi}_{10} = [0°, 1°, 0°]^{\mathrm{T}}$, $\boldsymbol{\varphi}_{20} = \boldsymbol{\varphi}_{30} = [0.5°, 1°, 0.5°]^{\mathrm{T}}$, and $\boldsymbol{\omega}_{10} = \boldsymbol{\omega}_{20} = \boldsymbol{\omega}_{30} = [0, 0, 0]^{\mathrm{T}}\text{rad/s}$. Suppose that the external disturbances in velocity and attitude motion are described as $\boldsymbol{d}_{v1} = \boldsymbol{d}_{v2} = \boldsymbol{d}_{v3} = [0.3\sin(t), 0.5\cos(2t),\ 0.2\cos(t)]^{\mathrm{T}}$ and $\boldsymbol{d}_{\omega2} = \boldsymbol{d}_{\omega3} = [0.3\cos(2t), 0.2\sin(3t),\ 0.5\sin(t)]^{\mathrm{T}}$. Meanwhile, the actuator faults are introduced in velocity channel as $\rho_{T_3} = 0.9$ during $t \in (2, 5]$ s. Assume that the virtual leader UAV is to complete low altitude penetration mission and track the desired trajectory as

Table 7.1 Structural and aerodynamic parameter values of UAV

Coefficient	Value	Unit	Coefficient	Value	Unit
m_l	20.64	kg	b_l	1.96	m
S_l	1.37	m^2	$\bar{c}_l$	0.76	m
ρ	1.29	$\text{kg}\cdot\text{m}^{-3}$	g	9.8	$\text{m}\cdot\text{s}^{-2}$
J_{xl}	1.607	$\text{kg}\cdot\text{m}^2$	J_{yl}	7.51	$\text{kg}\cdot\text{m}^2$
J_{zl}	7.18	$\text{kg}\cdot\text{m}^2$	J_{xzl}	0.59	$\text{kg}\cdot\text{m}^2$
c_{L0l}	0.1	/	$c_{L\alpha_l}$	0.25	rad^{-1}
c_{D0l}	0.5	/	$c_{Y\beta_l}$	−0.1	rad^{-1}
c_{l0l}	−0.001	/	$c_{l\beta_l}$	−0.038	rad^{-1}
c_{lpl}	−0.213	$\text{rad}^{-1}\cdot\text{s}$	c_{lrl}	0.114	$\text{rad}^{-1}\cdot\text{s}$
$c_{l\delta_{al}}$	−0.056	rad^{-1}	$c_{l\delta_{rl}}$	0.014	rad^{-1}
c_{m0l}	0.022	/	$c_{m\alpha_l}$	−0.473	rad^{-1}
c_{mql}	−3.449	$\text{rad}^{-1}\cdot\text{s}$	$c_{m\delta_{el}}$	−0.364	rad^{-1}
c_{n0l}	0.022	/	$c_{n\beta_l}$	0.036	rad^{-1}
c_{npl}	−0.151	$\text{rad}^{-1}\cdot\text{s}$	c_{nrl}	−0.195	$\text{rad}^{-1}\cdot\text{s}$
$c_{n\delta_{al}}$	−0.036	rad^{-1}	$c_{n\delta_{rl}}$	−0.055	rad^{-1}

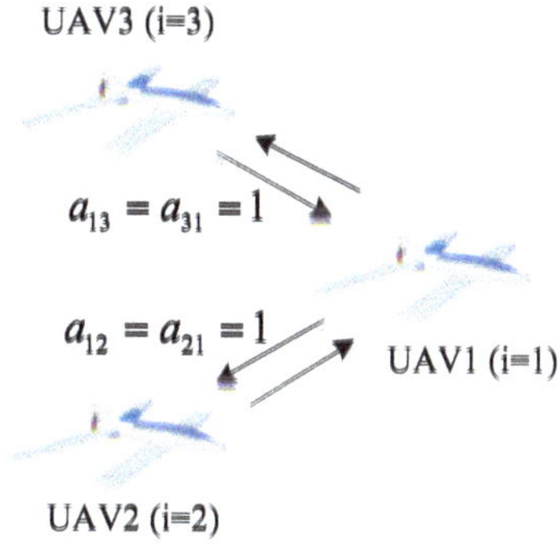

Fig. 7.2 Communication topology of UAVs

$$V_r = \begin{cases} 20 + \frac{30}{2s+1}, & 0s \le t < 7\,\text{s} \\ 20 + \frac{20}{0.4s+1}, & 7s \le t \le 12\,\text{s} \\ 20 + \frac{40}{0.4s+1}, & t > 12\,\text{s} \end{cases}$$

$$\boldsymbol{\varphi}_r = \begin{cases} [0, 0, 0]^{\mathrm{T}}, & 0s \le t < 5s \\ [0, \frac{-10}{0.1s+1}, 0]^{\mathrm{T}}, & 5s \le t \le 15s \\ [0, 0, 0]^{\mathrm{T}}, & t > 15s. \end{cases}$$

The controllers and disturbance observers are provided by (7.22), (7.23), (7.24), and (7.25) with design parameters as $\alpha_v = 0.5$, $\varsigma_{v1} = 0.1$, $c_{v1} = 0.1$, $c_{v2} = 0.2$, $\sigma_{v1} = \sigma_{v2} = 10\,(0s \le t \le 12s)\,;\,100\,(12s < t \le 14s)\,;\,150\,(t > 14s)$, $\sigma_{v3} = 10\,(0s \le t \le 12s)\,;\,100\,(12s < t \le 14s)\,;\,10\,(t > 14s)$, $k_{11} = k_{21} = k_{31} = 0.02(0s \le t \le 14s); 0.002\,(t > 14s)$, $k_{12} = k_{22} = k_{32} = 10\,(0s \le t \le 5s)\,;\,100\,(5s < t \le 12s)\,;\,20\,(12s < t \le 14s)\,;\,100\,(t > 14s)$, $\alpha_\varphi = 0.5$, $\varsigma_{\varphi 1} = 0.1$, $c_{\varphi 1} = 0.1$, $c_{\varphi 2} = 0.2$, $\sigma_{\varphi 1} = \sigma_{\varphi 2} = \sigma_{\varphi 3} = 1$, $\boldsymbol{K}_{11} = \boldsymbol{K}_{21} = \boldsymbol{K}_{31} = \boldsymbol{K}_{12} = \boldsymbol{K}_{22} = \boldsymbol{K}_{32} = \mathrm{diag}\{1, 1, 1\}$, $\boldsymbol{K}_{13} = \boldsymbol{K}_{23} = \boldsymbol{K}_{33} = \mathrm{diag}\{50, 60, 50\}$.

To comprehensively highlight the performance of the proposed approach, three situations are presented and discussed. The first case verifies the efficiency of proposed controller based on the above simulation scenario. The second one illustrates the superiority of dynamic event-triggered communication mechanism (DECM) over the periodic communication mechanism (PCM) and static event-triggered communication mechanism (SECM). The last case evaluates the sensitivity of proposed event-triggered conditions for different fault modes.

Case 1: In this case, the proposed controller is conducted based on the above simulation scenario. Figure 7.3 shows the trajectories of velocities and attitudes for the multi-UAV system. From Fig. 7.3a–d, it can be observed that the velocities and attitudes both converge availably to their desired trajectories, which validates the effectiveness of proposed control scheme. Additionally, the tracking errors of velocities and attitudes are depicted in Fig. 7.3e–h. The data release intervals and broadcasting instants of velocity and attitude for each UAV are presented in Figs. 7.4

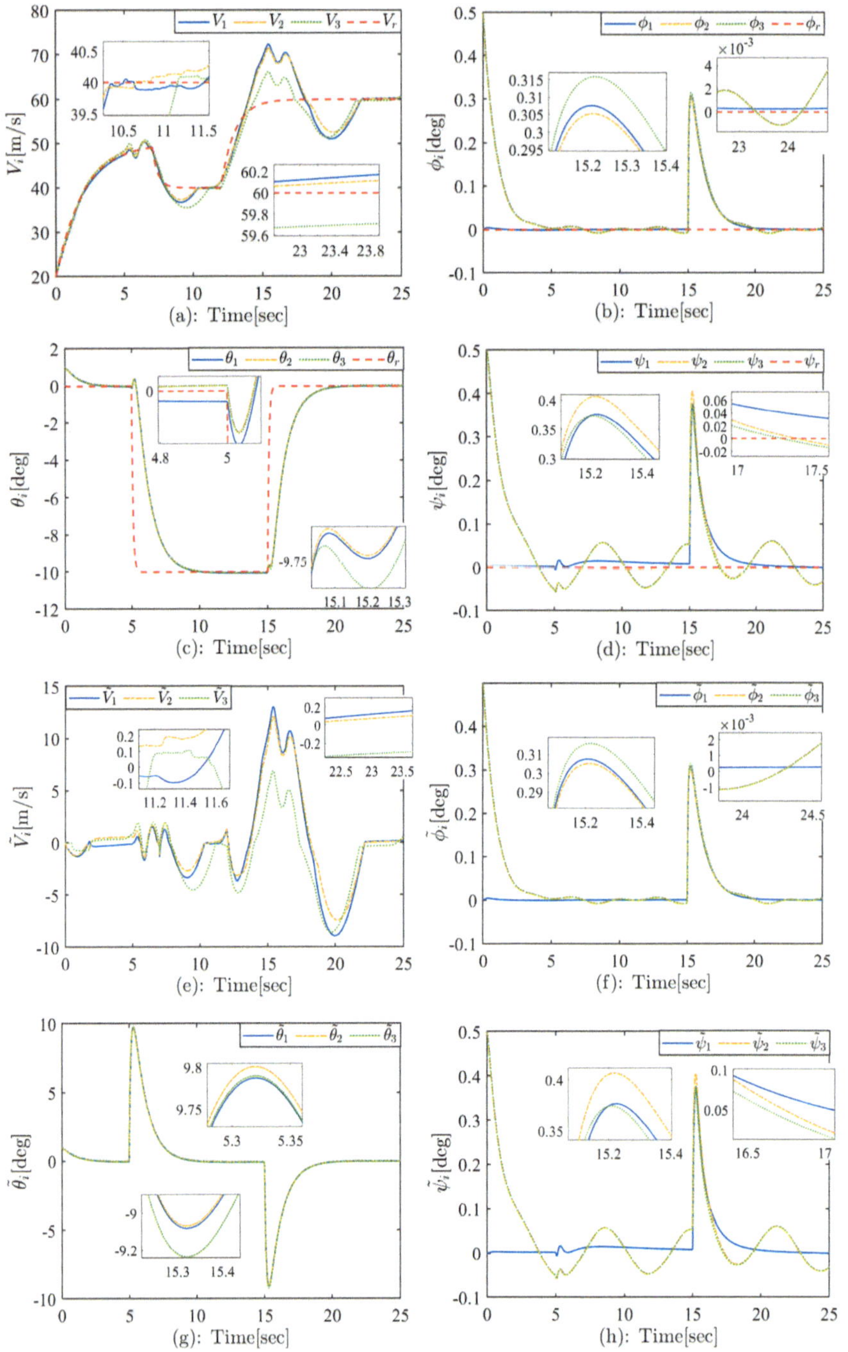

Fig. 7.3 Velocity and attitude tracking of multiple UAVs by the proposed controller. (**a**)–(**h**) show the curves of state responses and errors. (**a**) Velocity response. (**b**) Roll angle response. (**c**) Pitch angle response. (**d**) Yaw angle response. (**e**) Tracking error of velocity. (**f**) Tracking error of roll angle. (**g**) Tracking error of pitch angle. (**h**) Tracking error of yaw angle

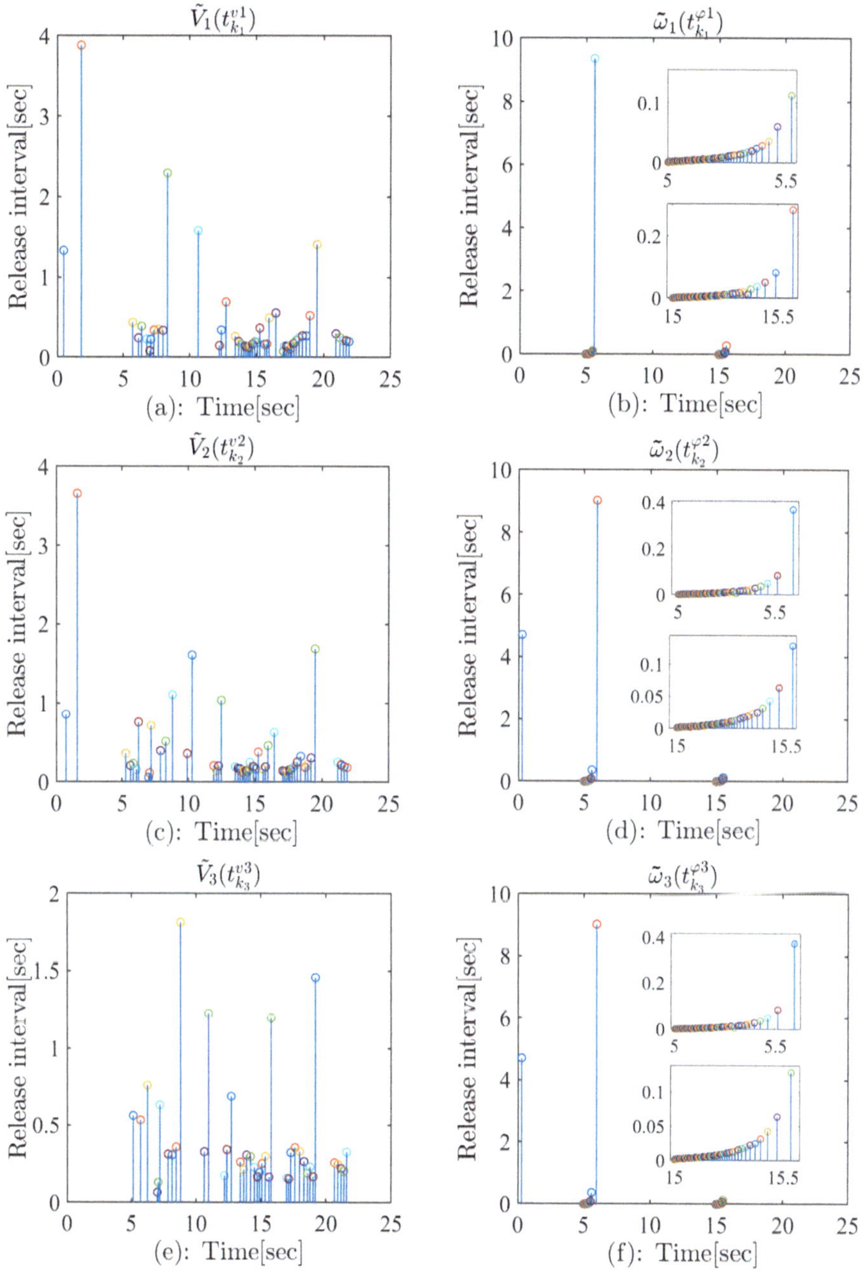

Fig. 7.4 Release intervals of velocity and attitude data for multi-UAV system. (**a**)–(**f**) show the curves of release intervals. (**a**) Data $\tilde{V}_1(t^{v1}_{k_1})$. (**b**) Data $\tilde{\boldsymbol{\omega}}_1(t^{\varphi 1}_{k_1})$. (**c**) Data $\tilde{V}_2(t^{v2}_{k_2})$. (**d**) Data $\tilde{\boldsymbol{\omega}}_2(t^{\varphi 2}_{k_2})$. (**e**) Data $\tilde{V}_3(t^{v3}_{k_3})$. (**f**) Data $\tilde{\boldsymbol{\omega}}_3(t^{\varphi 3}_{k_3})$

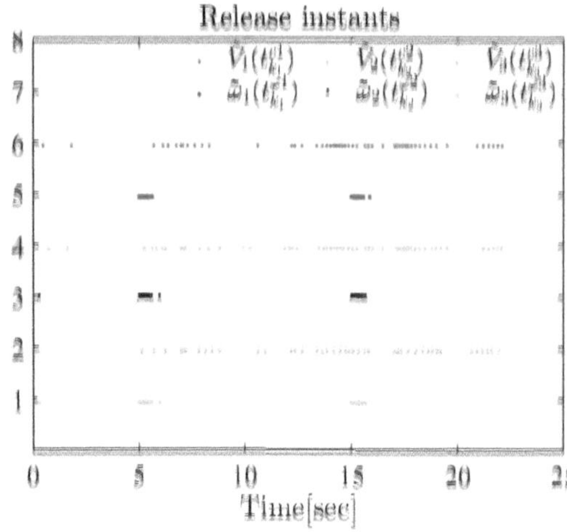

Fig. 7.5 Release instants of velocity and attitude data for multi-UAV system

Table 7.2 The number of data transmission for each UAV under different communication mechanisms

Transmitted data	$\tilde{V}_1\left(t_{k_1}^{v1}\right)$	$\tilde{\omega}_1\left(t_{k_1}^{\varphi 1}\right)$	$\tilde{V}_2\left(t_{k_2}^{v2}\right)$	$\tilde{\omega}_2\left(t_{k_2}^{\varphi 2}\right)$	$\tilde{V}_3\left(t_{k_3}^{v3}\right)$	$\tilde{\omega}_3\left(t_{k_3}^{\varphi 3}\right)$
PCM	2500	2500	2500	2500	2500	2500
SECM	187	583	148	582	146	584
DECM	51	143	52	149	42	146

and 7.5. Obviously, the communication among UAVs is intermittent, meanwhile no Zeno behavior exhibits. Thus we can see that the proposed event-based control scheme can maintain the expected tracking performances of velocities and attitudes despite the presence of actuator faults and external disturbances.

Case 2: To illustrate the advantages of event-triggered strategy in terms of DECM, two other communication mechanisms are evaluated in this comparative study: (i) PCM: Information exchanges among neighboring UAVs are performed in a periodic manner. (ii) SECM: Different from DECM in (7.20) and (7.21), the threshold parameter in SECM is fixed permanently as [20]. That is, $f_i\left(t, e_{vl}(t)\right) = \left|e_{vl}(t)\right| - \varsigma_{v2}$ and $f_i\left(t, \boldsymbol{e}_{\varphi l}(t)\right) = \left\|\boldsymbol{e}_{\varphi l}(t)\right\| - \varsigma_{\varphi 2}$. Choose $\varsigma_{v2} = 0.5$ and $\varsigma_{\varphi 2} = 0.2$.

The numbers of information transmission under different communication mechanisms are given in Table 7.2. While taking the tracking maneuver of UAV 3 for example, the comprehensive performances under SECM and DECM are plotted in Fig. 7.6. From Table 7.2, one can see that compared with PCM and SECM, DECM leads to least transmitted information, thus reducing the occupancy of limited communication resources. Even though with least communication transmission, the desired coordinated tracking under DECM can be still accomplished with scheduled precise requirements. This point is clearly shown in Fig. 7.6, where the responses of velocities and attitudes for DECM are almost similar to SECM. As a result, it can be concluded that the proposed dynamic event-based control scheme provides a

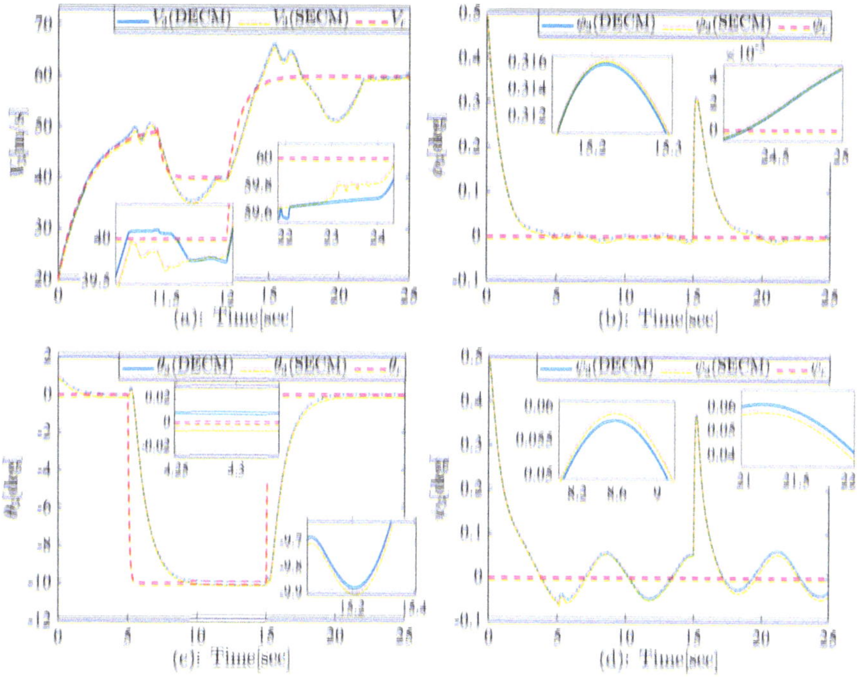

Fig. 7.6 Velocity and attitude tracking of UAV 3 under DECM and SECM. (**a**)–(**d**) show the curves of state responses. (**a**) Velocity response. (**b**) Roll angle response. (**c**) Pitch angle response. (**d**) Yaw angle response

better balance between saving communication resources and achieving well tracking performances than PCM and SECM.

Case 3: To further evaluate the sensitivity of proposed event-triggered conditions, we consider the following four different fault modes:

Fault mode 1: As defined in the above simulation scenario, the thrust of UAV 3 suffers the loss of effectiveness, that is, $\rho_{T_3} = 0.9$ during $t \in (2, 5]$ s.

Fault mode 2: Based on fault mode 1, the period of time when the thrust loses its effectiveness is enlarged, that is, $\rho_{T_3} = 0.9$ during $t \in (2, 8]$ s.

Fault mode 3: The aileron of UAV 3 loses its effectiveness described by $\boldsymbol{\rho}_{\delta_3} = \mathrm{diag}\{0.9, 1, 1\}$ during $t \in (5, 10]$ s.

Fault mode 4: The rudder of UAV 3 loses its effectiveness described by $\boldsymbol{\rho}_{\delta_3} = \mathrm{diag}\{1, 1, 0.9\}$ during $t \in (5, 10]$ s.

Based on the above fault modes, the fault-tolerant performances of UAV 3 for different fault modes are shown in Fig. 7.7. Robustness with proposed controller is demonstrated in case of the above four fault modes as presented in Fig. 7.8. The data release instants of velocity and attitude for UAV 3 with different faults are presented in Fig. 7.8. Besides, the corresponding amount of exchange of information under four fault modes is given in Table 7.3. From Figs. 7.7, 7.8, and Table 7.3, we can

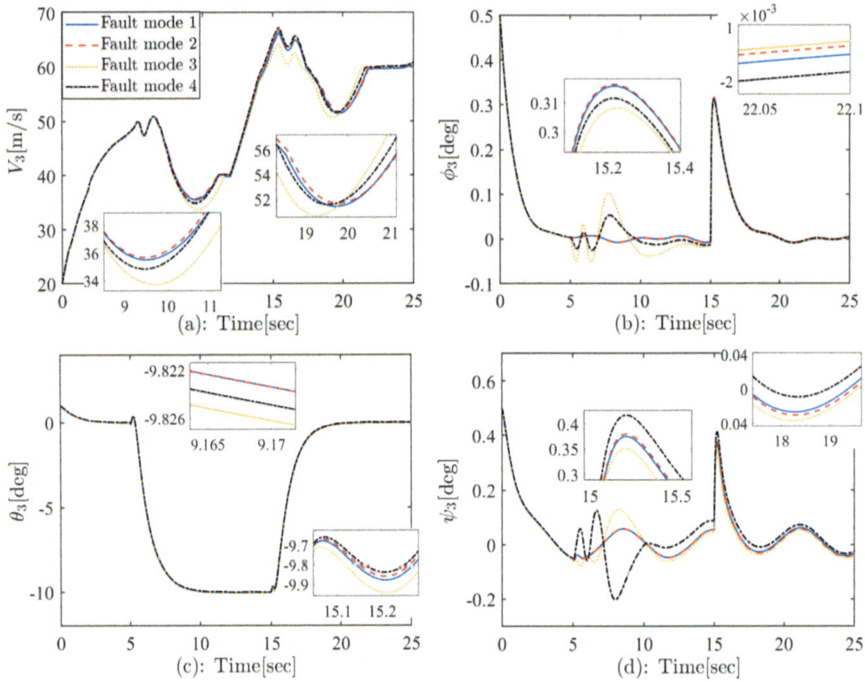

Fig. 7.7 Velocity and attitude tracking of UAV 3 under different fault modes. (**a**)–(**d**) show the curves of state responses. (**a**) Velocity response. (**b**) Roll angle response. (**c**) Pitch angle response. (**d**) Yaw angle response

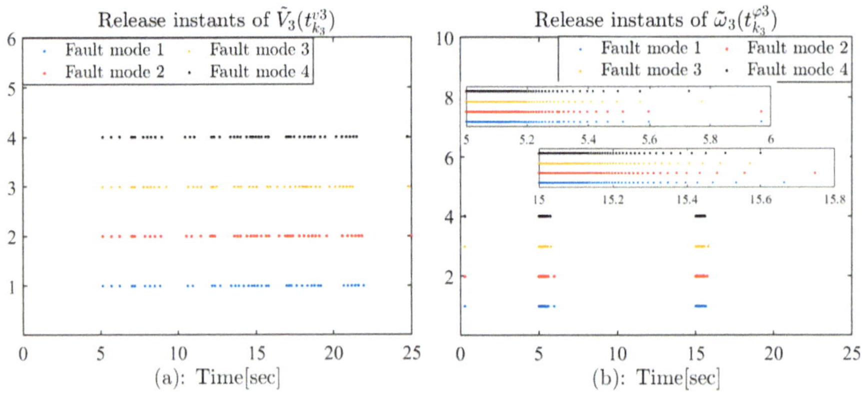

Fig. 7.8 Release instants of velocity and attitude data of UAV 3 under different fault modes

notice that the fault-tolerant property guarantees the stable tracking of UAV 3 in the presence of four different fault modes, and thus no additional communications are triggered in such cases.

Table 7.3 The number of data transmission under different fault modes

Transmitted data	$\tilde{V}_3\left(t_{k_3}^{v3}\right)$	$\tilde{\omega}_3\left(t_{k_3}^{\varphi 3}\right)$
Fault mode 1	42	146
Fault mode 2	45	147
Fault mode 3	46	146
Fault mode 4	47	145

7.5 Conclusions

This work investigates the event-based fault-tolerant coordinated control problem of multiple fixed-wing UAVs. A novel event-triggered communication mechanism has been presented, wherein the threshold parameter is not fixed permanently but determined by a dynamic rule. By imposing the proposed event-triggered mechanisms on translational and rotational subsystems, respectively, the inter-agent communication has been regulated to avoid the unnecessary communications and save on network resources. Furthermore, an event-based coordinated control scheme has been designed using only locally triggered state information at discretized instants. This work lays down the groundwork for other coordinated control problems of UAVs including formulating and maintaining a particular formation shape. Additionally, note that the extension of event-triggered mechanism to sensor-to-controller channel for UAV system remains open and deserves further investigation.

References

1. T. Kim, D. Qiao, Energy-efficient data collection for IoT networks via cooperative multi-hop UAV networks. IEEE Trans. Veh. Technol. **70**, 471–486 (2017)
2. H. Qiu, H. Duan, Multiple UAV distributed close formation control based on in-flight leadership hierarchies of pigeon flocks. Aerosp. Sci. Technol. **70**, 471–486 (2017)
3. Y. Chen, D. Chang, C. Zhang, Autonomous tracking using a swarm of UAVs: a constrained multi-agent reinforcement learning approach. IEEE Trans. Veh. Technol. **69**(11), 13702–13717 (2020)
4. Y. Wang, M. Shan, D. Wang, Motion capability analysis for multiple fixed-wing UAV formations with speed and heading rate constraints. IEEE Trans. Control Network Syst. **7**(2), 977–989 (2020)
5. Z. Yu, Y. Zhang, B. Jiang, C.Y. Su, J. Fu, Y. Jin, T. Chai, Fractional-order adaptive fault-tolerant synchronization tracking control of networked fixed-wing UAVs against actuator-sensor faults via intelligent learning mechanism. IEEE Trans. Neural Networks Learn. Syst. **32**(12), 5539–5553 (2021)
6. M. Lv, W. Yu, J. Cao, S. Baldi, Consensus in high-power multi-agent systems with mixed unknown control directions via hybrid Nussbaum-based control. IEEE Trans. Cybern. **52**(6), 5184–5196 (2022)
7. M. Lv, B.D. Schutter, C. Shi, S. Baldi, Logic-based distributed switching control for agents in power chained form with multiple unknown control directions. Automatica **137**, 110143 (2022)

8. Z. Yu, Y. Zhang, Z. Liu, Y. Qu, C.Y. Su, Distributed adaptive fractional-order fault tolerant cooperative control of networked unmanned aerial vehicles via fuzzy neural networks. IET Control Theory Appl. **13**(17), 2917–2929 (2019)
9. Z. Yu, Y. Qu, Y. Zhang, Distributed fault-tolerant cooperative control for multi-UAVs under actuator fault and input saturation. IEEE Trans. Control Syst. Technol. **27**(6), 2417–2429 (2019)
10. Y. Wang, T. Zhang, Z. Cai, J. Zhao, K. Wu, Multi-UAV coordination control by chaotic grey wolf optimization based distributed MPC with event-triggered strategy. Chin. J. Aeronaut. **33**(11), 2877–2897 (2020)
11. T.Z. Muslimov, R.A. Munasypov, Adaptive decentralized flocking control of multi-UAV circular formations based on vector fields and backstepping. ISA Trans. **107**, 143–159 (2020)
12. K. Guo, X. Li, L. Xie, Ultra-wideband and odometry-based cooperative relative localization with application to multi-UAV formation control. IEEE Trans. Cybern. **50**(6), 2590–2603 (2020)
13. J. Zhang, J. Yan, P. Zhang, Multi-UAV formation control based on a novel back-stepping approach. IEEE Trans. Veh. Technol. **69**(3), 2437–2448 (2020)
14. X. Dong, Y. Hua, Y. Zhou, Z. Ren, Y. Zhong, Theory and experiment on formation-containment control of multiple multirotor unmanned aerial vehicle systems. IEEE Trans. Autom. Sci. Eng. **16**(1), 229–240 (2019)
15. G.S. Seyboth, D.V. Dimarogonas, K.H. Hohansson, Event-based broadcasting for multi-agent average consensus. Automatica **49**(1), 245–252 (2013)
16. X. Yi, T. Yang, J. Wu, K.H. Johansson, Distributed event-triggered control for global consensus of multi-agent systems with input saturation. Automatica **100**, 1–9 (2019)
17. P. Duan, K. Liu, N. Huang, Z. Duan, Event-based distributed tracking control for second-order multiagent systems with switching networks. IEEE Trans. Syst. Man Cybern.: Syst. **50**(9), 3220–3230 (2020)
18. Y. Qian, L. Liu, G. Feng, Output consensus of heterogeneous linear multi-agent systems with adaptive event-triggered control. IEEE Trans. Autom. Control **64**(6), 2606–2613 (2019)
19. C. Peng, J. Zhang, Q. Han, Consensus of multiagent systems with nonlinear dynamics using an integrated sampled-data-based event-triggered communication scheme. IEEE Trans. Syst. Man Cybern.: Syst. **49**(3), 589–599 (2019)
20. C. Wang, C. Wen, L. Guo, L. Xing, Adaptive consensus control for nonlinear multiagent systems with unknown control directions using event-triggered communication. IEEE Trans. Cybern. **52**(5), 3057–3068 (2022)
21. X. Ge, Q.L. Han, Distributed formation control of networked multi-agent systems using a dynamic event-triggered communication mechanism. IEEE Trans. Ind. Electron. **64**(10), 8118–8127 (2017)
22. E. Garcia, Y. Cao, D.W. Casbeer, Decentralized event-triggered consensus with general linear dynamics. Automatica **50**(10), 2633–2640 (2014)
23. C. Viel, S. Bertrand, S. Kieffer, H. Piet-Lahanier, Distributed event-triggered control strategies for multi-agent formation stabilization and tracking. Automatica **106**, 110–116 (2019)
24. A. Girard, Dynamic triggering mechanisms for event-triggered control. IEEE Trans. Autom. Control **60**(7), 1992–1997 (2015)
25. V.S. Dolk, D.P. Borgers, W.P. Heemels, Dynamic event-triggered control: Tradeoffs between transmission intervals and performance, in *53rd IEEE Conference on Decision and Control* (2014), pp. 2764–2769
26. Z. Yu, Z. Liu, Y. Zhang, Y. Qu, C.Y. Su, Distributed finite-time fault-tolerant containment control for multiple unmanned aerial vehicles. IEEE Trans. Neural Networks Learn. Syst. **31**(6), 2077–2091 (2020)
27. A. Abbaspour, K.K. Yen, P. Forouzannezhad, A. Sargolzaei, A neural adaptive approach for active fault-tolerant control design in UAV. IEEE Trans. Syst. Man Cybern.: Syst. **50**(9), 3401–3411 (2020)

28. B. Zhang, X. Sun, S. Liu, M. Lv, X. Deng, Event-triggered adaptive fault-tolerant synchronization tracking control for multiple 6-DOF fixed-wing UAVs. IEEE Trans. Veh. Technol. **71**(1), 148–161 (2021)
29. H. Castaneda, O.S. Salas-Pena, J. Le-Morales, Extended observer based on adaptive second order sliding mode control for a fixed-wing UAV. ISA Trans. **66**, 226–232 (2017)
30. E. Oland, Nonlinear control of fixed-wing unmanned aerial vehicles, in *Norway: Norwegian University of Science and Technology* (2017)
31. M. Lv, W. Yu, J. Cao, S. Baldi, A separation-based methodology to consensus tracking of switched high-order nonlinear multi-agent systems. IEEE Trans. Neural Networks Learn. Syst. **33**(10), 5467–5479 (2021)
32. W. Ren, On consensus algorithms for double-integrator dynamics. IEEE Trans. Autom. Control **53**(6), 1503–1509 (2008)
33. W.E. Dixon, Adaptive regulation of amplitude limited robot manipulators with uncertain kinematics and dynamics. IEEE Trans. Autom. Control **52**(3), 488–493 (2007)
34. S. Zeghlache, H. Mekki, A. Bougueera, A. Djerioui, Actuator fault tolerant control using adaptive RBFNN fuzzy sliding mode controller for coaxial octorotor UAV. ISA Trans. **8**, 267–278 (2018)
35. F. Jiang, F. Pourpanah, Q. Hao, Design, implementation, and evaluation of a neural-network-based quadcopter UAV system. IEEE Trans. Ind. Electron. **67**(3), 2076–2085 (2020)
36. T. Dierks, S. Jagannathan, Output feedback control of a quadrotor UAV using neural networks. IEEE Trans. Neural Netw. **21**(1), 50–66 (2010)
37. Y. Song, L. He, D. Zhang, J. Qian, J. Fu, Neuroadaptive fault-tolerant control of quadrotor UAVs: a more affordable solution. IEEE Trans. Neural Networks Learn. Syst. **30**(7), 1975–1983 (2019)

Chapter 8
Control with Asynchronously Switching Topologies

The event-triggered communication is introduced in Chap. 7 as a means of reducing the communication resources required by multiple UAVs. Furthermore, communication interactions between UAVs can be reduced by switching between connected and unconnected topologies, thus reducing the wastage of communication resources. Nevertheless, communication topology switching gives rise to topology and controller switching that occur in a nonsynchronous manner, which is to say, asynchronous switching. In order to address this problem, an asynchronous switching observer is proposed in this chapter, which ensures that the follower UAVs are able to estimate the leader UAV state information under topology switching.

8.1 Introduction

Over the past decade, the field of control has witnessed a substantial surge in interest toward cooperative flight involving multi-UAVs [1–5]. Research in this domain can be broadly categorized into two main streams: the development of formation and path planning for multi-UAVs [1, 6, 7] and ensuring consensus in velocities and attitudes among multi-UAVs [8–10]. Notably, scholars have extensively explored the cooperative flight of fixed-wing multi-UAVs due to their inherent characteristics of high flexibility and endurance [11–16]. For example, in [13], a cooperative control scheme grounded in a fixed-time anti-saturation compensation system was introduced to address the distributed containment control problem in multi-UAV scenarios. Meanwhile, [15] presented a learning-based consensus protocol for fixed-wing multi-UAV systems exhibiting uncertain dynamics at predetermined altitudes. This protocol guarantees a bounded tracking error under external disturbances. Regrettably, existing results often focus solely on attitude/longitudinal models [13, 17], simplify the UAV system to a two-degree model [15, 16], or employ feedback linearization [18]. The model simplifications proposed in the aforementioned

M. Lv et al., *Unmanned Aerial Vehicles Formation Control and Safety Guarantee*, Cognitive Intelligence and Robotics, https://doi.org/10.1007/978-981-96-5829-9_8

studies neglect the nonlinear and coupling nature of the UAV systems, resulting in a simplified model that hinders the application of control algorithms to real fixed-wing UAV. As a result of this simplification, there is a difficulty in applying the designed control algorithms to the real-world UAV systems. In addition, each UAV being a six-degree-of-freedom (6-DoF) underdriven model [19], the nonlinear or coupled characteristics arising from the interaction of translational and rotational kinematics in the UAV system pose challenges in designing control algorithms. Evidently, there is a compelling need to delve into the control challenges specific to 6-DoF fixed-wing UAVs.

Considering the scenario of constrained communication resources, the communication topology of multiple UAV switching between connected and unconnected graphs can reduce the communication frequency between UAVs, thereby achieving the objective of reducing communication resources. Notably, considerable research has been dedicated to addressing multi-UAV systems under switching communication topologies [20–23]. In [20], a distributed collision avoidance formation controller ensures UAV stability in deployment scenarios with switching topologies. The work [21] proposes position and attitude robust controllers to address formation control challenges in quadrotor multi-UAVs, considering external disturbances and switching topologies. However, existing results often assume synchronized communication with topology switching when designing controllers. In reality, the identification time required for topology mode switching introduces delays due to factors such as transmission delays and network congestion. Consequently, a non-matching interval emerges between the topology and the controller/observer during the identification period, leading to inevitable asynchronous switching. Assuming synchronous topology switching may cause the designed controller/observer to induce system oscillations or instability. While a few results address asynchronous switching [24, 25], they face challenges as the Lyapunov function can only increase in the non-matching interval. Compensating for this requires decreasing the Lyapunov function in the matching interval, resulting in a conservative topological switching law aimed at ensuring Lyapunov function convergence (discussed in Remark 8.3). Moreover, to the best of our knowledge, the scenario of asynchronous topology switching is unexplored in the context of fixed-wing multi-UAVs in the existing literature.

Moreover, in real-world flight scenarios, the aging of the rudder can lead to severe damage to multi-UAV systems. To mitigate the impact of actuator failures, it is imperative to devise a set of rational fault-tolerant control strategies. Currently, various fault-tolerant cooperative control (FTCC) schemes for multi-UAVs have been proposed [26, 27]. However, all the aforementioned literature focuses on applying fault-tolerant control schemes exclusively to longitudinal or attitude dynamics, neglecting the intricate scenario involving 6-DoF and topology switching. Inspired by the considerations highlighted earlier, the key objectives of this work are delineated as follows.

- This work represents a pioneering effort in consensus design for 6-DoF fixed-wing multi-UAVs, addressing challenges posed by model uncertainty, actuator failures, and external disturbances within the context of asynchronous topology switching.
- In contrast to prior results [28, 29], the switching graph set under consideration in this study encompasses non-connected graphs that do not utilize the leader UAV as a spanning tree. Consequently, the frequency of inter-UAV communication is reduced, thereby further optimizing the utilization of communication resources. Furthermore, the designed asynchronous topology switching observer guarantees that this scenario still enables each follow-UAV to estimate the state information of the lead-UAV.
- To surmount the limitations of existing conservative asynchronous topology switching laws found in [24, 25], attributed to an inappropriate Lyapunov function, we adopt a new piecewise function. This function aligns with the inherent mechanism of asynchronous topology switching, ensuring that the system invokes the same Lyapunov function during the matching interval of the non-matching interval of the current topology mode and the previous topology mode. This guarantees a decrease in the Lyapunov function in the non-matching interval, resulting in a more lenient topology switching law.

Notations Let $\mathcal{R}$ represent the set of real numbers and $\mathcal{N}$ denote the set of nonnegative integers. $\lambda_{\min}(\boldsymbol{A})$ signifies the minimum eigenvalue of the matrix $\boldsymbol{A}$. The notation $\boldsymbol{A} \succeq \boldsymbol{B}$ indicates that $\boldsymbol{A} - \boldsymbol{B}$ is a semi-positive definite matrix. A continuous function $h(t)$ is categorized as a class K_∞ function if it is strictly increasing, satisfies $h(0) = 0$, and has a limit of $\lim_{t\to+\infty} h(t) = +\infty$.

Remark The main control schemes and contents of this chapter are from the published journal paper "Y. Y. Shi, J. Li, M. L. Lv, N. Wang, B. Y. Zhang, Distributed Consensus Control for 6-DoF Fixed-wing Multi-UAVs in Asynchronously Switching Topologies. IEEE Transactions on Vehicular Technology. 2024. https://doi.org/10.1109/TVT.2024.3520141." The authors appreciate the permission from Institute of Electrical and Electronics Engineers to reuse the results published in the relevant journal.

8.2 Problem Formulation and Preliminaries

8.2.1 UAV Kinematics and Dynamics

Contemplating a fleet of fixed-wing UAVs comprising a lead-UAV and N follow-UAVs, the 6-DoF dynamics of the jth UAV (where j belongs to the set $\{0, 1, \ldots, N\}$) expressed as per [30].

$$\begin{cases} \dot{\boldsymbol{p}}_j = \boldsymbol{R}_1\left(\boldsymbol{\varphi}_j\right)\boldsymbol{v}_j, \\ \dot{\boldsymbol{v}}_j = -\boldsymbol{S}\left(\boldsymbol{\omega}_j\right)\boldsymbol{v}_j + \frac{\boldsymbol{T}_j}{m_j} + \boldsymbol{R}_1^{\mathrm{T}}\left(\boldsymbol{\varphi}_j\right)\boldsymbol{g} + \frac{\boldsymbol{F}_j}{m_j} + \boldsymbol{d}_{vj}, \\ \dot{\boldsymbol{\varphi}}_j = \boldsymbol{R}_2^{-1}\left(\boldsymbol{\varphi}_j\right)\boldsymbol{\omega}_j, \\ \boldsymbol{J}_j\dot{\boldsymbol{\omega}}_j = -\boldsymbol{\omega}_j \times \boldsymbol{J}_j\boldsymbol{\omega}_j + \boldsymbol{N}_j + \boldsymbol{C}_j\boldsymbol{\delta}_j + \boldsymbol{J}_j\boldsymbol{d}_{wj}, \end{cases} \tag{8.1}$$

where $\boldsymbol{p}_j = [x_j, y_j, z_j]^{\mathrm{T}}$ represents the position in inertial frame, $\boldsymbol{v}_j = [u_j, v_j, w_j]^{\mathrm{T}}$ represents the linear velocity in the body fixed frame, and $\boldsymbol{\varphi}_j = [\phi_j, \theta_j, \psi_j]^{\mathrm{T}}$ represents the set of roll, pitch, and yaw angles. $\boldsymbol{T}_j = [T_{xj}, 0, 0]^{\mathrm{T}}$ and $\boldsymbol{\delta}_j = [\delta_{aj}, \delta_{ej}, \delta_{rj}]^{\mathrm{T}}$ are the system inputs. Furthermore, m_j is the mass of the UAV, $\boldsymbol{g} = [0, 0, g]^{\mathrm{T}}$ is the gravity acceleration in inertial frame, $\boldsymbol{R}_1(\cdot)$ denotes the matrix of transformation from the body fixed frame to inertial frame, $\boldsymbol{R}_2(\cdot)$ denotes the matrix mapping the time derivative of the Euler angle to the non-inertial angular velocity, and $\boldsymbol{J}_j$ denotes the inertia tensor matrix with the symmetric $x - z$ plane. Take the above matrices as

$$\boldsymbol{R}_1 = \begin{bmatrix} \mathrm{c}_{\psi j}\mathrm{c}_{\theta j} & -\mathrm{s}_{\psi j}\mathrm{s}_{\phi j} + \mathrm{c}_{\psi j}\mathrm{s}_{\theta j}\mathrm{s}_{\phi j} & \mathrm{s}_{\psi j}\mathrm{s}_{\phi j} + \mathrm{c}_{\psi j}\mathrm{s}_{\theta j}\mathrm{c}_{\phi j} \\ \mathrm{s}_{\psi j}\mathrm{c}_{\theta j} & \mathrm{c}_{\psi j}\mathrm{c}_{\phi j} + \mathrm{s}_{\psi j}\mathrm{s}_{\theta j}\mathrm{s}_{\phi j} & -\mathrm{c}_{\psi j}\mathrm{s}_{\phi j} + \mathrm{s}_{\psi j}\mathrm{s}_{\theta j}\mathrm{c}_{\phi j} \\ -\mathrm{s}_{\theta j} & \mathrm{c}_{\theta j}\mathrm{s}_{\phi j} & \mathrm{c}_{\theta j}\mathrm{c}_{\phi j} \end{bmatrix},$$

$$\boldsymbol{R}_2 = \begin{bmatrix} 1 & 0 & -s_{\psi j} \\ 0 & c_{\psi j} & s_{\psi j}c_{\theta j} \\ 0 & -s_{\theta j} & c_{\psi j}c_{\theta j} \end{bmatrix}, \quad \boldsymbol{J}_j = \begin{bmatrix} J_{xxj} & 0 & -J_{xzj} \\ 0 & J_{yyj} & 0 \\ -J_{zxj} & 0 & J_{zzj} \end{bmatrix}, \tag{8.2}$$

where c., s. represent $\cos(\cdot)$, $\sin(\cdot)$ functions, respectively. $\boldsymbol{F}_j$, and $\boldsymbol{N}_j$ denote the aerodynamic forces and aerodynamic moments, and their expressions are

$$\begin{aligned} \boldsymbol{F}_j &= \bar{q}_j S_j \boldsymbol{R}_3(\alpha_j, \beta_j)^{-1}[-C_{Dj}, C_{Yj}, -C_{Lj}]^{\mathrm{T}}, \\ \boldsymbol{N}_j &= \bar{q}_j S_j [bC_{lj}, \hat{c}C_{mj}, bC_{nj}]^{\mathrm{T}}, \end{aligned} \tag{8.3}$$

where b_j, $\hat{c}_j$, and S_j represent the wingspan, mean aerodynamic chord, and wing surface area, respectively. $\bar{q}_j = \frac{1}{2}\rho V_j^2$ is the dynamic pressure, where V_j is the total speed in body fixed frame satisfying $V_j = \|\boldsymbol{v}_j\|$, and ρ is the air density. the dimensionless coefficients $C_{kj}, k \in \{l, m, n, D, Y, L\}$, are presented in [30]. $\boldsymbol{R}_3(\cdot)$ maps the body fixed frame to a virtual wind frame defined based on the relative velocity of the UAV, which is displayed as

$$\boldsymbol{R}_3 = \begin{bmatrix} \mathrm{c}_{\alpha j}\mathrm{c}_{\beta j} & \mathrm{s}_{\beta j} & \mathrm{s}_{\alpha j}\mathrm{c}_{\beta j} \\ -\mathrm{c}_{\alpha j}\mathrm{s}_{\beta j} & \mathrm{c}_{\beta j} & -\mathrm{s}_{\alpha j}\mathrm{s}_{\beta j} \\ -\mathrm{s}_{\alpha j} & 0 & \mathrm{c}_{\alpha j} \end{bmatrix}, \tag{8.4}$$

where $\alpha_j = \arctan\left(\frac{w_j}{u_j}\right)$ and $\beta_j = \arcsin\left(\frac{v_j}{V_j}\right)$ denote the attack and sideslip angle, respectively. The control gain matrix $\boldsymbol{C}_j$ is defined by

$$\boldsymbol{C}_j = \overline{q}_j S_j \begin{bmatrix} b_j C_{l\delta aj} & 0 & b_j C_{l\delta rj} \\ 0 & \overline{c}_j C_{m\delta ej} & 0 \\ b_j C_{n\delta aj} & 0 & b_j C_{n\delta rj} \end{bmatrix}. \tag{8.5}$$

$\boldsymbol{S}\left(\boldsymbol{\omega}_j\right)$ is the cross-product matrix. For $i \in \mathcal{N}_{1:N}$, $\boldsymbol{d}_{vi}$ and $\boldsymbol{d}_{wi}$ denote the unknown and bounded external disturbances of the translational and rotational subsystem, respectively. For lead-UAV 0, we choose $\boldsymbol{d}_{v0} = \boldsymbol{d}_{w0} = 0$, and $\boldsymbol{T}_0$, $\boldsymbol{N}_0$ are known terms.

8.2.2 Switching Topology Theory

Consider a leader-following multi-UAV system in which UAVs exchange information based on communication topologies exchanged within a finite topology set $\hat{\mathcal{G}} = \{\mathcal{G}_1, \ldots, \mathcal{G}_l\}$. Define $\varsigma(t) : [0, +\infty) \to \Omega_{\text{st}}$ as a right continuous switching function, where Ω_{st} is a finite topology set. For $\mathcal{G}_{\varsigma(t)}$, the Laplacian matrix $\hat{\boldsymbol{\mathcal{L}}}_{\varsigma(t)}$ is expressed as

$$\hat{\boldsymbol{\mathcal{L}}}_{\varsigma(t)} = \boldsymbol{\mathcal{D}}_{\varsigma(t)} - \boldsymbol{\mathcal{A}}_{\varsigma(t)} = \begin{bmatrix} 0 & \boldsymbol{0}_{1\times N} \\ \tilde{\boldsymbol{\mathcal{L}}}_{\varsigma(t)} & \boldsymbol{\mathcal{L}}_{\varsigma(t)} \end{bmatrix},$$

where $\boldsymbol{\mathcal{A}}_{\varsigma(t)} = \left[a_{ij}^{\varsigma(t)}\right] \in \mathcal{R}^{N\times N}$, $i, j \in \mathcal{N}_{0:N}$, denotes the adjacency matrix of $\mathcal{G}_{\varsigma(t)}$. If UAV i cannot receive information from UAV j, then $a_{ij}^{\varsigma(t)} = 0$; else $a_{ij}^{\varsigma(t)} = 1$. $\boldsymbol{\mathcal{D}}_{\varsigma(t)} = \text{diag}\{\mathcal{D}_{0\varsigma(t)}, \ldots, \mathcal{D}_{N\varsigma(t)}\}$ denotes the degree matrix of $\mathcal{G}_{\varsigma(t)}$ with $\mathcal{D}_{i\varsigma(t)} = \sum_{j=1}^{N} a_{ij}^{\varsigma(t)}$, $i \in \mathcal{N}_{0:N}$. The switching topology is described in detail as follows.

Defining $\mathcal{U}_{\varsigma(t)}$ as the number of switches during $[0, t)$, the h-frequency of switching is expressed as

$$\varkappa(t) \triangleq \frac{\mathcal{U}_{\varsigma}(0, t)}{h(t)}, t \in [0, +\infty), \tag{8.6}$$

where $h(t)$ is class-K$_\infty$ function. Meanwhile, $\mathcal{W} \triangleq \{t_k | k \in \mathcal{N}_{0:\mathcal{U}_{\varsigma}}, t_0 = 0\}$ is defined as a sequence of topological moments sequence within $(0, t)$. Define the frequency of switching from $\mathcal{G}_a$ to $\mathcal{G}_b$ by

$$\iota_{ab}(t) \triangleq \frac{\Delta_{a\to b}}{\mathcal{U}_{\varsigma(t)}}, \ \forall (a, b) \in \mathcal{K}(\Omega_{\text{st}}), \ t \in [0, +\infty), \tag{8.7}$$

where (a, b) denotes switching from $\mathcal{G}_a$ to $\mathcal{G}_b$, $\mathcal{K}(\Omega_{\text{st}})$ denotes for (a, b) the set of all ordered, and $\Delta_{a\to b}$ denotes the number of switching from $\mathcal{G}_a$ to $\mathcal{G}_b$ during $(0, t)$. Define $\overline{\varkappa}, \overline{\iota}_{ab}$ to be the upper definite bounds of $\varkappa$, ι_{ab}, respectively, when $t \to +\infty$.

Topology switching is not synchronized based on the fact that it generates delay. Define $\varsigma_d(t) = \varsigma(t - d)$ as the time-delay switching signal, where d denotes the transition delay, $\mathcal{W}_d \triangleq \{t_k + d_k | k \in \mathcal{N}_{0:\mathcal{U}_\varsigma}, t_0 + d_0 = 0\}$ is the sequence of asynchronous topology switching moments in $[0, t)$. The switched delay d_k satisfies $0 \leq d_k < t_{k+1} - t_k$. For the sake of simple presentation, define $\mathbf{T}_{\perp(t_k, t_{k+1})}$ as the non-matching interval $[t_k, t_k + d_k)$ and $\mathbf{T}_{\top(t_k, t_{k+1})}$ as the matching interval $[t_k + d_k, t_{k+1})$. For $\mathcal{G}_l$, $i \in \Omega_{\text{st}}$, $t \in [0, +\infty)$, the h-frequency of the non-matching periods ϑ^*_{hl} and matching periods ϑ_{hl} as

$$\vartheta^*_{hl} \triangleq \sum_{k:\varsigma(t_k^-)=l} \frac{\mathbf{T}_{\perp(t_k, t_{k+1})}}{h(t)}, \quad \vartheta_{hl} \triangleq \sum_{k:\varsigma(t_k^-)=l} \frac{\mathbf{T}_{\top(t_k, t_{k+1})}}{h(t)} \tag{8.8}$$

and their asymptotic supremum and infimum are defined by

$$\overline{\vartheta}^*_{hl} \triangleq \lim_{t\to+\infty} \sup \vartheta^*_{hl}(t), \quad \underline{\vartheta}^*_{hl} \triangleq \lim_{t\to+\infty} \inf \vartheta^*_{hi}(t),$$

$$\overline{\vartheta}_{hl} \triangleq \lim_{t\to+\infty} \sup \vartheta_{hl}(t), \quad \underline{\vartheta}_{hl} \triangleq \lim_{t\to+\infty} \inf \vartheta_{hi}(t).$$

Differently from [31–33], this paper considers directed spanning tree Ω_{q} that does not contain the lead-UAV as the root, which is more generalizable. For $\forall i \in \Omega_{\text{P}}$, where Ω_{P} is the topological set containing the directed spanning tree rooted at the lead-UAV node, there exist a positive definite matrix $\mathcal{Q}_l = \text{diag}\{p_{l1}, \ldots, p_{lN}\}$ and a positive constant χ_l such that

$$\mathcal{Q}_l \mathcal{L}_l + \mathcal{L}_l^{\text{T}} \mathcal{Q}_l \succeq \chi_l \boldsymbol{I}_N. \tag{8.9}$$

8.2.3 *Fixed-Wing UAV Models with Actuator Failures and Model Uncertainty*

The whole follow-UAV systems can be decomposed into rotational and translational kinematics.

8.2.3.1 Translational Kinematics

For the ith fixed-wing UAV, the rotational dynamics gives $\dot{\omega}_l = \dot{\boldsymbol{R}}_2(\varphi_l)\dot{\varphi}_l + \boldsymbol{R}_2(\varphi_l)\ddot{\varphi}_l$. Substituting this into (8.1) further gives

$$\ddot{\boldsymbol{\varphi}}_l = (\boldsymbol{J}_l\boldsymbol{R}_2)^{-1}\boldsymbol{C}_l\boldsymbol{\delta}_l + (\boldsymbol{J}_l\boldsymbol{R}_2)^{-1}\left(-\boldsymbol{J}_l\dot{\boldsymbol{R}}_2\dot{\boldsymbol{\varphi}}_l + \boldsymbol{S}\left(\boldsymbol{J}_l\boldsymbol{R}_2\dot{\boldsymbol{\varphi}}_l\right)\boldsymbol{R}_2\dot{\boldsymbol{\varphi}}_l + \boldsymbol{N}_l + \boldsymbol{J}_l\boldsymbol{d}_{\varphi l}\right). \tag{8.10}$$

Considering the effect of the complex flight environment, there are uncertainties in the rotational inertia and aerodynamic parameters, $(\boldsymbol{J}_l\boldsymbol{R}_2)^{-1}\boldsymbol{S}\left(\boldsymbol{J}_l\boldsymbol{R}_2\dot{\boldsymbol{\varphi}}_l\right)\boldsymbol{R}_2\dot{\boldsymbol{\varphi}}_l$, $(\boldsymbol{J}_l\boldsymbol{R}_2)^{-1}\boldsymbol{N}_l$, $(\boldsymbol{J}_l\ \boldsymbol{R}_2)^{-1}\boldsymbol{C}_l\boldsymbol{\delta}_l$ are divided into certain terms $(\boldsymbol{J}_{0l}\boldsymbol{R}_2)^{-1}\ \boldsymbol{S}(\boldsymbol{J}_{0l}\ \boldsymbol{R}_2\dot{\boldsymbol{\varphi}}_l)\boldsymbol{R}_2\ \dot{\boldsymbol{\varphi}}_l$, $(\boldsymbol{J}_{0l}\boldsymbol{R}_2)^{-1}\boldsymbol{N}_l$, $(\boldsymbol{J}_{0l}\boldsymbol{R}_2)^{-1}\boldsymbol{C}_{0l}\boldsymbol{\delta}_l$ and uncertain terms $\Delta\left((\boldsymbol{J}_l\boldsymbol{R}_2)^{-1}\boldsymbol{S}(\boldsymbol{J}_l\boldsymbol{R}_2\dot{\boldsymbol{\varphi}}_l)\boldsymbol{R}_2\dot{\boldsymbol{\varphi}}_l\right)$, $\Delta(\boldsymbol{J}_l\boldsymbol{R}_2)^{-1}\ \boldsymbol{N}_l$, $\Delta\left((\boldsymbol{J}_l\boldsymbol{R}_2)^{-1}\boldsymbol{C}_l\right)\boldsymbol{\delta}_l$. Moreover, during actual flight, the UAV's rudder may experience failures. Failure models are established for the ailerons, elevator, and rudder, respectively.

$$\boldsymbol{\delta}_l = \boldsymbol{\varrho}_{\delta l}\boldsymbol{\delta}_{l0} + \boldsymbol{\delta}_{lf}, \tag{8.11}$$

where $\boldsymbol{\delta}_{l0} = [\delta_{al0}, \delta_{el0}, \delta_{rl0}]^{\mathrm{T}}$, $\boldsymbol{\varrho}_{\delta_l} = \mathrm{diag}\{\varrho_{\delta al}, \varrho_{\delta el}, \varrho_{\delta rl}\}$, $\boldsymbol{\delta}_{lf} = \left[\delta_{alf}, \delta_{elf}, \delta_{rlf}\right]^{\mathrm{T}}$, δ_{jl0} is the actual rudder input with $j \in \{a, e, r\}$, $\varrho_{\delta jl} \in [0, 1]$ is the unknown actuator efficiency factor, and δ_{jlf} is the bounded unknown bias fault. Equation (8.11) contains three cases as follows [34]:

(i) Fault-free: $\varrho_{\delta jl} \equiv 1, \delta_{jlf} = 0$
(ii) Partial loss of effectiveness: $\varrho_{\delta jl} \in (0, 1), \delta_{jlf} = 0$
(iii) Bias fault: $\varrho_{\delta jl} \equiv 1, \delta_{jlf} \neq 0$

Based on parameter uncertainties and rudder failures, the whole rotational dynamics (8.1) can be rewritten as

$$\begin{bmatrix}\dot{\boldsymbol{\varphi}}_l\\ \ddot{\boldsymbol{\varphi}}_l\end{bmatrix} = \begin{bmatrix}\mathbf{0}_{3\times3} & \boldsymbol{I}_3\\ \mathbf{0}_{3\times3} & \mathbf{0}_{3\times3}\end{bmatrix}\begin{bmatrix}\boldsymbol{\varphi}_l\\ \dot{\boldsymbol{\varphi}}_l\end{bmatrix} + \begin{bmatrix}\mathbf{0}_{3\times3}\\ \boldsymbol{B}_{l1}\boldsymbol{\varrho}_{\delta_l}\end{bmatrix}\boldsymbol{\delta}_{l0} + \begin{bmatrix}\mathbf{0}_{3\times3}\\ \boldsymbol{B}_{l2} + \boldsymbol{B}_{l3}\end{bmatrix}, \tag{8.12}$$

where $\boldsymbol{B}_{l1} = (\boldsymbol{J}_{l0}\boldsymbol{R}_2)^{-1}\boldsymbol{C}_{l0}$ and $\boldsymbol{B}_{l2} = (\boldsymbol{J}_{l0}\boldsymbol{R}_2)^{-1}\boldsymbol{F}_l - \boldsymbol{R}_2^{-1}\dot{\boldsymbol{R}}_2\dot{\boldsymbol{\varphi}}_l + (\boldsymbol{J}_{l0}\boldsymbol{R}_2)^{-1}\boldsymbol{S}\left(\boldsymbol{J}_{l0}\boldsymbol{R}_2\dot{\boldsymbol{\varphi}}_l\right)\boldsymbol{R}_2\dot{\boldsymbol{\varphi}}_l$ are known function terms, and $\boldsymbol{B}_{l3} = \Delta\left((\boldsymbol{J}_l\boldsymbol{R}_2)^{-1}\boldsymbol{S}\left(\boldsymbol{J}_l\boldsymbol{R}_2\dot{\boldsymbol{\varphi}}_l\right)\boldsymbol{R}_2\dot{\boldsymbol{\varphi}}_l\right) + \Delta(\boldsymbol{J}_l\boldsymbol{R}_2)^{-1}\boldsymbol{F}_l + \Delta\left((\boldsymbol{J}_l\boldsymbol{R}_2)^{-1}\boldsymbol{C}_l\right)\boldsymbol{\delta}_l + \boldsymbol{R}_2^{-1}\boldsymbol{d}_{\omega l} + (\boldsymbol{J}_{l0}\boldsymbol{R}_2)^{-1}\boldsymbol{C}_{l0}\boldsymbol{\delta}_{lf}$ is an uncertain function term.

8.2.3.2 Translational Kinematics

According to dynamic (8.1) and the definition of V_l, the derivative of scalar speed is obtained that

$$\dot{V}_l = \frac{\boldsymbol{v}_l^{\mathrm{T}}\dot{\boldsymbol{v}}_l}{V_l} = \frac{\boldsymbol{v}_l^{\mathrm{T}}}{V_l}\left(-\boldsymbol{S}(\boldsymbol{w}_l)\boldsymbol{v}_l + \frac{\boldsymbol{T}_l}{m_l} + \boldsymbol{R}_1^{\mathrm{T}}(\boldsymbol{\varphi}_l)\boldsymbol{g} + \frac{\boldsymbol{F}_l}{m_l} + \boldsymbol{d}_{vl}\right). \tag{8.13}$$

In order to achieve the engine thrust, (8.13) further obtains that

$$\dot{V}_i = \frac{\boldsymbol{v}_i^{\mathrm{T}}}{V_i}\left(\boldsymbol{R}_1^{\mathrm{T}}(\boldsymbol{\varphi}_i)\,\boldsymbol{g} + \frac{\boldsymbol{F}_i}{m_i} + \boldsymbol{d}_{vi}\right) + \frac{u_i T_{xi}}{m_i V_i}, \tag{8.14}$$

where $\boldsymbol{v}_i^{\mathrm{T}}\boldsymbol{S}(\boldsymbol{w}_i)\,\boldsymbol{v}_i = 0$. Similarly, the aerodynamic force $\boldsymbol{F}_i$ is decomposed into known term $\boldsymbol{F}_{i0}$ and uncertain term $\Delta\boldsymbol{F}_i$, and the engine failure is considered as

$$T_{xi} = \varrho_{Ti} T_{xi0} + T_{xif}, \tag{8.15}$$

where T_{xi0} is the desired thrust, $\varrho_{Ti} \in [0, 1]$ is an unknown effectiveness factor of the thrust, and T_{xif} is the unknown bias fault. Equation (8.14) contains three cases as follows:

(i) Fault-free: $\varrho_{Ti} = 1, T_{xif} = 0$
(ii) Partial loss of effectiveness: $\varrho_{Ti} \in (0, 1), T_{xif} = 0$
(iii) Bias fault: $\varrho_{Ti} = 1, T_{xif} \neq 0$

Thus, we can further obtain

$$\begin{aligned}\dot{V}_i =& \frac{\boldsymbol{v}_i^{\mathrm{T}}}{V_i}\left(\boldsymbol{R}_1^{\mathrm{T}}(\boldsymbol{\varphi}_i)\,\boldsymbol{g} + \frac{\boldsymbol{F}_{i0}}{m_i} + \boldsymbol{\Delta}_{vi} + \boldsymbol{d}_{vi}\right)\\ &+ \frac{u_i \varrho_{Ti} T_{xi0}}{m_i V_i} + \frac{u_i T_{xif}}{m_i V_i},\end{aligned} \tag{8.16}$$

where $\boldsymbol{\Delta}_{vi} = \frac{\Delta\boldsymbol{F}_i}{m_i}$.

Based on the above analysis, adopting the same handling, the rotational dynamic of the lead-UAV is constructed as

$$\begin{bmatrix}\dot{\boldsymbol{X}}_{01}\\ \dot{\boldsymbol{X}}_{02}\end{bmatrix} = \begin{bmatrix}\boldsymbol{0}_{3\times3} & \boldsymbol{I}_3\\ \boldsymbol{0}_{3\times3} & \boldsymbol{0}_{3\times3}\end{bmatrix}\begin{bmatrix}\boldsymbol{X}_{01}\\ \boldsymbol{X}_{02}\end{bmatrix} + \begin{bmatrix}\boldsymbol{0}_{3\times3}\\ \boldsymbol{G}_0(\boldsymbol{X}_0)\end{bmatrix}, \tag{8.17}$$

and the translational dynamic of the lead-UAV can be obtained that

$$\dot{\boldsymbol{X}}_{03} = \boldsymbol{H}_0(\boldsymbol{X}_0), \tag{8.18}$$

where $\boldsymbol{X}_{01} = \boldsymbol{\varphi}_0, \boldsymbol{X}_{02} = \dot{\boldsymbol{\varphi}}_0, \boldsymbol{X}_{03} = \boldsymbol{v}_0, \boldsymbol{G}_0 = (\boldsymbol{J}_0\boldsymbol{R}_2)^{-1}\left(\boldsymbol{C}_0\boldsymbol{\delta}_0 - \boldsymbol{S}(\boldsymbol{J}_0\boldsymbol{R}_2\dot{\boldsymbol{\varphi}}_0)\boldsymbol{R}_2\dot{\boldsymbol{\varphi}}_0 + \boldsymbol{N}_0 - \boldsymbol{J}_0\dot{\boldsymbol{R}}_2\dot{\boldsymbol{\varphi}}_0\right)$, $\boldsymbol{H}_0 = \boldsymbol{R}_1^{\mathrm{T}}\,\boldsymbol{g} - \boldsymbol{S}(\boldsymbol{\omega}_0)\,\boldsymbol{v}_0 + \frac{\boldsymbol{F}_{T0}}{m_0} + \frac{\boldsymbol{F}_{a0}}{m_0}$, $\boldsymbol{X}_0 = [\boldsymbol{X}_{01}^{\mathrm{T}}, \boldsymbol{X}_{02}^{\mathrm{T}}, \boldsymbol{X}_{03}^{\mathrm{T}}]^{\mathrm{T}}$.

Assumption 8.1 For the 6-DoF fixed-wing UAV model, both external disturbances and parameter uncertainties are bounded, i.e., $\|\boldsymbol{B}_{i3}\| \leq \bar{B}_{i3}$, $\|\boldsymbol{d}_{\omega i}\| \leq \bar{d}_{\omega i}$, $\|\boldsymbol{\delta}_{if}\| \leq \bar{\delta}_{if}$, $\|\boldsymbol{\Delta}_{vi}\| \leq \bar{\Delta}_{vi}$, $\|\boldsymbol{d}_{vi}\| \leq \bar{d}_{vi}$, $|T_{xif}| \leq \bar{F}_{Txif}$.

Assumption 8.2 For the lead-UAV, $\boldsymbol{F}_0$ and $\boldsymbol{H}_0$ satisfy the localized Lipschitz condition, i.e., $\|\boldsymbol{G}_0(\boldsymbol{X}_0)-\boldsymbol{G}_0(\boldsymbol{Y}_0)\| \le \gamma_1 \|\boldsymbol{X}_0-\boldsymbol{Y}_0\|$, $\|\boldsymbol{H}_0(\boldsymbol{X}_0)-\boldsymbol{H}_0(\boldsymbol{Y}_0)\| \le \gamma_2 \|\boldsymbol{X}_0-\boldsymbol{Y}_0\|$.

Remark 8.1 Assumption 8.1 stipulates that external disturbances and fault parameters are bounded, thereby ensuring the controllability of the follow-UAVs [35]. As the model of the lead-UAV is deterministic and stable, it aligns with Assumption 8.2. Consequently, Assumptions 8.1–8.2 are deemed reasonable.

8.2.4 Important Lemmas and Control Objective

The lemmas needed for this paper will be given as follows.

Lemma 8.1 ([36]) *The selected fuzzy rules are*

$$IF\ \theta_1\ is\ A_1^l,\ \ldots,\theta_n\ is\ A_n^l, THEN\ y\ \ is\ B^j\ (j=1,2,\ldots,N),$$

where $\boldsymbol{\Theta}=[\theta_1,\theta_2,\ldots,\theta_n]^{\mathrm{T}} \in \mathcal{R}^n$ *and* $\bar{y}(\boldsymbol{\Theta}) \in \mathcal{R}$ *denote the input and output of fuzzy system, and* A_i^l *and* B^j *are fuzzy sets in* $\mathcal{R}$. *The fuzzy output is*

$$\bar{y}(\boldsymbol{\Theta}) = \frac{\sum\limits_{j=1}^{N} \bar{w}_j \prod\limits_{i=1}^{n} \varpi_{A_i^j}(\theta_i)}{\sum\limits_{j=1}^{N} \left[\prod\limits_{i=1}^{n} \varpi_{A_i^j}(\theta_i)\right]},$$

where $\varpi_{A_i^j}, \varpi_{B_i^j}$ *are fuzzy membership functions, and* $\bar{w}_j = \max_{\omega\in\mathcal{R}} \varpi_{B^j}(\boldsymbol{\Theta})$. *Defining* $\bar{s}_j = \left(\prod_{i=1}^{n} \varpi_{A_i^j}(\theta_i)\right) / \left(\sum_{j=1}^{N} \left[\prod_{i=1}^{n} \varpi_{A_i^j}(\theta_i)\right]\right)$, *it holds* $\bar{y}(\boldsymbol{\Theta}) = \boldsymbol{W}^{\mathrm{T}}\boldsymbol{S}(\boldsymbol{\Theta})$, *where* $\boldsymbol{S}(\boldsymbol{\Theta}) = [\bar{s}_1(\boldsymbol{\Theta}), \bar{s}_2(\boldsymbol{\Theta}), \ldots, \bar{s}_n(\boldsymbol{\Theta})]^{\mathrm{T}}$ *is the basis function vector, and* $\boldsymbol{W} = [\bar{w}_1, \bar{w}_2, \ldots, \bar{w}_N]^{\mathrm{T}}$ *denotes the idealized weight matrix. For the nonlinear function* $y(\boldsymbol{\Theta}) : \mathcal{R}^n \to \mathcal{R}$ *within a given compact set* $\Omega(\boldsymbol{\Theta}) \subset \mathcal{R}^n$, *there exists a fuzzy logical system (FLS)* $\boldsymbol{W}^{\mathrm{T}}\boldsymbol{S}(\boldsymbol{\Theta})$ *satisfying*

$$y(\boldsymbol{\Theta}) = \boldsymbol{W}^{\mathrm{T}}\boldsymbol{S}(\boldsymbol{\Theta}) + \varepsilon(\boldsymbol{\Theta}), \tag{8.19}$$

where $\varepsilon(\boldsymbol{\Theta})$ *is the approximation error. There exists a positive constant* ε^* *satisfying* $|\varepsilon(\boldsymbol{\Theta})| \le \varepsilon^*$.

Lemma 8.2 ([37]) *For all* $b > 0$, *it has*

$$0 \le |a| - \frac{\sqrt{a^2+b}}{a^2} \le \sqrt{b},\ a \in \mathcal{R}.$$

Lemma 8.3 ([38]) *For all $b > 0$, it has*

$$0 \leq |b| - b \tanh(b/a) \leq 0.2785a, \ a \in \mathcal{R}.$$

Our objectives are to (i) develop an asynchronous switched observer for each follow-UAV to estimate the lead-UAV's velocity and attitude information and (ii) design an FTCC strategy to guarantee consensus attainment among the follow-UAVs with the lead-UAV.

8.3 Main Results

8.3.1 Asynchronous Switched Observer Design

For the follow-UAV i ($i = 1, \ldots, N$), the asynchronous switched observer is designed as

$$\begin{aligned}
\dot{\hat{X}}_{01}^{(i)} &= \hat{X}_{02}^{(i)} - \rho \langle \sigma_1^{(i)} \rangle_{\varsigma_d(t)} \\
\dot{\hat{X}}_{02}^{(i)} &= -\rho \langle \sigma_2^{(i)} \rangle_{\varsigma_d(t)} + G_0 \left(\hat{X}_0^{(i)} \right) \\
\dot{\hat{X}}_{03}^{(i)} &= -\rho \langle \sigma_3^{(i)} \rangle_{\varsigma_d(t)} + H_0 \left(\hat{X}_0^{(i)} \right),
\end{aligned} \tag{8.20}$$

where $\hat{X}_0^{(i)} = \left[\hat{X}_{01}^{(i)\mathrm{T}}, \hat{X}_{02}^{(i)\mathrm{T}}, \hat{X}_{03}^{(i)\mathrm{T}} \right]^{\mathrm{T}} \in \mathcal{R}^9$ denotes the observer state vector, $\hat{X}_{01}^{(i)}, \hat{X}_{02}^{(i)}, \hat{X}_{03}^{(i)}$ are estimates of lead-UAV attitude angle φ_0, attitude angle derivative $\dot{\varphi}_0$, and velocity v_0, respectively, ρ is a designed positive parameter presented later, and $\langle \sigma_l^{(i)} \rangle_{\varsigma_d(t)}$ represents the consensus error which is given as

$$\begin{aligned}
\langle \sigma_l^{(i)} \rangle_{\varsigma_d(t)} = &\sum_{j=1}^{N} \langle a_{ij} \rangle_{\varsigma_d(t)} \left(\hat{X}_{0l}^{(i)} - \hat{X}_{0l}^{(j)} \right) \\
&+ \langle a_{i0} \rangle_{\varsigma_d(t)} \left(\hat{X}_{0l}^{(i)} - X_{0l} \right), \ j \in \mathcal{N}_{0:N}.
\end{aligned}$$

Then, we present the following theorem as the main conclusion of this section.

Theorem 8.1 *Given the distributed asynchronous switched observer (8.20), assuming Assumption 8.2 holds, and the asynchronous topology switching signal $\varsigma_d(t)$ satisfies the following inequality:*

$$\sum_{l \in \Omega_q} \left(\underline{\lambda}_l \overline{\vartheta}_{hl}^{*} + \underline{\varpi}_l \overline{\vartheta}_{hl} \right) + \sum_{l \in \Omega_p} \left(-\overline{\lambda}_l \underline{\vartheta}_{hl}^{*} + \underline{\gamma}_l \overline{\vartheta}_{hl}^{*} - \overline{\varpi}_l \underline{\vartheta}_{hl} \right)$$

$$+\sum_{(a,b)\in\mathcal{K}(\Omega_{st})} \overline{\varkappa\iota}_{ab}c_{ab} < 0, \tag{8.21}$$

where $\underline{\lambda}_l, \overline{\lambda}_l, \underline{\gamma}_l, \overline{\varpi}_l, \underline{\varpi}_l, c_{ab}$ *are positive constants detailed later. Then the observer asymptotically estimates state information about lead-UAV attitude angle, derivative of attitude angle, and velocity.*

Proof Defining state estimation error $\tilde{\boldsymbol{X}}_0^{(i)} = \hat{\boldsymbol{X}}_0^{(i)} - \boldsymbol{X}_0$ and combining (8.17), (8.18), and (8.20), it has

$$\begin{aligned}
\dot{\tilde{\boldsymbol{X}}}_{01}^{(i)} &= \tilde{\boldsymbol{X}}_{02}^{(i)} - \rho\langle\boldsymbol{\sigma}_1^{(i)}\rangle_{\varsigma_d(t)},\\
\dot{\tilde{\boldsymbol{X}}}_{02}^{(i)} &= -\rho\langle\boldsymbol{\sigma}_2^{(i)}\rangle_{\varsigma_d(t)} + \tilde{\boldsymbol{G}}_0^{(i)}\left(\boldsymbol{X}_0, \hat{\boldsymbol{X}}_0^{(i)}\right),\\
\dot{\tilde{\boldsymbol{X}}}_{03}^{(i)} &= -\rho\langle\boldsymbol{\sigma}_3^{(i)}\rangle_{\varsigma_d(t)} + \tilde{\boldsymbol{H}}_0^{(i)}\left(\boldsymbol{X}_0, \hat{\boldsymbol{X}}_0^{(i)}\right),
\end{aligned} \tag{8.22}$$

where $\tilde{\boldsymbol{G}}_0^{(i)}\left(\boldsymbol{X}_0, \hat{\boldsymbol{X}}_0^{(i)}\right) = \boldsymbol{G}_0\left(\hat{\boldsymbol{X}}_0^{(i)}\right) - \boldsymbol{G}_0(\boldsymbol{X}_0)$, $\tilde{\boldsymbol{H}}_0^{(i)}\left(\boldsymbol{X}_0, \hat{\boldsymbol{X}}_0^{(i)}\right) = \boldsymbol{H}_0\left(\hat{\boldsymbol{X}}_0^{(i)}\right) - \boldsymbol{H}_0(\boldsymbol{X}_0)$. Under the influence of topology switching delay, we consider the following two cases.

Case I: During the matching interval $t \in \mathbf{T}_{\top(t_k,t_{k+1})}$, (8.22) can be rewritten as

$$\begin{aligned}
\dot{\boldsymbol{\xi}}_{1l} &= \boldsymbol{\xi}_{2l} - \rho\boldsymbol{\mathcal{L}}_{\varsigma(t_k)}\boldsymbol{\xi}_{1l},\\
\dot{\boldsymbol{\xi}}_{2l} &= -\rho\boldsymbol{\mathcal{L}}_{\varsigma(t_k)}\boldsymbol{\xi}_{2l} + \bar{\boldsymbol{G}}_{0l},\\
\dot{\boldsymbol{\xi}}_{3l} &= -\rho\boldsymbol{\mathcal{L}}_{\varsigma(t_k)}\boldsymbol{\xi}_{3l} + \bar{\boldsymbol{H}}_{0l}, l \in \mathcal{N}_{1:3},
\end{aligned} \tag{8.23}$$

where $\boldsymbol{\xi}_{jl} = \left[\tilde{\boldsymbol{X}}_{0jl}^{(1)}, \ldots, \tilde{\boldsymbol{X}}_{0jl}^{(N)}\right]^{\mathrm{T}}, j = 1, 2, 3$, $\bar{\boldsymbol{G}}_0 = \left[\tilde{\boldsymbol{G}}_0^{(1)}, \ldots, \tilde{\boldsymbol{G}}_0^{(N)}\right]^{\mathrm{T}}$, $\bar{\boldsymbol{H}}_0 = \left[\tilde{\boldsymbol{H}}_0^{(1)}, \ldots, \tilde{\boldsymbol{H}}_0^{(N)}\right]^{\mathrm{T}}$. Take the Lyapunov function as

$$L_{0\varsigma(t_k)} = \frac{1}{2}\sum_{j=1}^{3}\sum_{l=1}^{3} \boldsymbol{\xi}_{jl}^{\mathrm{T}}\boldsymbol{\mathcal{Q}}_{\varsigma(t_k)}\boldsymbol{\xi}_{jl}, \tag{8.24}$$

and the time derivative of $L_{0\varsigma(t_k)}$ is

$$\begin{aligned}
\dot{L}_{0\varsigma(t_k)} =& \sum_{l=1}^{3} \boldsymbol{\xi}_{1l}^{\mathrm{T}}\boldsymbol{\mathcal{Q}}_{\varsigma(t_k)}\left(\boldsymbol{\xi}_{2l} - \rho\boldsymbol{\mathcal{L}}_{\varsigma(t_k)}\boldsymbol{\xi}_{1l}\right)\\
&+ \sum_{l=1}^{3} \boldsymbol{\xi}_{2l}^{\mathrm{T}}\boldsymbol{\mathcal{Q}}_{\varsigma(t_k)}\left(-\rho\boldsymbol{\mathcal{L}}_{\varsigma(t_k)}\boldsymbol{\xi}_{2l} + \bar{\boldsymbol{G}}_{0l}\right)
\end{aligned}$$

$$+\sum_{l=1}^{3}\boldsymbol{\xi}_{3l}^{\mathrm{T}}\boldsymbol{\mathcal{Q}}_{\varsigma(t_k)}\left(-\rho\boldsymbol{\mathcal{L}}_{\varsigma(t_k)}\boldsymbol{\xi}_{3l}+\bar{\boldsymbol{H}}_{0l}\right). \tag{8.25}$$

According to Young's inequality and Assumption 8.2, it holds that

$$\sum_{l=1}^{3}\left(\boldsymbol{\xi}_{1l}^{\mathrm{T}}\boldsymbol{\mathcal{Q}}_{\varsigma(t_k)}\boldsymbol{\xi}_{2l}+\boldsymbol{\xi}_{2l}^{\mathrm{T}}\boldsymbol{\mathcal{Q}}_{\varsigma(t_k)}\bar{\boldsymbol{G}}_{0l}+\boldsymbol{\xi}_{3l}^{\mathrm{T}}\boldsymbol{\mathcal{Q}}_{\varsigma(t_k)}\bar{\boldsymbol{H}}_{0l}\right)$$
$$\leq\frac{\left\|\boldsymbol{\mathcal{Q}}_{\varsigma(t_k)}\right\|^2+\hat{\gamma}}{2}\sum_{j=1}^{3}\sum_{l=1}^{3}\|\boldsymbol{\xi}_{jl}\|^2+\frac{1}{2}\sum_{l=1}^{3}\|\boldsymbol{\xi}_{2l}\|^2, \tag{8.26}$$

where $\hat{\gamma}=3(\gamma_1^2+\gamma_2^2)$. When $\varsigma(t_k)\in\Omega_{\mathrm{p}}$, according to (8.9), it has

$$-\rho\sum_{j-1}^{3}\sum_{l-1}^{3}\boldsymbol{\xi}_{jl}^{\mathrm{T}}\boldsymbol{\mathcal{Q}}_{\varsigma(t_k)}\boldsymbol{\mathcal{L}}_{\varsigma(t_k)}\boldsymbol{\xi}_{jl}\leq-\frac{\rho}{2}\chi_{\varsigma(t_k)}\sum_{j=1}^{3}\sum_{l=1}^{3}\|\boldsymbol{\xi}_{jl}\|^2. \tag{8.27}$$

When $\varsigma(t_k)\in\Omega_{\mathrm{q}}$, it gets

$$-\rho\sum_{j=1}^{3}\sum_{l=1}^{3}\boldsymbol{\xi}_{jl}^{\mathrm{T}}\boldsymbol{\mathcal{Q}}_{\varsigma(t_k)}\boldsymbol{\mathcal{L}}_{\varsigma(t_k)}\boldsymbol{\xi}_{jl}$$
$$\leq\rho\|\boldsymbol{\mathcal{Q}}_{\varsigma(t_k)}\boldsymbol{\mathcal{L}}_{\varsigma(t_k)}\|^2\sum_{j=1}^{3}\sum_{l=1}^{3}\|\boldsymbol{\xi}_{jl}\|^2. \tag{8.28}$$

Substituting (8.26)–(8.28) into (8.25) yields

$$\dot{L}_{0\varsigma(t_k)}\leq\begin{cases}-\overline{\varpi}_{\varsigma(t_k)}L_{0\varsigma(t_k)}, & \varsigma(t)\in\Omega_{\mathrm{p}},\\ \underline{\varpi}_{\varsigma(t_k)}L_{0\varsigma(t_k)}, & \varsigma(t)\in\Omega_{\mathrm{q}},\end{cases} \tag{8.29}$$

where

$$\overline{\varpi}_{\varsigma(t_k)}=\frac{\rho\chi_{\varsigma(t_k)}-\left(\|\boldsymbol{\mathcal{Q}}_{\varsigma(t_k)}\|^2+1+\hat{\gamma}\right)}{\|\boldsymbol{\mathcal{Q}}_{\varsigma(t_k)}\|},$$
$$\underline{\varpi}_{\varsigma(t_k)}=\frac{\|\boldsymbol{\mathcal{Q}}_{\varsigma(t_k)}\boldsymbol{\mathcal{L}}_{\varsigma(t_k)}\|+(\|\boldsymbol{\mathcal{Q}}_{\varsigma(t_k)}\|^2+1+\hat{\gamma})}{\|\boldsymbol{\mathcal{Q}}_{\varsigma(t_k)}\|}.$$

This completes the design and analysis for the matching interval, and then the non-matching interval is discussed.

Case II: During the non-matching interval $t \in \mathbf{T}_{\perp(t_k,t_{k+1})}$, along the similar process of derivation as in (8.23) yields

$$\begin{aligned}\dot{\boldsymbol{\xi}}_{1l} &= \boldsymbol{\xi}_{2l} - \rho \boldsymbol{\mathcal{O}}_{\varsigma(t_k)} \boldsymbol{\xi}_{1l}, \\ \dot{\boldsymbol{\xi}}_{2l} &= -\rho \boldsymbol{\mathcal{O}}_{\varsigma(t_k)} \boldsymbol{\xi}_{2l} + \bar{\boldsymbol{G}}_{0l}, \\ \dot{\boldsymbol{\xi}}_{3l} &= -\rho \boldsymbol{\mathcal{O}}_{\varsigma(t_k)} \boldsymbol{\xi}_{3l} + \bar{\boldsymbol{H}}_{0l}, l \in \mathcal{N}_{1:3},\end{aligned} \tag{8.30}$$

where $\boldsymbol{\mathcal{O}}_{\varsigma(t_k)} = \left[\langle o_{ij} \rangle_{\varsigma(t_k)}\right] \in \mathcal{R}^{N \times N}$ with

$$\langle o_{ij} \rangle_{\varsigma(t_k)} = \begin{cases} -\langle a_{ij} \rangle_{\varsigma(t_k)}, & \text{if}(\langle a_{ij} \rangle_{\varsigma(t_k^-)} > 0) \\ & \quad \cap (\langle a_{ij} \rangle_{\varsigma(t_k)} > 0), \\ 0, & \text{otherwise} \end{cases} \tag{8.31}$$

for $i \neq j$, and $\langle o_{ij} \rangle_{\varsigma(t_k)} = \langle a_{i0} \rangle_{\varsigma(t_k^-)} \text{sgn}(\langle a_{i0} \rangle_{\varsigma(t_k)}) - \sum_{j=1}^{N} \langle o_{ij} \rangle_{\varsigma(t_k)}$, for $i = j$. It can further be seen that $\boldsymbol{\mathcal{O}}_{\varsigma(t_k)} = \boldsymbol{\mathcal{L}}_{\varsigma(t_k^-)} - \boldsymbol{\mathcal{B}}_{\varsigma(t_k)}$, where $\boldsymbol{\mathcal{B}}_{\varsigma(t_k)} = \left[\langle b_{ij} \rangle_{\varsigma(t_k)}\right] \in \mathcal{R}^{N \times N}$ with

$$\langle b_{ij} \rangle_{\varsigma(t_k)} = \begin{cases} -\langle a_{ij} \rangle_{\varsigma(t_k)}, & \text{if}(\langle a_{ij} \rangle_{\varsigma(t_k^-)} > 0) \\ & \quad \cap (\langle a_{ij} \rangle_{\varsigma(t_k)} > 0), \\ 0, & \text{otherwise} \end{cases} \tag{8.32}$$

for $i \neq j$, and $\langle b_{ij} \rangle_{\varsigma(t_k)} = \langle a_{i0} \rangle_{\varsigma(t_k^-)} \left(1 - \text{sgn}(\langle a_{i0} \rangle_{\varsigma(t_k)})\right) - \sum_{j=1}^{N} \langle b_{ij} \rangle_{\varsigma(t_k)}$, for $i = j$. Take the Lyapunov function as

$$L_{0\varsigma(t_k)} = \frac{1}{2} \sum_{j=1}^{3} \sum_{l=1}^{3} \boldsymbol{\xi}_{jl}^{\mathrm{T}} \boldsymbol{\mathcal{Q}}_{\varsigma(t_k^-)} \boldsymbol{\xi}_{jl}. \tag{8.33}$$

Similarly to (8.26), it has

$$\begin{aligned}&\sum_{l=1}^{3} \left(\boldsymbol{\xi}_{1l}^{\mathrm{T}} \boldsymbol{\mathcal{Q}}_{\varsigma(t_k^-)} \boldsymbol{\xi}_{2l} + \boldsymbol{\xi}_{2l}^{\mathrm{T}} \boldsymbol{\mathcal{Q}}_{\varsigma(t_k^-)} \bar{\boldsymbol{G}}_{0l} + \boldsymbol{\xi}_{3l}^{\mathrm{T}} \boldsymbol{\mathcal{Q}}_{\varsigma(t_k^-)} \bar{\boldsymbol{H}}_{0l} \right) \\ &\leq \frac{\left\| \boldsymbol{\mathcal{Q}}_{\varsigma(t_k^-)} \right\|^2 + \hat{\gamma}}{2} \sum_{j=1}^{3} \sum_{l=1}^{3} \|\boldsymbol{\xi}_{jl}\|^2 + \frac{1}{2} \sum_{l=1}^{3} \|\boldsymbol{\xi}_{2l}\|^2.\end{aligned} \tag{8.34}$$

When $\varsigma(t_k) \in \Omega_{\mathrm{p}}$, combining (8.9), (8.31), and (8.32), it obtains

$$
\begin{aligned}
&-\rho\sum_{j=1}^{3}\sum_{l=1}^{3}\boldsymbol{\xi}_{jl}^{\mathrm{T}}\boldsymbol{\mathcal{Q}}_{\varsigma(t_k^-)}\boldsymbol{\mathcal{B}}_{\varsigma(t_k)}\boldsymbol{\xi}_{jl} \\
&\le -\frac{\rho}{2}\chi_{\varsigma(t_k)}\sum_{j=1}^{3}\sum_{l=1}^{3}\|\boldsymbol{\xi}_{jl}\|^2+\rho\sigma_1\sum_{j=1}^{3}\sum_{l=1}^{3}\|\boldsymbol{\xi}_{jl}\|^2
\end{aligned}
\tag{8.35}
$$

where $\sigma_1 = \max\left\{\|\boldsymbol{\mathcal{Q}}_{\varsigma(t_k^-)}\boldsymbol{\mathcal{B}}_{\varsigma(t_k)}\|\,|\,\varsigma(t_k^-)\in\Omega_{\mathrm{p}}, k\in\mathcal{N}_{1:\mathcal{U}_\varsigma}\right\}$. Similarly, when $\varsigma(t_k)\in\Omega_{\mathrm{q}}$, we have

$$
-\rho\sum_{j=1}^{3}\sum_{l=1}^{3}\boldsymbol{\xi}_{jl}^{\mathrm{T}}\boldsymbol{\mathcal{Q}}_{\varsigma(t_k^-)}\boldsymbol{\mathcal{O}}_{\varsigma(t_k)}\boldsymbol{\xi}_{jl}\le\rho\sigma_2\sum_{j=1}^{3}\sum_{l=1}^{3}\|\boldsymbol{\xi}_{jl}\|^2,
\tag{8.36}
$$

where $\sigma_2=\max\left\{\|\boldsymbol{\mathcal{Q}}_{\varsigma(t_k^-)}\boldsymbol{\mathcal{O}}_{\varsigma(t_k)}\|\,|\,\varsigma(t_k^-)\in\Omega_{\mathrm{q}}, k\in\mathcal{N}_{1:\mathcal{U}_\varsigma}\right\}$.

Combining (8.34)–(8.36), one has

$$
\dot{L}_{0\varsigma(t_k)}\le\begin{cases}\left(\underline{\gamma}_{\varsigma(t_k^-)}-\overline{\lambda}_{\varsigma(t_k^-)}\right)L_{0\varsigma(t_k)}, & \varsigma(t_k)\in\Omega_{\mathrm{p}},\\ \underline{\lambda}_{\varsigma(t_k^-)}L_{0\varsigma(t_k)}, & \varsigma(t_k)\in\Omega_{\mathrm{q}},\end{cases}
\tag{8.37}
$$

where

$$
\begin{aligned}
\underline{\gamma}_{\varsigma(t_k^-)}&=\frac{2\rho\sigma_1}{\lambda_{\min}\left(\boldsymbol{\mathcal{Q}}_{\varsigma(t_k^-)}\right)},\\
\overline{\lambda}_{\varsigma(t_k^-)}&=\frac{\rho\chi_{\varsigma(t_k^-)}-\left(\|\boldsymbol{\mathcal{Q}}_{\varsigma(t_k^-)}\|^2+1+\hat{\gamma}\right)}{\|\boldsymbol{\mathcal{Q}}_{\varsigma(t_k^-)}\|},\\
\underline{\lambda}_{\varsigma(t_k^-)}&=\frac{\rho\sigma_2+\left(\|\boldsymbol{\mathcal{Q}}_{\varsigma(t_k^-)}\|^2+1+\hat{\gamma}\right)}{\|\boldsymbol{\mathcal{Q}}_{\varsigma(t_k^-)}\|}.
\end{aligned}
$$

To ensure convergence of observer error states, one has

$$
\rho>\max_{k\in\mathcal{N}_{1:\mathcal{U}_\varsigma}}\left\{\frac{\|\boldsymbol{\mathcal{Q}}_{\varsigma(t_k)}\|^2+1+\hat{\gamma}}{\chi_{\varsigma(t_k)}},\frac{\|\boldsymbol{\mathcal{Q}}_{\varsigma(t_k^-)}\|^2+1+\hat{\gamma}}{\chi_{\varsigma(t_k^-)}}\right\}.
$$

By analyzing the above two cases, for $t\in[t_k,t_{k+1}), k\in\mathcal{N}_{1:\mathcal{U}_\varsigma}$, take the piecewise Lyapunov function as

$$L_{0\varsigma(t_k)} = \begin{cases} \frac{1}{2}\sum_{j=1}^{3}\sum_{l=1}^{3} \boldsymbol{\xi}_{jl}^{\mathrm{T}} \boldsymbol{\mathcal{Q}}_{\varsigma(t_k^-)} \boldsymbol{\xi}_{jl}, & t \in \mathbf{T}_{\perp(t_k,t_{k+1})}, \\ \frac{1}{2}\sum_{j=1}^{3}\sum_{l=1}^{3} \boldsymbol{\xi}_{jl}^{\mathrm{T}} \boldsymbol{\mathcal{Q}}_{\varsigma(t_k)} \boldsymbol{\xi}_{jl}, & t \in \mathbf{T}_{\top(t_k,t_{k+1})}. \end{cases} \tag{8.38}$$

According to the above equation, there exists a positive constant $c_{\varsigma(t_k^-)\varsigma(t_k)} \geq \lambda_{\max}\left(\boldsymbol{\mathcal{Q}}_{\varsigma(t_k)}(\boldsymbol{\mathcal{Q}}_{\varsigma(t_k^-)})^{-1}\right)$ satisfying

$$L_{0\varsigma(t_k)}(t_k + d_k) \leq c_{\varsigma(t_k^-)\varsigma(t_k)} L_{0\varsigma(t_k)}(t_k^- + d_k). \tag{8.39}$$

Meanwhile, there exist two K_∞ functions such that $\alpha_1(\|\boldsymbol{\xi}\|) \leq L_{0\varsigma(t_k)}(t) \leq \alpha_2(\|\boldsymbol{\xi}\|)$, $k \in \mathcal{N}_{1:\mathcal{U}_\varsigma}$ with $\boldsymbol{\xi} = \left[\boldsymbol{\xi}_{11}^{\mathrm{T}}, \boldsymbol{\xi}_{12}^{\mathrm{T}}, \ldots, \boldsymbol{\xi}_{33}^{\mathrm{T}}\right]^{\mathrm{T}}$. Combining (8.29) and (8.37), it further has

$$\alpha_1(\boldsymbol{\xi}(t)) \leq \exp(\Lambda(t))\alpha_2(\boldsymbol{\xi}(0)), \tag{8.40}$$

where

$$\begin{aligned} \Lambda(t) &= \sum_{\substack{k=1 \\ \varsigma(t_k^-)\in\Omega_{\mathrm{q}}}}^{\mathcal{U}_\varsigma(0,t)} \underline{\lambda}_{\varsigma(t_k^-)} \mathbf{T}_{\perp(t_k,t_{k+1})} - \sum_{\substack{k=1 \\ \varsigma(t_k^-)\in\Omega_{\mathrm{p}}}}^{\mathcal{U}_\varsigma(0,t)} \overline{\lambda}_{\varsigma(t_k^-)} \mathbf{T}_{\perp(t_k,t_{k+1})} \\ &+ \sum_{\substack{k=1 \\ \varsigma(t_k^-)\in\Omega_{\mathrm{p}}}}^{\mathcal{U}_\varsigma(0,t)} \underline{\gamma}_{\varsigma(t_k^-)} \mathbf{T}_{\perp(t_k,t_{k+1})} + \sum_{\substack{k=0 \\ \varsigma(t_k)\in\Omega_{\mathrm{q}}}}^{\mathcal{U}_\varsigma(0,t)} \underline{\varpi}_{\varsigma(t_k)} \mathbf{T}_{\top(t_k,t_{k+1})} \\ &- \sum_{\substack{k=0 \\ \varsigma(t_k)\in\Omega_{\mathrm{p}}}}^{\mathcal{U}_\varsigma(0,t)} \overline{\varpi}_{\varsigma(t_k)} \mathbf{T}_{\top(t_k,t_{k+1})} + \sum_{k=1}^{\mathcal{U}_\varsigma(0,t)} \ln\left(c_{\varsigma(t_k^-)\varsigma(t_k)}\right). \end{aligned}$$

By the definition of $c_{\varsigma(t_k^-)\varsigma(t_k)}$ in (8.39), it further holds that

$$\sum_{k=1}^{\mathcal{U}_\varsigma(0,t)} \ln\left(c_{\varsigma(t_k^-)\varsigma(t_k)}\right) = \sum_{(a,b)\in\mathcal{K}(\Omega_{\mathrm{st}})} \ln\left(c_{\varsigma(t_k^-)\varsigma(t_k)}\right)\Delta_{a\to b}. \tag{8.41}$$

To facilitate the subsequent analysis, two functions are defined by

$$\Xi_1(t) = \sum_{l\in\Omega_{\mathrm{q}}} \underline{\lambda}_l \vartheta_{hl}^* - \sum_{l\in\Omega_{\mathrm{p}}} \overline{\lambda}_l \vartheta_{hl}^* + \sum_{l\in\Omega_{\mathrm{p}}} \underline{\gamma}_l \vartheta_{hl}^*$$

$$+\sum_{i\in\Omega_q}\underline{\varpi}_i\vartheta_{hi}-\sum_{i\in\Omega_p}\overline{\varpi}_i\vartheta_{hi}, \tag{8.42}$$

$$\Xi_2(t)=\sum_{(a,b)\in\mathcal{K}(\Omega_{st})}\ln\left(c_{\varsigma(t_k^-)\varsigma(t_k)}\right)\frac{\Delta_{a\to b}}{\mathcal{U}_\varsigma(0,t)}. \tag{8.43}$$

By the property of topological switching, combined with (8.41)–(8.43), we get $\Lambda(t)=h(t)\big(\Xi_1(t)+\varkappa(t)\ \Xi_2(t)\big)$. Together with (8.6)–(8.8), (8.21), and (8.40), it further yields

$$\|\boldsymbol{\xi}(t)\|\le\alpha_1^{-1}\left(\exp(\Lambda(t))\alpha_2(\|\boldsymbol{\xi}(0)\|)\right). \tag{8.44}$$

Based on the above inequation, it can be obtained that $\boldsymbol{\xi}$ is bounded and converges to 0 for $t\to+\infty$. Eventually, the observer error is given to converge asymptotically to 0, i.e., the observer can asymptotically estimate all the signals of the lead-UAV, completing the proof of Theorem 8.1.

Remark 8.2 For asynchronous topology switching, the consensus error is reconstructed. The current state-of-the-art results [24, 25] employ the Laplacian matrix construction of the previous topology $\mathcal{G}_{\varsigma(t_{k-1})}$ during the non-matching interval. While this approach may be effective in other contexts, it is not consistent with the intrinsic mechanism of asynchronous topology switching. Specifically, during the non-matching interval, each UAV receives information only from its common neighbors before and after the topology switch. As illustrated in Fig. 8.1, UAV 3 is only receiving information from UAV 0, rather than from both UAV 0 and UAV 2, during $t\in\mathbf{T}_{\perp(t_k,t_{k+1})}$. In light of the aforementioned evidence, we proceeded to construct the transition matrix $\boldsymbol{\mathcal{O}}_{\varsigma(t_k)}$ during the non-matching interval such that $\langle\boldsymbol{\sigma}_j\rangle_{\varsigma_d(t)}=\boldsymbol{\mathcal{O}}_{\varsigma(t_k)}\boldsymbol{\xi}_{jl}$.

Remark 8.3 Figure 8.2 illustrates the distinction between the Lyapunov function as formulated in [24] and the approach presented in this paper. In Fig. 8.2a, the conventional Lyapunov function undergoes a transformation during topology switching. On the other hand, Fig. 8.2b showcases the innovative piecewise Lyapunov function devised in this paper, aligning with the inherent mechanism of asynchronous topology switching. Consequently, it ensures continuity at the moment of topology

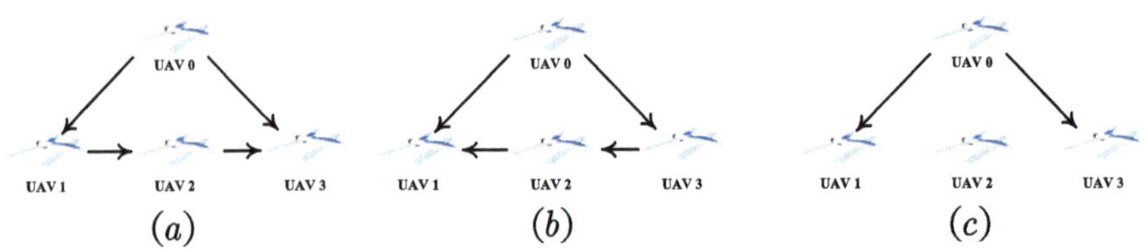

Fig. 8.1 Topology switching rule diagram. (**a**) Communication topology before switching $\mathcal{G}_{\varsigma(t_{k-1})}$. (**b**) Communication topology to complete switching $\mathcal{G}_{\varsigma(t_k)}$. (**c**) Communication topology during topology identification $\mathcal{G}_{\varsigma_d(t)}$

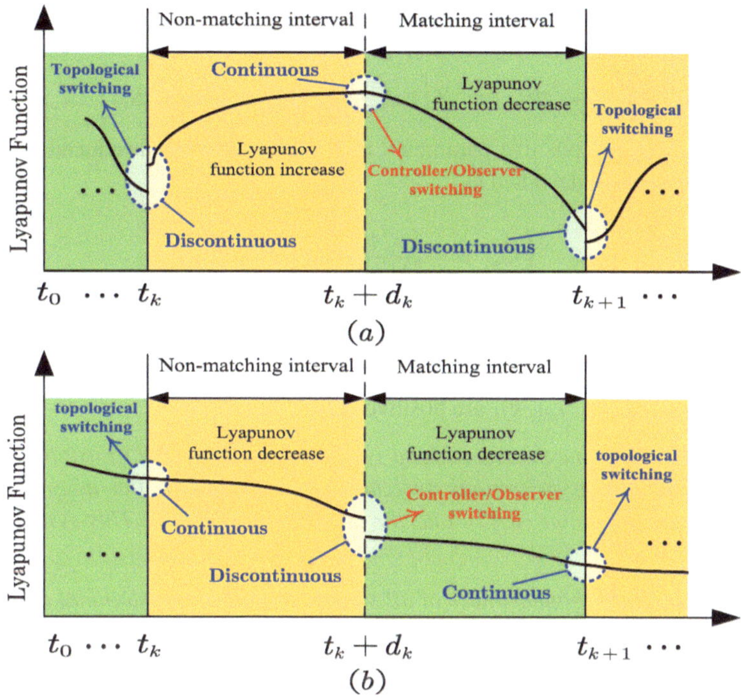

Fig. 8.2 Comparison of Lyapunov functions. (**a**) The Lyapunov function proposed in [24]. (**b**) The Lyapunov function proposed in this paper

switching while introducing discontinuity during controller/observer switching. The advantage of this design lies in the piecewise Lyapunov function (8.38), featuring the same positive definite diagonal matrix in both the matching interval of the previous topology and the non-matching interval of the current topology. Furthermore, by adjusting the parameter $\overline{\lambda}_{\varsigma(t_k^-)}$ in (8.37), the Lyapunov function in the non-matching interval can still be guaranteed to decrease if $\varsigma(t_k^-) \in \Omega_{\mathrm{p}}$ is within the prescribed range. In comparison to [24, 25], this results in a more lenient topological switching law.

8.3.2 Distributed Controller Design for Velocity Subsystem

In order to obtain the desired control objective, an FLS-based adaptive fault-tolerant speed controller is proposed as

$$T_{xi0} = -\hat{\aleph}_{Fi} \tanh\left(\frac{\tilde{V}_i \hat{\aleph}_{Fi} \varpi_{vi}}{\hat{l}_{vi}}\right) - \frac{\tilde{V}_i \hat{\aleph}_{Ti}^2 \eta_{vi}^2}{\varpi_{vi}\sqrt{\tilde{V}_i^2 \hat{\aleph}_{Ti}^2 \eta_{vi}^2 + \overline{\varepsilon}_{vi}}}, \tag{8.45}$$

where $\eta_{vl} = K_{vl}\tilde{V}_l + \frac{\hat{\Phi}_{vl}\boldsymbol{v}_l^{\mathrm{T}}\boldsymbol{\Psi}_{vl}}{2b_{l1}V_l} + \frac{\tilde{V}_l\boldsymbol{v}_l^{\mathrm{T}}\boldsymbol{v}_l}{2b_{l2}V_l^2} + \frac{\tilde{V}_l}{2V_0^{(l)2}}\hat{\boldsymbol{X}}_{03}^{(l)\mathrm{T}}\hat{\boldsymbol{X}}_{03}^{(i)} + \frac{\boldsymbol{v}_l^{\mathrm{T}}}{V_l}\left(\boldsymbol{R}_1^{\mathrm{T}}\boldsymbol{g} + \frac{\boldsymbol{F}_{l0}}{m_l}\right) - \frac{\hat{\boldsymbol{X}}_{03}^{(l)\mathrm{T}}\boldsymbol{H}_0}{V_0^{(l)}}$, $\varpi_{vl} = \frac{u_l}{m_l V_l}$, $\boldsymbol{\Psi}_{vl} = \frac{\tilde{V}_l}{V_l}\left[u_l\boldsymbol{S}_{vl}^{\mathrm{T}}\boldsymbol{S}_{vl}, v_l\boldsymbol{S}_{vl}^{\mathrm{T}}\boldsymbol{S}_{vl}, w_i\boldsymbol{S}_{vi}^{\mathrm{T}}\boldsymbol{S}_{vl}\right]^{\mathrm{T}}$, $\hat{l}_{vl}$, $\overline{\varepsilon}_{vl}$, b_{l1}, b_{l2} are the designed positive parameter, and $\tilde{V}_l$, K_{vl} will be introduced later. The adaptive laws can be designed as

$$\begin{cases} \dot{\hat{\Phi}}_{vl} = \dfrac{r_{1l}\tilde{V}_l\boldsymbol{v}_l^{\mathrm{T}}\boldsymbol{\Psi}_{vl}}{2b_{l1}V_l} - k_{1l}\hat{\Phi}_{vl}, \\ \dot{\hat{\aleph}}_{Fl} = r_{2l}|\tilde{V}_l|\varpi_{vl} - k_{2l}\hat{\aleph}_{Fl}, \\ \dot{\hat{\aleph}}_{Tl} = r_{3l}\tilde{V}_l\eta_{vl} - k_{3l}\hat{\aleph}_{Tl}, \end{cases} \tag{8.46}$$

where $r_{1l}, r_{2l}, r_{3l}, k_{1l}, k_{2l}, k_{3l}$ are positive design parameters.

Theorem 8.2 *Examine the closed-loop system, comprising UAV translational kinematics (8.16), the asynchronous switched observer (8.20), the adaptive fault-tolerant thrust controller (8.45), and the adaptive laws (8.46). This system facilitates:*

(1) Maintaining the boundedness of all signals related to translational kinematics with the controller designed to handle actuator failures, model uncertainty, and external disturbances

(2) Achieving consensus speed across the entire multi-UAV system, as the speed of the follow-UAV precisely tracks the speed of the lead-UAV with bounded error

Proof First define the speed error $\tilde{V}_l = V_l - V_0^{(l)}$, $V_0^{(l)} = \left\|\hat{\boldsymbol{X}}_{03}^{(l)}\right\|$. Take the Lyapunov function as $L_v = \sum_{l=1}^{N}\frac{1}{2}\tilde{V}_l^2$, and the time derivative of it is

$$\begin{aligned} \dot{L}_v =& \sum_{l=1}^{N}\tilde{V}_l\left(\frac{\boldsymbol{v}_l^{\mathrm{T}}}{V_l}\left(\boldsymbol{R}_1^{\mathrm{T}}\boldsymbol{g} + \frac{\boldsymbol{F}_{l0}}{m_l} + \Delta_{vl} + \boldsymbol{d}_{vl}\right) + \frac{u_l T_{xlf}}{m_l V_l} + \frac{u_l Q_{Tl} T_{xl0}}{m_l V_l}\right) \\ & - \sum_{l=1}^{N}\frac{\tilde{V}_l\hat{\boldsymbol{X}}_{03}^{(l)\mathrm{T}}}{V_0^{(l)}}\left(-\rho\langle\boldsymbol{\sigma}_3^{(l)}\rangle_{\varsigma_d(t)} + \boldsymbol{H}_0\right). \end{aligned} \tag{8.47}$$

For the uncertainty term Δ_{vl} of (8.47), FLS approximates it according to Lemma 8.1. Define $\boldsymbol{\Theta}_{vl} = \left[\boldsymbol{v}_l^{\mathrm{T}}, \boldsymbol{\varphi}_l^{\mathrm{T}}, \dot{\boldsymbol{\varphi}}_l^{\mathrm{T}}, \boldsymbol{\delta}_{l0}^{\mathrm{T}}\right]^{\mathrm{T}}$ as the input to the FLS. The fuzzy membership function is chosen as

$$\varpi_{A_l^j}(\theta_l) = \exp\left(-\frac{(\theta_l + 2 - 0.5j)}{2}\right), j = 1, 2, \ldots, 7,$$

where θ_l represents the elements in $\boldsymbol{\Theta}_{vl}$. It has

$$\sum_{l=1}^{N}\frac{\tilde{V}_l}{V_l}\boldsymbol{v}_l^{\mathrm{T}}\boldsymbol{W}_{v_l}^{\mathrm{T}}\boldsymbol{S}_{v_l}\le\sum_{l=1}^{N}\frac{\tilde{V}_l\Phi_{vl}\boldsymbol{v}_l^{\mathrm{T}}\boldsymbol{\Psi}_{vl}}{2b_{l1}V_l}+\sum_{l=1}^{N}\frac{b_{l1}}{2}\boldsymbol{l}^{\mathrm{T}}\boldsymbol{l},$$

$$\sum_{l=1}^{N}\frac{\tilde{V}_l}{V_l}\boldsymbol{v}_l^{\mathrm{T}}\left(\boldsymbol{\varepsilon}_{vl}+\boldsymbol{d}_{vl}\right)\le\sum_{l=1}^{N}\frac{\tilde{V}_l^2\boldsymbol{v}_l^{\mathrm{T}}\boldsymbol{v}_l}{2b_{l2}V_l^2}+\sum_{l=1}^{N}\frac{b_{l2}}{2}\boldsymbol{\tau}_{vl}^{\mathrm{T}}\boldsymbol{\tau}_{vl},\tag{8.48}$$

where $\boldsymbol{\tau}_{vl}=[\tau_{vl1},\tau_{vl2},\tau_{vl3}]^{\mathrm{T}}$, $\tau_{vlj}\ge|\varepsilon_{vlj}+d_{vlj}|$, $\Phi_{vl}=\max_{j=1,2,3}\left\{\boldsymbol{W}_{vlj}^{\mathrm{T}}\boldsymbol{W}_{vlj}\right\}$.

In addition, it gets

$$\sum_{l=1}^{N}\frac{\tilde{V}_l\hat{\boldsymbol{X}}_{03}^{(l)\mathrm{T}}}{V_0^{(l)}}\rho\langle\boldsymbol{\sigma}_3^{(l)}\rangle_{\varsigma_d(t)}\le\sum_{l=1}^{N}\frac{\tilde{V}_l^2}{2V_0^{(l)2}}\hat{\boldsymbol{X}}_{03}^{(l)\mathrm{T}}\hat{\boldsymbol{X}}_{03}^{(l)}+\frac{1}{2}\overline{\beta}.\tag{8.49}$$

Substituting (8.48) and (8.49) into (8.47) yields

$$\begin{aligned}\dot{L}_v\le&\sum_{l=1}^{N}\tilde{V}_l\left(\frac{u_lT_{xlf}}{m_lV_l}+\frac{u_l\varrho_{Tl}T_{xl0}}{m_lV_l}-\frac{\hat{\boldsymbol{X}}_{03}^{(l)\mathrm{T}}\boldsymbol{H}_0}{V_0^{(l)}}+\frac{\boldsymbol{v}_l^{\mathrm{T}}}{V_l}\left(\boldsymbol{R}_1^{\mathrm{T}}\boldsymbol{g}+\frac{\boldsymbol{F}_{l0}}{m_l}\right)\right.\\&\left.+\frac{\tilde{V}_l}{2V_0^{(l)2}}\hat{\boldsymbol{X}}_{03}^{(l)\mathrm{T}}\hat{\boldsymbol{X}}_{03}^{(l)}+\frac{\Phi_{vl}\boldsymbol{v}_l^{\mathrm{T}}\boldsymbol{\Psi}_{vl}}{2b_{l1}V_l}+\frac{\tilde{V}_l\boldsymbol{v}_l^{\mathrm{T}}\boldsymbol{v}_l}{2b_{l2}V_l^2}\right)+\frac{1}{2}\overline{\beta}\\&+\sum_{l=1}^{N}\left(\frac{b_{l1}}{2}\boldsymbol{l}^{\mathrm{T}}\boldsymbol{l}+\frac{b_{l2}}{2}\boldsymbol{\tau}_{vl}^{\mathrm{T}}\boldsymbol{\tau}_{vl}\right).\end{aligned}\tag{8.50}$$

Define $\aleph_{Tl}=1/\varrho_{Tl},\aleph_{Fl}=\aleph_{Tl}\sup_{t\ge0}T_{xlf}$, and according to Lemma 8.3, it further holds that

$$\begin{aligned}\sum_{l=1}^{N}\tilde{V}_l\frac{u_lT_{xlf}}{m_lV_l}&\le\sum_{l=1}^{N}\varrho_{T_l}|\tilde{V}_l|\varpi_{vl}\aleph_{Fl}\\&\le\sum_{l=1}^{N}\varrho_{Tl}\left(\tilde{V}_l\hat{\aleph}_{Fl}\varpi_{vl}\tanh\left(\frac{\tilde{V}_l\hat{\aleph}_{Fl}\varpi_{vl}}{\hat{\iota}_{vl}}\right)+0.2785\hat{\iota}_{vl}\right)\\&+\sum_{l=1}^{N}\varrho_{Tl}|\tilde{V}_l|\varpi_{vl}\tilde{\aleph}_{Fl}.\end{aligned}\tag{8.51}$$

Next, take the final Lyapunov function of the translation system as

$$L_2=L_v+\sum_{l=1}^{N}\frac{\tilde{\Phi}_{vl}^2}{2r_{1l}}+\sum_{l=1}^{N}\frac{\tilde{\aleph}_{Fl}^2}{2r_{2l}}+\sum_{l=1}^{N}\frac{\tilde{\aleph}_{Tl}^2}{2r_{3l}},\tag{8.52}$$

where $\tilde{\Phi}_{vi}, \tilde{\aleph}_{Fi}, \tilde{\aleph}_{Ti}$ denote the estimation error of $\Phi_{vi}, \aleph_{Fi}, \aleph_{Ti}$, respectively, satisfying $\tilde{\Phi}_{vi} = \Phi_{vi} - \hat{\Phi}_{vi}, \tilde{\aleph}_{Fi} = \aleph_{Fi} - \hat{\aleph}_{Fi}, \tilde{\aleph}_{Ti} = \aleph_{Ti} - \hat{\aleph}_{Ti}$ with $\hat{\Phi}_{vi}, \hat{\aleph}_{Fi}, \hat{\aleph}_{Ti}$ being the estimates.

Based on the thrust controller (8.45) and the adaptive laws (8.46), the time derivative of (8.52) denotes

$$
\begin{aligned}
\dot{L}_2 \leq & -\sum_{i=1}^{N} K_{vi} \tilde{V}_i^2 - \sum_{i=1}^{N} \frac{k_{1i}}{2r_{1i}} \tilde{\Phi}_{vi}^2 - \sum_{i=1}^{N} \frac{k_{2i}}{2r_{2i}} \tilde{\aleph}_{Fi}^2 - \sum_{i=1}^{N} \frac{k_{3i}}{2r_{3i}} \tilde{\aleph}_{Ti}^2 + \sum_{i=1}^{N} \frac{k_{1i}}{2r_{1i}} \Phi_{vi}^2 \\
& + \sum_{i=1}^{N} \frac{k_{2i}}{2r_{2i}} \aleph_{Fi}^2 + \sum_{i=1}^{N} \left(\frac{b_{i1}}{2} \boldsymbol{l}^{\mathrm{T}} \boldsymbol{l} + \frac{b_{i2}}{2} \boldsymbol{\tau}_{vi}^{\mathrm{T}} \boldsymbol{\tau}_{vi} + 0.2785 \hat{l}_{vi} \varrho_{Ti} \right) \\
& + \sum_{i=1}^{N} \frac{k_{3i}}{2r_{3i}} \aleph_{Ti}^2 + \frac{1}{2} \overline{\beta} \\
\leq & -\kappa_{v1} L_2 + \kappa_{v2},
\end{aligned}
\tag{8.53}
$$

where

$$
\begin{aligned}
\kappa_{v1} = & \min_{1 \leq i \leq N} \{2K_{vi}, k_{1i}, k_{2i}, k_{3i}\}, \\
\kappa_{v2} = & \sum_{i=1}^{N} \left(\frac{b_{i1}}{2} \boldsymbol{l}^{\mathrm{T}} \boldsymbol{l} + \frac{b_{i2}}{2} \boldsymbol{\tau}_{vi}^{\mathrm{T}} \boldsymbol{\tau}_{vi} + 0.2785 \hat{l}_{vi} \varrho_{Ti} \right) \\
& + \sum_{i=1}^{N} \frac{k_{1i}}{2r_{1i}} \Phi_{vi}^2 + \sum_{i=1}^{N} \frac{k_{2i}}{2r_{2i}} \aleph_{Fi}^2 + \sum_{i=1}^{N} \frac{k_{3i}}{2r_{3i}} \aleph_{Ti}^2 + \frac{1}{2} \overline{\beta}.
\end{aligned}
$$

Finally, it gets

$$
0 \leq L_2(t) \leq \left[\frac{\kappa_{v2}}{\kappa_{v1}} + L_2(0) \right] \exp(-\kappa_{v1} t) + \frac{\kappa_{v2}}{\kappa_{v1}}, \quad t \geq 0.
$$

From the above, we can conclude that all signals in the UAV translational error $\tilde{V}_i$ are bounded. Furthermore, the consensus velocity tracking error approaches arbitrarily small values, denoted by a positive constant $\hat{\imath}_2$, ensuring that $\lim_{t\to\infty} |V_i - V_0| \leq \hat{\imath}_2$. This concludes the proof of Theorem 8.2.

8.3.3 Distributed Controller Design for Attitude Subsystem

Two reasonable assumptions are posited to fulfill the intended control objectives.

Assumption 8.3 ([39]) For all $t \in (0, +\infty)$, it is supposed that $\partial \boldsymbol{B}_{i3}/\partial \boldsymbol{\delta}_{i0} + (\boldsymbol{J}_{i0}\boldsymbol{R}_2)^{-1} \boldsymbol{C}_0 \boldsymbol{\varrho}_{\delta_i} \neq 0$ and $\|\Delta \boldsymbol{J}_i^{-1} \boldsymbol{C}_i \boldsymbol{\varrho}_{\delta_i} (\boldsymbol{J}_{i0}^{-1} \boldsymbol{C}_{i0} \boldsymbol{\varrho}_{\delta_i})^{-1}\|_\infty < 1$.

Assumption 8.4 ([40]) There exists a positive number $g_{i,\min}$ such that $\lambda_{\min}(\boldsymbol{J}_{i0}^{-1} \boldsymbol{C}_0 \boldsymbol{\varrho}_{\delta_i} (\boldsymbol{J}_{i0}^{-1} \boldsymbol{C}_0)^{\mathrm{T}}) \geq g_{i,\min}$.

Remark 8.4 Assumptions 8.3 and 8.4 ensure the controllability of the rotational dynamics (8.12) and assert that the controller $\boldsymbol{\delta}_{i0}$ can generate adequate output even in the presence of faults. Additionally, since $g_{i,\min}$ is solely employed in the analysis process, both of the aforementioned assumptions are deemed reasonable.

The FLS-based adaptive fault-tolerant attitude controller is formulated as

$$\boldsymbol{\delta}_{i0} = -\left(\boldsymbol{J}_{i0}^{-1} \boldsymbol{C}_{i0}\right)^{\mathrm{T}} \boldsymbol{R}_2 \frac{\boldsymbol{\hbar}_i \hat{\aleph}_{\varphi i}^2 \boldsymbol{\eta}_{\varphi i}^{\mathrm{T}} \boldsymbol{\eta}_{\varphi i}}{\sqrt{\hat{\aleph}_{\varphi i}^2 \boldsymbol{\hbar}_i^{\mathrm{T}} \boldsymbol{\hbar}_i \boldsymbol{\eta}_{\varphi i}^{\mathrm{T}} \boldsymbol{\eta}_{\varphi i} + \overline{\varepsilon}_{\varphi_i}}}, \tag{8.54}$$

where a_{i1}, a_{i2}, $\overline{\varepsilon}_{\varphi_i}$ are positive parameters, $\boldsymbol{\hbar}_i$, $\boldsymbol{\epsilon}_{i1}$, $\boldsymbol{\epsilon}_{i2}$, and $\boldsymbol{K}_{\varphi i}$ and $\boldsymbol{K}_{\hbar i}$ will be introduced later. $\boldsymbol{\eta}_{\varphi i} = \boldsymbol{K}_{\hbar i} \boldsymbol{\hbar}_i + \boldsymbol{B}_{i2} - \boldsymbol{G}_0 \left(\hat{\boldsymbol{X}}_0^{(i)}\right) + \boldsymbol{\hbar}_i / (2a_{i1}) + \hat{\Phi}_{\varphi i} \boldsymbol{\Psi}_{\varphi i} / (2a_{i2}) + \boldsymbol{K}_{\varphi i} \boldsymbol{\epsilon}_{i2} + \boldsymbol{\epsilon}_{i1}$, $\boldsymbol{\Psi}_{\varphi_i} = \left[\hbar_{i1} \boldsymbol{S}_{\varphi_{i1}}^{\mathrm{T}} \boldsymbol{S}_{\varphi_{i1}}, \hbar_{i2} \boldsymbol{S}_{\varphi_{i2}}^{\mathrm{T}} \boldsymbol{S}_{\varphi_{i2}}, \hbar_{i3} \boldsymbol{S}_{\varphi_{i3}}^{\mathrm{T}} \boldsymbol{S}_{\varphi_{i3}}\right]^{\mathrm{T}}$. The adaptive laws can be designed as

$$\begin{cases} \dot{\hat{\Phi}}_{\varphi i} = \dfrac{r_{4i} \boldsymbol{\hbar}_i^{\mathrm{T}} \Psi_{\varphi_i}}{2a_{i2}} - k_{4i} \hat{\Phi}_{\varphi i}, \\ \dot{\hat{\aleph}}_{\varphi i} = r_{5i} \boldsymbol{\hbar}_i^{\mathrm{T}} \boldsymbol{\eta}_{\varphi i} - k_{5i} \hat{\aleph}_{\varphi i}, \end{cases} \tag{8.55}$$

where $r_{4i}, r_{5i}, k_{4i}, k_{5i}$ are positive design parameters.

Theorem 8.3 *Examine the rotational dynamics (8.10), comprising the asynchronous switched observer (8.20), the adaptive fault-tolerant controller (8.54), and adaptive update laws (8.55). Assumptions 8.1, 8.3, and 8.4 are valid, ensuring satisfaction of the following for multi-UAVs:*

(1) The controller, designed to handle actuator failures, model uncertainty, and external disturbances, maintains the boundedness of all signals related to rotational kinematics.

(2) The attitude angle of the follow-UAV precisely follows the attitude angle of the lead-UAV, with bounded error, facilitating uniformity across the entire multi-UAV system.

Proof At first, the UAV attitude state and estimated state error is defined by

$$\begin{aligned} \boldsymbol{\epsilon}_{i1} &= \boldsymbol{\varphi}_i - \hat{\boldsymbol{X}}_{01}^{(i)}, \\ \boldsymbol{\epsilon}_{i2} &= \dot{\boldsymbol{\varphi}}_i - \hat{\boldsymbol{X}}_{02}^{(i)}, \quad i \in \mathcal{N}_{1:N}. \end{aligned} \tag{8.56}$$

The time derivatives of $\boldsymbol{\epsilon}_{l1}, \boldsymbol{\epsilon}_{l2}$ along (8.12) and (8.20) are

$$\begin{aligned}\dot{\boldsymbol{\epsilon}}_{l1} &= \boldsymbol{\epsilon}_{l2} + \rho\langle\boldsymbol{\sigma}_1^{(l)}\rangle_{\varsigma d(t)},\\ \dot{\boldsymbol{\epsilon}}_{l2} &= \boldsymbol{B}_{l1}\boldsymbol{Q}_{\delta l}\boldsymbol{\delta}_{l0} + \boldsymbol{B}_{l2} + \boldsymbol{B}_{l3} - \boldsymbol{G}_0 + \rho\langle\boldsymbol{\sigma}_2^{(i)}\rangle_{\varsigma d(t)}.\end{aligned} \tag{8.57}$$

Take the Lyapunov function as $L_\epsilon = \sum_{l=1}^{N} \boldsymbol{\epsilon}_{l1}^{\mathrm{T}}\boldsymbol{\epsilon}_{l1}$, and the time derivative of L_ϵ is

$$\dot{L}_\epsilon = \sum_{l=1}^{N} \boldsymbol{\epsilon}_{l1}^{\mathrm{T}} \left(-\boldsymbol{K}_{\varphi l}\boldsymbol{\epsilon}_{l1} + \boldsymbol{\hbar}_l + \rho\langle\boldsymbol{\sigma}_1^{(l)}\rangle_{\varsigma d(t)}\right), \tag{8.58}$$

where $\boldsymbol{\hbar}_l = \boldsymbol{\epsilon}_{l2} - \boldsymbol{\pi}_l$, $\boldsymbol{\pi}_l = -\boldsymbol{K}_{\varphi l}\boldsymbol{\epsilon}_{l1}$ is the virtual controller, and $\boldsymbol{K}_{\varphi l} \in \mathcal{R}^{3\times 3}$ is the designed positive definite matrix with $\lambda_{\min}(\boldsymbol{K}_{\varphi l}) > 0.5$.

Further, the Lyapunov function is designed to be $L_\hbar = \sum_{i=1}^{N} \boldsymbol{\hbar}_l^{\mathrm{T}}\boldsymbol{\hbar}_l$ derived as

$$\dot{L}_\hbar = \sum_{l=1}^{N} \boldsymbol{\hbar}_l^{\mathrm{T}}\left(\boldsymbol{B}_{l1}\boldsymbol{Q}_{\delta l}\boldsymbol{\delta}_{l0} + \boldsymbol{B}_{l2} + \boldsymbol{B}_{l3} - \boldsymbol{G}_0 + \rho\langle\boldsymbol{\sigma}_2^{(l)}\rangle_{\varsigma d(t)} - \dot{\boldsymbol{\pi}}_l\right). \tag{8.59}$$

For the uncertainty term $\boldsymbol{B}_{l3}$ of (8.59), FLS approximates it according to Lemma 8.1. Define $\boldsymbol{\Theta}_{\varphi l} = \left[\boldsymbol{v}_l^{\mathrm{T}}, \boldsymbol{\varphi}_l^{\mathrm{T}}, \dot{\boldsymbol{\varphi}}_l^{\mathrm{T}}, \boldsymbol{\delta}_{l0}^{\mathrm{T}}\right]^{\mathrm{T}}$ as the input to the FLS. Since $\boldsymbol{\Theta}_{\varphi l}$ contains the control input $\boldsymbol{\delta}_{l0}$, the problem of algebraic loops occurs if FLS is used directly to approximate it. To solve this problem, the control input is processed with a low-pass filter to obtain $\boldsymbol{B}_{l3} = \boldsymbol{B}'_{l3} + \boldsymbol{\Delta}_{l3}$. The fuzzy membership function is chosen as

$$\varpi_{A_l^j}(\theta_l) = \exp\left(-\frac{(\theta_l + 2 - 0.5j)}{2}\right), j = 1, 2, \ldots, 7,$$

where θ_l represents the elements in $\boldsymbol{\Theta}_{\varphi l}$. Using Young's inequality, we get

$$\begin{aligned}\hbar_{lj}\left(\varepsilon_{\varphi lj} + \Delta_{l3,j}\right) &\le \frac{\hbar_{lj}^2}{2a_{l1}} + \frac{a_{l1}}{2}\tau_{lj}^2,\\ \hbar_{lj}\boldsymbol{W}_{\varphi lj}^{\mathrm{T}}\boldsymbol{S}_{\varphi lj} &\le \frac{\hbar_{lj}^2\boldsymbol{\Phi}_{\varphi l}\boldsymbol{S}_{\varphi lj}^{\mathrm{T}}\boldsymbol{S}_{\varphi lj}}{2a_{l2}} + \frac{a_{l2}}{2}.\end{aligned} \tag{8.60}$$

Defining $\boldsymbol{l} = [1, 1, 1]^{\mathrm{T}}$, $\boldsymbol{\tau}_l = [\tau_{l1}, \tau_{l2}, \tau_{l3}]^{\mathrm{T}}$, it further holds that

$$\sum_{l=1}^{N} \boldsymbol{\hbar}_l^{\mathrm{T}}\boldsymbol{B}_{l3} \le \sum_{l=1}^{N} \frac{\boldsymbol{\hbar}_l^{\mathrm{T}}\boldsymbol{\hbar}_l}{2a_{l1}} + \sum_{l=1}^{N} \frac{\Phi_{\varphi l}\boldsymbol{\hbar}_l^{\mathrm{T}}\boldsymbol{\Psi}_{\varphi l}}{2a_{l2}} + \sum_{l=1}^{N} \frac{a_{l1}}{2}\boldsymbol{\tau}_l^{\mathrm{T}}\boldsymbol{\tau}_l + \sum_{l=1}^{N} \frac{a_{l2}}{2}\boldsymbol{l}^{\mathrm{T}}\boldsymbol{l}. \tag{8.61}$$

In addition, following the definition of the consensus error and (8.44), it has

$$\sum_{l=1}^{N} \boldsymbol{\epsilon}_{l1}^{\mathrm{T}} \rho \langle \boldsymbol{\sigma}_1^{(l)} \rangle_{\varsigma_d(t)} + \sum_{l=1}^{N} \boldsymbol{\hbar}_l^{\mathrm{T}} \rho \langle \boldsymbol{\sigma}_2^{(l)} \rangle_{\varsigma_d(t)} \leq \frac{1}{2} \sum_{l=1}^{N} \left(\boldsymbol{\epsilon}_{l1}^{\mathrm{T}} \boldsymbol{\epsilon}_{l1} + \boldsymbol{\hbar}_l^{\mathrm{T}} \boldsymbol{\hbar}_l \right) + \frac{\bar{\beta}}{2}, \tag{8.62}$$

where $\sum_{l=1}^{N} \rho^2 (\| \langle \boldsymbol{\sigma}_1^{(l)} \rangle_{\varsigma_d(t)} \|^2 + \| \langle \boldsymbol{\sigma}_2^{(l)} \rangle_{\varsigma_d(t)} \|^2) \leq \beta \| \boldsymbol{\xi}(t) \|^2 \leq \beta \left(\alpha_1^{-1} (\exp(\Lambda(t)) \alpha_2 (\| \boldsymbol{\xi}(0) \|)) \right)^2 \triangleq \bar{\beta}$, $\beta = \rho^2 \max\{ \| \mathcal{O}_{\varsigma(t_k)} \|^2, \| \mathcal{L}_{\varsigma(t_k)} \|^2 \mid k \in \mathcal{N}_{0:\mathcal{U}_\varsigma} \}$.

According to the definition of $\boldsymbol{\pi}_l$, it has

$$\begin{aligned} -\sum_{l=1}^{N} \boldsymbol{\hbar}_l^{\mathrm{T}} \dot{\boldsymbol{\hbar}}_l &= \sum_{l=1}^{N} \boldsymbol{\hbar}_l^{\mathrm{T}} \boldsymbol{K}_{\varphi_l} \left(\boldsymbol{\epsilon}_{l2} + \rho \langle \boldsymbol{\sigma}_1^{(l)} \rangle_{\varsigma_d(t)} \right) \\ &\leq \sum_{l=1}^{N} \boldsymbol{\hbar}_l^{\mathrm{T}} \left(\boldsymbol{K}_{\varphi_l} \boldsymbol{\epsilon}_{l2} + \frac{1}{2} \boldsymbol{\hbar}_l \right) + \frac{\lambda_{\max} \left(\boldsymbol{K}_{\varphi_l}^{\mathrm{T}} \boldsymbol{K}_{\varphi_l} \right)}{2} \bar{\beta}. \end{aligned} \tag{8.63}$$

To compensate for the influence of actuator failures, the parameter boundaries are estimated, defining $\aleph_{\varphi l} = \inf_{t \geq 0} \lambda_{\min} \left(\boldsymbol{J}_{l0}^{-1} \boldsymbol{C}_0 \boldsymbol{\varrho}_{\delta_l} \left(\boldsymbol{J}_{l0}^{-1} \boldsymbol{C}_0 \right)^{\mathrm{T}} \right)$. Take the ultimate Lyapunov of the rotational system as

$$L_1 = L_\epsilon + L_\hbar + \sum_{l=1}^{N} \frac{1}{2 r_{4l}} \tilde{\Phi}_{\varphi l}^2 + \sum_{l=1}^{N} \frac{g_{\varphi l}}{2 r_{5l}} \tilde{\aleph}_{\varphi l}^2, \tag{8.64}$$

where $\tilde{\Phi}_{\varphi l}, \tilde{\aleph}_{\varphi l}$ represent the estimation error of $\Phi_{\varphi l}, \aleph_{\varphi l}$, respectively, satisfying $\tilde{\Phi}_{\varphi l} = \Phi_{\varphi l} - \hat{\Phi}_{\varphi l}, \tilde{\aleph}_{\varphi l} = \aleph_{\varphi l} - \hat{\aleph}_{\varphi l}$ with $\hat{\Phi}_{\varphi l}, \hat{\aleph}_{\varphi l}$ being the estimates.

Combining (8.61) and (8.62), the time derivative of (8.64) is

$$\begin{aligned} \dot{L}_1 \leq & -\sum_{l=1}^{N} \left(\lambda_{\min}(\boldsymbol{K}_{\varphi l}) - \frac{1}{2} \right) \boldsymbol{\epsilon}_{l1}^{\mathrm{T}} \boldsymbol{\epsilon}_{l1} - \sum_{l=1}^{N} \frac{k_{4l}}{2 r_{4l}} \tilde{\Phi}_{\varphi l}^2 - \sum_{l=1}^{N} \left(\lambda_{\min}(\boldsymbol{K}_{\hbar l}) - 1 \right) \boldsymbol{\hbar}_l^{\mathrm{T}} \boldsymbol{\hbar}_l \\ & + \sum_{l=1}^{N} \frac{k_{4l}}{2 r_{4l}} \Phi_{\varphi l}^2 + \frac{\lambda_{\max} \left(\boldsymbol{K}_{\varphi l}^{\mathrm{T}} \boldsymbol{K}_{\varphi l} \right) + 1}{2} \bar{\beta} + \sum_{l=1}^{N} \frac{a_{l1}}{2} \boldsymbol{\tau}_l^{\mathrm{T}} \boldsymbol{\tau}_l + \sum_{l=1}^{N} \frac{a_{l2}}{2} \boldsymbol{l}^{\mathrm{T}} \boldsymbol{l} \\ & + \sum_{l=1}^{N} \boldsymbol{\hbar}_l^{\mathrm{T}} \left(\boldsymbol{B}_{l1} \boldsymbol{\varrho}_{\delta l} \boldsymbol{\delta}_{l0} + \boldsymbol{\eta}_{\varphi l} \right) - \sum_{l=1}^{N} \frac{g_{\varphi l}}{r_{5l}} \tilde{\aleph}_{\varphi l} \dot{\hat{\aleph}}_{\varphi l}, \end{aligned} \tag{8.65}$$

where $\boldsymbol{K}_{\hbar i}$ is the designed positive definite parameter matrix with $\lambda_{\min}(\boldsymbol{K}_{\hbar i}) > 1$. For the penultimate term of (8.65), combining the controller with Lemma 8.2, (8.54) and (8.55) gives

$$
\begin{aligned}
\sum_{i=1}^{N} \boldsymbol{\hbar}_i^{\mathrm{T}} \left(\boldsymbol{B}_{i1}\boldsymbol{\varrho}_{\delta_i}\delta_{i0} + \boldsymbol{\eta}_{\varphi i}\right) &\leq -\sum_{i=1}^{N} \frac{g_{\varphi i}\boldsymbol{\hbar}_i^{\mathrm{T}}\boldsymbol{\hbar}_i\hat{\aleph}_{\varphi i}^2\boldsymbol{\eta}_{\varphi i}^{\mathrm{T}}\boldsymbol{\eta}_{\varphi i}}{\sqrt{\hat{\aleph}_{\varphi i}^2\boldsymbol{\hbar}_i^{\mathrm{T}}\boldsymbol{\hbar}_i\boldsymbol{\eta}_{\varphi i}^{\mathrm{T}}\boldsymbol{\eta}_{\varphi i} + \overline{\varepsilon}_{\varphi i}}} + \sum_{i=1}^{N} \boldsymbol{\hbar}_i^{\mathrm{T}}\boldsymbol{\eta}_{\varphi i} \\
&\leq \sum_{i=1}^{N} g_{\varphi i}\sqrt{\overline{\boldsymbol{\varepsilon}}_{\varphi i}} + \sum_{i=1}^{N} g_{\varphi i}\tilde{\aleph}_{\varphi i}\boldsymbol{\hbar}_i^{\mathrm{T}}\boldsymbol{\eta}_{\varphi i},
\end{aligned}
\tag{8.66}
$$

where $\overline{\boldsymbol{\varepsilon}}_{\varphi i}$ is a positive real parameter. Moreover, using (8.66) and adaptive laws (8.55), (8.65) is rewritten as

$$
\begin{aligned}
\dot{L}_1 \leq & -\sum_{i=1}^{N}\left(\lambda_{\min}(\boldsymbol{K}_{\varphi i}) - \frac{1}{2}\right)\boldsymbol{\epsilon}_{i1}^{\mathrm{T}}\boldsymbol{\epsilon}_{i1} - \sum_{i=1}^{N}\frac{k_{4i}}{2r_{4i}}\tilde{\Phi}_{\varphi i}^2 + \sum_{i=1}^{N}\frac{a_{i2}}{2}\boldsymbol{l}^{\mathrm{T}}\boldsymbol{l} \\
& -\sum_{i=1}^{N}\frac{g_{\varphi i}k_{5i}}{2r_{5i}}\tilde{\aleph}_{\varphi i}^2 + \frac{\lambda_{\max}\left(\boldsymbol{K}_{\varphi i}^{\mathrm{T}}\boldsymbol{K}_{\varphi i}\right)+1}{2}\bar{\beta} + \sum_{i=1}^{N}\frac{a_{i1}}{2}\boldsymbol{\tau}_i^{\mathrm{T}}\boldsymbol{\tau}_i \\
& +\sum_{i=1}^{N}\frac{a_{i2}}{2}\boldsymbol{l}^{\mathrm{T}}\boldsymbol{l} + \sum_{i=1}^{N}\frac{k_{4i}}{2r_{4i}}\Phi_{\varphi i}^2 + \sum_{i=1}^{N}\frac{g_{\varphi i}k_{5i}}{2r_{5i}}\tilde{\aleph}_{\varphi i}^2 + \sum_{i=1}^{N} g_{\varphi i}\sqrt{\overline{\boldsymbol{\varepsilon}}_{\varphi i}} \\
& \leq -\kappa_{\varphi 1}L_1 + \kappa_{\varphi 2},
\end{aligned}
\tag{8.67}
$$

where

$$
\begin{aligned}
\kappa_{\varphi 1} = &\min_{1\leq i\leq N}\left\{2\lambda_{\min}(\boldsymbol{K}_{\varphi i}) - 1, 2\lambda_{\min}(\boldsymbol{K}_{\hbar i}) - 2, k_{4i}, k_{5i}\right\}, \\
\kappa_{\varphi 2} = &\frac{\lambda_{\max}\left(\boldsymbol{K}_{\varphi i}^{\mathrm{T}}\boldsymbol{K}_{\varphi i}\right)+1}{2}\bar{\beta} + \sum_{i=1}^{N}\frac{a_{i1}}{2}\boldsymbol{\tau}_i^{\mathrm{T}}\boldsymbol{\tau}_i + \sum_{i=1}^{N}\frac{a_{i2}}{2}\boldsymbol{l}^{\mathrm{T}}\boldsymbol{l} \\
& + \sum_{i=1}^{N}\frac{k_{4i}}{2r_{4i}}\boldsymbol{\Phi}_{\varphi i}^2 + \sum_{i=1}^{N}\frac{g_{\varphi i}k_{5i}}{2r_{5i}}\tilde{\aleph}_{\varphi i}^2 + \sum_{i=1}^{N} g_{\varphi i}\sqrt{\overline{\boldsymbol{\varepsilon}}_{\varphi i}}.
\end{aligned}
$$

Finally, it obtains that

$$
0 \leq L_1(t) \leq \left[\frac{\kappa_{\varphi 2}}{\kappa_{\varphi 1}} + L_1(0)\right]\exp\left(-\kappa_{\varphi 1}t\right) + \frac{\kappa_{\varphi 2}}{\kappa_{\varphi 1}},\ t \geq 0.
$$

The results indicate that all the signals in the rotational error system (8.57) are semi-globally uniformly ultimately bounded (SGUUB). Thus, it is obtained that each follow-UAV can track the attitude of the lead-UAV under adjustable error, i.e., there exists a positive constant $\hat{\iota}_1$ such that $\lim_{t\to\infty} \|\boldsymbol{\varphi}_i - \boldsymbol{\varphi}_0\| \le \lim_{t\to\infty} \|\boldsymbol{\epsilon}_{i1}\| + \|\tilde{\boldsymbol{X}}_{01}^{(i)}\| \le \hat{\iota}_1$. This completes the proof of Theorem 8.3.

Remark 8.5 Based on the above analysis, we can see that the consensus error $\|\boldsymbol{\varphi}_i - \boldsymbol{\varphi}_0\|$ can be adjusted to any value by increasing $\lambda_{\min}(\boldsymbol{K}_{\varphi i})$, $\lambda_{\min}(\boldsymbol{K}_{\hbar i})$ or decreasing $a_{i1}, a_{i2}, \varepsilon_{\varphi i}$, but the adjusted result produces a large control input. Therefore the balance between the tracking performance and the control inputs in designing the controller should be made.

Remark 8.6 The iterative derivation of the virtual controller increases the computational complexity of the algorithm. This paper does not consider the issue of "complexity explosion" due to the simplicity of the structure of the virtual controller π_i for the attitude system and the fact that only its first-order derivatives are required. Furthermore, an investigation into the structure of the controller reveals that the controller (8.54), (8.45) in question is relatively simple and does not entail a multitude of complex computations. On the other hand, the deployment of the FLS has an impact on the computational complexity of the algorithm. However, the selected FLS is relatively simple and comprises a limited number of iterations, which results in a low computational burden for the overall algorithm.

8.4 Simulation and Analyses

In this section, we focus on a multi-UAV system comprising one lead-UAV (UAV 0) and four follow-UAVs (UAV 1–4). The goal is to design asynchronous switched observers and distributed consensus controllers to ensure that the speeds and attitude angles of the follow-UAVs synchronize with those of the lead-UAV within the desired error. This synchronization is achieved under the communication switching topology $\overline{\mathcal{G}} = \{\mathcal{G}_1, \mathcal{G}_2, \mathcal{G}_3\}$, where $\Omega_{\mathrm{st}} = \{1, 2, 3\}$, $\Omega_{\mathrm{p}} = \{1, 2\}$, and $\Omega_{\mathrm{q}} = \{3\}$. The Laplacian matrices for the three topological graphs can be designed as follows:

$$\mathcal{L}_1=\begin{bmatrix} 1 & 0 & 0 & 0 \\ -1 & 1 & 0 & 0 \\ -1 & 0 & 1 & 0 \\ 0 & -1 & 0 & 1 \end{bmatrix}, \quad \mathcal{L}_2=\begin{bmatrix} 1 & 0 & 0 & 0 \\ -1 & 1 & 0 & 0 \\ 0 & -1 & 1 & 0 \\ 0 & 0 & -1 & 1 \end{bmatrix}, \quad \mathcal{L}_3=\begin{bmatrix} 1 & 0 & 0 & 0 \\ -1 & 1 & 0 & 0 \\ 0 & 0 & 1 & -1 \\ 0 & 0 & -1 & 1 \end{bmatrix}.$$

Then, with respect to the topology switching signal $\varsigma(t)$, we choose $h(t) = t$, $\mathcal{U}_\varsigma(0, t) = 0.14t + t^{0.25}$, $\vartheta_{h1} = 0.55 - t^{-0.4}$, $\vartheta_{h2} = 0.7 - t^{-0.4}$, $\vartheta_{h3} = 0.1 + t^{-0.55}$, $\vartheta^*_{h1} = \vartheta^*_{h2} = 0.002 + t^{-0.4}/3$, $\vartheta^*_{h3} = 0.001 + 0.25t^{-0.4}$, $\iota_{12} = 0.2$, $\iota_{21} = 0.2$, $\iota_{13} = 0.1$, $\iota_{31} = 0.1$, $\iota_{32} = 0.2$, $\iota_{23} = 0.2$. According to (8.6)–(8.8), it has $\overline{\vartheta}_{h1} = \underline{\vartheta}_{h1} = 0.55$, $\overline{\vartheta}_{h2} = \underline{\vartheta}_{h2} = 0.7$, $\overline{\vartheta}_{h2} = \underline{\vartheta}_{h2} = 0.1$, $\overline{\vartheta}^*_{h1} = \underline{\vartheta}^*_{h1} = \overline{\vartheta}^*_{h2} = \underline{\vartheta}^*_{h2} =$

0.002, $\overline{\vartheta}^*_{h3} = \underline{\vartheta}^*_{h3} = 0.001$. To verify the effectiveness of the simulation, we set the simulation time $t = 50$s, which yields $\mathcal{U}_\varsigma = 10$, $\vartheta_{h1} = 0.3409$, $\vartheta_{h2} = 0.4409$, $\vartheta_{h3} = 0.2163$, $\vartheta^*_{h1} = \vartheta^*_{h2} = 0.0717$, $\vartheta^*_{h3} = 0.0533$. Moreover, the calculation from (8.9) gives $\chi_1 = \chi_2 = 0.3$, $\rho = 1.5$, $\mathcal{Q}_3 = \text{diag}\{3, 3, 3, 3\}$, and

$$\mathcal{Q}_1 = \begin{bmatrix} 0.5938 & 0.2188 & 0.1250 & 0.0625 \\ 0.2188 & 0.3750 & 0 & 0.1250 \\ 0.1250 & 0 & 0.2500 & 0 \\ 0.0625 & 0.1250 & 0 & 0.2500 \end{bmatrix},$$

$$\mathcal{Q}_2 = \begin{bmatrix} 1.0938 & 0.5938 & 0.2500 & 0.0625 \\ 0.5938 & 0.9375 & 0.4375 & 0.1250 \\ 0.2500 & 0.4375 & 0.7500 & 0.2500 \\ 0.0625 & 0.1250 & 0.2500 & 0.5000 \end{bmatrix}.$$

Figure 8.3a illustrates the topology switching signals and the topology delay signals for observer/controller asynchronous switching due to topology recognition.

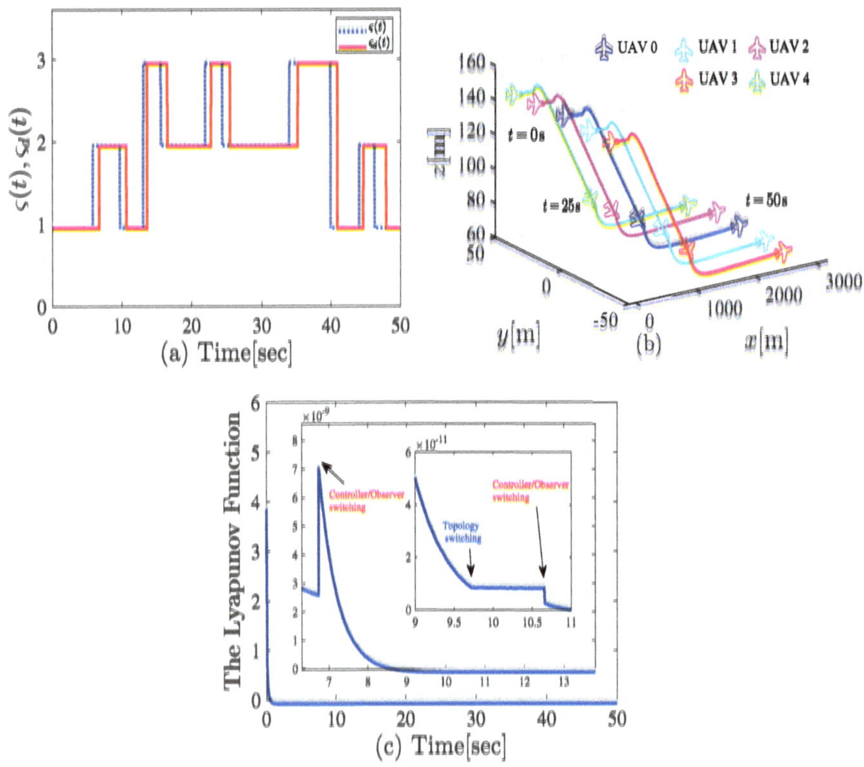

Fig. 8.3 (**a**) Topology switching and controller/observer switching trajectories. (**b**) Multi-UAV trajectory. (**c**) The Lyapunov function trajectory

Table 8.1 Airframe parameter values of UAV

Coefficient	Value	Unit	Coefficient	Value	Unit
g	9.8	$m \cdot s^{-2}$	$\bar{\rho}$	1.29	$kg \cdot m^2$
m_l	20.64	kg	S_l	1.37	m^2
b_l	1.96	m	$\bar{c}_l$	0.76	m
c_{L0l}	0.1	/	$c_{L\alpha l}$	0.25	rad^{-1}
c_{l0l}	−0.001	/	$c_{l\beta l}$	−0.038	rad^{-1}
c_{D0l}	0.5	/	$c_{Y\beta l}$	−0.1	rad^{-1}
c_{m0l}	0.022	/	$c_{m\alpha l}$	−0.473	rad^{-1}
c_{n0l}	0.022	/	$c_{n\alpha l}$	0.036	rad^{-1}
c_{lpl}	−0.213	$rad^{-1}\cdot s$	c_{lrl}	0.114	$rad^{-1}\cdot s$
$c_{l\delta al}$	−0.056	rad^{-1}	$c_{l\delta rl}$	0.014	rad^{-1}
$c_{n\delta al}$	−0.036	rad^{-1}	$c_{n\delta rl}$	−0.055	rad^{-1}
c_{mql}	−3.449	$rad^{-1}\cdot s$	$c_{m\delta l}$	−0.364	rad^{-1}
c_{npl}	−0.151	$rad^{-1}\cdot s$	c_{nrl}	−0.195	$rad^{-1}\cdot s$
J_{xxl}	1.607	$kg \cdot m^2$	J_{yyl}	7.51	$kg \cdot m^2$
J_{zzl}	7.18	$kg \cdot m^2$	J_{xzl}	0.59	$kg \cdot m^2$

Table 8.2 Initial state values of UAV

	$x_l(0)$	$y_l(0)$	$z_l(0)$	$\phi_l(0)$	$\theta_l(0)$	$\psi_l(0)$	$V_l(0)$
UAV0	40 m	0 m	150 m	0°	1°	0°	20 m/s
UAV1	10 m	−20 m	150 m	0.2°	1°	0°	25 m/s
UAV2	10 m	20 m	150 m	1°	1°	0°	20 m/s
UAV3	−10 m	−35 m	150 m	0°	1°	− 0.5°	19.5 m/s
UAV4	−10 m	35 m	150 m	0.2°	1.5°	− 0.2°	15 m/s

The values of multi-UAV parameters are shown in Table 8.1. The UAVs 2-4 are influenced by external disturbances, which are assumed to be $\boldsymbol{d}_{\omega l} = [0.2\sin(0.1t), 0.3\sin(0.2t), 0.5\sin(0.2t)]^{\mathrm{T}}$ and $\boldsymbol{d}_{vl} = [0.3\sin(t), 0.5\cos(2t), 0.2\cos(t)]^{\mathrm{T}}$. In addition, for UAV 3, there is the failure model for the rudder, elevator, and ailerons at $t \in [10, 20)$, i.e. $\boldsymbol{\delta}_{3f} = [0, 1, 0]^{\mathrm{T}}$, $\boldsymbol{\varrho}_{\delta 3} = \mathrm{diag}\{0.8, 0.8, 0.9\}$. The initial states of the multi-UAVs are shown in Table 8.2.

Assuming that the multi-UAV needs to complete the low-altitude crossing task and track the required trajectory as

$$V_r = \begin{cases} 20 + \frac{30}{2s+1}, & 0\mathrm{s} \le t < 7\mathrm{s}, \\ 20 + \frac{20}{0.4s+1}, & 7\mathrm{s} \le t < 15\mathrm{s}, \\ 20 + \frac{40}{s+1}, & t \ge 15\mathrm{s}, \end{cases}$$

$$\boldsymbol{\varphi}_r = \begin{cases} [0, 0, 0]^{\mathrm{T}}, & 0\mathrm{s} \le t < 5\mathrm{s}, \\ \left[0, \frac{-5}{1.5s+1}, 0\right]^{\mathrm{T}}, & 5\mathrm{s} \le t < 25\mathrm{s}, \\ [0, 0, 0]^{\mathrm{T}}, & t \ge 25\mathrm{s}. \end{cases}$$

Next, the design parameters are chosen as $K_{v1} = 200$, $K_{v2} = 300$, $K_{v3} = 400$, $K_{v4} = 270$, $\boldsymbol{K}_{\hbar 1} = \boldsymbol{K}_{\hbar 2} = \boldsymbol{K}_{\hbar 3} = \boldsymbol{K}_{\hbar 4} = \text{diag}\{50, 55, 50\}(0\text{s} \le t < 5\text{s})$; $\text{diag}\{15, 30, 15\}(t \ge 5\text{s})$, $\boldsymbol{K}_{\varphi 1} = \text{diag}\{60, 50, 200\}(0\text{s} \le t < 5\text{s}, t \ge 25\text{s})$; $\text{diag}\{105, 200, 270\}(5\text{s} \le t < 15\text{s})$; $\text{diag}\{150, 600, 400\}(15\text{s} \le t < 25\text{s})$, $\boldsymbol{K}_{\varphi 2} = \text{diag}\{40, 50, 45\}(0\text{s} \le t < 5\text{s})$; $\text{diag}\{100, 100, 100\}(5\text{s} \le t < 15\text{s})$; $\text{diag}\{120, 150, 120\}(15\text{s} \le t < 25\text{s})$; $\text{diag}\{120, 120, 120\}(t \ge 25\text{s})$; $\boldsymbol{K}_{\varphi 3} = \text{diag}\{40, 50, 45\}(0\text{s} \le t < 5\text{s})$; $\text{diag}\{100, 100, 100\}(5\text{s} \le t < 25\text{s})$; $\text{diag}\{110, 110, 110\}(t \ge 25\text{s})$, $\boldsymbol{K}_{\varphi 4} = \text{diag}\{40, 50, 45\}(0\text{s} \le t < 5\text{s})$; $\text{diag}\{90, 90, 90\}(5\text{s} \le t < 15\text{s})$; $\text{diag}\{100, 100, 100\}(t \ge 25\text{s})$, for $i = 1, \ldots, 4$, $a_{1i} = a_{2i} = b_{1i} = b_{2i} = 1$, $\bar{\varepsilon}_{\varphi i} = 0.01$, $r_{1i} = r_{4i} = r_{5i} = 0.01$, $r_{3i} = 1$, $k_{1i} = 1$, $k_{3i} = k_{4i} = k_{5i} = 25$, $\hat{l}_{vi} = 2$, $r_{21} = r_{22} = r_{24} = 1$, $r_{23} = 0.0003$, $k_{21} = k_{22} = k_{24} = 25$, $k_{23} = 22$, $\bar{\varepsilon}_{v1} = \bar{\varepsilon}_{v2} = \bar{\varepsilon}_{v4} = 0.05$, $\bar{\varepsilon}_{v3} = 0.03$.

Figure 8.3b illustrates the trajectories of the multi-UAV system, demonstrating that the followers can maintain their original formation despite the influence of topology switching. In Fig. 8.3c, the continuity of the piecewise Lyapunov function during topology switching and its discontinuity during controller/observer

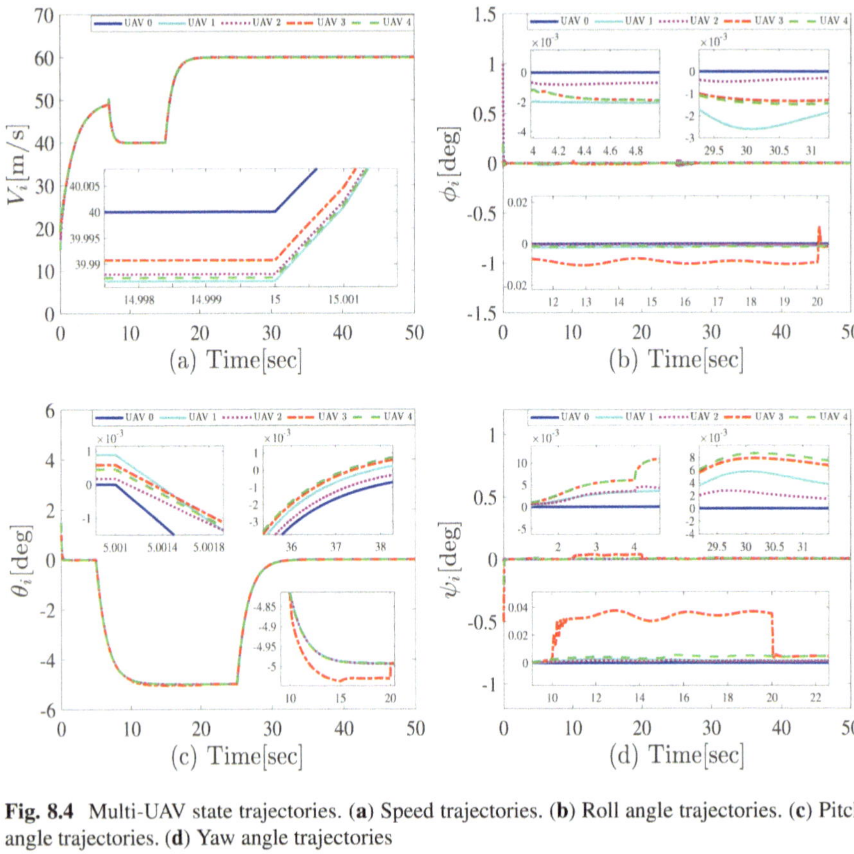

Fig. 8.4 Multi-UAV state trajectories. (**a**) Speed trajectories. (**b**) Roll angle trajectories. (**c**) Pitch angle trajectories. (**d**) Yaw angle trajectories

switching are evident. Importantly, the Lyapunov function does not increase in the non-matching interval. The convergence of velocities and attitudes to the desired commands is confirmed by Fig. 8.4a–d, providing validation for the effectiveness of the proposed algorithm. Figure 8.4b–d illustrates the trajectory of the attitude angle of the multi-UAS. Although an error is present in the attitude angle of the multi-UAV, it is minimal and does not significantly affect the overall flight. Furthermore, due to the coupling relationship between the three angles (specifically, the roll and yaw angles are highly coupled), the impact of the error is further compensated, enabling the multi-UAVs to remain in formation and fly cooperatively under the assigned mission, as illustrated in Fig. 8.3b. To demonstrate the superiority of the proposed control algorithm and the advantages of asynchronous switching observer, in Figs. 8.5, 8.6, 8.7, and 8.8, the proposed control scheme is compared with the observerless approach presented in [35]. To ensure compatibility with the method in [35], it is assumed to be implemented under a fixed topology $\mathcal{G}_1$. As shown in

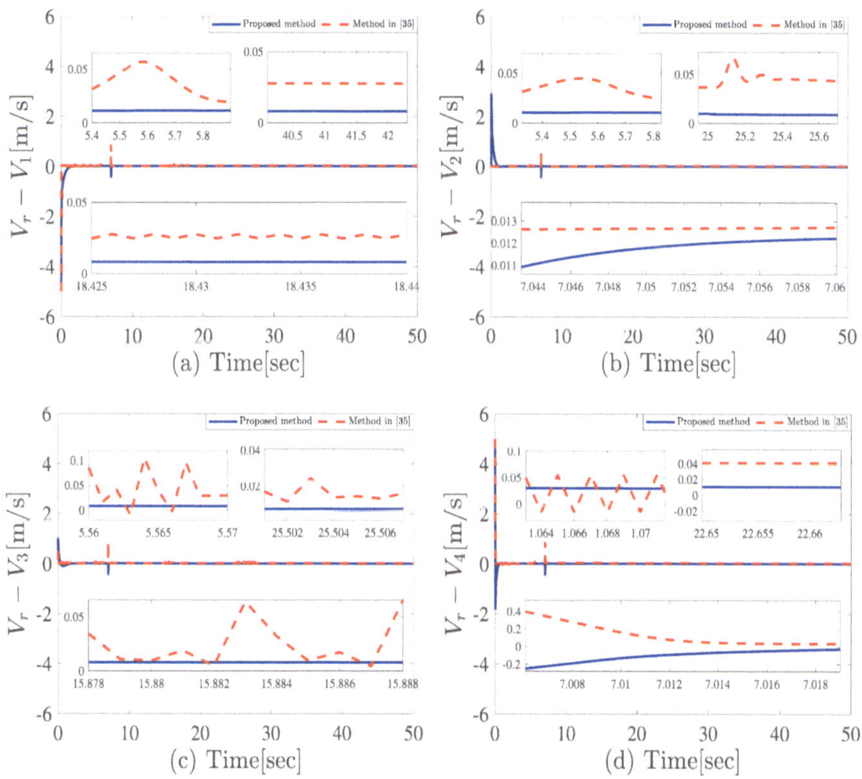

Fig. 8.5 Comparison of speed tracking error trajectories. (**a**) Error trajectories of UAV 1. (**b**) Error trajectories of UAV 2. (**c**) Error trajectories of UAV 3. (**d**) Error trajectories of UAV 4

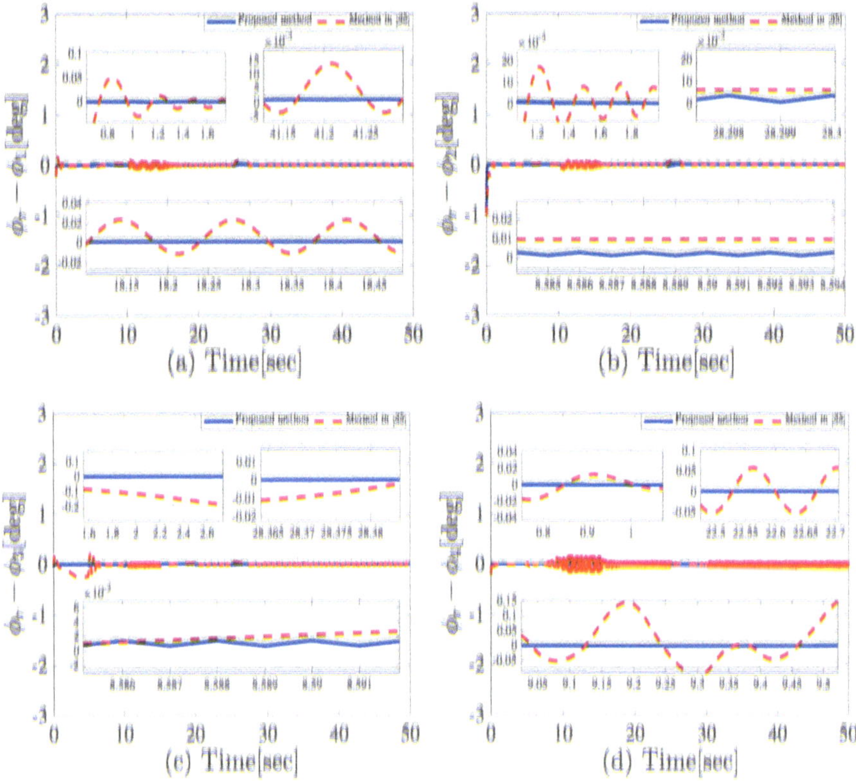

Fig. 8.6 Comparison of roll angle tracking error trajectories. (**a**) Error trajectories of UAV 1. (**b**) Error trajectories of UAV 2. (**c**) Error trajectories of UAV 3. (**d**) Error trajectories of UAV 4

Figs. 8.5, 8.6, 8.7, and 8.8, the method in this paper exhibits minor tracking errors under topology switching. Notably, the oscillation amplitude is significantly smaller than that observed in [35], highlighting the advantage of the proposed algorithm presented in this paper.

8.5 Conclusion

In this study, an asynchronous switched observer is introduced for each follow-UAV, facilitating the asymptotic estimation of lead-UAV state information under switching topologies. To achieve this, a novel piecewise Lyapunov function is adopted, resulting in a more lenient topological switching law. Concurrently, an

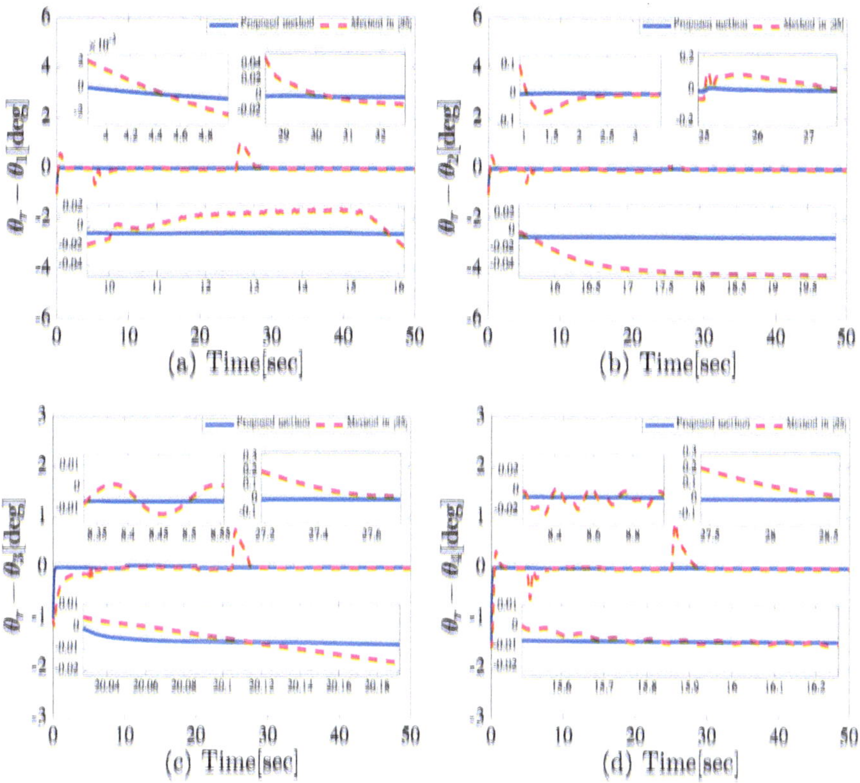

Fig. 8.7 Comparison of pitch angle tracking error trajectories. (a) Error trajectories of UAV 1. (b) Error trajectories of UAV 2. (c) Error trajectories of UAV 3. (d) Error trajectories of UAV 4

FLS-based FTCC scheme is devised for 6-DoF fixed-wing multi-UAVs, accounting for parameter certainty, external disturbances, and actuator failures. This scheme ensures that follow-UAVs can effectively track the observer states, enabling them to synchronize with the lead-UAV's speed and angles. The proposed algorithm's effectiveness is validated through numerical and semi-physical simulations. Considering the significance of communication resource conservation in multi-UAV systems [5], event-triggered communication emerges as an essential tool. Consequently, extending the multi-UAV control strategy presented in this paper to incorporate an event-triggered communication mechanism stands as an intriguing avenue for further investigation.

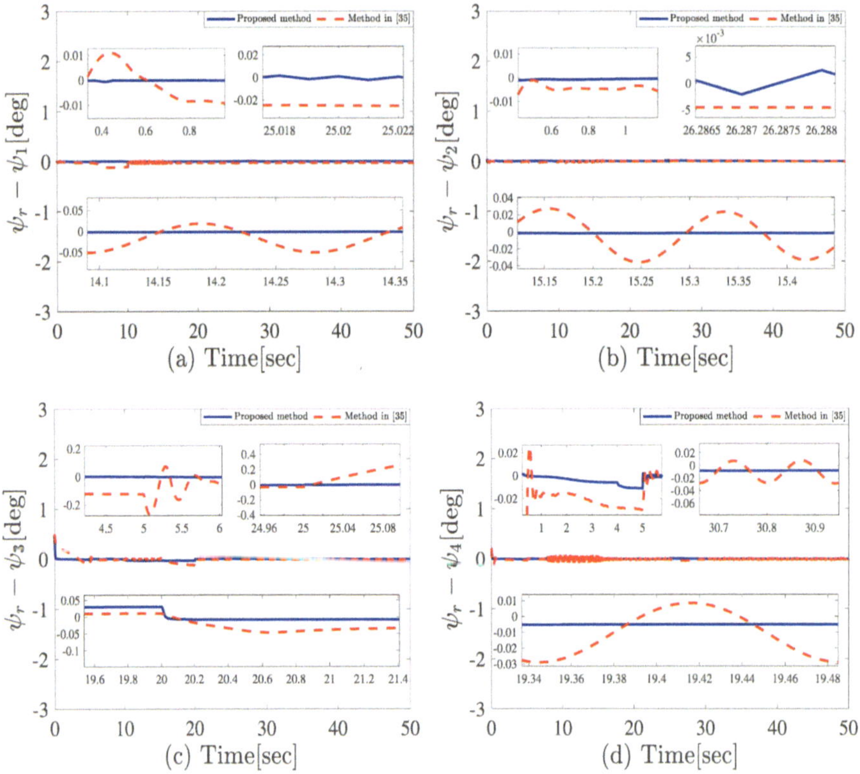

Fig. 8.8 Comparison of yaw angle tracking error trajectories. (**a**) Error trajectories of UAV 1. (**b**) Error trajectories of UAV 2. (**c**) Error trajectories of UAV 3. (**d**) Error trajectories of UAV 4

References

1. Y. Ding, B. Xin, L. Dou, J. Chen, B. M. Chen, A memetic algorithm for curvature-constrained path planning of messenger UAV in air-ground coordination. IEEE Trans. Autom. Sci. Eng. **19**(4), 3735–3749 (2022)
2. J. Wu, J. Zhang, Y. Sun, X. Li, L. Gao, G. Han, Multi-UAV collaborative dynamic task allocation method based on ISOM and attention mechanism. IEEE Trans. Veh. Technol. **73**(5), 6225–6235 (2024)
3. C. Yuan, Y. Zhang, Z. Liu, A survey on technologies for automatic forest fire monitoring, detection, and fighting using unmanned aerial vehicles and remote sensing techniques. Can. J. Forest Res. **45**(7), 783–792 (2015)
4. T. Kopfstedt, M. Mukai, M. Fujita, C. Ament, Control of formations of UAVs for surveillance and reconnaissance missions. IFAC Proc. Vol. **41**(2), 5161–5166 (2008)
5. M. Lv, C.K. Ahn, P. Wang, R. Zuo, J. Cao, Event-based distributed bipartite consensus of high-order nonlinear multi-agent network: applications to hypersonic flight vehicles. IEEE Trans. Autom. Sci. Eng. https://doi.org/10.1109/TASE.2024.3408650
6. J. Zhang, J. Yan, P. Zhang, Multi-UAV formation control based on a novel back-stepping approach. IEEE Trans. Veh. Technol. **69**(3), 2437–2448 (2020)

7. T. Oktay, Y. Eraslan, Autonomous flight performance optimization of fixed-wing unmanned aerial vehicle with morphing wingtip. Aircraft Eng. Aerospace Technol. **96**(3), 475–482 (2024)
8. Z. Yu, Y. Zhang, Z. Liu, Y. Qu, C.-Y. Su, B. Jiang, Decentralized finite-time adaptive fault-tolerant synchronization tracking control for multiple UAVs with prescribed performance. J. Franklin Inst. **357**(16), 11830–11862 (2020)
9. O. Kose, T. Oktay, Simultaneous quadrotor autopilot system and collective morphing system design. Aircraft Eng. Aerospace Technol. **92**(7), 1093–1100 (2020)
10. O. Kose, T. Oktay, E. Özen, Simultaneous arm morphing quadcopter and autonomous flight system design. Aircraft Eng. Aerospace Technol. **95**(10), 1624–1632 (2023)
11. S. Hu, W. Ni, X. Wang, A. Jamalipour, Disguised tailing and video surveillance with solar-powered fixed-wing unmanned aerial vehicle. IEEE Trans. Veh. Technol. **71**(5), 5507–5518 (2022)
12. M. Silic, K. Mohseni, An experimental evaluation of radio models for localizing fixed-wing UAVs in rural environments. IEEE Trans. Veh. Technol. **72**(5), 5576–5586 (2023)
13. L. Cui, Q. Zhou, D. Huang, Fixed-time disturbance observer based distributed cooperative containment control with fixed-time anti-saturation compensator for multi-unmanned aerial vehicles. Int. J. Robust Nonlinear Control **33**(17), 10584–10605 (2023)
14. D. Yin, X. Yang, C. Wang, H. Yu, S. Chen, A two-layer communication relay planning method for a fixed-wing UAVs swarm. IEEE Trans. Veh. Technol. **73**(5), 7140–7156 (2024)
15. Q. Wang, S. Zhao, X. Wang, Distributed control for coordinated tracking of fixed-wing unmanned aerial vehicles under model uncertainty and disturbances. Appl. Sci. **11**(21), 9830 (2021)
16. J. Wang, M. Xin, Integrated optimal formation control of multiple unmanned aerial vehicles. IEEE Trans. Control Syst. Technol. **21**(5), 1731–1744 (2012)
17. E.M. Coates, D. Reinhardt, T.I. Fossen, Reduced-attitude control of fixed-wing unmanned aerial vehicles using geometric methods on the two-sphere. IFAC-PapersOnLine **53**(2), 5749–5756 (2020)
18. Z. Jin, L. Bai, Z. Wang, P. Zhang, Self-triggered distributed formation control of fixed-wing unmanned aerial vehicles subject to velocity and overload constraints. IEEE Trans. Autom. Sci. Eng. **21**(3), 4082–4093 (2024)
19. B.L. Stevens, F.L. Lewis, E.N. Johnson, *Aircraft Control and Simulation: Dynamics, Controls Design, and Autonomous Systems* (John Wiley & Sons, London, 2015)
20. J. Wu, C. Luo, Y. Luo, K. Li, Distributed UAV swarm formation and collision avoidance strategies over fixed and switching topologies. IEEE Trans. Cybern. **52**(10), 10969–10979 (2021)
21. H. Liu, Y. Wang, F.L. Lewis, K.P. Valavanis, Robust formation tracking control for multiple quadrotors subject to switching topologies. IEEE Trans. Control of Netw. Syst. **7**(3), 1319–1329 (2020)
22. Z. Wang, T. Liu, Z.-P. Jiang, Cooperative formation control under switching topology: An experimental case study in multirotors. IEEE Trans. Cybern. **51**(12), 6141–6153 (2020)
23. X. Dong, Y. Li, C. Lu, G. Hu, Q. Li, Z. Ren, Time-varying formation tracking for UAV swarm systems with switching directed topologies. IEEE Trans. Neural Netw. Learn. Syst. **30**(12), 3674–3685 (2019)
24. K. Li, C.-C. Hua, X. You, Distributed asynchronous consensus control for nonlinear multiagent systems under switching topologies. IEEE Trans. Autom. Control **66**(9), 4327–4333 (2020)
25. K. Li, C.-C. Hua, X. You, C. K. Ahn, Leader-following asynchronous consensus for multiagent systems with unknown control and output directions. Automatica **132**, 109832 (2021)
26. M. Lv, C.K. Ahn, B. Zhang, A. Fu, Fixed-time antisaturation cooperative control for networked fixed-wing unmanned aerial vehicles considering actuator failures. IEEE Trans. Aerospace Electron. Syst. **59**(6), 8812–8825 (2023)
27. Z. Yu, Y. Zhang, B. Jiang, J. Fu, Y. Jin, T. Chai, Composite adaptive disturbance observer-based decentralized fractional-order fault-tolerant control of networked UAVs. IEEE Trans. Syst. Man Cybern. Syst. **52**(2), 799–813 (2020)

28. Y. Qi, S. Zhou, Y. Kang, S. Yan, Formation control for unmanned aerial vehicles with directed and switching topologies. Int. J. Aerospace Eng. **2016**, 7657452 (2016)
29. Z. Zuo, M. Defoort, B. Tian, Z. Ding, Distributed consensus observer for multiagent systems with high-order integrator dynamics. IEEE Trans. Autom. Control **65**(4), 1771–1778 (2019)
30. E. Oland, R. Schlanbusch, R. Kristiansen, Underactuated waypoint tracking of a fixed-wing UAV. IFAC Proc. Vol. **46**(30), 126–133 (2013)
31. X. Wang, G.-H. Yang, Fault-tolerant consensus tracking control for linear multiagent systems under switching directed network. IEEE Trans. Cybern. **50**(5), 1921–1930 (2020)
32. Z. Liu, H. Huang, J. H. Park, J. Huang, X. Wang, M. Lv, Adaptive fuzzy control for unknown nonlinear multiagent systems with switching directed communication topologies. IEEE Trans. Fuzzy Syst. **31**(7), 2487–2494 (2023)
33. M. Lv, W. Yu, J. Cao, S. Baldi, Consensus in high-power multiagent systems with mixed unknown control directions via hybrid Nussbaum-based control, IEEE Trans. Cybern. **52**(6), 5184–5196 (2022)
34. Z. Yu, Y. Zhang, B. Jiang, C.-Y. Su, J. Fu, Y. Jin, T. Chai, Nussbaum-based finite-time fractional-order backstepping fault-tolerant flight control of fixed-wing UAV against input saturation with hardware-in-the-loop validation. Mech. Syst. Signal Proc. **153**, 107406 (2021)
35. B. Zhang, X. Sun, M. Lv, S. Liu, L. Li, Distributed adaptive fixed-time fault-tolerant control for multiple 6-DOF UAVs with full-state constraints guarantee. IEEE Syst. J. **16**(3), 4792–4803 (2022)
36. W. Sun, S.-F. Su, Y. Wu, J. Xia, Adaptive fuzzy event-triggered control for high-order nonlinear systems with prescribed performance. IEEE Trans. Cybern. **52**(5), 2885–2895 (2020)
37. L. Xing, C. Wen, Y. Zhu, H. Su, Z. Liu, Output feedback control for uncertain nonlinear systems with input quantization. Automatica **65**, 191–202 (2016)
38. E. Cruz-Zavala, J. A. Moreno, L. M. Fridman, Uniform robust exact differentiator. IEEE Trans. Autom. Control **56**(11), 2727–2733 (2011)
39. G. Lai, Z. Liu, Y. Zhang, C. P. Chen, Adaptive position/attitude tracking control of aerial robot with unknown inertial matrix based on a new robust neural identifier. IEEE Trans. Neural Netw. Learn. Syst. **27**(1), 18–31 (2015)
40. S. Tong, Y. Li, P. Shi, Observer-based adaptive fuzzy backstepping output feedback control of uncertain MIMO pure-feedback nonlinear systems. IEEE Trans. Fuzzy Syst. **20**(4), 771–785 (2012)

Part IV
Safety and Security in UAV Operations

Chapter 9
Optimal Formation Tracking Control

Except for the velocity and attitude tracking control problem discussed in the previous six chapters, the cooperative formation is another core aspect of multi-UAV control and serves as the foundation of executing flight missions [1–3]. In complex flight environment, UAV formation would be affected by external disturbances, such as gusts and electromagnetic interference, as well as model uncertainty, which significantly influences the safety and stability of cooperative flight among multiple UAVs. In response to this challenge, this chapter designs a learning-based optimal control methodology to ensure the performance robustness and optimality of multi-UAV system in the face of strong disturbances.

9.1 Introduction

In recent years, the utilization of unmanned aerial vehicle (UAV) formations has proven successful in various applications such as load transport [1], surveillance [2], and target enclosing [3]. Numerous formation control methodologies have been proposed to address these applications. For instance, in [4], a decentralized sliding mode controller was introduced to achieve consensus in altitude and heading angle for UAV formations. Kang et al. [5] developed a time-varying formation controller that leveraged local status information sharing to solve the consensus control problem in UAV formations. Several studies [6–9] directed their attention toward coordinated formation stabilization and formation tracking of UAVs, considering fixed topologies and switching topologies, respectively. However, it is important to note that most existing results [4–9] fail to guarantee performance optimality. Complicating matters further, UAV formations are often exposed to intricate disturbances when operating in complex environments [10–13]. This reality poses challenges to ensuring optimality, thus greatly limiting the performance of UAV formations.

M. Lv et al., *Unmanned Aerial Vehicles Formation Control and Safety Guarantee*,
Cognitive Intelligence and Robotics, https://doi.org/10.1007/978-981-96-5829-9_9

Consequently, there is a critical need to pursue both optimality and robustness to achieve stable formation control in the presence of complex disturbances. This objective falls under the domain of mixed robust control and optimal control [14–17].

Primarily, the current focus in the realm of mixed robust control and optimal control design lies in the framework of the two-player zero-sum game. This framework offers a viable solution to the robust control problem, where the controller acts as the minimizing player, while the disturbance represents the maximizing counterpart [18, 19]. Recent advancements have presented various approaches for addressing two-player zero-sum games through online learning of control and disturbance policies [20, 21]. These methods involve the adaptation of control and disturbance policies using reinforcement learning, with neural networks employed to identify the corresponding value function.

Reinforcement learning provides an optimal design technique for control systems [22]. Different from the traditional optimal control designs, controllers incorporating reinforcement learning technique are capable of learning the approximate solution to optimal control from the feedback of surroundings [23, 24]. With respect to learning-based optimal control designs [25–29], online approximate solutions based on policy iteration are developed, whereby the neural networks are deployed to approximate the value function and control policy. This advantage has motivated researchers to develop several learning-based control approaches for the path planning [30], obstacle avoidance [31], and resource allocation [32] of UAV formation.

Taking inspiration from foregoing literature, a feedforward-feedback learning-based optimal control is developed for multiple UAVs. The objective is to seek both the optimality and robustness of cooperative formation tracking. To this end, a backstepping-based feedback control technique is employed to transform the UAV formation tracking problem into an equivalent optimal regulation problem. Subsequently, a learning-based optimal control is derived by utilizing a two-player zero-sum game framework based on Echo State Network (ESN) approximation. The main contributions of this chapter can be summarized as follows:

- In response to the challenge posed by the inability of most UAV formation control methods [4–9] to ensure performance optimality, a feedforward-feedback learning-based optimal control scheme is devised. The proposed control scheme addresses the complex disturbances encountered in cooperative UAV formation tracking while also guaranteeing performance optimality.
- In addition to the model-based backstepping controllers [5–9], the integration of reinforcement learning offers the potential to achieve robust control actions. Through the introduction of the cooperative policy iteration algorithm based on two-player zero-sum game, the tracking performances of UAV formation are further optimized.
- In contrast to most learning-based control methods, such as [25–28, 32], which satisfy the persistency of excitation condition by introducing probing noise to

the system dynamics or through a replay strategy involving the collection and training of large amounts of recorded data, our proposed algorithm takes a different way. The proposed algorithm attempts to eliminate the persistency of excitation condition by introducing innovative compensation terms into the ESN tuning law. Furthermore, our proposed algorithm addresses the requirement for initial admissible control by incorporating a novel piecewise compensation term into the ESN tuning law, which is based on a newly proposed Lyapunov function.

The rest of chapter is organized as follows: Sect. 9.2 presents the problem formulation and preliminaries. In Sect. 9.3, an ESN-based approximation for the value function and the stability analysis are provided. Experiment and simulation validations are discussed in Sect. 9.4. Section 9.5 draws the conclusion.

Remark The main control schemes and contents of this chapter are from the published journal paper "B. Y. Zhang, M. L. Lv, S. H. Cui, X. W. Bu, J. H. Park, Learning-based optimal cooperative formation tracking control for multiple UAVs: A feedforward-feedback design framework, IEEE Transactions on Automation Science and Engineering, DOI: 10.1109/TASE.2023.3322028." The authors appreciate the permission from Institute of Electrical and Electronics Engineers to reuse the results published in the relevant journal.

9.2 Problem Formulation

9.2.1 *Basic Graph Theory*

Given a graph $\mathcal{G} = (\mathcal{V}, \mathcal{E}, \mathcal{A})$ to describe the connections among the multiple UAVs, it consists of nodes $\mathcal{V} = \{v_1, \cdots, v_n\}$ and the sets of edges $\mathcal{E} = \{(i, j), \ i, j \in \mathcal{V}, \text{and } i \neq j\}$, and $\mathcal{A} = [a_{ij}] \in \Re^{n \times n}$ denotes the weighted adjacency matrix of $\mathcal{G}$. If there exists an edge between the ith UAV and jth UAV, then $a_{ij} = a_{ji} \neq 0$, and otherwise $a_{ij} = a_{ji} = 0$. Moreover, the neighbors of ith UAV is denoted by the set $\mathcal{N}_i = \{v_j : \ (v_j, \ v_i) \in \mathcal{E}\}$. The Laplacian matrix is defined by $\mathcal{L} = \mathcal{D} - \mathcal{A}$, where $\mathcal{D} = \text{diag}\{d_1, \cdots, d_n\}$ with $d_i = \sum_{j=1}^{n} a_{ij}$. $\mathcal{G}$ is connected if there is a path from each UAV to others.

For the sake of multitask need, a switching graph is considered for UAV formation. Specifically, define an infinite sequence of time intervals $[t_k, t_{k+1})$, where $t_0 = 0$ and $0 < t_\tau \leq t_{k+1} - t_k$ with t_τ denoting the dwell time. The graph keeps fixed during $[t_k, t_{k+1})$ and switches at time t_{k+1}. Let $\sigma(t) : [0, +\infty) \rightarrow \{1, 2, \ldots, n\}$ stand for a switching signal, where n represents the number of all probable graphs. The value of $\sigma(t)$ is the index of switching topology. Define $a_{ij}^{\sigma(t)}$ as the weight of $\mathcal{A}$ for $\sigma(t)$. Let $\mathcal{N}_i^{\sigma(t)}$, $\mathcal{G}^{\sigma(t)}$, and $\mathcal{L}^{\sigma(t)}$ be the neighbor set, the topology, and the Laplacian matrix for $\sigma(t)$, respectively.

9.2.2 UAV Dynamic Model

Consider a formation of N identical fixed wing UAVs. By introducing the auxiliary dynamics with a path parameter θ_i, the kinematic model for each UAV is expressed as [33, 34]

$$\begin{cases} \dot{x}_i = V_i \cos\psi_i \cos\gamma_i + w_{xi} \\ \dot{y}_i = V_i \sin\psi_i \cos\gamma_i + w_{yi} \\ \dot{z}_i = V_i \sin\gamma_i + w_{zi} \\ \dot{\psi}_i = g \tan\phi_{ci} / V_i \\ \dot{\gamma}_i = \kappa \left(\gamma_{ci} - \gamma_i\right) \\ \dot{\theta}_i = \varphi_i \\ \dot{\varphi}_i = \mu_i, \end{cases} \tag{9.1}$$

where x_i, y_i, and z_i denote the position in the inertial frame, V_i is the airspeed, ψ_i is the heading angle, and γ_i is the air-relative flight path angle. The control inputs are selected as ϕ_{ci} and γ_{ci}, which represent the commands of roll angle and flight path angle, respectively. w_{xi}, w_{yi}, and w_{zi} are complicated unknowns along x-, y-, and z-axes, respectively, and μ_i is a virtual control law to generate the path. Note that we follow the coordinated turn assumption to achieve the motion of ψ_i. And the dynamics of γ_i is described using a first-order system with a time constant κ to accommodate the relatively slow response set up by the autopilot. θ_i is a path parameter, which can be any physical quantity, and we take $\dot{\theta}_i = \varphi_i$ and $\dot{\varphi}_i = \mu_i$.

Definition 9.1 (Cooperative UAV Formation Tracking) Multiple UAVs are said to achieve cooperative formation tracking if

$$\lim_{t\to\infty} \left(\boldsymbol{p}_i(t) - \Delta\boldsymbol{p}_i(t) - \boldsymbol{p}_0(t)\right) = \mathbf{0}, \tag{9.2}$$

where $\boldsymbol{p}_i(t) = [x_i(t), y_i(t), z_i(t)]^{\mathrm{T}}$ denotes the position of the ith UAV, $\Delta\boldsymbol{p}_i(t) = [\Delta x_i(t), \Delta y_i(t), \Delta z_i(t)]^{\mathrm{T}}$ represents the relative position with respect to the virtual leader which specifies the expected time-varying formation, and $\boldsymbol{p}_0(t)$ is the position of virtual leader, expressed as

$$\begin{aligned} \boldsymbol{p}_0(t) = \{&x_0, y_0, z_0 \in \Re \,|\, \theta_0 \in [\theta_{\min}, \theta_{\max}] \\ &\mapsto x_0 = r_x(\theta_0), y_0 = r_y(\theta_0), z_0 = r_z(\theta_0)\} \end{aligned} \tag{9.3}$$

where $\dot{\theta}_0 = f(\theta_0, t)$ with function $f : \Re^n \to \Re^n$, and $\theta_{\min}$ and $\theta_{\max}$ are the minimum and maximum values of θ_0, respectively.

From (9.2), define $e_{xi} = x_i - \Delta x_i - x_0$, $e_{yi} = y_i - \Delta y_i - y_0$, and $e_{zi} = z_i - \Delta z_i - z_0$. Taking the time derivative of e_{xi}, e_{yi}, and e_{zi} along (9.1) yields

$$\begin{cases} \dot{e}_{xi} = f_{xi}\left(\boldsymbol{X}_i\right) + w_{xi} - \Delta\dot{x}_i \\ \dot{e}_{yi} = f_{yi}\left(\boldsymbol{X}_i\right) + w_{yi} - \Delta\dot{y}_i \\ \dot{e}_{zi} = f_{zi}\left(\boldsymbol{X}_i\right) + w_{zi} - \Delta\dot{z}_i, \end{cases} \tag{9.4}$$

where $f_{xi}\left(\boldsymbol{X}_i\right) = V_i \cos\psi_i \cos\gamma_i - \frac{\partial x_0}{\partial\theta_0}\varphi_0$, $f_{yi}\left(\boldsymbol{X}_i\right) = V_i \sin\psi_i \cos\gamma_i - \frac{\partial y_0}{\partial\theta_0}\varphi_0$, and $f_{zi}\left(\boldsymbol{X}_i\right) = V_i \sin\gamma_i - \frac{\partial z_0}{\partial\theta_0}\varphi_0$, $\boldsymbol{X}_i = [x_i, y_i, z_i, \psi_i, \gamma_i, \theta_0, \varphi_0]^{\mathrm{T}}$.

For the sake of simplification, let $u_{1i} = g\tan\phi_{ci}/V_i$ and $u_{2i} = \kappa\left(\gamma_{ci} - \gamma_i\right)$. Then the implementable control laws can be calculated as $\phi_{ci} = \arctan\left(u_{1i}V_i/g\right)$ and $\gamma_{ci} = \gamma_i + u_{2i}/\kappa$. Define $\boldsymbol{e}_i = \left[e_{xi}, e_{yi}, e_{zi}\right]^{\mathrm{T}}$, $\boldsymbol{d}_i = \left[w_{xi}, w_{yi}, w_{zi}\right]^{\mathrm{T}}$, $\boldsymbol{u}_i = [u_{i1}, u_{i2}, \mu_0]^{\mathrm{T}}$, and $\boldsymbol{f}_i = \left[f_{xi}, f_{yi}, f_{zi}\right]^{\mathrm{T}}$. It follows from (9.4) that

$$\begin{cases} \dot{\boldsymbol{e}}_i = \boldsymbol{f}_i + \boldsymbol{d}_i - \Delta\dot{\boldsymbol{p}}_i \\ \dot{\boldsymbol{f}}_i = \boldsymbol{F}_i + \boldsymbol{G}_i\boldsymbol{u}_i, \end{cases} \tag{9.5}$$

where

$$\boldsymbol{G}_i = \begin{bmatrix} -V_i \sin\psi_i \cos\gamma_i & -V_i \cos\psi_i \sin\gamma_i & -\frac{\partial x_0}{\partial\theta_0} \\ V_i \cos\psi_i \cos\gamma_i & -V_i \sin\psi_i \sin\gamma_i & -\frac{\partial y_0}{\partial\theta_0} \\ 0 & V_i \cos\gamma_i & -\frac{\partial z_0}{\partial\theta_0} \end{bmatrix},$$

$$\boldsymbol{F}_i = -\left[\frac{\partial\left(\frac{\partial x_0}{\partial\theta_0}\right)}{\partial\theta_0}\varphi_0^2, \frac{\partial\left(\frac{\partial y_0}{\partial\theta_0}\right)}{\partial\theta_0}\varphi_0^2, \frac{\partial\left(\frac{\partial z_0}{\partial\theta_0}\right)}{\partial\theta_0}\varphi_0^2\right]^{\mathrm{T}}.$$

9.2.3 Control Objective

The objective of this chapter is to design a feedforward-feedback learning-based optimal control scheme such that the multiple UAVs follow the virtual leader and maintain a prescribed formation configuration in the presence of complicated disturbances while respecting the switching topologies $\mathcal{G}^{\sigma(t)}$.

This section presents a mixed feedforward-feedback based optimal control scheme to achieve the cooperative UAV formation tracking (see Definition 1) with complicated disturbances. Firstly, the feedback control law, denoted by $\boldsymbol{u}_i^a$, is designed using the backstepping control method to obtain a new transformed error dynamics. Subsequently, an optimal feedforward control action, denoted by $\boldsymbol{u}_i^*$ is derived. As a result, the mixed feedforward-feedback based control signal is given as $\boldsymbol{u}_i = \boldsymbol{u}_i^a + \boldsymbol{u}_i^*$. The overall control configuration is shown in Fig. 9.1.

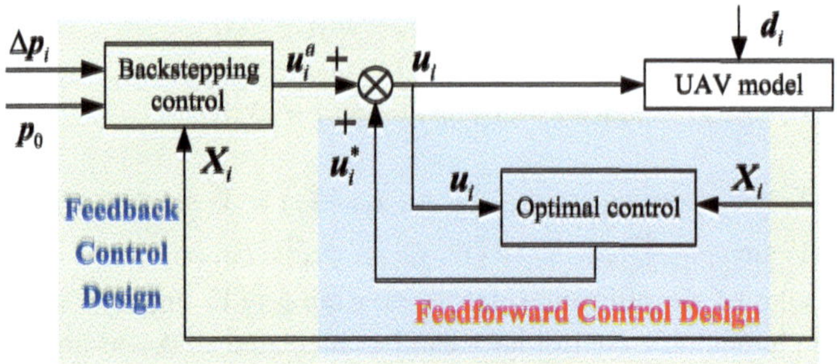

Fig. 9.1 Configuration of feedforward-feedback control scheme

9.3 Main Results

9.3.1 *Backstepping-Based Feedback Control Design*

A backstepping-based feedback control design is proposed as follows:

Step 1: To carry out the controller design, some notations are defined as follows: $\tilde{\boldsymbol{f}}_l = \boldsymbol{f}_l - \boldsymbol{f}_{ld}$, $\boldsymbol{f}_{ld}$ is the virtual control such that $\boldsymbol{f}_{ld} = \boldsymbol{f}_{ld}^a + \boldsymbol{f}_{ld}^*$, where $\boldsymbol{f}_{ld}^a$ is the feedback virtual control and $\boldsymbol{f}_{ld}^*$ is the feedforward optimal term. Then it follows from (9.5) that

$$\dot{\boldsymbol{e}}_l = \tilde{\boldsymbol{f}}_l + \boldsymbol{f}_{ld}^a + \boldsymbol{f}_{ld}^* + \boldsymbol{d}_l - \Delta\dot{\boldsymbol{p}}_l. \tag{9.6}$$

Consider the following Lyapunov function candidate:

$$L_{l1} = \frac{1}{2}\boldsymbol{e}_l^{\mathrm{T}}\boldsymbol{e}_l. \tag{9.7}$$

Taking the time derivative of L_{l1} along (9.6) yields

$$\dot{L}_{l1} = \boldsymbol{e}_l^{\mathrm{T}}\left(\tilde{\boldsymbol{f}}_l + \boldsymbol{f}_{ld}^a + \boldsymbol{f}_{ld}^* + \boldsymbol{d}_l - \Delta\dot{\boldsymbol{p}}_l\right). \tag{9.8}$$

The feedback virtual control $\boldsymbol{f}_{ld}^a$ is chosen as

$$\boldsymbol{f}_{ld}^a = -k_{2l}\sum_{j\in\mathcal{N}_l^{\sigma(t)}} a_{lj}^{\sigma(t)}\left(\boldsymbol{e}_l - \boldsymbol{e}_j\right) - k_{1l}\boldsymbol{e}_l + \Delta\dot{\boldsymbol{p}}_l, \tag{9.9}$$

where k_{1l} and k_{2l} are positive design parameters.

Substituting (9.9) into (9.8) gives

$$\dot{L}_{l1} = -k_{2l}\boldsymbol{e}_l^{\mathrm{T}}\sum_{j\in\mathcal{N}_l^{\sigma(t)}} a_{lj}^{\sigma(t)}\left(\boldsymbol{e}_l - \boldsymbol{e}_j\right) + \boldsymbol{e}_l^{\mathrm{T}}\tilde{\boldsymbol{f}}_l$$

$$- k_{1l} \|e_l\|^2 + e_l^{\mathrm{T}} d_l + e_l^{\mathrm{T}} f_{ld}^*. \tag{9.10}$$

Step 2: Choose the following Lyapunov function candidate:

$$L_{l2} = \frac{1}{2} \tilde{f}_l^{\mathrm{T}} \tilde{f}_l. \tag{9.11}$$

As shown in Fig. 9.1, the control input u_l is taken as $u_l = u_l^a + u_l^*$. The time derivative of L_{l2} along (9.5) is

$$\dot{L}_{l2} = \tilde{f}_l^{\mathrm{T}} \left\{ F_l + G_l \left(u_l^a + u_l^* \right) - \dot{f}_{ld} \right\}. \tag{9.12}$$

The feedback control u_l^a is designed as

$$\begin{aligned} u_l^a &= - k_{2l} G_l^{-1} \sum_{j \in \mathcal{N}_l^{\sigma(t)}} \left\{ a_{lj}^{\sigma(t)} \left(e_l - e_j \right) + a_{lj}^{\sigma(t)} \left(\tilde{f}_l - \tilde{f}_j \right) \right\} \\ &= k_{2l} G_l^{-1} \frac{\tilde{f}_l}{\|\tilde{f}_l\|^2} e_l^{\mathrm{T}} \sum_{j \in \mathcal{N}_l^{\sigma(t)}} a_{lj}^{\sigma(t)} \left(\tilde{f}_l - \tilde{f}_j \right) \\ &= G_l^{-1} \left(F_l + k_{3l} \tilde{f}_l + e_l - \dot{f}_{ld} \right), \end{aligned} \tag{9.13}$$

where k_{3l} is a positive design parameter.

Substituting (9.13) into (9.12) yields

$$\begin{aligned} \dot{L}_{l2} &= - k_{2l} \tilde{f}_l^{\mathrm{T}} \sum_{j \in \mathcal{N}_l^{\sigma(t)}} \left\{ a_{lj}^{\sigma(t)} \left(e_l - e_j \right) + a_{lj}^{\sigma(t)} \left(\tilde{f}_l - \tilde{f}_j \right) \right\} \\ &= k_{2l} e_l^{\mathrm{T}} \sum_{j \in \mathcal{N}_l^{\sigma(t)}} \left\{ a_{lj}^{\sigma(t)} \left(\tilde{f}_l - \tilde{f}_j \right) \right\} \\ &= k_{3l} \|\tilde{f}_l\|^2 - \tilde{f}_l^{\mathrm{T}} e_l + \tilde{f}_l^{\mathrm{T}} G_l u_l^*. \end{aligned} \tag{9.14}$$

Theorem 9.1 Given the error dynamics (9.5), define the augmented control input vector $\tau_l = \left[f_{ld}^{\mathrm{T}}, u_l^{\mathrm{T}} \right]^{\mathrm{T}}$ such that $\tau_l = \tau_l^a + \tau_l^*$, where $\tau_l^a = \left[f_{ld}^{a\mathrm{T}}, u_l^{a\mathrm{T}} \right]^{\mathrm{T}}$ is the augmented feedback control input in (9.9) and (9.13), and $\tau_l^* = \left[f_{ld}^{*\mathrm{T}}, u_l^{*\mathrm{T}} \right]^{\mathrm{T}}$ is the augmented feedforward control action which optimally stabilizes the transformed error dynamics

$$\dot{\mathfrak{S}}_l = C_l \tau_l^* + K_l d_l, \tag{9.15}$$

where $C_l = \begin{bmatrix} 1_{3\times3} & 0_{3\times3} \\ 0_{3\times3} & G_l \end{bmatrix}$, $K_l = \begin{bmatrix} 1_{3\times3} \\ 0_{3\times3} \end{bmatrix}$, and $\mathfrak{S}_l = \left[e_l^{\mathrm{T}}, \tilde{f}_l^{\mathrm{T}} \right]^{\mathrm{T}}$.

Then the optimal formation tracking of UAVs is accomplished by forcing $\Im_i$ to converge to an arbitrarily small region around origin in an optimal manner.

Proof Based on (9.7) and (9.11), let us consider the entire Lyapunov function

$$L = \sum_{i=1}^{N} (L_{i1} + L_{i2}). \tag{9.16}$$

Taking the time derivative of L along (9.10) and (9.14) gives

$$\begin{aligned}\dot{L} =& -k_{2i}\sum_{i=1}^{N}(\boldsymbol{e}_i+\widetilde{\boldsymbol{f}}_i)^{\mathrm{T}}\sum_{j\in\mathcal{N}_i^{\sigma(t)}}\left\{a_{ij}^{\sigma(t)}(\boldsymbol{e}_i-\boldsymbol{e}_j)+a_{ij}^{\sigma(t)}(\widetilde{\boldsymbol{f}}_i-\widetilde{\boldsymbol{f}}_j)\right\}\\ &+\sum_{i=1}^{N}\left(\boldsymbol{e}_i^{\mathrm{T}}\boldsymbol{f}_{id}^{*}+\widetilde{\boldsymbol{f}}_i^{\mathrm{T}}\boldsymbol{G}_i\boldsymbol{u}_i^{*}+\boldsymbol{e}_i^{\mathrm{T}}\boldsymbol{d}_i\right)\\ &-\sum_{i=1}^{N}\left(k_{1i}\|\boldsymbol{e}_i\|^2+k_{3i}\|\widetilde{\boldsymbol{f}}_i\|^2\right).\end{aligned} \tag{9.17}$$

Let $\boldsymbol{e} = [\boldsymbol{e}_1^{\mathrm{T}}, \boldsymbol{e}_2^{\mathrm{T}}, \cdots, \boldsymbol{e}_N^{\mathrm{T}}]^{\mathrm{T}}$, $\widetilde{\boldsymbol{f}} = [\widetilde{\boldsymbol{f}}_1^{\mathrm{T}}, \widetilde{\boldsymbol{f}}_2^{\mathrm{T}}, \cdots, \widetilde{\boldsymbol{f}}_N^{\mathrm{T}}]^{\mathrm{T}}$, and $\boldsymbol{\Lambda}^{\sigma(t)} = \begin{bmatrix}\mathcal{L}^{\sigma(t)} & \mathcal{L}^{\sigma(t)}\\ \mathcal{L}^{\sigma(t)} & \mathcal{L}^{\sigma(t)}\end{bmatrix}$. Then, the following equality holds:

$$\begin{aligned}&\sum_{i=1}^{N}(\boldsymbol{e}_i+\widetilde{\boldsymbol{f}}_i)^{\mathrm{T}}\sum_{j\in\mathcal{N}_i^{\sigma(t)}}\left\{a_{ij}^{\sigma(t)}(\boldsymbol{e}_i-\boldsymbol{e}_j)+a_{ij}^{\sigma(t)}(\widetilde{\boldsymbol{f}}_i-\widetilde{\boldsymbol{f}}_j)\right\}\\ &=\begin{bmatrix}\boldsymbol{e}\\ \widetilde{\boldsymbol{f}}\end{bmatrix}^{\mathrm{T}}\left(\boldsymbol{\Lambda}^{\sigma(t)}\otimes\boldsymbol{I}_3\right)\begin{bmatrix}\boldsymbol{e}\\ \widetilde{\boldsymbol{f}}\end{bmatrix}.\end{aligned} \tag{9.18}$$

Assuming that graph $\mathcal{G}^{\sigma(t)}$ is uniformly connected, $\mathcal{L}^{\sigma(t)}$ is symmetric and positive semi-definite, and further $\boldsymbol{\Lambda}^{\sigma(t)}$ is positive semi-definite. Then, it holds that

$$-k_{2i}\begin{bmatrix}\boldsymbol{e}\\ \widetilde{\boldsymbol{f}}\end{bmatrix}^{\mathrm{T}}\left(\boldsymbol{\Lambda}^{\sigma(t)}\otimes\boldsymbol{I}_3\right)\begin{bmatrix}\boldsymbol{e}\\ \widetilde{\boldsymbol{f}}\end{bmatrix}\le 0. \tag{9.19}$$

Substituting (9.19) into (9.17) gives rise to

$$\dot{L} \le \sum_{i=1}^{N}\left(\boldsymbol{e}_i^{\mathrm{T}}\boldsymbol{f}_{id}^{*}+\widetilde{\boldsymbol{f}}_i^{\mathrm{T}}\boldsymbol{G}_i\boldsymbol{u}_i^{*}+\boldsymbol{e}_i^{\mathrm{T}}\boldsymbol{d}_i\right)$$

$$
\begin{aligned}
&- \sum_{i=1}^{N} \left(k_{1i} \| \boldsymbol{e}_i \|^2 + k_{3i} \| \widetilde{\boldsymbol{f}}_i \|^2 \right) \\
&\leq \sum_{i=1}^{N} \left\{ \Im_i^{\mathrm{T}} \left(\begin{bmatrix} \mathbf{1}_{3\times 3} & \mathbf{0}_{3\times 3} \\ \mathbf{0}_{3\times 3} & \boldsymbol{G}_i \end{bmatrix} \boldsymbol{\tau}_i^* + \begin{bmatrix} \mathbf{1}_{3\times 3} \\ \mathbf{0}_{3\times 3} \end{bmatrix} \boldsymbol{d}_i \right) \right\} \\
&- c_i \sum_{i=1}^{N} \| \Im_i \|^2 .
\end{aligned} \tag{9.20}
$$

It follows from (9.20) that when $\boldsymbol{\tau}_i^*$ stabilizes the system (9.15), the first term in the right-hand side of (9.20) turns negative.

This completes the proof of Theorem 9.1.

9.3.2 Learning-Based Optimal Feedforward Control Design

In this section, a zero-sum game-based optimal controller is designed to stabilize the transformed error dynamics (9.15), which forces $\Im_i$ to converge to an arbitrarily small region around origin in an optimal manner. The complete design framework is shown in Fig. 9.2.

Given the feedforward control policies $\boldsymbol{\tau}_i^*$ and $\boldsymbol{d}_i$, the infinite horizon integral cost is defined as

$$
V_i = \int_t^{\infty} \left(Q\left(\Im_i \right) + \boldsymbol{\tau}_i^{*\mathrm{T}} \boldsymbol{R}_\tau \boldsymbol{\tau}_i^* - \boldsymbol{d}_i^{\mathrm{T}} \boldsymbol{R}_d \boldsymbol{d}_i \right) dt, \tag{9.21}
$$

where $Q\left(\Im_i \right) = \Im_i^{\mathrm{T}} \boldsymbol{Q} \Im_i$ is a penalty on the error $\Im_i$, and $\boldsymbol{R}_\tau \in \Re^{6\times 6}$ and $\boldsymbol{R}_d \subset \Re^{6\times 6}$ arc positive design matrixes.

Then, the optimal cost for two-player zero-sum game can be obtained

$$
V_i^* = \min_{\boldsymbol{\tau}_i^*} \max_{\boldsymbol{d}_i} \int_t^{\infty} \left(Q\left(\Im_i \right) + \boldsymbol{\tau}_{i*\mathrm{T}} \boldsymbol{R}_\tau \boldsymbol{\tau}_i^* - \boldsymbol{d}_i^{\mathrm{T}} \boldsymbol{R}_d \boldsymbol{d}_i \right) dt. \tag{9.22}
$$

In view of the value function (9.21), the Hamiltonian function associated with feedforward control policies $\boldsymbol{\tau}_i^*$ and $\boldsymbol{d}_i$ is defined as

$$
H_i = Q\left(\Im_i \right) + \boldsymbol{\tau}_i^{*\mathrm{T}} \boldsymbol{R}_\tau \boldsymbol{\tau}_i^* - \boldsymbol{d}_i^{\mathrm{T}} \boldsymbol{R}_d \boldsymbol{d}_i + \nabla V_i^{\mathrm{T}} \left(\boldsymbol{C}_i \boldsymbol{\tau}_i^* + \boldsymbol{K}_i \boldsymbol{d}_i \right), \tag{9.23}
$$

where $\nabla V_i = \partial V_i / \partial \Im_i \in \Re^6$.

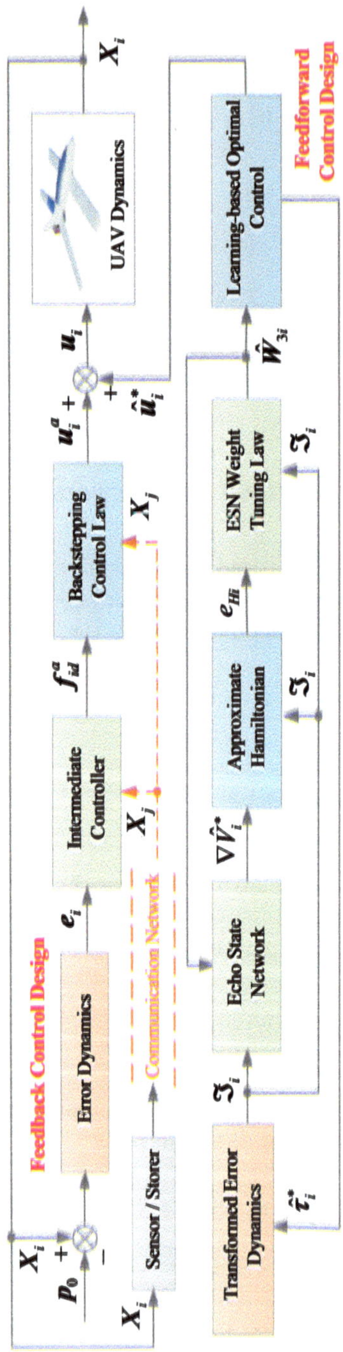

Fig. 9.2 Overall design framework of proposed control methodology

Based on the stationarity condition of Nash equilibrium for (9.23), the optimal feedforward control policies are obtained by

$$\bar{\boldsymbol{\tau}}_i^* = -\frac{1}{2}\boldsymbol{R}_\tau^{-1}\boldsymbol{C}_i^{\mathrm{T}}\nabla V_i^* \tag{9.24}$$

$$\boldsymbol{d}_i^* = \frac{1}{2}\boldsymbol{R}_d^{-1}\boldsymbol{K}_i^{\mathrm{T}}\nabla V_i^*, \tag{9.25}$$

where the optimal value function V_i^* is the solution of the following Hamilton-Jacobi-Isaacs (HJI) equation

$$\begin{aligned} 0 = &-\frac{1}{4}\nabla V_i^{*\mathrm{T}}\boldsymbol{C}_i\boldsymbol{R}_\tau^{-1}\boldsymbol{C}_i^{\mathrm{T}}\nabla V_i^* + \frac{1}{4}\nabla V_i^{*\mathrm{T}}\boldsymbol{K}_i\boldsymbol{R}_d^{-1}\boldsymbol{K}_i^{\mathrm{T}}\nabla V_i^* \\ &+ Q\left(\Im_i\right), \end{aligned} \tag{9.26}$$

where $\nabla V_i^*(0) = 0$.

If the HJI equation can be solved by V_i^*, the optimal feedforward control policies $\bar{\boldsymbol{\tau}}_i^*$ and $\boldsymbol{d}_i^*$ can be implemented by (9.24) and (9.25). However, the analytic solution to the HJI equation is generally hard to obtain due to its inherently nonlinear feature. To overcome this difficulty, a learning-based control algorithm is proposed to learn the solution to the HJI equation online using a critic ESN in order to get the optimal feedforward control policies $\bar{\boldsymbol{\tau}}_i^*$ and $\boldsymbol{d}_i^*$.

To implement the feedforward control policies, an online critic ESN is used to approximate the optimal value function as [35]

$$\begin{aligned} \dot{\boldsymbol{b}}_i(t) &= \frac{1}{\beta}\left(-\alpha\boldsymbol{b}_i(t) + \phi\left(\boldsymbol{W}_{1i}\Im_i(t) + \boldsymbol{W}_{2i}\boldsymbol{b}_i(t)\right)\right) \\ \nabla V_i^*(t) &= \boldsymbol{W}_{3i}\left[\Im_i(t);\boldsymbol{b}_i(t)\right] + \boldsymbol{\varepsilon}_i\left(\Im_i(t),\boldsymbol{b}_i(t)\right), \end{aligned} \tag{9.27}$$

where α is a positive leaky rate, β is a positive time constant, and $\boldsymbol{W}_{1i}$, $\boldsymbol{W}_{2i}$, and $\boldsymbol{W}_{3i}$ are weighted matrices for input, reservoir states, and output of ESN, respectively. Thereinto, $\boldsymbol{W}_{1i}$ and $\boldsymbol{W}_{2i}$ are always sparse matrices generated stochastically and do not need to be tuned. Only $\boldsymbol{W}_{3i}$ is needed to be trained. The training of $\boldsymbol{W}_{3i}$ uses the gradient descent method in the following. $\phi(\cdot)$ is the reservoir unit function, $\Im_i(t)$ is the input of ESN, and $\boldsymbol{\varepsilon}_i(\Im_i(t),\boldsymbol{b}_i(t))$ is the approximation error. The output activation function is taken as an identity function.

Remark 9.1 The main reasons of using ESN to approximate the optimal value functions are that compared with the existing NNs [36], the ESN uses a dynamical reservoir to replace the hidden layer of the recurrent NNs. Hence, we only need to train the output weight $\boldsymbol{W}_{3i}$, which reduces the online computation of weight tuning for UAV system.

Using (9.27), the optimal feedforward control policies are given by

$$\bar{\boldsymbol{\tau}}_i^* = -\frac{1}{2}\boldsymbol{R}_\tau^{-1}\boldsymbol{C}_i^{\mathrm{T}}\boldsymbol{W}_{3i}\left[\Im_i(t);\boldsymbol{b}_i(t)\right] + \boldsymbol{\varepsilon}_{\tau i} \tag{9.28}$$

$$\boldsymbol{d}_i^* = \frac{1}{2}\boldsymbol{R}_d^{-1}\boldsymbol{K}_i^{\mathrm{T}}\boldsymbol{W}_{3i}\left[\Im_i(t);\boldsymbol{b}_i(t)\right] + \boldsymbol{\varepsilon}_{di}, \tag{9.29}$$

where $\boldsymbol{\varepsilon}_{\tau i} = -\frac{1}{2}\boldsymbol{R}_\tau^{-1}\boldsymbol{C}_i^{\mathrm{T}}\boldsymbol{\varepsilon}_i$ and $\boldsymbol{\varepsilon}_{di} = \frac{1}{2}\boldsymbol{R}_d^{-1}\boldsymbol{K}_i^{\mathrm{T}}\boldsymbol{\varepsilon}_i$.

Assumption 9.1 ([37]) The approximation error $\boldsymbol{\varepsilon}_i$ is bounded by $\|\boldsymbol{\varepsilon}_i\| \le \kappa_\varepsilon$ with κ_ε a positive constant. Then it follows that $\|\boldsymbol{\varepsilon}_{\tau i}\| \le \kappa_{\varepsilon\tau}$ and $\|\boldsymbol{\varepsilon}_{di}\| \le \kappa_{\varepsilon d}$ with $\kappa_{\varepsilon\tau}$ and $\kappa_{\varepsilon d}$ positive constants.

Substituting (9.27) into the HJI equation (9.26) gives

$$\begin{aligned}\varepsilon_i^{\mathrm{HJI}} =& -\frac{1}{4}(\boldsymbol{W}_{3i}\left[\Im_i(t);\boldsymbol{b}_i(t)\right])^{\mathrm{T}}\boldsymbol{P}_{\tau i}\boldsymbol{W}_{3i}\left[\Im_i(t);\boldsymbol{b}_i(t)\right]\\ &+\frac{1}{4}(\boldsymbol{W}_{3i}\left[\Im_i(t);\boldsymbol{b}_i(t)\right])^{\mathrm{T}}\boldsymbol{P}_{di}\boldsymbol{W}_{3i}\left[\Im_i(t);\boldsymbol{b}_i(t)\right]\\ &+Q(\Im_i),\end{aligned} \tag{9.30}$$

where $\boldsymbol{P}_{\tau i} = \boldsymbol{C}_i\boldsymbol{R}_\tau^{-1}\boldsymbol{C}_i^{\mathrm{T}}$, $\boldsymbol{P}_{di} = \boldsymbol{K}_i\boldsymbol{R}_d^{-1}\boldsymbol{K}_i^{\mathrm{T}}$, and $\varepsilon_i^{\mathrm{HJI}}$ is the residual error arising from ESN approximation.

Since the ideal $\boldsymbol{W}_{3i}$ is unknown, we let $\widehat{\boldsymbol{W}}_{3i}$ be the estimate of $\boldsymbol{W}_{3i}$, and we define $\widetilde{\boldsymbol{W}}_{3i} = \boldsymbol{W}_{3i} - \widehat{\boldsymbol{W}}_{3i}$. Then, $\nabla V_i^*(\Im_i(t))$ can be estimated as

$$\nabla\widehat{V}_i^*(t) = \widehat{\boldsymbol{W}}_{3i}\left[\Im_i(t);\boldsymbol{b}_i(t)\right]. \tag{9.31}$$

Substituting (9.31) into (9.24) and (9.25), the estimated optimal feedforward control policies are given by

$$\widehat{\boldsymbol{\tau}}_i^* = -\frac{1}{2}\boldsymbol{R}_\tau^{-1}\boldsymbol{C}_i^{\mathrm{T}}\widehat{\boldsymbol{W}}_{3i}\left[\Im_i(t);\boldsymbol{b}_i(t)\right] \tag{9.32}$$

$$\widehat{\boldsymbol{d}}_i = \frac{1}{2}\boldsymbol{R}_d^{-1}\boldsymbol{K}_i^{\mathrm{T}}\widehat{\boldsymbol{W}}_{3i}\left[\Im_i(t);\boldsymbol{b}_i(t)\right]. \tag{9.33}$$

The approximate Hamiltonian function is obtained as

$$\begin{aligned}\widehat{H}_i =&\left(\widehat{\boldsymbol{W}}_{3i}\left[\Im_i(t);\boldsymbol{b}_i(t)\right]\right)^{\mathrm{T}}\left(\boldsymbol{C}_i\widehat{\boldsymbol{\tau}}_i^*+\boldsymbol{K}_i\widehat{\boldsymbol{d}}_i\right)+Q(\Im_i)\\ &+\widehat{\boldsymbol{\tau}}_i^{*\mathrm{T}}\boldsymbol{R}_\tau\widehat{\boldsymbol{\tau}}_i^* - \widehat{\boldsymbol{d}}_i^{\mathrm{T}}\boldsymbol{R}_d\widehat{\boldsymbol{d}}_i \overset{\Delta}{=} e_{Hi}.\end{aligned} \tag{9.34}$$

Then, it follows along (9.30)–(9.33) that

$$
\begin{aligned}
e_{Hi} = & \left(\widetilde{\boldsymbol{W}}_{3i}\left[\Im_i(t);\boldsymbol{b}_i(t)\right]\right)^{\mathrm{T}}\left(\boldsymbol{C}_i\widehat{\boldsymbol{\tau}}_i^* + \boldsymbol{K}_i\widehat{\boldsymbol{d}}_i\right) - \varepsilon_i^{\mathrm{HJI}} \\
& - \frac{1}{4}\left(\widetilde{\boldsymbol{W}}_{3i}\left[\Im_i(t);\boldsymbol{b}_i(t)\right]\right)^{\mathrm{T}}\boldsymbol{P}_{\tau i}\widetilde{\boldsymbol{W}}_{3i}\left[\Im_i(t);\boldsymbol{b}_i(t)\right] \\
& + \frac{1}{4}\left(\widetilde{\boldsymbol{W}}_{3i}\left[\Im_i(t);\boldsymbol{b}_i(t)\right]\right)^{\mathrm{T}}\boldsymbol{P}_{di}\widetilde{\boldsymbol{W}}_{3i}\left[\Im_i(t);\boldsymbol{b}_i(t)\right].
\end{aligned}
\tag{9.35}
$$

It is desired to select $\widetilde{\boldsymbol{W}}_{3i}$ to minimize the squared residual error $E_{Hi} = \frac{1}{2}e_{Hi}^2$ by the gradient descent algorithm. Then, the tuning law of $\widehat{\boldsymbol{W}}_{3i}$ is designed as

$$
\dot{\widehat{\boldsymbol{W}}}_{3i} = -\lambda_i\frac{\partial E_{Hi}}{\partial\widehat{\boldsymbol{W}}_{3i}} = -\lambda_i\left(\boldsymbol{C}_i\widehat{\boldsymbol{\tau}}_i^* + \boldsymbol{K}_i\widehat{\boldsymbol{d}}_i\right)\left[\Im_i(t);\boldsymbol{b}_i(t)\right]^{\mathrm{T}}e_{Hi}, \tag{9.36}
$$

where λ_i is a positive tuning parameter.

Remark 9.2 Before starting the training process, an initial admissible control signal is necessary. In studies [38] (Definition 1), [39] (Definition 1), and [40] (Assumption 1), the initial admissible control solution is assumed to be existent a priori in the stability analysis and is chosen manually in simulation and experimental examples. This priori assumption is practically difficult or even impossible to be satisfied since the output weights of the ESNs generally have dozens of dimensions [35]. We remove the initial admissible control condition by proposing new piecewise adaptation laws (9.38) for the critic ESN with the help of a newly proposed operator Γ_i (9.39) which is selected based on Lyapunov's sufficient condition (9.37). This operation is desired to pull the controller back to the admissible range when the control is not admissible.

Remark 9.3 In terms of tuning law (9.36), the persistency of excitation (PE) condition is needed to guarantee the convergence of $\widehat{\boldsymbol{W}}_{3i}$. Generally, the PE condition is directly assumed to be satisfied [38] (eq. (27)) and is always satisfied by exerting probing noise on the system dynamics [39] (eq. (28)), while in [40] (Section III. C), the PE condition is solved by so-called experience replay strategy in the sense of collecting and training massive amounts of recorded data. In contrast with aforementioned methods, we attempt to remove the PE assumption via an adaptive manner by introducing appropriate compensation terms (9.40), (9.41) in the adaptation laws (9.36) of the critic ESN. On the one hand, to stabilize the instability terms of (9.36) caused by the absence of the PE assumption, we introduce the compensation terms (9.40) in the adaptation law (9.36). On the other hand, to offset the superfluous terms of $\widehat{\boldsymbol{W}}_{3i}$ in (9.36), the compensation terms (9.41) are designed combined with the adaptation laws (9.36).

To remove the requirement of initial admissible control pair, we introduce a Lyapunov function candidate L_{i3} satisfying

$$
\dot{L}_{i3} = -\frac{1}{2}\nabla L_{i3}^{\Im_i\mathrm{T}}\boldsymbol{P}_{\tau i}\widehat{\boldsymbol{W}}_{3i}\left[\Im_i(t);\boldsymbol{b}_i(t)\right]
$$

$$+\frac{1}{2}\nabla L_{l3}^{\Im_l \mathrm{T}} \boldsymbol{P}_{dl} \widehat{\boldsymbol{W}}_{3l} \left[\boldsymbol{\Im}_l(t); \boldsymbol{b}_l(t)\right], \tag{9.37}$$

where $\nabla L_{l3}^{\Im_l}$ denotes the partial derivative of L_{l3} with respect to $\boldsymbol{\Im}_l$.

Lemma 9.1 ([41]) *Given the transformed error dynamics (9.15) with associated value function* (9.21) *and optimal control policies (9.24) and (9.25), it is supposed that there exists a continuously differentiable Lyapunov function L_{l3} such that $\dot{L}_{l3} = \nabla L_{l3}^{\Im_l \mathrm{T}} \left(\boldsymbol{C}_l \bar{\boldsymbol{\tau}}_l^* + \boldsymbol{K}_l \boldsymbol{d}_l^*\right) < 0$. Let $\boldsymbol{R}_{\Im_l}$ be an appropriate positive definite matrix. Then it holds that $\nabla L_{l3}^{\Im_l \mathrm{T}} \left(\boldsymbol{C}_l \bar{\boldsymbol{\tau}}_l^* + \boldsymbol{K}_l \boldsymbol{d}_l^*\right) = -\nabla L_{l3}^{\Im_l \mathrm{T}} \boldsymbol{R}_{\Im_l} \nabla L_{l3}^{\Im_l}$.*

Based on (9.37), a compensation term is designed as

$$\begin{aligned}\boldsymbol{T}_{1l} =& - \Gamma_l \lambda_l \frac{\partial \dot{L}_{l3}}{\partial \widehat{\boldsymbol{W}}_{3l}} = \frac{1}{2} \Gamma_l \lambda_l \boldsymbol{P}_{\tau l} \nabla L_{l3}^{\Im_l} \left[\boldsymbol{\Im}_l(t); \boldsymbol{b}_l(t)\right]^{\mathrm{T}} \\ & - \frac{1}{2} \Gamma_l \lambda_l \boldsymbol{P}_{dl} \nabla L_{l3}^{\Im_l} \left[\boldsymbol{\Im}_l(t); \boldsymbol{b}_l(t)\right]^{\mathrm{T}},\end{aligned} \tag{9.38}$$

where the operator Γ_l is defined as

$$\Gamma_l = \begin{cases} 0, & \text{if } \nabla L_{l3}^{\Im_l \mathrm{T}} \left(\boldsymbol{C}_l \widehat{\boldsymbol{\tau}}_l^* + \boldsymbol{K}_l \widehat{\boldsymbol{d}}_l\right) < 0 \\ 1, & \text{else.} \end{cases} \tag{9.39}$$

Remark 9.4 The operator Γ_l is chosen based on Lyapunov's sufficient condition for stability. This operation is desired to pull the controller back to the admissible range when the control is not admissible. If the closed-loop system is unstable, the operator $\Gamma_l = 1$, (9.38) will be activated, which turns $\dot{L}_{l3} < 0$ hold true. Otherwise, the operator $\Gamma_l = 0$ and (9.38) do not take effect.

To relax the PE condition, the following compensation terms are introduced in the tuning law for $\widehat{\boldsymbol{W}}_{3l}$:

$$\boldsymbol{T}_{2l} = \lambda_l \sigma_l \boldsymbol{M}_l \left[\boldsymbol{\Im}_l(t); \boldsymbol{b}_l(t)\right]^{\mathrm{T}} \tag{9.40}$$

$$\begin{aligned}\boldsymbol{T}_{3l} =& - \frac{1}{4} \lambda_l \boldsymbol{M}_l^{\mathrm{T}} \boldsymbol{P}_{\tau l} \boldsymbol{M}_l \boldsymbol{H}_l \left[\boldsymbol{\Im}_l(t); \boldsymbol{b}_l(t)\right]^{\mathrm{T}} \\ & + \frac{1}{4} \lambda_l \boldsymbol{M}_l^{\mathrm{T}} \boldsymbol{P}_{dl} \boldsymbol{M}_l \boldsymbol{H}_l \left[\boldsymbol{\Im}_l(t); \boldsymbol{b}_l(t)\right]^{\mathrm{T}},\end{aligned} \tag{9.41}$$

where $\boldsymbol{M}_l = \widehat{\boldsymbol{W}}_{3l} \left[\boldsymbol{\Im}_l(t); \boldsymbol{b}_l(t)\right]$, $\boldsymbol{H}_l = \boldsymbol{C}_l \widehat{\boldsymbol{\tau}}_l^* + \boldsymbol{K}_l \widehat{\boldsymbol{d}}_l$, and σ_l is a learning rate.

Based on (9.38)–(9.41), the new tuning law of $\widehat{\boldsymbol{W}}_{3l}$ denotes

$$\begin{aligned}\dot{\widehat{\boldsymbol{W}}}_{3l} =& - \lambda_l \boldsymbol{H}_l \left[\boldsymbol{\Im}_l(t); \boldsymbol{b}_l(t)\right]^{\mathrm{T}} e_{Hl} + \lambda_l \sigma_l \boldsymbol{M}_l \left[\boldsymbol{\Im}_l(t); \boldsymbol{b}_l(t)\right]^{\mathrm{T}} \\ & - \frac{1}{4} \lambda_l \boldsymbol{M}_l^{\mathrm{T}} \boldsymbol{P}_{\tau l} \boldsymbol{M}_l \boldsymbol{H}_l \left[\boldsymbol{\Im}_l(t); \boldsymbol{b}_l(t)\right]^{\mathrm{T}}\end{aligned}$$

$$+\frac{1}{4}\lambda_l \boldsymbol{M}_l^{\mathrm{T}} \boldsymbol{P}_{dl} \boldsymbol{M}_l \boldsymbol{H}_l [\Im_l(t); \boldsymbol{b}_l(t)]^{\mathrm{T}}$$
$$+\frac{1}{2}\Gamma_l \lambda_l \boldsymbol{P}_{\tau l} \nabla L_{l3}^{\Im_l} [\Im_l(t); \boldsymbol{b}_l(t)]^{\mathrm{T}}$$
$$-\frac{1}{2}\Gamma_l \lambda_l \boldsymbol{P}_{dl} \nabla L_{l3}^{\Im_l} [\Im_l(t); \boldsymbol{b}_l(t)]^{\mathrm{T}}. \tag{9.42}$$

9.3.3 Stability Analysis

Theorem 9.2 *Consider the transformed error dynamics* (9.15). *Let the ESN weight tuning law be provided by (9.42). The feedforward control policies are given by (9.32) and (9.33). With Assumption 9.1, the tracking error* $\Im_l$ *and the weight estimation error* $\widetilde{\boldsymbol{W}}_{3l}$ *are uniformly ultimately bounded (UUB).*

Proof Consider the following Lyapunov function candidate:

$$L_{l4} = L_{l3} + \frac{1}{2\lambda_l}\mathrm{tr}\left(\widetilde{\boldsymbol{W}}_{3l}^{\mathrm{T}} \widetilde{\boldsymbol{W}}_{3l}\right). \tag{9.43}$$

Taking the time derivative of (9.43) gives

$$\dot{L}_{l4} = \nabla L_{l3}^{\Im_l} \boldsymbol{H}_l + \frac{1}{\lambda_l}\mathrm{tr}\left(\dot{\widetilde{\boldsymbol{W}}}_{3l}^{\mathrm{T}} \widetilde{\boldsymbol{W}}_{3l}\right). \tag{9.44}$$

Using (9.35) and (9.42), it follows that

$$\dot{\widetilde{\boldsymbol{W}}}_{3l} = -\lambda_l \boldsymbol{H}_l [\Im_l(t); \boldsymbol{b}_l(t)]^{\mathrm{T}} \left(\widetilde{\boldsymbol{M}}_l^{\mathrm{T}} \boldsymbol{H}_l - \frac{1}{4}\widetilde{\boldsymbol{M}}_l^{\mathrm{T}} \boldsymbol{P}_{\tau l} \widetilde{\boldsymbol{M}}_l\right.$$
$$\left.+\frac{1}{4}\widetilde{\boldsymbol{M}}_l^{\mathrm{T}} \boldsymbol{P}_{dl} \widetilde{\boldsymbol{M}}_l - \varepsilon_l^{\mathrm{HJI}}\right) - \lambda_l \sigma_l \boldsymbol{M}_l [\Im_l(t); \boldsymbol{b}_l(t)]^{\mathrm{T}}$$
$$+\frac{1}{4}\lambda_l \boldsymbol{H}_l [\Im_l(t); \boldsymbol{b}_l(t)]^{\mathrm{T}} \boldsymbol{M}_l^{\mathrm{T}} \boldsymbol{P}_{\tau l} \boldsymbol{M}_l$$
$$-\frac{1}{4}\lambda_l \boldsymbol{H}_l [\Im_l(t); \boldsymbol{b}_l(t)]^{\mathrm{T}} \boldsymbol{M}_l^{\mathrm{T}} \boldsymbol{P}_{dl} \boldsymbol{M}_l$$
$$-\frac{1}{2}\Gamma_l \lambda_l \boldsymbol{P}_{\tau l} \nabla L_{l3}^{\Im_l} [\Im_l(t); \boldsymbol{b}_l(t)]^{\mathrm{T}}$$
$$+\frac{1}{2}\Gamma_l \lambda_l \boldsymbol{P}_{dl} \nabla L_{l3}^{\Im_l} [\Im_l(t); \boldsymbol{b}_l(t)]^{\mathrm{T}}, \tag{9.45}$$

where $\widetilde{\boldsymbol{M}}_l = \widetilde{\boldsymbol{W}}_{3l} [\Im_l(t); \boldsymbol{b}_l(t)]$.

Substituting (9.45) into (9.44), the term $\mathrm{tr}\left(\dot{\widetilde{W}}_{3i}^{\mathrm{T}}\widetilde{W}_{3i}\right)$ becomes

$$
\begin{aligned}
&\frac{1}{\lambda_i}\mathrm{tr}\left(\dot{\widetilde{W}}_{3i}^{\mathrm{T}}\widetilde{W}_{3i}\right) = -\mathrm{tr}\left(\widetilde{M}_i^{\mathrm{T}}H_iH_i^{\mathrm{T}}\widetilde{M}_i\right) + \mathrm{tr}\left(\sigma_i\widetilde{M}_i^{\mathrm{T}}\widetilde{M}_i\right)\\
&+\frac{1}{4}\mathrm{tr}\left((W_{3i}\left[\Im_i(t);b_i(t)\right])^{\mathrm{T}}P_{\tau i}W_{3i}\left[\Im_i(t);b_i(t)\right]H_i^{\mathrm{T}}\widetilde{M}_i\right)\\
&-\frac{1}{4}\mathrm{tr}\left((W_{3i}\left[\Im_i(t);b_i(t)\right])^{\mathrm{T}}P_{di}W_{3i}\left[\Im_i(t);b_i(t)\right]H_i^{\mathrm{T}}\widetilde{M}_i\right)\\
&-\mathrm{tr}\left(\sigma_i(W_{3i}\left[\Im_i(t);b_i(t)\right])^{\mathrm{T}}\widetilde{M}_i\right) + \mathrm{tr}\left(\varepsilon_i^{\mathrm{HJI}}H_i^{\mathrm{T}}\widetilde{M}_i\right)\\
&-\frac{1}{2}\mathrm{tr}\left(\widetilde{M}_i^{\mathrm{T}}P_{\tau i}W_{3i}\left[\Im_i(t);b_i(t)\right]H_i^{\mathrm{T}}\widetilde{M}_i\right)\\
&+\frac{1}{2}\mathrm{tr}\left(\widetilde{M}_i^{\mathrm{T}}P_{di}W_{3i}\left[\Im_i(t);b_i(t)\right]H_i^{\mathrm{T}}\widetilde{M}_i\right)\\
&-\frac{1}{2}\mathrm{tr}\left(\Gamma_i\left[\Im_i(t);b_i(t)\right]\nabla L_{i3}^{\Im_i\mathrm{T}}P_{\tau i}^{\mathrm{T}}\widetilde{W}_{3i}\right)\\
&+\frac{1}{2}\mathrm{tr}\left(\Gamma_i\left[\Im_i(t);b_i(t)\right]\nabla L_{i3}^{\Im_i\mathrm{T}}P_{di}^{\mathrm{T}}\widetilde{W}_{3i}\right).
\end{aligned}
\tag{9.46}
$$

Define

$$
\begin{aligned}
N_i &= H_iH_i^{\mathrm{T}} + \frac{1}{2}P_{\tau i}W_{3i}\left[\Im_i(t);b_i(t)\right]H_i^{\mathrm{T}}\\
&\quad-\frac{1}{2}P_{di}W_{3i}\left[\Im_i(t);b_i(t)\right]H_i^{\mathrm{T}} - \sigma_iI_6
\end{aligned}
\tag{9.47}
$$

$$
\begin{aligned}
S_i &= \frac{1}{4}(W_{3i}\left[\Im_i(t);b_i(t)\right])^{\mathrm{T}}P_{\tau i}W_{3i}\left[\Im_i(t);b_i(t)\right]H_i^{\mathrm{T}}\\
&\quad-\frac{1}{4}(W_{3i}\left[\Im_i(t);b_i(t)\right])^{\mathrm{T}}P_{di}W_{3i}\left[\Im_i(t);b_i(t)\right]H_i^{\mathrm{T}}\\
&\quad-\sigma_i(W_{3i}\left[\Im_i(t);b_i(t)\right])^{\mathrm{T}}+\varepsilon_i^{\mathrm{HJI}}H_i^{T}.
\end{aligned}
\tag{9.48}
$$

Combining (9.46)–(9.48), (9.44) can be rewritten as

$$
\begin{aligned}
\dot{L}_{i4} =&\nabla L_{i3}^{\Im_i\mathrm{T}}H_i - \mathrm{tr}\left(\widetilde{M}_i^{\mathrm{T}}N_i\widetilde{M}_i\right) + \mathrm{tr}\left(S_i\widetilde{M}_i\right)\\
&-\frac{1}{2}\mathrm{tr}\left(\Gamma_i\left[\Im_i(t);b_i(t)\right]\nabla L_{i3}^{\Im_i\mathrm{T}}P_{\tau i}^{\mathrm{T}}\widetilde{W}_{3i}\right)\\
&+\frac{1}{2}\mathrm{tr}\left(\Gamma_i\left[\Im_i(t);b_i(t)\right]\nabla L_{i3}^{\Im_i\mathrm{T}}P_{di}^{\mathrm{T}}\widetilde{W}_{3i}\right).
\end{aligned}
\tag{9.49}
$$

It can be concluded from [41] that $\boldsymbol{S}_i$ is bounded by $\|\boldsymbol{S}_i\| \leq \kappa_s$. Let the parameters $\boldsymbol{P}_{\tau i}$, $\boldsymbol{P}_{\tau i}$, and σ_i be chosen such that $\lambda_{\min}(\boldsymbol{N}_i) > 0$. Then it follows that

$$\begin{aligned}\dot{L}_{i4} \leq & \nabla L_{i3}^{\Im_i \mathrm{T}} \boldsymbol{H}_i - \lambda_{\min}(\boldsymbol{N}_i) \|\widetilde{\boldsymbol{M}}_i\|^2 + \kappa_s \|\widetilde{\boldsymbol{M}}_i\| \\ & - \frac{1}{2}\mathrm{tr}\left(\Gamma_i [\Im_i(t); \boldsymbol{b}_i(t)] \nabla L_{i3}^{\Im_i \mathrm{T}} \boldsymbol{P}_{\tau i}^{\mathrm{T}} \widetilde{\boldsymbol{W}}_{3i}\right) \\ & + \frac{1}{2}\mathrm{tr}\left(\Gamma_i [\Im_i(t); \boldsymbol{b}_i(t)] \nabla L_{i3}^{\Im_i \mathrm{T}} \boldsymbol{P}_{di}^{\mathrm{T}} \widetilde{\boldsymbol{W}}_{3i}\right). \end{aligned} \tag{9.50}$$

According to (9.39), two cases of $\Gamma_i = 0$ and $\Gamma_i = 1$ are analyzed as follows.

Case 1: $\Gamma_i = 0$, which indicates that $\nabla L_{i3}^{\Im_i \mathrm{T}} \boldsymbol{H}_i < 0$. It follows from (9.50) that

$$\begin{aligned}\dot{L}_{i4} \leq & -\kappa_{\Im} \left\|\nabla L_{i3}^{\Im_i}\right\| - \lambda_{\min}(\boldsymbol{N}_i) \|\widetilde{\boldsymbol{M}}_i\|^2 + \kappa_s \|\widetilde{\boldsymbol{M}}_i\| \\ = & -\kappa_{\Im} \left\|\nabla L_{i3}^{\Im_i}\right\| - \lambda_{\min}(\boldsymbol{N}_i) \left(\|\widetilde{\boldsymbol{M}}_i\| - \frac{\kappa_s}{2\lambda_{\min}(\boldsymbol{N}_i)}\right)^2 \\ & + \frac{\kappa_s^2}{4\lambda_{\min}(\boldsymbol{N}_i)}. \end{aligned} \tag{9.51}$$

If the following inequalities

$$\left\|\nabla L_{i3}^{\Im_i}\right\| > \frac{\kappa_s^2}{4\kappa_{\Im}\lambda_{\min}(\boldsymbol{N}_i)} \tag{9.52}$$

and

$$\|\widetilde{\boldsymbol{M}}_i\| > \frac{\kappa_s}{\lambda_{\min}(\boldsymbol{N}_i)} \tag{9.53}$$

hold true, then $\dot{L}_{i4} < 0$ is satisfied.

Case 2: $\Gamma_i = 1$, which means that $\nabla L_{i3}^{\Im_i \mathrm{T}} \boldsymbol{H}_i \geq 0$. Then (9.50) further satisfies

$$\begin{aligned}\dot{L}_{i4} \leq & \nabla L_{i3}^{\Im_i \mathrm{T}} \boldsymbol{H}_i - \lambda_{\min}(\boldsymbol{N}_i) \|\widetilde{\boldsymbol{M}}_i\|^2 + \kappa_s \|\widetilde{\boldsymbol{M}}_i\| \\ & - \frac{1}{2}\mathrm{tr}\left([\Im_i(t); \boldsymbol{b}_i(t)] \nabla L_{i3}^{\Im_i \mathrm{T}} \boldsymbol{P}_{\tau i}^{\mathrm{T}} \widetilde{\boldsymbol{W}}_{3i}\right) \\ & + \frac{1}{2}\mathrm{tr}\left([\Im_i(t); \boldsymbol{b}_i(t)] \nabla L_{i3}^{\Im_i \mathrm{T}} \boldsymbol{P}_{di}^{\mathrm{T}} \widetilde{\boldsymbol{W}}_{3i}\right). \end{aligned} \tag{9.54}$$

By adding and subtracting $\nabla L_{l3}^{\Im_l \mathrm{T}} \boldsymbol{C}_l \bar{\boldsymbol{\tau}}_l^*$ and $\nabla L_{l3}^{\Im_l \mathrm{T}} \boldsymbol{K}_l \boldsymbol{d}_l^*$ to the right-hand side of (9.54), one arrives at

$$\begin{aligned}
\dot{L}_{l4} \leq & \nabla L_{l3}^{\Im_l \mathrm{T}} \left(\boldsymbol{C}_l \bar{\boldsymbol{\tau}}_l^* + \boldsymbol{K}_l \boldsymbol{d}_l^* \right) + \kappa_s \left\| \widetilde{\boldsymbol{M}}_l \right\| - \lambda_{\min} \left(\boldsymbol{N}_l \right) \left\| \widetilde{\boldsymbol{M}}_l \right\|^2 \\
& + \nabla L_{l3}^{\Im_l \mathrm{T}} \left\{ \boldsymbol{C}_l \left(\widehat{\boldsymbol{\tau}}_l^* - \bar{\boldsymbol{\tau}}_l^* \right) + \boldsymbol{K}_l \left(\widehat{\boldsymbol{d}}_l - \boldsymbol{d}_l^* \right) \right\} \\
& - \frac{1}{2} \mathrm{tr} \left(\left[\boldsymbol{\Im}_l (t) ; \boldsymbol{b}_l (t) \right] \nabla L_{l3}^{\Im_l \mathrm{T}} \boldsymbol{P}_{\tau l}^{\mathrm{T}} \widetilde{\boldsymbol{W}}_{3l} \right) \\
& + \frac{1}{2} \mathrm{tr} \left(\left[\boldsymbol{\Im}_l (t) ; \boldsymbol{b}_l (t) \right] \nabla L_{l3}^{\Im_l \mathrm{T}} \boldsymbol{P}_{dl}^{\mathrm{T}} \widetilde{\boldsymbol{W}}_{3l} \right).
\end{aligned} \tag{9.55}$$

Substituting (9.28), (9.29), (9.32), and (9.33) into (9.55), it follows

$$\begin{aligned}
\dot{L}_{l4} \leq & \nabla L_{l3}^{\Im_l \mathrm{T}} \left(\boldsymbol{C}_l \bar{\boldsymbol{\tau}}_l^* + \boldsymbol{K}_l \boldsymbol{d}_l^* \right) - \nabla L_{l3}^{\Im_l \mathrm{T}} \left(\boldsymbol{C}_l \boldsymbol{\varepsilon}_{\tau i} + \boldsymbol{K}_l \boldsymbol{\varepsilon}_{\tau l} \right) \\
& - \lambda_{\min} \left(\boldsymbol{N}_l \right) \left\| \widetilde{\boldsymbol{M}}_l \right\|^2 + \kappa_s \left\| \widetilde{\boldsymbol{M}}_l \right\|.
\end{aligned} \tag{9.56}$$

Recalling Assumption 9.1, it can be concluded that $\left\| \boldsymbol{C}_l \boldsymbol{\varepsilon}_{\tau i} + \boldsymbol{K}_l \boldsymbol{\varepsilon}_{\tau l} \right\| \leq \kappa_c$ with κ_c a positive parameter.

$$\begin{aligned}
\dot{L}_{l4} \leq & - \lambda_{\min} \left(\boldsymbol{R}_{\Im l} \right) \left\| \nabla L_{l3}^{\Im_l} \right\|^2 + \kappa_c \left\| \nabla L_{l3}^{\Im_l} \right\| \\
& - \lambda_{\min} \left(\boldsymbol{N}_l \right) \left\| \widetilde{\boldsymbol{M}}_l \right\|^2 + \kappa_s \left\| \widetilde{\boldsymbol{M}}_l \right\| \\
\leq & - \lambda_{\min} \left(\boldsymbol{R}_{\Im l} \right) \left(\left\| \nabla L_{l3}^{\Im_l} \right\| - \frac{\kappa_c}{2 \lambda_{\min} \left(\boldsymbol{R}_{\Im l} \right)} \right)^2 \\
& - \lambda_{\min} \left(\boldsymbol{N}_l \right) \left(\left\| \widetilde{\boldsymbol{M}}_l \right\| - \frac{\kappa_s}{2 \lambda_{\min} \left(\boldsymbol{N}_l \right)} \right)^2 + \zeta_l,
\end{aligned} \tag{9.57}$$

where $\zeta_l = \frac{\kappa_s^2}{4 \lambda_{\min} (\boldsymbol{N}_l)} + \frac{\kappa_c^2}{4 \lambda_{\min} (\boldsymbol{R}_{\Im l})}$.

If the following inequalities

$$\left\| \nabla L_{l3}^{\Im_l} \right\| > \frac{\kappa_c}{2 \lambda_{\min} \left(\boldsymbol{R}_{\Im l} \right)} + \sqrt{\frac{\zeta_l}{\lambda_{\min} \left(\boldsymbol{R}_{\Im l} \right)}} \tag{9.58}$$

and

$$\left\| \widetilde{\boldsymbol{M}}_l \right\| > \frac{\kappa_s}{2 \lambda_{\min} \left(\boldsymbol{N}_l \right)} + \sqrt{\frac{\zeta_l}{\lambda_{\min} \left(\boldsymbol{N}_l \right)}} \tag{9.59}$$

hold true, there exists $\dot{L}_{l4} < 0$.

This completes the proof of Theorem 9.2.

Theorem 9.3 *With Assumption 9.1, the feedback policy pair* $(\hat{\boldsymbol{\tau}}_l^*, \hat{\boldsymbol{d}}_l)$ *converges to the approximate Nash equilibrium solution of the zero-sum game, i.e.,* $\left\|\bar{\boldsymbol{\tau}}_l^* - \hat{\boldsymbol{\tau}}_l^*\right\|$ *and* $\|\boldsymbol{d}_l^* - \hat{\boldsymbol{d}}_l\|$ *are UUB.*

Proof From (9.28), (9.29), (9.32), and (9.33), it follows that

$$\left\|\bar{\boldsymbol{\tau}}_l^* - \hat{\boldsymbol{\tau}}_l^*\right\| \leqq \frac{1}{2}\|-\boldsymbol{R}_\tau^{-1}\boldsymbol{C}_l^{\mathrm{T}}\widetilde{\boldsymbol{W}}_{3l}\left[\boldsymbol{\Im}_l(t); \boldsymbol{b}_l(t)\right]\| \tag{9.60}$$

$$\|\boldsymbol{d}_l^* - \hat{\boldsymbol{d}}_l\| \leqq \frac{1}{2}\|-\boldsymbol{R}_d^{-1}\boldsymbol{K}_l^{\mathrm{T}}\widetilde{\boldsymbol{W}}_{3l}\left[\boldsymbol{\Im}_l(t); \boldsymbol{b}_l(t)\right]\|. \tag{9.61}$$

Invoking Theorem 9.2, $[\boldsymbol{\Im}_l(t); \boldsymbol{b}_l(t)]$ is bounded by $\|[\boldsymbol{\Im}_l(t); \boldsymbol{b}_l(t)]\| \leq b_z$; then it follows that

$$\left\|\bar{\boldsymbol{\tau}}_l^* - \hat{\boldsymbol{\tau}}_l^*\right\| \leq \frac{1}{2}\lambda_{\max}\left(\boldsymbol{R}_\tau^{-1}\boldsymbol{C}_l^{\mathrm{T}}\right) b_z \widetilde{\boldsymbol{W}}_{3l} \tag{9.62}$$

$$\|\boldsymbol{d}_l^* - \hat{\boldsymbol{d}}_l\| \leq \frac{1}{2}\lambda_{\max}\left(\boldsymbol{R}_d^{-1}\boldsymbol{K}_l^{\mathrm{T}}\right) b_z \widetilde{\boldsymbol{W}}_{3l}. \tag{9.63}$$

Recalling the boundedness of $\widetilde{\boldsymbol{W}}_{3l}$, it is obvious that $\left\|\bar{\boldsymbol{\tau}}_l^* - \hat{\boldsymbol{\tau}}_l^*\right\|$ and $\|\boldsymbol{d}_l^* - \hat{\boldsymbol{d}}_l\|$ are UUB. This completes the proof of Theorem 9.3.

9.4 Semi-physical Experiments and Analyses

To showcase the effectiveness of our proposed control algorithm, comprehensive flight tests encompassing both experimental and simulated validations are conducted in the presence of complex disturbances. In Section A, an experimental test utilizing the Links-RT UAV Platform is constructed to validate the effectiveness of proposed controller. Additionally, in Section B, the comparisons with state-of-the-art approaches are presented through simulation validation, thereby demonstrating the superiority of our proposed controller. These two sections encompass two representative flight tasks: low altitude penetration with formation reconfiguration and coordinated turn with formation keeping.

To demonstrate the feasibility of proposed controller, the experimental validation is carried out based on the Links-RT UAV Platform (supported by Beijing Links Co., Ltd.). The overall experiment setup and detailed hardcore parts of Links-RT UAV Platform are presented in Figs. 9.3 and 9.4. The interactive relationship between functional units is depicted in Fig. 9.5. The proposed controller is firstly implemented in MATLAB. Then the code generation by Links-Auto Coder is used to convert the MATLAB language into C code, which is downloaded by Pixhawk. The Pixhawk is responsible for executing the proposed control algorithm with

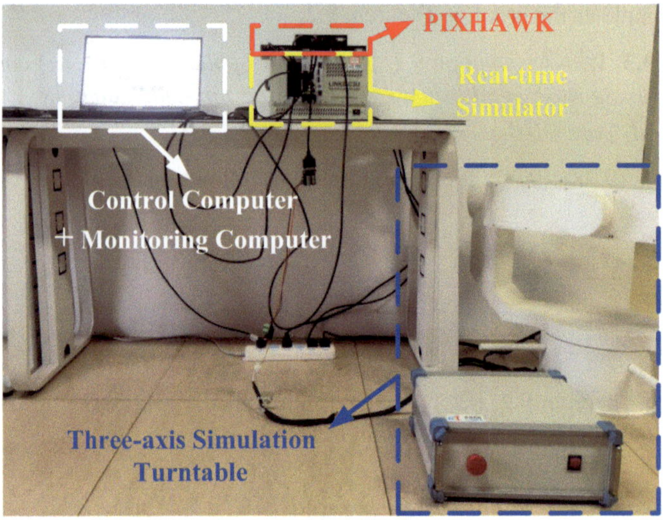

Fig. 9.3 Experiment prototype of the Links-RT UAV Platform

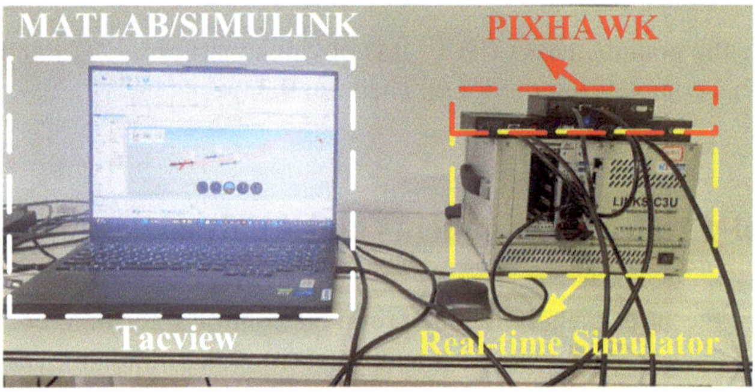

Fig. 9.4 Hardcore parts of the Links-RT UAV Platform

a sampling time of 2 ms and generating pulse-width modulation (PWM) signals, which are sent to the real-time simulator, where the multi-UAV dynamics model is compiled and loaded into the real-time simulator, such that the velocity and attitude motion are calculated and transmitted to Pixhawk and monitoring software, while presenting the 3D visual scenes in Tacview and plotting the velocity and attitude tracking curves by MATLAB.

In the experiment test, the performance of our proposed control scheme is assessed through a low altitude penetration task for UAV formation reconfiguration. The planner path profile of references is expressed as $x_0(\theta_0) = 0$, $y_0(\theta_0) = 32\theta_0$, $z_0(\theta_0) = 500 - 20\theta_0$ $(t \le 5)$, and $z_0(\theta_0) = 400\,(5 < t \le 14)$. The initial values of the path parameters are taken as $\theta_0 = 0\,\text{rad}$ and $\dot{\theta}_0 = 1\,\text{rad}/\text{s}$. The initial airspeed, heading angle, and pitch angle of each UAV are taken as $V_0 = 30\,\text{m/s}$,

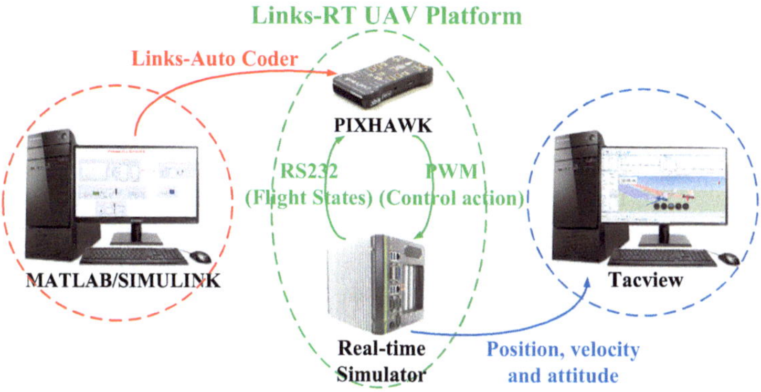

Fig. 9.5 Interactive relationship between functional units

Table 9.1 Formation reconfiguration and switching topologies

	Δx(m)	Δy(m)	Δz(m)	$a_{ij}^{\sigma(t)}$
UAV1	0	$-2t$	0	$a_{12}^{\sigma(t)}=1\,(t\in[0,14]\,\mathrm{s})$
				$a_{13}^{\sigma(t)}=1\,(t\in[0,14]\,\mathrm{s})$
UAV2	$-20+0.5t$	0	0	$a_{21}^{\sigma(t)}=1\,(t\in[0,14]\,\mathrm{s})$
				$a_{23}^{\sigma(t)}=\begin{cases}1\,(t\in[0,5]\,\mathrm{s})\\0\,(t\in(5,14]\,\mathrm{s})\end{cases}$
UAV3	$20-0.5t$	0	0	$a_{31}^{\sigma(t)}=1\,(t\in[0,14]\,\mathrm{s})$
				$a_{32}^{\sigma(t)}=\begin{cases}0\,(t\in[0,5]\,\mathrm{s})\\1\,(t\in(5,14]\,\mathrm{s})\end{cases}$

$\psi_0=\pi/2\,\mathrm{rad}$, and $\gamma_0=-\arctan(2/3)\,\mathrm{rad}$, respectively. The time constant of pitch dynamics is chosen by $\kappa=1$. The desired relative positions of UAVs with respect to virtual leader and the preset topologies are listed in Table 9.1. Moreover, the flight test is performed for the cases of complicated disturbances as follows:

$$\begin{aligned}\omega_x&=5\sin(0.1t)+\xi_x\\ \omega_y&=3\cos(0.1t)+\xi_y\\ \omega_z&=3\cos(0.5t)+\xi_z,\end{aligned}\tag{9.64}$$

where ξ_x, ξ_y, and ξ_z represent random noise signals with a normal distribution.

Remark 9.5 Different from the constant disturbances in [42, 43] and low-frequency disturbance with small-value amplitude in [44, 45], the assumed disturbances (9.64) take into account the factors of electromagnetic interferences and gusts among disturbances, which integrates the characteristics of high-frequency vibration and large-value amplitude. This disturbance set fits in the complex flight environment.

The control parameters of proposed feedback controller are chosen as $k_{1l} = k_{2l} = k_{3l} = 90$, $l = 1, 2, 3$. For a learning-based feedforward controller, we take $\boldsymbol{Q} = \boldsymbol{I}_6$, $\boldsymbol{R}_\tau = 10^8 \boldsymbol{I}_6$, and $\boldsymbol{R}_d = 10^8 \boldsymbol{I}_3$. For ESN, the reservoir is selected as $\boldsymbol{W}_2 \in \Re^{100\times100}$, the internal unit function ϕ is defined as an identity function, and we select $\alpha = 10$ and $\beta = 10$. To relax the initial admissible control, the Lyapunov function L_{l3} is chosen as $L_{l3} = \Im_l^{\mathrm{T}} \boldsymbol{Q}_{\Im l} \Im_l$ with $\boldsymbol{Q}_{\Im l} = \boldsymbol{I}_6$; then it follows $\dot{L}_{l3} = 2\Im_l^{\mathrm{T}} \boldsymbol{Q}_{\Im l}(\boldsymbol{C}_l \hat{\boldsymbol{\tau}}_l^* + \boldsymbol{K}_l \hat{\boldsymbol{d}}_l)$. The initial output weights are set to be zero. The learning rates are selected by $\lambda_l = \sigma_l = 10$, $l = 1, 2, 3$.

The 3D visual scenes displayed by Tacview are presented in Fig. 9.6a–b. And the corresponding experimental results drawn by MATLAB are exhibited in Figs. 9.7,

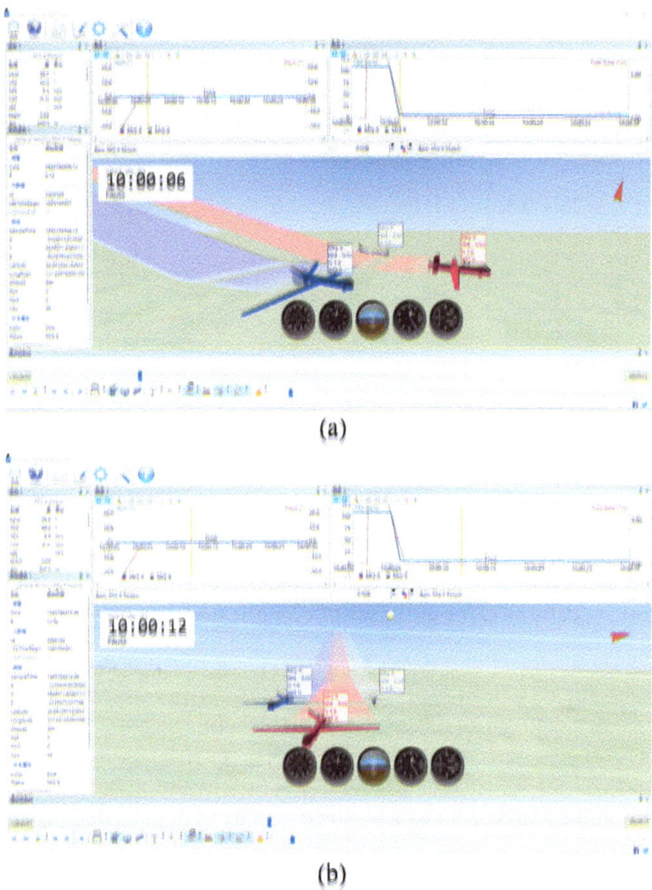

Fig. 9.6 3D visual scenes displayed by Tacview. (**a**) 3D visual scene at 6th second. (**b**) 3D visual scene at 12th second

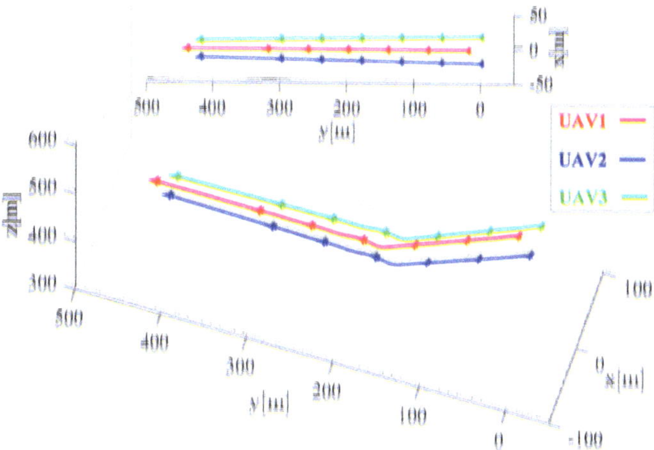

Fig. 9.7 Position trajectories of UAV formation

9.8, 9.9, 9.10, and 9.11. Figure 9.7 shows the position trajectories of UAV formation, where the initial configuration is described by a loose "V" shape, and the final positions of UAV formation are realized by a close "V" one. Figure 9.8 presents the state responses for each UAV during flight. And Fig. 9.9 describes the tracking errors for each UAV. It can be observed from Fig. 9.9 that cooperative formation tracking is well achieved by our proposed control algorithm. Taking UAV1 for example, Fig. 9.10 gives the control signals, and the evolution of operator Γ_i is shown in Fig. 9.11.

To highlight the disturbance rejection ability of our proposed method, our proposed control algorithm is compared with state-of-the-art approaches [3, 34] by simulation validation. The first comparative method is taken from [3], where a conventional backstepping controller is considered. The second comparative approach is a disturbance observer based backstepping controller in [34].

In this section, the superiority of our proposed control scheme is emphasized through a coordinated turn with formation keeping. The path profile of virtual leader is expressed as $x_0(\theta_0) = 1000\cos\theta_0$, $y_0(\theta_0) = 1500\sin(2\theta_0)$, and $z_0(\theta_0) = 1000$. The initial values of path parameters are taken as $\theta_0 = 0$ rad and $\dot{\theta}_0 = 0.01$ rad/s. The initial states for each UAV are taken as $V_0 = 30$ m/s, $\psi_0 = \pi/2$ rad, and $\gamma_0 = 0$ rad. The time constant of pitch dynamics is chosen as $\kappa = 1$. The preset desired relative positions of UAVs 1–3 with respect to virtual leader and the fixed topologies are shown in Table 9.2. The complicated disturbances ω_x, ω_y, and ω_z are the same as (9.64).

The control parameters of proposed feedback controller are chosen as $k_{1i} = 100$, $k_{2i} = 0.1$, $k_{3i} = 100$, $i = 1, 2, 3$. For learning-based feedforward controller, we

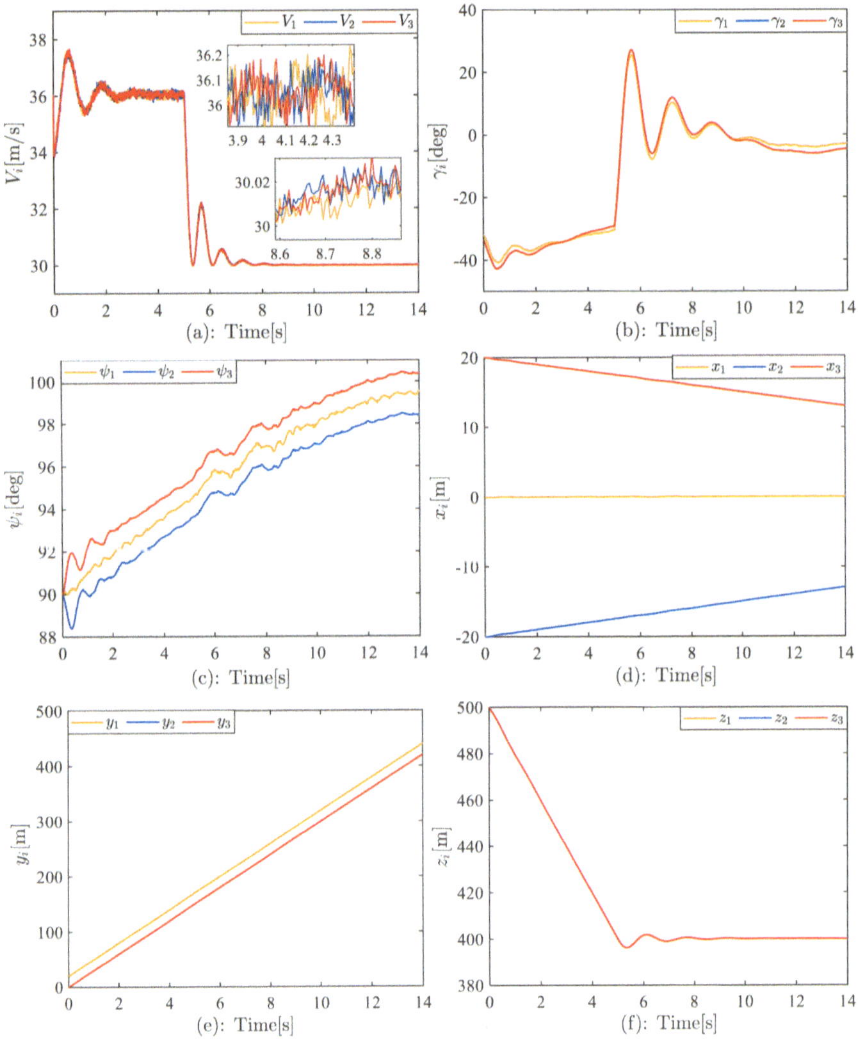

Fig. 9.8 Steady-state responses under proposed controller for low altitude penetration task: (**a**) V_i response. (**b**) γ_i response. (**c**) ψ_i response. (**d**) x_i response. (**e**) y_i response. (**f**) z_i response

take $\boldsymbol{Q} = \boldsymbol{I}_6$, $\boldsymbol{R}_\tau = \boldsymbol{I}_6$, and $\boldsymbol{R}_d = \boldsymbol{I}_3$. For ESN, the reservoir is selected as $\boldsymbol{W}_2 \in \Re^{1000\times1000}$, the internal unit function ϕ is defined as an identity function, and we select $\alpha = 10$ and $\beta = 10$. To relax the initial admissible control, the Lyapunov function L_{i3} is chosen as $L_{i3} = \Im_i^{\mathrm{T}} \boldsymbol{Q}_{\Im i} \Im_i$ with $\boldsymbol{Q}_{\Im i} = \boldsymbol{I}_6$; then one has $\dot{L}_{i3} = 2\Im_i^{\mathrm{T}} \boldsymbol{Q}_{\Im i}(\boldsymbol{C}_i \hat{\boldsymbol{\tau}}_i^* + \boldsymbol{K}_i \hat{\boldsymbol{d}}_i)$. The initial output weights are set to be zero. The learning rates are selected by $\lambda_i = \sigma_i = 10$, $i = 1, 2, 3$.

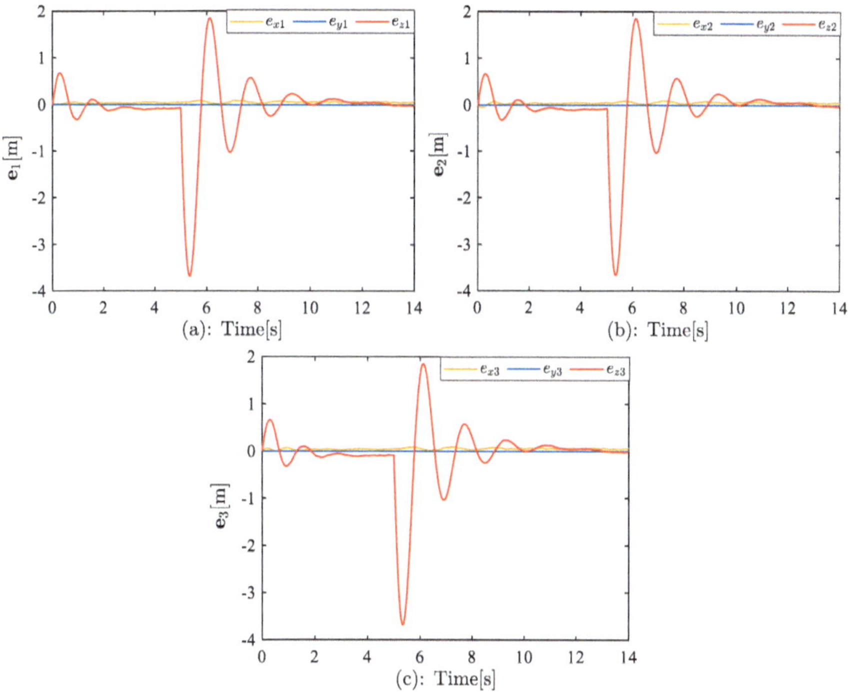

Fig. 9.9 Tracking errors under proposed controller for low altitude penetration task: (**a**) Tracking errors for UAV1. (**b**) Tracking errors for UAV2. (**c**) Tracking errors for UAV3

Taking the UAV1 for example, Fig. 9.12 shows the curves of formation tracking errors under three different controllers. It is observed from Fig. 9.12 that with the help of feedback compensation of the learning-based algorithm, our proposed method exhibits stronger disturbance rejection ability in the presence of complicated disturbances. To visually evaluate the tracking performances of different controllers, the indices of the maximum error fluctuations, offset errors, and integral of absolute errors are listed in Table 9.3, which validates the superiority of our proposed control scheme.

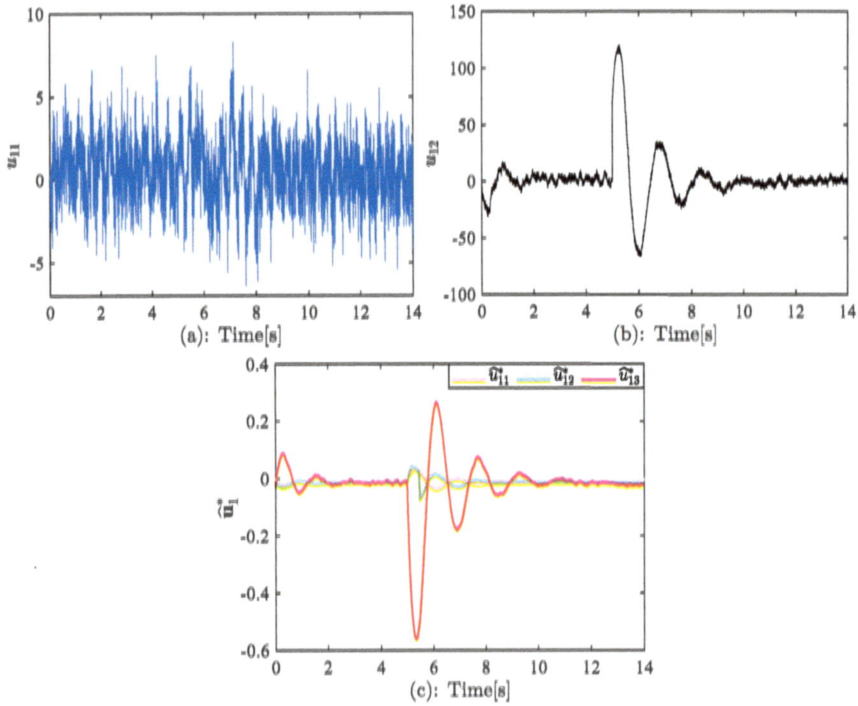

Fig. 9.10 Control actions for UAV1: (**a**) Control input u_{11}. (**b**) Control input u_{12}. (**c**) Estimated optimal control policy $\widehat{\boldsymbol{u}}_1^*$

9.5 Conclusions

This chapter has presented a feedforward-feedback learning-based optimal control scheme to provide the cooperative UAV formation tracking in the presence of complicated disturbances. A two-player zero-sum game framework is designed, where the critic ESN is derived to approximate the optimal feedforward policies. To remove the PE condition and the requirement of initial admissible control in weight tuning law, appropriate compensation terms and a new Lyapunov function have been introduced into adaptive tuning laws. Simulation results have validated the effectiveness and superiority of our proposed control algorithm over the state-of-the-art approaches. One of the future researches would be the extension of our control design to complex networked systems.

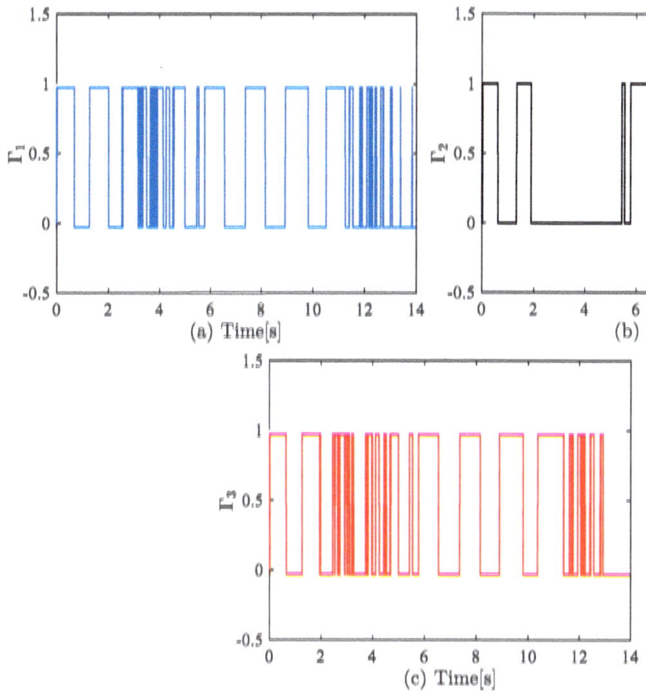

Fig. 9.11 Evolution of the operators Γ_i: (**a**) Γ_1. (**b**) Γ_2. (**c**) Γ_3

Table 9.2 Formation keeping and fixed topologies

	Δx(m)	Δy(m)	Δz(m)	$a_{ij}^{\sigma(t)}$
UAV1	0	0	0	$a_{12}^{\sigma(t)} = 1\ (t \in [0, 12]\ \mathrm{s})$
				$a_{13}^{\sigma(t)} = 1\ (t \in [0, 12]\ \mathrm{s})$
UAV2	− 20	0	0	$a_{21}^{\sigma(t)} = 1\ (t \in [0, 12]\ \mathrm{s})$
				$a_{23}^{\sigma(t)} = 1\ (t \in [0, 12]\ \mathrm{s})$
UAV3	20	0	0	$a_{31}^{\sigma(t)} = 1\ (t \in [0, 12]\ \mathrm{s})$
				$a_{32}^{\sigma(t)} = 1\ (t \in [0, 12]\ \mathrm{s})$

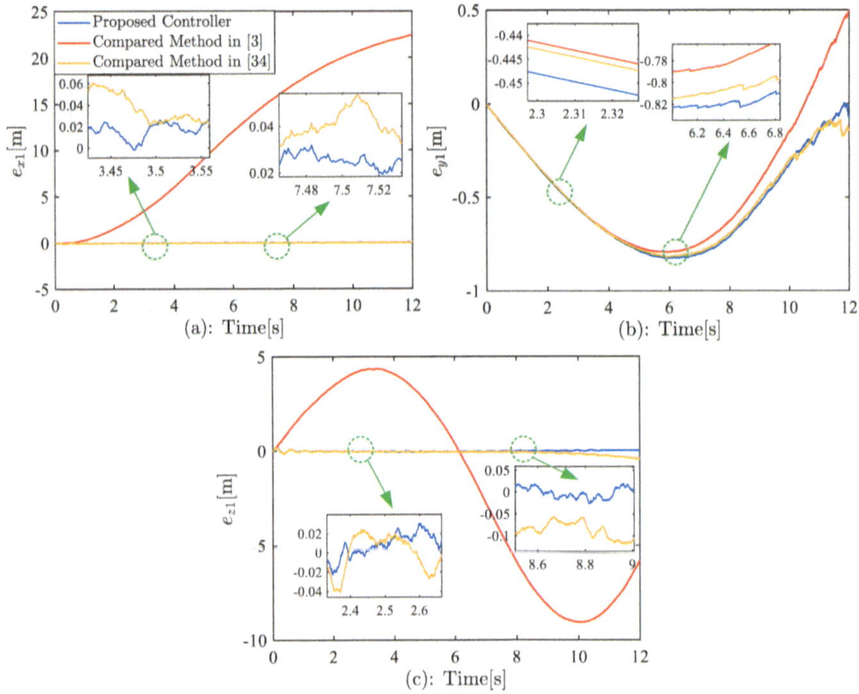

Fig. 9.12 Tracking errors under the proposed controller and comparative controllers. (**a**) e_{x1} response. (**b**) e_{y1} response. (**c**) e_{z1} response

Table 9.3 Transient and steady-state performances under different controllers

	Controller	Maximum error fluctuations	Offset error	Integral of absolute error
e_{x1}	**Proposed controller**	**0.0836**	**0.0496**	**0.3773**
	Compared method in [3]	22.4448	24.4430	136.4567
	Compared method in [34]	0.7999	0.277014	0.4535
e_{y1}	**Proposed controller**	**0.8233**	**−0.0724**	**5.9935**
	Compared method in [3]	0.7916	0.4951	5.7112
	Compared method in [34]	0.8148	−0.1372	5.9513
e_{z1}	**Proposed controller**	**0.2035**	**0.0294**	**0.2813**
	Compared method in [3]	9.0657	−5.7929	54.5148
	Compared method in [34]	0.4535	−0.4531	0.9986

References

1. M. Mozaffari, W. Saad, M. Bennis, M. Debbah, Wireless communication using unmanned aerial vehicles (UAVs): optimal transport theory for hover time optimization. IEEE Trans Wirel Commun **16**(12), 8052–8066 (2017)
2. B. Angus, A. David, Trajectory optimization for high-altitude long endurance UAV maritime radar surveillance. IEEE Trans. Aerospace Electron. Syst. **56**(3), 2406–2421 (2020)
3. X. Dong, Y. Li, C. Lu, G. Hu, Q. Li, Z. Ren, Time-varying formation tracking for UAV swarm systems with switching directed topologies. IEEE Trans. Neural Netw. Learn. Syst. **30**(12), 3674–3685 (2019)
4. S. Rao, D. Ghose, Sliding mode control-based autopilots for leaderless consensus of unmanned aerial vehicles. IEEE Trans. Control Syst. Technol. **22**(5), 1964–1972 (2014)
5. Y. Kang, Y. Kuang, J. Cheng, B. Zhang, Y. Qi, S. Zhou, K. Mao, Robust leaderless time-varying formation control for unmanned aerial vehicle swarm system with Lipschitz nonlinear dynamics and directed switching topologies. Chin. J. Aeronaut. **35**(1), 124–136 (2022)
6. M. Lv, C.K. Ahn, B. Zhang, A. Fu, Fixed-time anti-saturation cooperative control for networked fixed-wing unmanned aerial vehicles considering actuator failures. IEEE Trans. Aerospace Electro. Syst. (2023). https://doi.org/10.1109/TAES.2023.3311420
7. Y. Zou, Z. Zhou, X. Dong, Z. Meng, Distributed formation control for multiple vertical takeoff and landing UAVs with switching topologies. IEEE/ASME Trans. Mechatron. **23**(4), 1750–176 (2018)
8. X. Dong, B. Yu, S. Z, Y. Zhong, Time-varying formation control for unmanned aerial vehicles: theories and applications. IEEE Trans. Control Syst. Technol. **23**(1), 340–348 (2015)
9. X. Dong, Y. Zhou, Z. Ren, Y. Zhong, Time-varying formation tracking for second-order multi-agent systems subjected to switching topologies with application to quadrotor formation flying. IEEE Trans. Ind. Electron. **64**(6), 5014–5024 (2017)
10. M. Lv, B.D. Schutter, S. Baldi, Nonrecursive control for formation-containment of HFV swarms with dynamic event-triggered communication. IEEE Trans. Ind. Inform. **19**(3), 3188–3197 (2023)
11. M. Lv, Z. Chen, B.D. Schutter, S. Baldi, Prescribed-performance tracking for high-power nonlinear dynamics with time-varying unknown control coefficients. Automatica **146**, 110584 (2022)
12. Y. Ke, K. Wang, B.M. Chen, Design and implementation of a hybrid UAV with model-based flight capabilities. IEEE/ASME Trans. Mechatron. **23**(3), 1114–1125 (2018)
13. P. Hemakumara, S. Sukkarieh, Learning UAV stability and control derivatives using Gaussian processes. IEEE Trans. Robot. **29**(4), 813–824 (2013)
14. F. Santoso, M.A. Garratt, S.G. Anavatti, State-of-the-art intelligent flight control systems in unmanned aerial vehicles. IEEE Trans. Autom. Sci. Eng. **15**(2), 613C627 (2018)
15. Z. Zhen, Y. Chen, L. Wen, B. Han, An intelligent cooperative mission planning scheme of UAV swarm in uncertain dynamic environment. Aerospace Sci. Technol. **100**, 105826 (2020)
16. L. He, N. Aouf, B. Song, Explainable deep reinforcement learning for UAV autonomous path planning. Aerospace Sci. Technol. **118**, 107052 (2021)
17. C. Wang, J. Wang, Y. Shen, X. Zhang, Autonomous navigation of UAVs in large-scale complex environments: a deep reinforcement learning approach. IEEE Trans. Veh. Technol. **68**(3), 2124–2136 (2019)
18. Y. Fu, T. Chai, Online solution of two-player zero-sum games for continuous-time nonlinear systems with completely unknown dynamics. IEEE Trans. Neural Netw. Learn. Syst. **27**(12), 2577–2587 (2016)
19. J. Zhao, Y. Lv, Z. Zhao, Adaptive learning based output-feedback optimal control of CT two-player zero-sum games. IEEE Trans. Circuits Syst. II: Express Briefs **69**(3), 1437–1441 (2022)
20. H. Li, D. Liu, D. Wang, Integral reinforcement learning for linear continuous-time zero-sum games with completely unknown dynamics. IEEE Trans. Autom. Sci. Eng. **11**(3), 706–714 (2014)

21. S. Mehraeen, T. Dierks, S. Jagannathan, M.L. Crow, Zero-sum two-player game theoretic formulation of affine nonlinear discrete-time systems using neural networks. IEEE Trans. Cybern. **43**(6), 1641–1655 (2013)
22. V. Saxena, J. Jalden, H. Klessig, Optimal UAV base station trajectories using flow-level models for reinforcement learning. IEEE Trans. Cogn. Commun. Netw. **5**(4), 1101–1112 (2019)
23. X. Liu, Y. Liu, Y. Chen, Reinforcement learning in multiple-UAV networks: deployment and movement design. IEEE Trans. Veh. Technol. **68**(8), 8036–8049 (2019)
24. T. Guo, N. Jiang, B. Li, X. Zhu, Y. Wang, W. Du, UAV navigation in high dynamic environments: a deep reinforcement learning approach. Chin. J. Aeronaut. **34**(2), 479–489 (2021)
25. H. Zargarzadeh, T. Dierks, S. Jagannathan, Optimal control of nonlinear continuous-time systems in strict-feedback form. IEEE Trans. Neural Netw. Learn. Syst. **26**(10), 2535–2549 (2015)
26. S.R. Moghadam, M. Farrokhi, Optimal output feedback control of a class of uncertain systems with input constraints using parallel feedforward compensator. J. Franklin Inst. **357**(18), 13449–13476 (2020)
27. B. Zhao, D. Liu, C. Luo, Reinforcement learning-based optimal stabilization for unknown nonlinear systems subject to inputs with uncertain constraints. IEEE Trans. Neural Netw. Learn. Syst. **31**(10), 4330–4340 (2020)
28. N. Wang, Y. Gao, H. Zhao, C.K. Ahn, Reinforcement learning based optimal tracking control of an unknown unmanned surface vehicle. IEEE Trans. Neural Netw. Learn. Syst. **32**(7), 3034–3045 (2021)
29. X. Yang, H. He, Decentralized event-triggered control for a class of nonlinear-interconnected systems using reinforcement learning. IEEE Trans. Cybern. **51**(2), 635–648 (2021)
30. L. Wang, K. Wang, C. Pan, W. Xu, N. Aslam, L. Hanzo, Multi-agent deep reinforcement learning-based trajectory planning for multi-UAV assisted mobile edge computing. IEEE Trans. Cogn. Commun. Netw. **7**(1), 73–84 (2021)
31. A. Singla, S. Padakandla, S. Bhatnagar, Memory-based deep reinforcement learning for obstacle avoidance in UAV with limited environment knowledge. IEEE Trans. Intell. Transp. Syst. **22**(1), 107–118 (2021)
32. J. Cui, Y. Liu, A. Nallanathan, Multi-agent reinforcement learning- based resource allocation for UAV networks. IEEE Trans. Wirel. Commun. **19**(2), 729–743 (2020)
33. J. Yang, C. Liu, M. Coombes, Y. Yan, W.H. Chen, Optimal path following for small fixed-wing UAVs under wind disturbances. IEEE Trans. Control Syst. Technol. **29**(3), 996–1008 (2021)
34. X. Yu, J. Yang, S. Li, Finite-time path following control for small- scale fixed-wing UAVs under wind disturbances. J. Franklin Inst. **357**(12), 7879–7903 (2020)
35. H. Jaeger, M. Lukosevicius, D. Popovici, Optimization and applications of echo state networks with leaky-integrator neurons. Neural Netw. **20**(3), 335–352 (2007)
36. J.M. Keller, D. Liu, D.B. Fogel, *Recurrent Neural Networks* (Wiley-IEEE Press, 2016)
37. Z. Shi, M. Han, Support vector echo-state machine for chaotic time-series prediction. IEEE Trans. Neural Netw. **18**(2), 359–372 (2007)
38. K.G. Vamvoudakis, F.L. Lewis, Online actor-critic algorithm to solve the continuous-time infinite horizon optimal control problem. Automatica **46**, 878–888 (2010)
39. J. Qin, M. Li, Y. Shi, Q. Ma, W. Zheng, Optimal synchronization control of multiagent systems with input saturation via off-policy reinforcement learning.IEEE Trans. Neural Netw. Learn. Syst. **30**(1), 85–96 (2019)
40. X. Yang, H. He, Adaptive critic learning and experience replay for decentralized event-triggered control of nonlinear interconnected systems. IEEE Trans. Syst. Man Cybern. Syst. **50**(11), 4043–4055 (2020)
41. H. Zhang, L. Cui, Y. Luo, Near-optimal control for nonzero-sum differential games of continuous-time nonlinear systems using single-network ADP. IEEE Trans. Cybern. **43**(1), 206–216 (2013)

42. Z. Yu, Y. Qu, Y. Zhang, Distributed fault-tolerant cooperative control for multi-UAVs under actuator fault and input saturation. IEEE Trans. Control Syst. Technol. **27**(6), 2417–2429 (2019)
43. K. Klausen, C. Meissen, T.I. Fossen, M. Arcak, T.A. Johansen, Cooperative control for multirotors transporting an unknown suspended load under environmental disturbances. IEEE Trans. Control Syst. Technol. **28**(2), 653–660 (2020)
44. S. Shao, M. Chen, Y. Zhang, Adaptive discrete-time flight control using disturbance observer and neural networks. IEEE Trans. Neural Netw. Learn. Syst. **30**(12), 3708–3721 (2019)
45. Y. Chen, R. Yu, Y. Zhang, C. Liu, Circular formation flight control for unmanned aerial vehicles with directed network and external disturbance. IEEE/CAA J. Autom. Sin. **7**(2), 505–516 (2020)

Chapter 10
Secure Control Under Asynchronous DoS

In addition to the safety issues related to UAV formation caused by complex strong disturbances mentioned in Chap. 9, the data transmission links among multiple UAVs in open communication networks are susceptible to potential malicious attacks. This leads to the interruptions of data transmission, resulting in a loss of stable formation control due to the cascading effects. In response to aperiodic denial-of-service (DoS) attacks, this chapter introduces a resilient fixed-time observer and proposes an attack-oriented fixed-time security consensus controller.

10.1 Introduction

In recent years, with the rapid development of wireless communication, distributed computation, and advanced control schemes, networked unmanned aerial vehicles (UAVs) have become indispensable and widely applied in various engineering and industrial operations due to the flexibility, scalability, and remote accessibility [1]. In the field of control for networked UAVs, considerable attention has been paid in state-of-the-art research results in terms of both theory and practice. Liao et al. [2] investigated the distributed formation and reconfiguration control of vertical takeoff and landing (VTOL) UAVs. For nonlinear multi-UAVs with unmodeled dynamics, Yu et al. [3] proposed a neural adaptive distributed formation control method. Du et al. [4] proposed a model predictive formation tracking-containment control strategy for multi-UAVs with obstacle avoidance ability. Considering clutter environment, Ze et al. [5] focused on the time-varying formation planning and distributed control for UAVs with time-varying formation size. Ma et al. [6] proposed an adaptive fault-tolerant formation control method for heterogeneous UAVs-UGVs systems with multiple actuator faults. Moreover, recently, a learning-based optimal cooperative formation tracking control strategy is proposed by Zhang et al. [7], using a novel feedforward-feedback design framework. Nevertheless, we

M. Lv et al., *Unmanned Aerial Vehicles Formation Control and Safety Guarantee*, Cognitive Intelligence and Robotics, https://doi.org/10.1007/978-981-96-5829-9_10

should take notice that the aforementioned state-of-the-art consensus control design methodologies of the networked UAVs are based on the secure communication environment with no cyberattacks.

However, the widespread adoption of networked UAVs has also made them vulnerable to various kinds of cyberattacks (e.g., DoS attack, deception attack, and replay attack), posing significant threats to system security and stability. Denial-of-service (DoS) attacks are particularly challenging because they can disrupt communication channels, preventing timely data exchange and normal communication between system components [8, 9]. To mitigate the threats from DoS attacks, security control strategies have become a critical research field, aiming to ensure the robust operation of distributed systems even under malicious attacks [10–12]. Recently, several state-of-the-art results have emerged, primarily focusing on various attack modes. Various effective security control protocols have been proposed in [13–18] to ensure the security consensus of distributed formation systems under DoS attacks. For instance, Ma et al. [14] investigated the resilient control of networked unmanned surface vehicles in face of aperiodic DoS attacks by developing a dynamic event-triggered mechanism. Considering the security of multiarea power systems under hybrid cyberattacks, Cheng et al. [15] and Zhang et al. [16] proposed the resilient distributed coordination control scheme and H_∞ load frequency control scheme, respectively. Although these contributions have significantly advanced the field, they have focused solely on synchronous attacks, where the attack time periods across communication channels are synchronized. Actually, DoS attacks often occur asynchronously, making them more difficult to detect and mitigate. Asynchronous DoS attacks, which target distributed formation systems at irregular intervals, pose a significant challenge to the resilience of these systems [19–23]. To list a few, Zhao et al. [21] designed an active security control strategy for disturbed switched systems under asynchronous DoS attacks. Zhang et al. [22] proposed an adaptive security practical regulation scheme of multi-agent systems with DoS attacks with asynchronous communication. To solve the consensus issue of networked systems under asynchronous DoS attack, Zhang et al. [23] developed a hybrid dynamic event-triggered control method. Despite the growing recognition of this threat, comprehensive studies on security of networked UAV systems facing asynchronous DoS attacks are still lacking. This leads to the first inspiration of our work: How to achieve the resilient security control of networked UAVs formation against the asynchronous DoS cyberattacks?

Moreover, the existing literature on resilient security control under DoS attacks has often neglected the impact of stable convergence speed of the system. The convergence speed, referring to the time required for the system to converge to the desired state trajectory, is a crucial metric for evaluating the consensus control performance. Most state-of-the-art results have only considered the exponential stability of the system without imposing constraints on convergence time

[24–27]. Zhao et al. [25] investigated the event-triggered control of distributed networked systems under DoS attacks with exponential convergence, and the practical applications based on an Internet-based three-tank system are conducted. Li et al. [27] proposed a stability analysis method for H_∞-controlled active quarter-vehicle suspension systems under periodic DoS attacks via a resilient event-triggered scheme. In recent years, numerous researches on time-constrained consensus control of distributed networked systems like networked UAVs have been conducted. For example, Ren et al. [28] proposed an event-triggered control scheme under asynchronous switching with finite-time convergence. Bai et al. [29] developed a neural adaptive fixed-time control strategy to achieve the semi-globally uniform stability of nonlinear multi-agent non-affine systems based on the mean value theorem. Recently, a novel specified-time affine formation maneuver control strategy of multi-agent systems over directed communication networks is proposed by Zhao et al. [30]. Although the time-constrained stability has been achieved in abovementioned researches, the issue of system control security was not taken into account. This leads to the second inspiration of our work: How to achieve the fast and stable convergence of networked UAV formation with existence of asynchronous DoS cyberattacks?

Inspired by the preceding discourse, this chapter delves into the fixed-time resilient security control design for networked UAVs in the face of asynchronous DoS attacks. The proposed security control strategy ensures system stability within a predetermined settling time, even amidst asynchronous DoS attack conditions. The key contributions of this work are outlined as follows:

- Contrasting with existing state observation methods [13–18], this chapter introduces an innovative resilient fixed-time observer, which is capable of precisely determining the leader's state information, even under the scenarios where connectivity is compromised due to DoS attacks.
- Divergent from the design methodologies presented in [25–27], this study introduces an observer-based fixed-time resilient security control protocol, which ensures rapidly stable convergence for networked UAV systems, even in the presence of asynchronous DoS attacks.
- By examining network resilience across various attack patterns, this study innovatively derives the boundary conditions necessary for the critical stability of networked UAV systems against connectivity-sustained/broken DoS attacks.

The remainder of this chapter is organized as follows: Sect. 10.2 presents the problem formulation and preliminaries. Sections 10.3.1 and 10.3.2 give the fixed-time resilient security control protocol under only connectivity-sustained DoS attacks and switching connectivity-sustained/broken DoS attacks, respectively. Simulation analyses are conducted in Sect. 10.4. Finally, the conclusion of research is given in Sect. 10.5.

10.2 Problem Formulation

10.2.1 UAV Dynamic Model

The dynamic model of each UAV among the network is formulated as

$$\begin{cases} \dot{x}_i = V_i \cos\theta_i \cos\phi_i \\ \dot{y}_i = V_i \cos\theta_i \sin\phi_i \\ \dot{z}_i = V_i \sin\theta_i \\ \dot{V}_i = (T_i - D_i)/m_i - g\sin\theta_i \\ \dot{\theta}_i = -\dfrac{g\cos\theta_i}{V_i} + \dfrac{a_{\theta i}}{V_i} \\ \dot{\phi}_i = \dfrac{a_{\phi i}}{V_i \cos\theta_i}, \quad i = 1, 2, \ldots, N, \end{cases} \tag{10.1}$$

where $p_i = [x_i, y_i, z_i]^{\top}$ and V_i are the position vector and velocity value, respectively. The control input signals $a_{Ti} = (T_i - D_i)/m_i$, $a_{\phi i} = L_i \sin\phi_i / m_i$, $a_{\theta i} = L_i \cos\phi_i / m_i$ with T_i, D_i, L_i are thrust, drag force, and lift, and θ_i and ϕ_i are the flight path angle and heading angle. Based on the above discussions, the velocity vector $v_i = [v_{xi}, v_{yi}, v_{zi}]^{\mathrm{T}}$ with $v_{xi} = V_i \cos\theta_i \cos\phi_i$, $v_{yi} = V_i \cos\theta_i \sin\phi_i$, and $v_{zi} = V_i \sin\theta_i$.

Then, dynamics (10.1) can be further expressed in the following second-order form:

$$\begin{cases} \dot{p}_i(t) = v_i(t) \\ \dot{v}_i(t) = R_i u_i(t) + G_i \quad i = 1, 2, \ldots, N, \end{cases} \tag{10.2}$$

where $R_i = \begin{bmatrix} \cos\phi_i \cos\theta_i & -\sin\phi_i & -\sin\theta_i \cos\phi_i \\ \sin\phi_i \cos\theta_i & \cos\phi_i & -\sin\phi_i \sin\theta_i \\ \sin\theta_i & 0 & \cos\theta_i \end{bmatrix}$, $G_i = [0, 0, -g]^{\mathrm{T}}$, and the input vector $u_i = [a_{Ti}, a_{\phi i}, a_{\theta i}]^{\mathrm{T}}$.

10.2.2 Attack Graph Theory

The evolution of the network topology under DoS attack is denoted by a switching signal $\sigma(t)$: $[0, +\infty) \to \mathcal{P} = \{1, 2, \ldots, M\}$, where M is the number of possible topologies. The communication graph $\mathcal{G}_{\sigma(t)} = (\mathcal{V}, \mathcal{E}_{\sigma(t)}, \mathcal{A}_{\sigma(t)})$, with $\mathcal{V}$ and $\mathcal{E}_{\sigma(t)}$ being the sets of nodes and edges, respectively. $(\kappa_i, \kappa_j) \in \mathcal{E}_{\sigma(t)}$ holds if and only if node κ_i is accessible to node κ_j under attack topology $\sigma(t)$. The adjacency matrix $\mathcal{A}_{\sigma(t)} = [a_{ij}^{\sigma(t)}] \in \mathbb{R}^{N\times N}$, with $a_{ij}^{\sigma(t)} = 1$ if edge (κ_i, κ_j) is not attacked; otherwise $a_{ij}^{\sigma(t)} = 0$. The degree matrix $\mathcal{D}_{\sigma(t)} = \text{diag}\{d_1^{\sigma(t)}, \ldots, d_N^{\sigma(t)}\}$, with $d_i^{\sigma(t)} = \sum_{j=1}^{N} a_{ij}^{\sigma(t)}$. Then, the Laplacian matrix is given as $\mathcal{L}_{\sigma(t)} = \mathcal{D}_{\sigma(t)} - \mathcal{A}_{\sigma(t)}$. Besides, the communication relationship between followers and the leader can be described by $\mathcal{B}_{\sigma(t)} = \text{diag}\{b_1^{\sigma(t)}, \ldots, b_N^{\sigma(t)}\}$, where $b_i^{\sigma(t)} = 1$ if edge (κ_0, κ_i) is not attacked, and $b_i^{\sigma(t)} = 0$, otherwise. Moreover, the information exchange matrix is calculated as $\bar{\mathcal{L}}_{\sigma(t)} = \mathcal{L}_{\sigma(t)} + \mathcal{B}_{\sigma(t)}$.

10.2.3 Mathematical Preliminaries

Lemma 10.1 ([24]) *For a connected graph $\mathcal{G}_r$, $r \in \mathcal{P}$, there exists a diagonal matrix $\Psi_r = \text{diag}\{\psi_{r1}, \psi_{r2}, \ldots, \psi_{rN}\}$, $\psi_{ri} > 0$ $(i \in \mathbb{N}_{1:N})$ such that $\Psi_r \mathcal{L}_r + (\mathcal{L}_r)^\top \Psi_r \succeq \omega_r I_N$, with ω_r a positive constant.*

Lemma 10.2 ([26]) *For a nonlinear system $\dot{x}(t) = f(t, x)$, with $x \in \mathbb{R}^n$, if there exists a radially unbounded continuous function $V(x,t)$ such that $\dot{V}(x,t) \leq -\zeta_1 V^\alpha(x,t) - \zeta_2 V^\beta(x,t) + \delta$, where ζ_1, ζ_2, δ are positive constants, and $0 < \alpha < 1$, $\beta > 1$, the origin of the system is semi-global practical fixed-time stable (SGPFTS). The settling time is bounded by $T \leq T_{\max} = [1/(\zeta_1\delta(1-\alpha))] + [1/(\zeta_2\delta(\beta-1))]$. Specifically, if $\dot{V}(x,t) \leq -\zeta_1 V^\alpha(x,t) - \zeta_2 V^\beta(x,t)$, the origin of the system is globally fixed-time stable (GFTS). The settling time is bounded by $T \leq T_{\max} = [1/(\zeta_1(1-\alpha))] + [1/(\zeta_2(\beta-1))]$.*

10.2.4 Modeling of Asynchronous DoS Attacks

Definition 10.1 (Connectivity-Sustained DoS Attack) The network connectivity is maintained, and the network topology $\mathcal{G}_S$ contains at least one directed spanning tree (CS-DoS).

Definition 10.2 (Connectivity-Broken DoS Attacks) The network connectivity is broken with no directed spanning tree among the network topology $\mathcal{G}_U$ (CB-DoS).

Define the set of possible topologies under asynchronous DoS attacks as $\overline{\mathcal{G}} = \{\mathcal{G}_1, \ldots, \mathcal{G}_M\}$; then we have $\overline{\mathcal{G}} = \overline{\mathcal{G}}_S \cup \overline{\mathcal{G}}_U$ and $\mathcal{P} = \mathcal{P}_S \cup \mathcal{P}_U$. Besides, let $\mathcal{K}(\mathcal{P})$ be the set of combinations (i, j) and that there exists the switching graph from $\mathcal{G}_i$ to $\mathcal{G}_j$ $(i, j \in \mathcal{P})$. $M_\sigma(0, t)$ denotes the number of switches during $(0, t]$, and $0 < \tau_0 < \tau_1 < \ldots < \tau_{M_\sigma(0,t)}$ are the switching instants. The possible network topologies under different DoS attack modes are depicted in Fig. 10.1.

For a time interval $(t_1, t_2]$, the activation period of a specific attack topology $\mathcal{G}_j (j \in \mathcal{P})$ is given by $\Xi_j(t_1, t_2) = (t_1, t_2] \cap (\cup_{\sigma(\tau_i)=j}(\tau_i, \tau_{i+1}])$. Besides, the

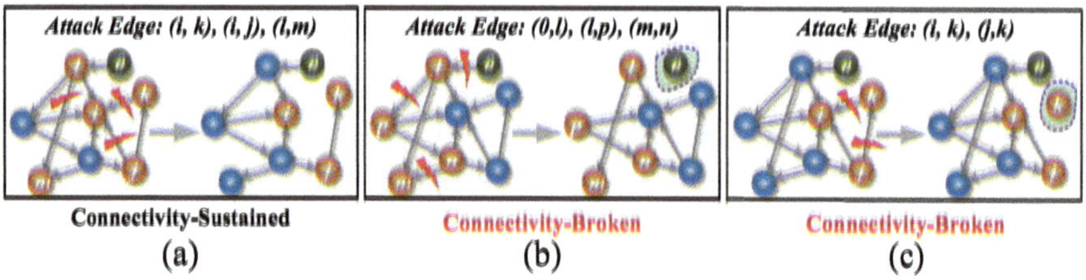

Fig. 10.1 Schematic of possible network topologies under different DoS attack modes: (**a**) CS-DoS attack. (**b**) CB-DoS attack with isolated leader κ_0. (**c**) CB-DoS attack with isolated follower agent κ_k

DoS attack strength is usually limited, which can be described by the following assumptions:

Assumption 10.1 (DoS Attack Duration) There exist constants $\bar{\varXi}_{j[S]}$, $\bar{\varXi}_{j[U]}$ and functions $\rho_{j[S]}(t_0, t)$, $\rho_{j[U]}(t_0, t)$, such that the duration of CS-DoS attack (including initial topology $\mathcal{G}_0$) and CB-DoS attack satisfies

$$\varXi_j\left(t_0, t\right) \leq \bar{\varXi}_{j[U]} + \rho_{j[U]}\left(t_0, t\right), \ j \in \mathcal{P}_U \tag{10.3}$$

$$\varXi_k\left(t_0, t\right) \geq \bar{\varXi}_{k[S]} + \rho_{k[S]}\left(t_0, t\right), \ k \in \mathcal{P}_S. \tag{10.4}$$

Assumption 10.2 (DoS Attack Frequency) There exist constants $\bar{M}_{ij[U]}$, $\bar{M}_{ij[S]}$ and functions $\rho_{ij[U]}$, $\rho_{ij[S]}$, such that the mode switching frequency under DoS attacks is bounded by

$$M_{ij}\left(t_0, t\right) \leq \bar{M}_{ij[U]} + \rho_{ij[U]}\left(t_0, t\right), \ (i, j) \in \mathcal{K}(\mathcal{P}_U) \tag{10.5}$$

$$M_{ij}\left(t_0, t\right) \geq \bar{M}_{ij[S]} + \rho_{ij[S]}\left(t_0, t\right), \ (i, j) \in \mathcal{K}(\mathcal{P}_S). \tag{10.6}$$

10.2.5 Attack and Defense Motivations

The DoS attacker aims to disrupt the normal communication among agents by compromising network connectivity. Conversely, the defender's objective is to sustain acceptable consensus control performance despite asynchronous DoS attacks. This is achieved by devising effective defensive strategies. The comprehensive security control design framework of this study is illustrated in Fig. 10.2.

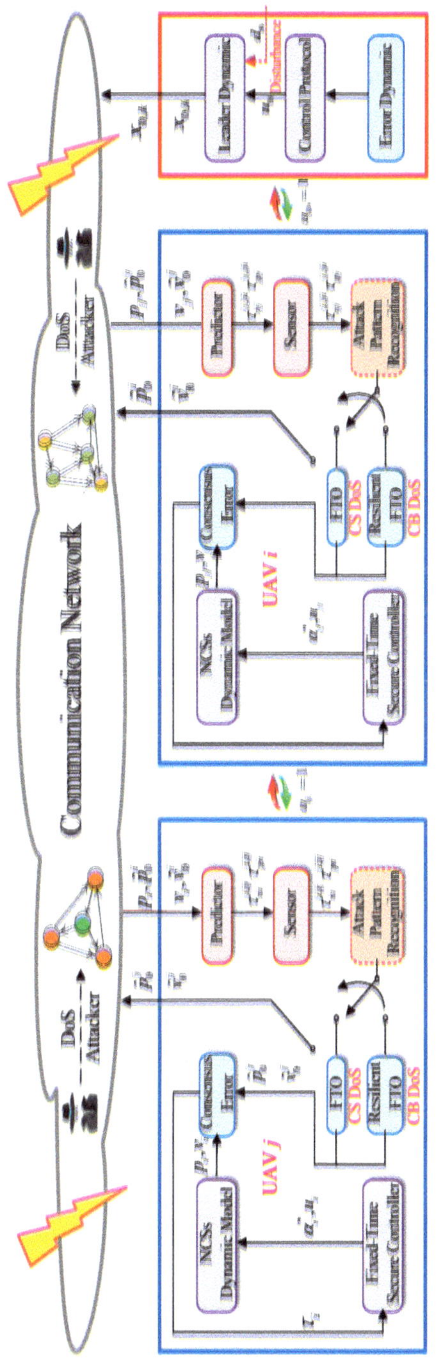

Fig. 10.2 Design framework of fixed-time resilient security control protocol under asynchronous DoS attacks

10.3 Main Results

10.3.1 Fixed-Time Security Consensus Control Under Only Connectivity-Sustained DoS Attacks

In this section, we focus on a scenario involving solely connectivity-sustained DoS attacks, where the attack topology $\mathcal{G}_\sigma(t)$ consistently contains a directed spanning tree, with the leader that is the root across the entire time horizon.

10.3.1.1 Design of Fixed-Time Observer

The main purpose is to design an observer $\hat{p}_0^i$ for follower κ_i, such that $\hat{p}_0^i \to p_0$, $\hat{v}_0^i \to v_0$ when $t \to T_1$, where $\hat{p}_0^i$ is the estimate of p_0. Define the observation error $\tilde{p}_0^i = \hat{p}_0^i - p_0$, $\tilde{v}_0^i = \hat{v}_0^i - v_0$ and the distributed observation error under DoS attack topology $\sigma(t)$ as $e_{pi}^{\sigma(t)} = \sum_{j=1}^N a_{ij}^{\sigma(t)}(\hat{p}_0^j - \hat{p}_0^i) + b_i^{\sigma(t)}(p_0 - \hat{p}_0^i)$, $e_{vi}^{\sigma(t)} = \sum_{j=1}^N a_{ij}^{\sigma(t)}(\hat{v}_0^j - \hat{v}_0^i) + b_i^{\sigma(t)}(v_0 - \hat{v}_0^i)$. Then, the fixed-time observer is designed as

$$\begin{cases} \dot{\hat{p}}_0^i = \hat{v}_0^i + \phi_{pi}(e_{pi}^{\sigma(t)}) \\ \dot{\hat{v}}_0^i = \phi_{vi}(e_{vi}^{\sigma(t)}) + G_i, \end{cases} \tag{10.7}$$

where $\phi_{pi}(\cdot) = \theta_{p1}\lceil e_{pi}^{\sigma(t)}\rfloor^\alpha + \theta_{p2}\lceil e_{pi}^{\sigma(t)}\rfloor^\beta$, and $\phi_{vi}(\cdot)(\cdot) = \theta_{v1}\lceil e_{vi}^{\sigma(t)}\rfloor^\alpha + \theta_{v2}\lceil e_{vi}^{\sigma(t)}\rfloor^\beta + \theta_{v3}\mathrm{sig}(e_{vi}^{\sigma(t)})$, with $0 < \alpha < 1$, $\beta > 1$. $\theta_{p1}, \theta_{p2}, \theta_{v3}$ are positive design parameters for the observer which will be determined later.

Theorem 10.1 *For the networked UAV systems (10.2), if the parameters* $\min\{\theta_{p1}, \theta_{v1}\} > ((\Phi + 1)2^{\alpha+1/2} + \mu)/\Gamma(N)^\alpha 2^{\alpha+1/2}$, $\min\{\theta_{p2}, \theta_{v2}\} > ((\Phi + 1)2^{\beta+1/2}+\mu)/\Gamma(N)^\beta (Nn)^{1-\beta/2}2^{\beta+1/2}$, $\theta_{v3} > N(2N-1)(\bar{u}_0 + \bar{d}_0)/\Gamma(N)$ *with* $\mu > 0$*, then observer (10.7) drives the observation errors* $\tilde{p}_0^i$ *and* $\tilde{v}_0^i$ *which converge to zero within a fixed settling time* T_1*, with the settling time given by*

$$T_1 \le T_{1\max} := 2/\mu\,(1-\alpha) + 2/\mu\,(\beta - 1). \tag{10.8}$$

Proof Choose Lyapunov function $V_{0\sigma(t)} = \frac{1}{2}(\tilde{p}_0^\top \tilde{p}_0 + \tilde{v}_0^\top \tilde{v}_0)$, $\sigma(t) \in \mathcal{P}$. Combining (10.1) and (10.7), one has

$$\begin{aligned} \dot{V}_{0\sigma(t)} = {} & \tilde{p}_0^\top \Big(\tilde{v}_0 - \theta_{p1}\left\lceil \bar{\mathcal{L}}_{\sigma(t)}\tilde{p}_0\right\rfloor^\alpha - \theta_{p2}\left\lceil \bar{\mathcal{L}}_{\sigma(t)}\tilde{p}_0\right\rfloor^\beta \\ & + \tilde{v}_0^\top\big(-\theta_{v1}\left\lceil \bar{\mathcal{L}}_{\sigma(t)}\tilde{v}_0\right\rfloor^\alpha - \theta_{v2}\left\lceil \bar{\mathcal{L}}_{\sigma(t)}\tilde{v}_0\right\rfloor^\beta \\ & -\theta_{v3}\mathrm{sig}\big(\bar{\mathcal{L}}_{\sigma(t)}\tilde{v}_0\big) + G - \mathbf{1}_N(u_0 + d_0)\big), \end{aligned} \tag{10.9}$$

where $\tilde{p}_0$, $\tilde{v}_0$, G are augmented forms of $\tilde{p}_0^i$, $\tilde{v}_0^i$, G_i, respectively.

According to Lemma 10.2 and the basic inequality, one has

$$\tilde{p}_0^\top \tilde{v}_0 \leq \left(2V_{0\sigma(t)}\right)^{\frac{\alpha_v+1}{2}} + \left(2V_{0\sigma(t)}\right)^{\frac{\beta_v+1}{2}} \tag{10.10}$$

$$\tilde{v}_0^\top G \leq \Phi((2V_{0\sigma(t)})^{\frac{\alpha_v+1}{2}} + (2V_{0\sigma(t)})^{\frac{\beta_v+1}{2}}), \tag{10.11}$$

where Φ is a positive constant.

The maximum and minimum singular values of Laplacian matrix are global time-varying due to the switching of DoS attack patterns. According to [31], a lower bound of the minimum singular value is given by $\lambda_{\min}(\bar{\mathcal{L}}_{\sigma(t)}) \geq ((N-1)/N)^{(N-1)/2}/(N^2+N-1) \triangleq \Gamma(N)$. Moreover, according to the Gershgorin theorem [25], all the singular values of $\bar{\mathcal{L}}_{\sigma(t)}$ satisfy $|\lambda| \leq |l_{i,i}| + \sum_{j=1,j\neq i}^N |l_{i,j}| \leq 2N-1$, which gives an upper bound of the maximum singular value of $\bar{\mathcal{L}}_{\sigma(t)}$, i.e., $\lambda_{\max}(\bar{\mathcal{L}}_{\sigma(t)}) \leq 2N-1$. Thus, we have

$$-\sum_{k=1}^{n} \tilde{p}_0^\top \tilde{\theta}_{p1} \lceil \bar{\mathcal{L}}_{\sigma(t)} \tilde{p}_0 \rfloor^{\alpha} \leq -\tilde{\theta}_{1,\min} \Gamma(N)^{\alpha} \left(2V_{0\sigma(t)}\right)^{\frac{\alpha+1}{2}} \tag{10.12}$$

$$-\sum_{k=1}^{n} \tilde{p}_0^\top \tilde{\theta}_{p2} \lceil \bar{\mathcal{L}}_{\sigma(t)} \tilde{p}_0 \rfloor^{\beta} \leq -\tilde{\theta}_{2,\min} \tilde{\Gamma}(N)^{\beta} \left(2V_{0\sigma(t)}\right)^{\frac{\beta+1}{2}} \tag{10.13}$$

$$\begin{aligned} &\tilde{v}_0^\top \left(-\theta_{v3}\mathrm{sig}(\bar{\mathcal{L}}_{\sigma(t)}\tilde{v}_0) - \mathbf{1}_N \left(u_0(t) + d_0(t)\right)\right) \\ &\leq -\left(\theta_{v3}\frac{\Gamma(N)}{2N-1} - N\left(\bar{u}_0 + \bar{d}_0\right)\right) \|\tilde{v}_0\|, \end{aligned} \tag{10.14}$$

where $\tilde{\Gamma}(N) = \Gamma(N)(Nn)^{(1-\beta)/2}$.

Substituting (10.10)–(10.14) into (10.9) yields

$$\begin{aligned} \dot{V}_{0\sigma(t)} \leq &-\left(\theta_{1,\min}\Gamma(N)^{\alpha}2^{\frac{\alpha+1}{2}} - (\Phi+1)\,2^{\frac{\alpha+1}{2}}\right) V_{0\sigma(t)}^{\frac{\alpha+1}{2}} \\ &-\left(\theta_{2,\min}\Gamma(N)^{\beta}(Nn)^{\frac{1-\beta}{2}}2^{\frac{\beta+1}{2}} - (\Phi+1)\,2^{\frac{\beta+1}{2}}\right) \\ &\times V_{0\sigma(t)}^{\frac{\beta+1}{2}} - \left(\theta_{v3}\frac{\Gamma(N)}{2N-1} - N\left(\bar{u}_0+\bar{d}_0\right)\right)\|\tilde{v}_0\|. \end{aligned} \tag{10.15}$$

Define $\min\{\theta_{p1}, \theta_{v1}\} > ((\Phi+1)2^{\alpha+1/2} + \mu)/\Gamma(N)^{\alpha}2^{\alpha+1/2}$, $\min\{\theta_{p2}, \theta_{v2}\} > ((\Phi+1)2^{\beta+1/2} + \mu)/\Gamma(N)^{\beta}(Nn)^{1-\beta/2}2^{\beta+1/2}$, and $\theta_{v3} > N(2N-1)$ $(\bar{u}_0 + \bar{d}_0)/\Gamma(N)$ with $\mu > 0$ be observer gains; then inequality (10.15) is further formulated as

$$\dot{V}_{0\sigma(t)} \leq -\mu V_{0\sigma(t)}^{\frac{\alpha+1}{2}} - \mu V_{0\sigma(t)}^{\frac{\beta+1}{2}}, \tag{10.16}$$

which implies that the observation error system is globally fixed-time stable under the asynchronous CS-DoS attacks with the settling time bounded by $T_1 \leq T_{1\max}$. This completes the proof of Theorem 10.1.

10.3.1.2 Design of Fixed-Time Security Control Protocol

Define the tracking error $\tau_l = [\tau_{pl}, \tau_{vl}]^{\top}$, with $\tau_{pl} = p_l - \hat{p}_0^l$, $\tau_{vl} = v_l - \hat{v}_0^l$. Then, the following error dynamic is obtained:

$$\begin{cases} \dot{\tau}_{pl} = \tau_{vl} + \Delta_{pl}, \\ \dot{\tau}_{vl} = u_l + d_l + \Delta_{vl}, \end{cases} \tag{10.17}$$

where $\Delta_{pl} = -\theta_{p1}\lceil e_{pl}^{\sigma(t)} \rfloor^{\alpha} - \theta_{p2}\lceil e_{pl}^{\sigma(t)} \rfloor^{\beta}$ and $\Delta_{vl} = -\theta_{v1}\lceil e_{vl}^{\sigma(t)} \rfloor^{\alpha} - \theta_{v2}\lceil e_{vl}^{\sigma(t)} \rfloor^{\beta} - \theta_{v3}\text{sig}(e_{vl}^{\sigma(t)})$. Before continuing, we make the following coordinate transformation for the ith follower ($i \in \mathbb{N}_{1:N}$): $\bar{\tau}_l = (\tau_l)^{\frac{1}{\mu}} - (\alpha_l^*)^{\frac{1}{\mu}}$, where α_l^* is the virtual control signal to be designed later, and $\mu_1 = 1 > \cdots > \mu = (4n + 3 - 2k)/(4n + 1) > 1/2$.

The security control protocol against only CS-DoS attacks is presented in Table 10.1, with the control parameters given by $\wp_{l,1} = \xi_l + \rho_l + \rho_l + \bar{\beta}_l + 2\tilde{\Phi} + \frac{\hbar}{\kappa}$, and $\wp_{l2} = \xi_l + 2\tilde{\Phi} - \epsilon_l$, with $\rho_l = 2^{1-\mu}\frac{2-\mu_{n-1}}{\kappa}$, $\varpi = \frac{2-\mu}{3-\mu}(\frac{1}{3-\mu})^{\frac{1}{2-\mu}}$, and $\xi_l = 2^{\frac{\kappa+\delta}{2}}(Nn)^{\frac{\kappa+\delta-2}{2}}\xi$, with ξ a positive design constant.

Theorem 10.2 *For the networked UAV systems (10.1), execute the fixed-time observer (10.7) with gains θ_{p1}, θ_{p2}, θ_{v1}, θ_{v2}, and θ_{v3} given in Theorem 10.1 and fixed-time security control protocol given in Table 10.1. Then the networked UAV systems (10.1) is semi-global, practical, and fixed-time stable in spite of*

Table 10.1 Control protocol against only CS-DoS attacks

Virtual control for Step k ($2 \leq k \leq n-1$) under CS-DoS attacks
$\alpha_l^* = -\varpi^2 \frac{\bar{\tau}_{pl}^{4-\mu}}{\sqrt{\bar{\tau}_{pl}^{6-2\mu} + \varepsilon_{l,1}}} - \epsilon_l \bar{\tau}_{pl}^{\kappa+\delta+\mu-2}\tanh\left(\frac{\bar{\tau}_{pl}^{\kappa+\delta}}{\varepsilon_{l,2}}\right) - \wp_{l,1}\bar{\tau}_{pl}^{\mu} - \wp_{l,2}\bar{\tau}_{pl}^{\mu+\delta}.$
Security control protocol for Step under CS-DoS attacks
$u_l = -\varpi^2 \frac{\bar{\tau}_{vl}^{4-\mu}}{\sqrt{\bar{\tau}_{vl}^{6-2\mu} + \varepsilon_{l,1}}} - \epsilon_l \bar{\tau}_{vl}^{\kappa+\delta+\mu-2}\tanh\left(\frac{\bar{\tau}_{vl}^{\kappa+\delta}}{\varepsilon_{l,2}}\right) - \wp_{l,1}\bar{\tau}_{vl}^{\mu} - \wp_{l,2}\bar{\tau}_{vl}^{\mu+\delta} + \dot{\hat{p}}_0^l - \theta_{v3}\text{sig}(\bar{\tau}_{pl}).$

asynchronous CS-DoS attacks, with the settling time bounded by $T := T_1 + T_2$, *where* $T_2 \leq T_{2\max} := 2/\tilde{\xi}\phi\,(2-\kappa) + 2/\tilde{\xi}\phi\,(\kappa+\delta-2)$.

Proof The proof process of Theorem 10.2 is conducted based on the backstepping design procedure.

Step 1: In this step, the integral-type Lyapuonv candidate function $V_1 = \sum_{i=1}^{N} Q_l$ is chosen, where the nonnegative C^1 function Q_l is

$$Q_{pl} = \int_{\alpha_l^*}^{\tau_{pl}} \left[\theta^{\frac{1}{\mu}} - (\alpha_l^*)^{\frac{1}{\mu}}\right]^{2-\mu} d\theta, \tag{10.18}$$

which satisfies $\frac{\partial Q_{pl}}{\partial \alpha_l^*} = \int_{\alpha_l^*}^{\tau_{vl}} [\theta^{\frac{1}{\mu}} - (\alpha_l^*)^{\frac{1}{\mu}}]^{1-\mu} d\theta (\mu-2) \frac{d\alpha_l^{*1/\mu}}{d\alpha_l^*}$, $\frac{\partial Q_{pl}}{\partial \tau_{vl}} = \bar{\tau}_{pl}^{2-\mu}$, $Q_{pl} \leq 2\bar{\tau}_{pl}^2$.

Using Lemma 10.1 and Lemma 10.2, and $(|\bar{n}| + |\bar{m}|)^{\frac{1}{\gamma}} \leq |\bar{n}|^{\frac{1}{\gamma}} + |\bar{m}|^{\frac{1}{\gamma}}$, $|\bar{n}^{\frac{1}{\gamma}} - \bar{m}^{\frac{1}{\gamma}}| \leq 2^{1-\frac{1}{\gamma}} |\bar{n} - \bar{m}|^{\frac{1}{\gamma}}$, with $\gamma \geq 1$ is a constant, $\bar{m} \in \mathbb{R}$, and $\bar{n} \in \mathbb{R}$, we have

$$\begin{aligned}\bar{\tau}_{pl}^{2-\mu} \dot{\tau}_{pl} \leq\ & \bar{\tau}_{pl}^{2-\mu} \alpha_{l+1}^* + \rho_{vl}\left(\bar{\tau}_{pl}^{\kappa} + \bar{\tau}_{vl}^{\kappa}\right) + \bar{\beta}_{vl} \bar{\tau}_{pl}^{\kappa} \\ & + \tilde{\Phi}\left(\bar{\tau}_{pl}^{\kappa} + \bar{\tau}_{pl}^{\kappa+\delta}\right) + \varpi_k \left|\bar{\tau}_{pl}\right|^{3-\mu} \\ & + 2\left(\bar{\tau}_{pl}^{\kappa} + \bar{\tau}_{pl}^{\kappa+\delta}\right) + |\Delta_l|^{3-\mu},\end{aligned} \tag{10.19}$$

where $\tilde{\Phi} = \frac{2-\mu}{2-\mu+\mu_r}(\frac{\mu_r}{2-\mu+\mu_r})^{\frac{\mu_r}{2-\mu}}(2^{1-\mu_r}\Phi)^{\frac{2-\mu+\mu_r}{2-\mu}}$, and $\rho_{vl} = 2^{1-\mu}\frac{2-\mu}{\kappa}$, $\varpi = \frac{2-\mu}{3-\mu}(\frac{1}{3-\mu})^{\frac{1}{2-\mu}}$.

According to the expression of Q_l, it holds that

$$(\mu-2)\int_{\alpha_l^*}^{\tau_{pl}} \left[\theta^{\frac{1}{\mu}} - (\alpha_l^*)^{\frac{1}{\mu}}\right]^{1-\mu} d\theta \leq \hbar \bar{\tau}_{pl}, \tag{10.20}$$

where $\hbar = (2-\mu)2^{1-\mu}$. Besides, noting $d(\alpha_l^*)^{1/\mu}/dt$ is a continuous function, and assuming $|d(\alpha_l^*)^{1/\mu}/dt| \leqq \Lambda_{vl}$ with $\Lambda_{vl} > 0$, which leads to

$$\begin{aligned}&(\mu-2)\frac{d\left(\alpha_l^*\right)^{\frac{1}{\mu}}}{dt}\int_{\alpha_l^*}^{\tau_{vl}} \left[\theta^{\frac{1}{\mu}} - (\alpha_l^*)^{\frac{1}{\mu}}\right]^{1-\mu} d\theta \\ &\leqq \frac{\hbar \bar{\tau}_l^{\kappa}}{\kappa} + \frac{\hbar \Lambda_{vl}^{\kappa/(\kappa-1)}(\kappa-1)}{\kappa}.\end{aligned} \tag{10.21}$$

Using the inequalities $0 \leq |z| - \frac{z^2}{\sqrt{z^2+\varepsilon}} \leq \sqrt{\varepsilon}$, with $\varepsilon > 0$, , $z \in \mathbb{R}$, and $0 \leq |\vartheta| - \vartheta \tanh\left(\frac{\vartheta}{\varepsilon}\right) \leq 0.2785\varepsilon$, with $\varepsilon > 0$, $\vartheta \in \mathbb{R}$, the following results are derived:

$$-\bar{\tau}_{pi}^{2-\mu} \frac{\varpi^2 \bar{\tau}_{pi}^{4-\mu}}{\sqrt{\bar{\tau}_{pi}^{6-2\mu} + \varepsilon_{i,1}}} \leq -\varpi \left|\bar{\tau}_{pi}\right|^{3-\mu} + \sqrt{\varepsilon_{i,1}} \tag{10.22}$$

$$-\epsilon_i \bar{\tau}_{pi}^{\kappa+\delta} \tanh\left(\frac{\bar{\tau}_{pi}^{\kappa+\delta}}{\varepsilon_{i,2}}\right) \leq -\epsilon_i \bar{\tau}_{pi}^{\kappa+\delta} + 0.2785\tilde{\varepsilon}_{i,2}, \tag{10.23}$$

where $\tilde{\varepsilon}_{i,2} = \epsilon_i \varepsilon_{i,2}$ is a positive constant.

Combining (10.19)–(10.23) and substituting the virtual control signal α_i^* into V_{k-1}, we have

$$\begin{aligned}\dot{V} \leq &-\sum_{i=1}^{N}\sum_{j=p}\left[\xi_{i,j}+2(n-k)\right]\left(\bar{\tau}_{i,j}^{\kappa}+\bar{\tau}_{i,j}^{\kappa+\delta}\right)+\sum_{i=1}^{N}\sum_{j=p}\sqrt{\varepsilon_{i,j,1}} \\ &+\sum_{i=1}^{N}\sum_{j=p}\Delta_{i,j}^{3-\mu_j}+\sum_{i=1}^{N}\sum_{j=p}0.2785\tilde{\varepsilon}_{i,j,2}+\sum_{i=1}^{N}\rho_i\bar{\tau}_{vi}^{\kappa} \\ &+\sum_{i=1}^{N}\sum_{j=p}\frac{\hbar_j \Lambda_{i,j}^{\kappa/(\kappa-1)}(\kappa-1)}{\kappa}.\end{aligned} \tag{10.24}$$

Step 2: Consider the integral-type Lyapunov candidate function $V_n = V_{n-1} + \sum_{i=1}^{N} Q_i$, where the nonnegative C^1 function Q_i is defined as

$$Q_{vi} = \int_{\alpha_{i,n}^*}^{\tau_{vi}} \left[\theta^{\frac{1}{\mu}} - \left(\alpha_i^*\right)^{\frac{1}{\mu}}\right]^{2-\mu} d\theta. \tag{10.25}$$

Then, along the similar derivation process of Step 1, one can obtain

$$\begin{aligned}\dot{V}_n \leq &-\sum_{i=1}^{N}\sum_{j=p,v}\xi_{i,j}\left(\bar{\tau}_{i,j}^{\kappa}+\bar{\tau}_{i,j}^{\kappa+\delta}\right)+\sum_{i=1}^{N}\sum_{j=p,v}\sqrt{\varepsilon_{i,j,1}} \\ &+\sum_{i=1}^{N}\sum_{j=p,v}\Delta_{i,j}^{3-\mu_j}+\sum_{i=1}^{N}\sum_{j=p,v}0.2785\tilde{\varepsilon}_{i,j,2} \\ &+\sum_{i=1}^{N}\sum_{j=v}\frac{\hbar_j \Lambda_{i,j}^{\kappa/(\kappa-1)}(\kappa-1)}{\kappa}.\end{aligned} \tag{10.26}$$

Now, consider the total Lyapunov energy function for the networked UAVs $\bar{V} = V_1 + V_2$, and one has

$$\begin{aligned}\dot{\bar{V}} \leq & -2^{\frac{\kappa+\delta}{2}}(Nn)^{\frac{\kappa+\delta-2}{2}}\xi\sum_{i=1}^{N}\sum_{j=p,v}\bar{\tau}_{i,j}^{\kappa}+\sum_{i=1}^{N}\sum_{j=p,v}\Delta_{i,j}^{3-\mu_j} \\ & -2^{\frac{\kappa+\delta}{2}}(Nn)^{\frac{\kappa+\delta-2}{2}}\xi\sum_{i=1}^{N}\sum_{j=p,v}\bar{\tau}_{i,j}^{\kappa+\delta}+\tilde{\aleph}_{i,j},\end{aligned} \quad (10.27)$$

where $\aleph_{i,j}$ is a positive constant, and $\aleph_{i,j} = \sum_{i=1}^{N}\sum_{j=v}\frac{\hbar_j\Lambda_{i,j}^{\kappa/(\kappa-1)}(\kappa-1)}{\kappa} + \sum_{i=1}^{N}\sum_{j=p,v}\sqrt{\varepsilon_{i,j,1}} + 0.2785\sum_{i=1}^{N}\sum_{j=p,v}\tilde{\varepsilon}_{i,j,2}$.

According to Lemma 10.2 and the fact $Q_{vi} \leq 2\bar{\tau}_{vi}^2$, it can be deduced that

$$\sum_{i=1}^{N}\sum_{j=p,v}\left(\bar{\tau}_{pi}^2\right)^{\frac{\kappa}{2}} \geq \left(\frac{\bar{V}}{2}\right)^{\frac{\kappa}{2}} \quad (10.28)$$

$$\sum_{i=1}^{N}\sum_{j=p,v}\left(\bar{\tau}_{pi}^2\right)^{\frac{\kappa+\delta}{2}} \geq (Nn)^{\frac{2-\kappa-\delta}{2}}\left(\frac{\bar{V}}{2}\right)^{\frac{\kappa+\delta}{2}}. \quad (10.29)$$

Substituting (10.28) and (10.29) into (10.27), we have

$$\dot{\bar{V}} \leq -\tilde{\xi}\bar{V}^{\frac{\kappa}{2}} - \tilde{\xi}\bar{V}^{\frac{\kappa+\delta}{2}} + \sum_{i=1}^{N}\sum_{j=p,v}\Delta_{i,j}^{3-\mu_j} + \sum_{j=p,v}\aleph_{i,j}, \quad (10.30)$$

where $\tilde{\xi} = 2^{\frac{\kappa+\delta}{2}}(Nn)^{\frac{\kappa+\delta-2}{2}}\xi$ is a positive constant.

According to Theorem 10.1, the observation errors for all followers reach origin when $t \to T_1$; thus

$$\begin{aligned}\lim_{t\to T_1}\Delta_{i,j} = & -\lim_{t\to T_1}\tilde{\theta}_{p1}\left\lceil\bar{\mathcal{L}}_{\sigma(t)}\tilde{p}_0\right\rfloor^{\alpha_v} \\ & -\lim_{t\to T_1}\tilde{\theta}_{p2}\left\lceil\bar{\mathcal{L}}_{\sigma(t)}\tilde{p}_0\right\rfloor^{\beta_v} = 0,\end{aligned} \quad (10.31)$$

which implies that $\sum_{i=1}^{N}\sum_{j=1}^{n}\Delta_{i,j}^{3-\mu_j} = 0$, when $t \geq T_1$.

Combining (10.30) and (10.31), one can derive that for $t \in [T_1, +\infty)$, (10.28) further becomes

$$\dot{\bar{V}} \leq -\tilde{\xi}\bar{V}^{\frac{\kappa}{2}} - \tilde{\xi}\bar{V}^{\frac{\kappa+\delta}{2}} + \tilde{\aleph}. \quad (10.32)$$

By using Lemma 10.2, we can determine that (10.17) is semi-global practical fixed-time stable (SGPFTS) in the sense of (10.32), and the settling time is given by $T_{2\max} := 2/\tilde{\xi}\phi\,(2-\kappa) + 2/\tilde{\xi}\phi\,(\kappa+\delta-2)$. From the above discussions, it can be summarized that the networked UAV systems (1) is fixed-time stable when $t \geq T_{\max} = T_1 + T_2$. Specifically, the limiting behavior of the error τ_l as $t \to T_{\max}$ is given by $\lim_{t \to T_{\max} := T_1 + T_2} \tau_l = \varepsilon$, where ε is a small enough constant. This completes the proof of Theorem 10.2.

10.3.2 *Fixed-Time Resilient Security Consensus Control Under Connectivity-Sustained/Broken DoS Attacks*

In this section, we explore the scenario of switching CS/CB-DoS attacks, focusing on a more intricate and widely applicable case from the perspective of network connectivity.

10.3.2.1 Design of Resilient Fixed-Time Observer

Clearly, the observer (10.7) proposed in Sect. 10.3.1 becomes infeasible under the scenario of switching connectivity-sustained/broken DoS attacks, due to the unbounded estimate of leader's information for attack-affected nodes when the connectivity of the network is disrupted.

To address this limitation and mitigate the adverse effects of such attacks on network performance, we develop a novel resilient fixed-time observer, which is formulated as (10.7), with the resilient observer gain parameters given later.

Theorem 10.3 *The resilient fixed-time observer (10.7) can accurately estimate the leader's state for each agent within a fixed settling time T_3, if the resilient observer gains satisfy* $\min\{\theta_{k,1}\} > \frac{1}{h_1\alpha}(\Phi+1)2^{\frac{\alpha+1}{2}}$, $\min\{\theta_{k,2}\} > \frac{1}{h_1\beta}(\Phi+1)2^{\frac{\beta+1}{2}}$,

$\theta_{k,3} > \max\left\{N(2N-1)(\bar{u}_0+\bar{d}_0)/\Gamma(N),\, N(\bar{u}_0+\bar{d}_0)/\|\bar{\mathcal{L}}_{\sigma(t)}\|\right\}$, and the following boundary condition holds:

$$\frac{1-\vartheta}{2}\sum_{j\in\mathcal{P}_U}\frac{d}{dt}\tilde{\rho}_{j[U],\vartheta}(0,t) - \frac{1-\vartheta}{2}\sum_{k\in\mathcal{P}_S}\frac{d}{dt}\tilde{\rho}_{k[S],\vartheta}(0,t)$$
$$+\frac{d}{dt}\tilde{\rho}_{mn,\vartheta}(0,t) \leq 0, \quad \vartheta \in \{\alpha,\beta\}, \tag{10.33}$$

where $\tilde{\rho}_{mn,\theta}(0,t)$, $\tilde{\rho}_{j[U],\theta}(0,t)$, and $\tilde{\rho}_{k[S],\theta}(0,t)$ are $\mathcal{FK}_\infty$ functions which will be defined later.

Proof Choose the candidate Lyapunov function as

$$V_{0\sigma(t)} = \frac{1}{2}\sum_{l=1}^{n} z_l^{\top}\Psi_{\sigma(t)} z_l,\ \sigma(t) \in \mathcal{P}, \tag{10.34}$$

where $z_l = [\tilde{p}_l^0, \tilde{v}_l^0]^{\top}$, and $\Psi_{\sigma(t)}$ is a positive-definite diagonal matrix.

According to Lemma 10.1, one has

$$\Psi_{\sigma(t)}\bar{\mathcal{L}}_{\sigma(t)} + \left(\bar{\mathcal{L}}_{\sigma(t)}\right)^{\top}\Psi_{\sigma(t)} \geq \omega_{\sigma(t)} I_N \tag{10.35}$$

when $\sigma(t) \in \mathcal{P}_S$, and $\omega_{\sigma(t)}$ is a positive constant.

Moreover, we have

$$\begin{aligned} &\tilde{v}_0^{\top}\left(-\phi_3 \mathrm{sig}(\bar{\mathcal{L}}_{\sigma(t)}\tilde{v}_0) - \mathbf{1}_N\left(u_0 + d_0\right)\right) \\ \leq &\begin{cases} -\left(\phi_3 \frac{\Gamma(N)}{2N-1} - N\left(\bar{u}_0 + \bar{d}_0\right)\right)\|\tilde{v}_0\|, & \sigma(t) \in \mathcal{P}_S \\ -\left(\phi_3 \left\|\bar{\mathcal{L}}_{\sigma(t)}\right\| - N\left(\bar{u}_0 + \bar{d}_0\right)\right)\|\tilde{v}_0\|, & \sigma(t) \in \mathcal{P}_U. \end{cases} \end{aligned} \tag{10.36}$$

Then, combining (10.34)–(10.36) and using the similar idea of [21] (a detailed derivation is omitted here due to page limitations), one can deduce that

$$\dot{V}_{0\sigma(t)}\left(z\left(t\right)\right) \leq -\bar{\vartheta}_{1\sigma(t)} V_{0\sigma(t)}^{\frac{1+\alpha}{2}} - \bar{\vartheta}_{2\sigma(t)} V_{0\sigma(t)}^{\frac{1+\beta}{2}} \tag{10.37}$$

with the time-varying switching design parameters $\bar{\vartheta}_{1\sigma(t)}$ and $\bar{\vartheta}_{2\sigma(t)}$ given by

$$\bar{\vartheta}_{k,1} = \begin{cases} \dfrac{2\hat{\vartheta}_{k,1}}{\|\Psi_{\sigma(t)}\|^{\frac{1+\alpha_v}{2}}}, & \sigma(t) \in \mathcal{P}_S \\ -\dfrac{2\check{\vartheta}_{k,1}}{\left(\vartheta_{\min}(\Psi_{\sigma(t)})\right)^{\frac{1+\alpha_v}{2}}}, & \sigma(t) \in \mathcal{P}_U \end{cases} \tag{10.38}$$

and

$$\bar{\vartheta}_{k,2} = \begin{cases} \dfrac{2\hat{\vartheta}_{k,2}}{\|\Psi_{\sigma(t)}\|^{\frac{1+\beta_v}{2}}}, & \sigma(t) \in \mathcal{P}_S \\ -\dfrac{2\check{\vartheta}_{k,2}}{\left(\vartheta_{\min}(\Psi_{\sigma(t)})\right)^{\frac{1+\beta_v}{2}}}, & \sigma(t) \in \mathcal{P}_U. \end{cases} \tag{10.39}$$

Remark 10.1 In fact, (10.37) delineates an essential prerequisite for ensuring the fixed-time stability of the observation error dynamics. When $\sigma(t) \in \mathcal{P}_S$, the network is connected and $\hat{\bar{\vartheta}}_{1\sigma(t)} > 0$, $\hat{\bar{\vartheta}}_{2\sigma(t)} > 0$, causing a decrease in $V_{0\sigma(t)}(z(t))$. When $\sigma(t) \in \mathcal{P}_U$, the connectivity is broken and $\hat{\bar{\vartheta}}_{1\sigma(t)} < 0$, $\hat{\bar{\vartheta}}_{2\sigma(t)} < 0$, which leads to an increase in $V_{0\sigma(t)}(z(t))$. Besides, it is imperative to recognize that if $\sigma(t) \in \mathcal{P}_S$, there are two possible cases: (i) If $|(\tau_{\sigma(t)}, \tau_{\sigma(t)+1})| \geq T_{\sigma(t)}$, where

$T_{\sigma(t)} \triangleq 2/(\bar{\vartheta}_{1\sigma(t)}(1-\alpha_v)) + 2/(\bar{\vartheta}_{2\sigma(t)}(\beta_v - 1))$, the system's observation error is steered toward stability for $t \geq T_{\sigma(t)}$. (ii) If $|(\tau_{\sigma(t)}, \tau_{\sigma(t)+1})| < T_{\sigma(t)}$, the value of $V_{0\sigma(t)}$ starts to rebound after $T_{\sigma(t)}$, and the system would not be stable within the current interval. In the following, we prove that (10.33) is a critical condition for the convergence of the networked UAV systems over the entire time horizon.

Case 1: If $V_{0\sigma(t)} \geq 1$, $\sigma(t) \in \mathcal{P}$: Define $z = [z_1^\top, \ldots, \tilde{v}_0^\top]^\top$. By integrating (10.37) and noting $V_{0\sigma(t)}^{\frac{1+\alpha_v}{2}} \leq V_{0\sigma(t)}^{\frac{1+\beta_v}{2}}$, we get $\int_{\tau_{M_\sigma(0,t)}}^{\top} V_{0\sigma(t)}^{-\frac{1+\alpha}{2}}(z(t))dV_{0\sigma(t)}(z(t)) \leq \int_{\tau_{M_\sigma(0,t)}}^{\top} -(\bar{\vartheta}_{1\sigma(t)} + \bar{\vartheta}_{2\sigma(t)})dt$, which leads to when $t \in [\tau_{M_\sigma(0,t)}, \tau_{M_\sigma(0,t)+1})$

$$V_{0\sigma(t)}(z(t)) \leq \Big(V_{0\sigma(t)}^{\frac{1-\alpha}{2}}\left(z\left(\tau_{M_\sigma(0,t)}\right)\right) - \frac{1-\alpha}{2} \times \left(\bar{\vartheta}_{1\sigma(t)}+\bar{\vartheta}_{2\sigma(t)}\right)\left(t-\tau_{M_\sigma(0,t)}\right)\Big)^{\frac{2}{1-\alpha}}. \tag{10.40}$$

From the definition of $V_{0\sigma(t)}$, there exists a positive constant $\mu_{ij} \geq \lambda_{\max}(\Psi_j(\Psi_i)^{-1})$ such that

$$V_{0j}(z(t)) \leq \mu_{ij} V_{0i}(z(t)),\ i, j \in \mathcal{P}. \tag{10.41}$$

Iterating (10.40) and (10.41) yields

$$\begin{aligned} V_{0\sigma(t)} \leq{} & \Bigg(\mu_{\sigma(t)\sigma(\tau_{M_\sigma(0,t)})}^{\frac{1-\alpha_v}{2}} V_{0\sigma(\tau_{M_\sigma(0,t)})}^{\frac{1-\alpha_v}{2}}\left(z\left(\tau_{M_\sigma(0,t)}\right)\right) \\ & - \frac{(1-\alpha_v)\left(\bar{\vartheta}_{1\sigma(t)}+\bar{\vartheta}_{2\sigma(t)}\right)}{2}\left(t-\tau_{M_\sigma(0,t)}\right)\Bigg)^{\frac{2}{1-\alpha_v}} \\ \leq{} & \cdots \\ \leq{} & \Bigg(\Gamma_1(t)\, V_{0\sigma(\tau_0)}^{\frac{1-\alpha_v}{2}}(z(\tau_0)) - \sum_{i=1}^{M_\sigma(0,t)+1}\Big((1-\alpha_v) \\ & \times \Gamma_2(t)\frac{\bar{\vartheta}_{1\sigma(t)}+\bar{\vartheta}_{2\sigma(t)}}{2}\Delta t_i\Big)\Bigg)^{\frac{1-\alpha_v}{2}}, \end{aligned} \tag{10.42}$$

where $\Gamma_1(t) = \prod_{i=1}^{M_\sigma(0,t)+1} \mu_{\sigma(\tau_i)\sigma(\tau_{i-1})}^{(1-\alpha)/2}$, and $\Gamma_2(t) = \prod_{j=i+1}^{M_\sigma(0,t)+1} \mu_{\sigma(\tau_j)\sigma(\tau_{j-1})}^{(1-\alpha)/2}$.

Moreover, noting the definition of Δt, we have

$$\sum_{i=1}^{M_\sigma(0,t)+1}\left(\frac{(1-\alpha)}{2}\Gamma_2(t)\left(\bar{\vartheta}_{1\sigma(t)}+\bar{\vartheta}_{2\sigma(t)}\right)\Delta t_i\right)$$

$$= -\frac{(1-\alpha)}{2}\sum_{j\in\mathcal{P}_U}\left|\bar{\vartheta}_{1j}+\bar{\vartheta}_{2j}\right|\left|(0,t]\cap\left(\bigcup_{\sigma(\tau_i)=j}(\tau_i,\tau_{i+1}]\right)\right|$$

$$\times\exp\sum_{(m,n)\in\mathcal{E}(\mathcal{P})}\ln\mu_{mn}^{\frac{1-\alpha}{2}}M_{mn}\left(\tau_{j-1},\tau_{M_\sigma(0,t)+1}\right)$$

$$+\frac{(1-\alpha)}{2}\sum_{k\in\mathcal{P}_S}\left|\bar{\vartheta}_{1k}+\bar{\vartheta}_{2k}\right|\left|(0,t]\cap\left(\bigcup_{\sigma(\tau_i)=k}(\tau_i,\tau_{i+1}]\right)\right|$$

$$\times\exp\sum_{(m,n)\in\mathcal{E}(\mathcal{P})}\ln\mu_{mn}^{\frac{1-\alpha}{2}}M_{mn}\left(\tau_{k-1},\tau_{M_\sigma(0,t)+1}\right). \tag{10.43}$$

Substituting (10.2)–(10.4) and (10.43) into (10.42) leads to

$$V_{0\sigma(t)}(z(t)) \le \left(\tilde{\rho}_{mn}(0,t)+\frac{(1-\alpha)}{2}\sum_{j\in\mathcal{P}_U}\tilde{\rho}_{j[U]}(0,t)\right.$$
$$\left.-\frac{(1-\alpha)}{2}\sum_{j\in\mathcal{P}_S}\tilde{\rho}_{j[S]}(0,t)\right)^{\frac{1-\alpha}{2}}, \tag{10.44}$$

where

$$\tilde{\rho}_{mn,\theta}(0,t) = -(1-\theta)/2\sum\nolimits_{k\in\mathcal{P}_S}\hat{\Gamma}_\theta(\tau_{k-1},t)|\tilde{\vartheta}|\bar{\Xi}_{k[S]}+(1-\theta)/2,$$

$$\sum\nolimits_{j\in\mathcal{P}_U}\hat{\Gamma}_\theta(\tau_{j-1},t)|\tilde{\vartheta}_j|\bar{\Xi}_{j[U]}+\hat{\Gamma}_\theta(0,t)V_{0\sigma(0)}^{\frac{1-\alpha}{2}}(0),$$

$$\tilde{\rho}_{j[U],\theta}(0,t) = \sum\nolimits_{j\in\mathcal{P}_U}\hat{\Gamma}_\theta(\tau_{j-1},t)|\tilde{\vartheta}_j|\rho_{j[U]}(0,t),$$

$$\tilde{\rho}_{k[S],\theta}(0,t) = \sum\nolimits_{k\in\mathcal{P}_S}\hat{\Gamma}_\theta(\tau_{k-1},t)|\tilde{\vartheta}|\rho_{k[S]}(0,t)$$

with $\hat{\Gamma}_\theta(\tau_a,t)=\exp(\sum_{(m,n)\in\mathcal{E}(\mathcal{P})}\ln\mu_{mn}^{\frac{1-\theta}{2}}(\bar{M}_{mn}+\rho_{mn}(\tau_a,t)))$.

It is obvious that $V_{0\sigma(0)} = \frac{1}{2}\sum_{i=1}^n z_i^\top\Psi_{\sigma(0)}z_i > 0$, $\rho_{k[S]}(0,t)$, and $\rho_{j[U]}(0,t)$ are $\mathcal{FK}_\infty$ functions, and the former grows faster than the later in terms of (10.31), thus the right side of inequality (10.44) is decreasing by t, which implies that there must be a solution T_{0,α_v} to the equation $\tilde{\rho}_{mn}(0,t)+(1-\alpha_v)/2(\sum_{j\in\mathcal{P}_U}\tilde{\rho}_{j[U]}(0,t)-\sum_{j\in\mathcal{P}_S}\tilde{\rho}_{j[S]}(0,t))=0$, which implies that $\lim_{t \to T_{0,\alpha_V}} V_{0\sigma(t)}(z(t)) \to 0$. From the above discussions, we have

$$\lim_{t \to T_{3,\alpha}} |z(t)| \to 0, \; T_{3,\alpha} > 0, \tag{10.45}$$

which indicates that $z(t)$ is bounded and converges to the origin within a fixed settling time $T_{3,\alpha}$.

Case 2: If $0 < V_{0\sigma(t)} < 1$, $\sigma(t) \in \mathcal{P}$: $V_{0\sigma(t)}^{\frac{1+\alpha}{2}} > V_{0\sigma(t)}^{\frac{1+\beta}{2}}$. Integrating (10.8) yields that $\int_{\tau_{M_\sigma}(0,t)}^{\top} V_{0\sigma(t)}^{-\frac{1+\beta}{2}}(z(t))dV_{0\sigma(t)}(z(t)) \le \int_{\tau_{M_\sigma}(0,t)}^{\top} -\left(\bar{\vartheta}_{1\sigma(t)} + \bar{\vartheta}_{2\sigma(t)}\right)dt$, which leads to

$$V_{0\sigma(t)}(z(t)) \le \Bigg(V_{0\sigma(t)}^{\frac{1-\beta}{2}}\left(z\left(\tau_{M_\sigma}(0,t)\right)\right) - \frac{1-\beta}{2} \times \left(\bar{\bar{\vartheta}}_{1\sigma(t)} + \bar{\bar{\vartheta}}_{2\sigma(t)}\right)\left(t - \tau_{M_\sigma}(0,t)\right)\Bigg)^{\frac{2}{1-\beta}}, \tag{10.46}$$

where $t \in [\tau_{M_\sigma}(0,t), \tau_{M_\sigma}(0,t)+1})$.

Similarly, we have

$$V_{0\sigma(t)}(z(t)) \le \Bigg(\tilde{\rho}_{mn}(0,t) + \frac{(1-\beta)}{2}\sum_{j \in \mathcal{P}_U}\tilde{\rho}_{j[U]}(0,t) - \frac{(1-\beta)}{2}\sum_{j \in \mathcal{P}_S}\tilde{\rho}_{j[S]}(0,t)\Bigg)^{\frac{1-\beta}{2}}. \tag{10.47}$$

Using the similar derivation process of Case 1, one can also conclude that $\lim_{t \to T_{3,\beta}} |z(t)| \to 0$, where $T_{3,\beta} > 0$. Therefore, it can be concluded that the proposed resilient observer can achieve the stable convergence within a fixed settling time $T_3 = \max\{T_{3,\alpha}, T_{3,\beta}\}$. In other words, $\hat{p}_0^i \to x_0$, for $t \ge T_3$. This completes the proof of Theorem 10.3.

10.3.2.2 Design of Resilient Security Control Protocol

The resilient security control protocol for the case of switching CS/CB-DoS attacks is the same as given in Table 10.1, based on the proposed resilient fixed-time observer. Therefore, the proof is omitted here to avoid repetition.

Remark 10.2 (Network Resilience Analysis) The CB-DoS attack results in an increase in $V_{0\sigma(t)}(z(t))$ (*performance bounce*). However, attack strength boundary condition (10.33) ensures a decrease in the upper boundary of $V_{0\sigma(t)}(z(t))$ (*performance recovery*). The energy function increases during the interval $[\tau_i, \tau_{i+1})$ for $i \in \mathcal{P}_U$ and decreases for $i \in \mathcal{P}_S$. The constraints on the duration and frequency of attacks help regulate the escalation of the energy function, ensuring that

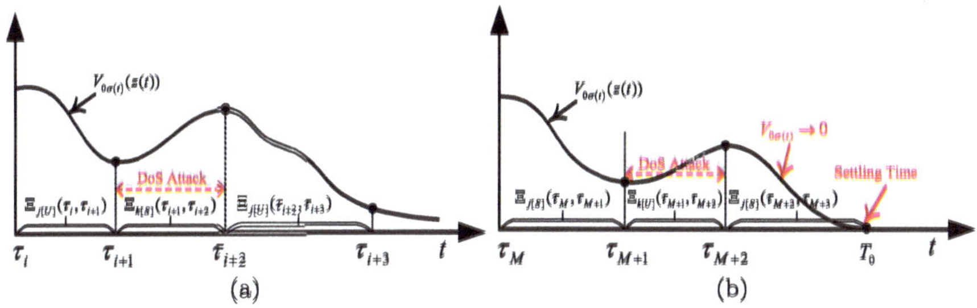

Fig. 10.3 Sketch of the evolution of $V_{0\sigma(t)}(z(t))$ under the scenarios of connectivity-sustained/broken switching DoS attacks. (**a**) The value of $V_{0\sigma(t)}(z(t))$ decreases during the DoS-inactive intervals and CS-DoS attack intervals. Conversely, it increases during CB-DoS attacks. (**b**) Over the entire time horizon, $V_{0\sigma(t)}(z(t))$ exhibits a resilient attenuation trend and ultimately decreases to the origin after T_0

$V_{0\sigma(t)}(z(t)) \to 0$ as t approaches the settling time T_0. The evolution of $V_{0\sigma(t)}(z(t))$ under the switching connectivity-sustained/broken DoS attacks is illustrated in Fig. 10.3.

Remark 10.3 Although existing state-of-the-art resilient security control strategies have effectively tackled synchronous periodic DoS attacks [13–18], they usually presuppose predictable attack patterns. Our proposed strategy aims to achieve the secure consensus of networked UAV systems by addressing the challenge of asynchronous DoS attacks with stochastic combinations of paralyzed edges, where each communication edge is susceptible to independent disruption. It is obvious that our designed control scheme is of more practical significance, from the perspective of dealing with various complex attack scenarios.

10.4 Simulation and Analysis

In this section, we conduct simulations and analyses to validate the effectiveness of the proposed secure control method against asynchronous DoS attacks. Moreover, we present the comparative simulations with the finite-time security control method proposed in [27].

In order to assess the control capabilities of our proposed resilient secure control protocol in the face of dynamic CS/CB DoS attacks, we have conducted a comprehensive simulation, which is based on a network of UAVs. The connectivity-sustained DoS attack and connectivity-broken DoS attack pattern are depicted in Fig. 10.4.

As shown in Fig. 10.4, two cases of attack communication are considered: (i) If the edge (κ_1, κ_3) is disrupted due to a DoS attack, the formation remains connected (CS-DoS attack). (ii) If the edge (κ_3, κ_4) is disrupted under a DoS attack, the formation becomes unconnected (CB-DoS attack). The reference trajectory is

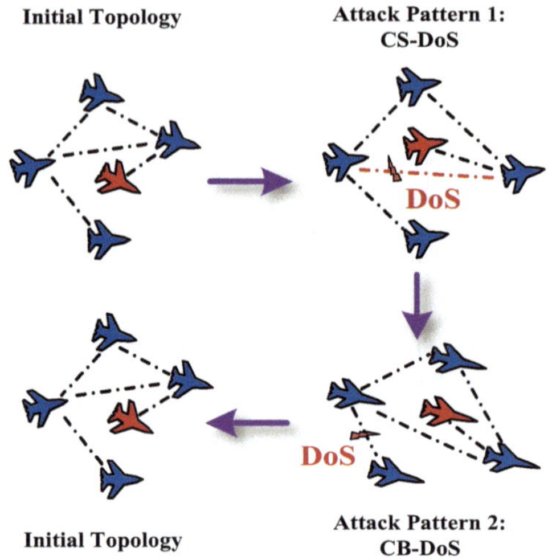

Fig. 10.4 Flight formation topologies under different attack patterns

Table 10.2 Parameters of resilient observer and controller

Parameters of resilient fixed-time observer								
Φ	α	β	μ	$\bar{u}_0$	$\bar{d}_0$	$\theta_{1,\min}$	$\theta_{2,\min}$	$\theta_{n,3}$
12	0.7	0.8	1.75	10	3	35.66	38.42	15.33
Parameters of resilient security control protocol								
μ_1	μ_2	ϖ_1	ϖ_2	$\epsilon_{i,1}$	$\epsilon_{i,2}$	κ	δ	ξ
1	7/9	0.25	0.29	3.5	8.1	1.78	4/9	6.5
$\rho_{i,1}$	$\rho_{i,2}$	$\xi_{i,1}$	$\xi_{i,2}$	$\wp_{i,1,1}$	$\wp_{i,1,2}$	$\wp_{i,2,1}$	$\wp_{i,2,2}$	
0.73	0.68	5.02	5.02	13.42	10.95	25.16	22.28	

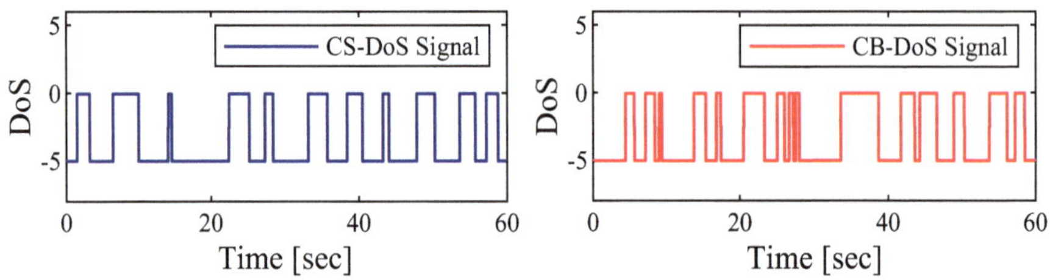

Fig. 10.5 Asynchronous aperiodic attack signals of CS-DoS and CB-DoS

defined by $p_0 = [50\sin(0.5t), 30\cos(0.5t), 2t]^\top$. Table 10.2 lists the parameters of resilient fixed-time observer and security controller. Moreover, the duration of CS-DoS signal and CB-DoS signal is depicted in Fig. 10.5. Different from the case of only connectivity-sustained DoS attack, the connectivity-sustained DoS attack and connectivity-broken DoS attack are active alternately in this scenario, which means that the connectivity of the formation network may be disrupted in some specific attack time intervals.

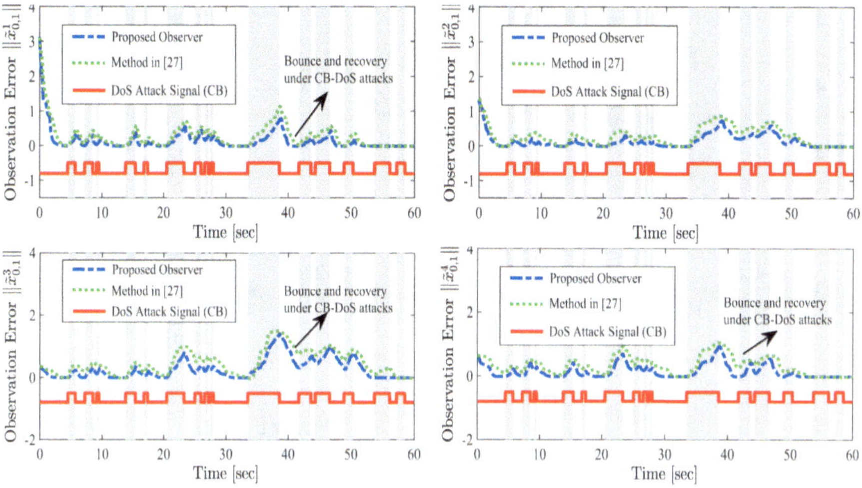

Fig. 10.6 Responses of observation errors under asynchronous switching CS/CB-DoS attacks

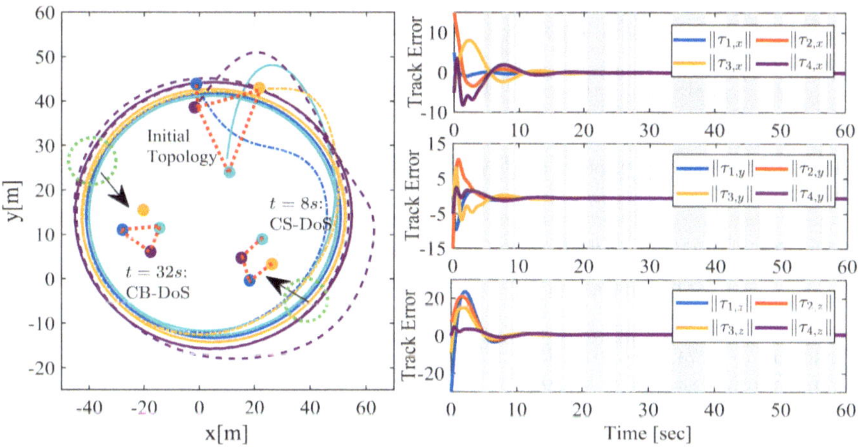

Fig. 10.7 Flight trajectories of all UAVs and responses of tracking errors under CS/SB-DoS attacks

The simulation outcomes, when subjected to alternating CS/CB-DoS attacks, are presented in Figs. 10.6, 10.7, and 10.8. Figure 10.6 displays the evolution of the observation errors $\hat{x}_{0,1}^{i}$, demonstrating the convergence performance achieved by the proposed resilient fixed-time observer. Figure 10.7 presents the flight trajectories of the formation along with the tracking errors τ_i, showcasing the efficacy of the proposed resilient security control protocol in countering the dynamic CS/CB-DoS attacks. Figure 10.8 captures the positional snapshots of all UAVs within the formation at four distinct time points, highlighting the network communication integrity under the influence of alternating CS/CB-DoS attacks. Furthermore,

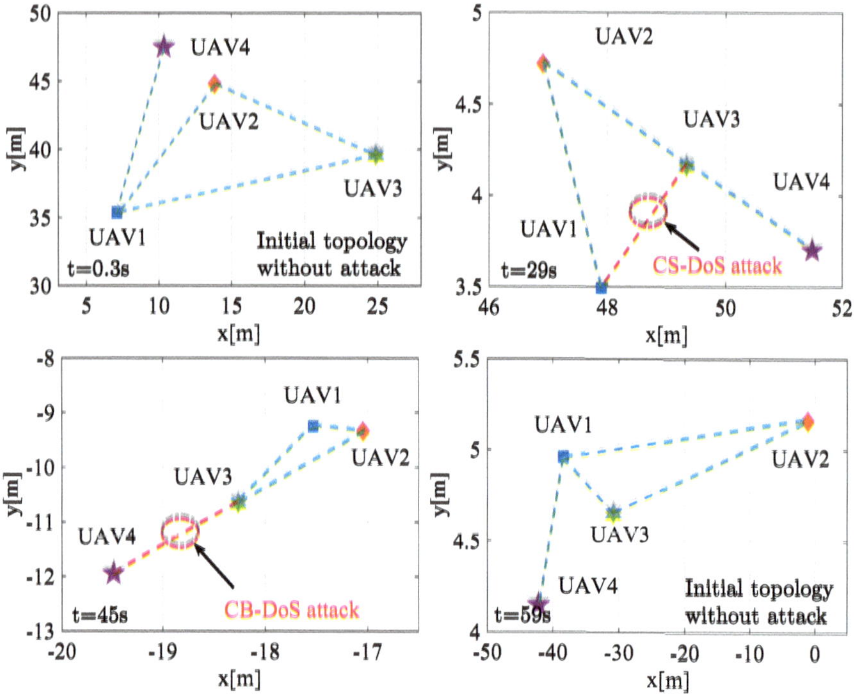

Fig. 10.8 Positional snapshots of all UAVs within the formation at four distinct simulation time points

Table 10.3 Settling time of the formation under different methods

Agent	AFTO method in [24]			Proposed FTETO method		
	T_1	T_2	T_{max}	T_1	T_2	T_{max}
Follower 1	74.82s	25.75s	100.55s	32.33s	17.42s	49.75s$_{\downarrow 50.52\%}$
Follower 2	66.08s	20.35s	86.43s	20.92s	13.43s	34.35s$_{\downarrow 60.26\%}$
Follower 3	72.84s	27.37s	100.21s	29.96s	14.55s	44.51s$_{\downarrow 55.58\%}$
Follower 4	58.09s	17.32s	75.41s	15.78s	15.99s	31.77s$_{\downarrow 57.87\%}$

Table 10.3 compares the settling time of the formation under the proposed control scheme with that of the approach in [24], indicating the rapid convergence ability of the proposed resilient security control protocol.

10.5 Conclusions

In this study, we tackle the challenge of fixed-time resilient security control for networked control systems of UAVs in the face of asynchronous DoS attacks.

We introduce a resilient fixed-time observer capable of precisely estimating the leader's information. Subsequently, we develop an innovative fixed-time security consensus control protocol to ensure rapid and stable convergence of UAVs, even when subjected to asynchronous DoS attacks. Additionally, we derive the boundary condition for system stability under scenarios of connectivity-sustained/broken DoS attacks. Looking ahead, our future research will delve deeper into the development of resilient security control protocols that not only maintain system performance but also provide guaranteed performance levels in the presence of cyber threats.

References

1. Z.Y. Wu, Y. Wang, J.L. Xiong, M. Xie, Switching periodic event-triggered H_∞ control for NCSs and its application to UAV. IEEE Trans. Veh. Technol. **74**(4), 5078–5088 (2024)
2. F. Liao, R. Teo, J.L. Wang, X.X. Dong, F. Lin, K.M. Peng, Distributed formation and reconfiguration control of VTOL UAVs. IEEE Trans. Control Syst. Technol. **25**(1), 270–277 (2018)
3. Y.J. Yu, J. Guo, C.K. Ahn, Z.R. Xiang, Neural adaptive distributed formation control of nonlinear multi-UAVs with unmodeled dynamics. IEEE Trans. Neural Netw. Learn. Syst. **34**(11), 9555–9561 (2023)
4. Z.X. Du, H. Zhang, Z.P. Wang, H.C. Yan, Model predictive formation tracking-containment control for multi-UAVs with obstacle avoidance. IEEE Trans. Syst. Man Cybern. Syst. **54**(6), 3404–3414 (2024)
5. K.R. Ze, W. Wang, K.X. Liu, J.H. Liu, Time-varying formation planning and distributed control for multiple UAVs in clutter environment. IEEE Trans. Ind. Electron. **71**(9), 11305–11315 (2024)
6. Y.J. Ma, B. Jiang, J. Wang, J.Y. Gong, Adaptive fault-tolerant formation control for heterogeneous UAVs-UGVs systems with multiple actuator faults. IEEE Trans. Aerospace Electron. Syst. **59**(5), 6705–6716 (2023)
7. B.Y. Zhang, M.L. Lv, S.H. Cui, X.W. Bu, J.H. Park, Learning-based optimal cooperative formation tracking control for multiple UAVs: a feedforward-feedback design framework. IEEE Trans. Autom. Sci. Eng. https://doi.org/10.1109/TASE.2023.3322028
8. J.H. Lu, H.Y. Jiang, X.C. Hou, P. Wang, Distributed edge-based event-triggered control for voltage restoration and current sharing in DC microgrids under DoS attacks. IEEE Trans. Ind. Electron. **70**(8), 8053–8063 (2023)
9. Z. Han, W. Wang, J.S. Huang, Z.Q. Wang, Distributed adaptive formation tracking control of mobile robots with event-triggered communication and denial-of-service attacks. IEEE Trans. Ind. Electron. **70**(4), 4077–4087 (2023)
10. C. Peng, H.T. Sun, Switching-like event-triggered control for networked control systems under malicious denial of service attacks. IEEE Trans. Autom. Control **65**(9), 3943–3949 (2020)
11. F.H. Li, Z.S. Hou, Learning-based model-free adaptive control for nonlinear discrete-time networked control systems under hybrid cyber attacks. IEEE Trans. Cybern. **54**(3), 1560–1570 (2024)
12. X. Cai, K.B. Shi, S. Kun, S.M. Zhong, Y.C. Soh, Y. Yu, Performance degradation estimation mechanisms for networked control systems under DoS attacks and its application to autonomous ground vehicle. IEEE Trans. Cybern. **54**(5), 2992–3002 (2024)
13. S.L. Hu, D. Yue, X.L. Chen, Z.H. Chen, X.P. Xie, Resilient H_∞ filtering for event-triggered networked systems under nonperiodic DoS jamming attacks. IEEE Trans. Cybern. **51**(3), 1392–1403 (2021)

14. Y. Ma, X. Qi, Z.X. Li, S.L. Hu, M.A. Sotelo, Resilient control for networked unmanned surface vehicles with dynamic event-triggered mechanism under aperiodic DoS attacks. IEEE Trans. Veh. Technol. **73**(4), 5824–5833 (2024)
15. Z.H. Cheng, S.L. Hu, D. Yue, C.X. Dou, S.G. Shen, Resilient distributed coordination control of multiarea power systems under hybrid attacks. IEEE Trans. Syst. Man Cybern. Syst. **52**(1), 7–18 (2022)
16. G.D. Zhang, J. Li, O. Bamisile, Y.K. Xing, D.S. Cai, Q. Huang, An H_∞ load frequency control scheme for multi-area power system under cyber-attacks and time-varying delays. IEEE Trans. Power Syst. **38**(2), 1336–1349 (2023)
17. R. Merco, F. Ferrante, P. Pisu, A hybrid controller for DoS-resilient string-stable vehicle platoons. IEEE Trans. Intell. Transp. Syst. **22**(3), 1697–1707 (2021)
18. L. Qiu, S.Y. Zhou, Y.D. Wen, M. Najariyan, C.X. Liu, Stability and stabilization of semi-Markov jump linear system with DoS attacks. IEEE Trans. Circuits Syst.-II: Express Briefs **71**(6), 3126–3130 (2024)
19. Y.F. Zhang, Z.G. Wu, Asynchronous control of Markov jump systems under aperiodic DoS attacks. IEEE Trans. Circuits Syst.-II: Express Briefs **70**(2), 685–689 (2023)
20. S.L. Dong, M.Q. Liu, Adaptive fuzzy asynchronous control for nonhomogeneous Markov jump power systems under hybrid attacks. IEEE Trans. Fuzzy Syst. **31**(3), 1009–1019 (2023)
21. R. Zhao, Z.Q. Zuo, Y.J. Wang, W.T. Zhang, Active control strategy for disturbed switched systems under asynchronous DoS attacks. Syst. Control Lett. **6**, 2701–2706 (2022)
22. J. Zhang, D.S. Yang, H.G. Zhang, H.G. Su, Adaptive secure practical fault-tolerant output regulation of multiagent systems with DoS attacks by asynchronous communications. IEEE Trans. Netw. Sci. Eng. **10**(6), 4046–4055 (2023)
23. J. Zhang, C. Zhao, F.C. Liu, C.C. Wang, Q. Geng, Hybrid dynamic event-triggered control for networked control systems subject to asynchronous DoS attacks. IEEE Syst. J. **17**(2), 1974–1985 (2023)
24. W.W. Zhan, Y.J. Chen, Z.G. Wu, Y.N. Wang, Event-triggered finite-time formation control for networked nonholonomic mobile robots under denial-of-service attacks. IEEE Trans. Netw. Sci. Eng. **10**(6), 3754–3766 (2023)
25. N. Zhao, P. Shi, W. Xing, C.P. Lim, Event-triggered control for networked systems under denial of service attacks and applications. IEEE Trans. Circuits Syst.-I: Regul. Pap. **69**(2), 811–820 (2022)
26. H.J. Yang, D. Ye, Observer-based fixed-time secure tracking consensus for networked high-order multiagent systems against DoS attacks. IEEE Trans. Cybern. **52**(4), 2018–2031 (2022)
27. W.X. Li, H.P. Du, Z.G. Feng, L. Deng, D.H. Ning, W.H. Li, Stability analysis for H_∞-controlled active quarter-vehicle suspension systems with a resilient event-triggered scheme under periodic DoS attacks. IEEE Trans. Cybern. **54**(4), 2358–2368 (2024)
28. H.L. Ren, G.D. Zong, T.S. Li, Event-triggered finite-time control for networked switched linear systems with asynchronous switching. IEEE Trans. Syst. Man Cybern. Syst. **48**(11), 1874–1884 (2018)
29. W. Bai, P.X.P. Liu, H.Q. Wang, Neural-network-based adaptive fixed-time control for nonlinear multiagent non-affine systems. IEEE Trans. Neural Netw. Learn. Syst. **35**(1), 570–583 (2024)
30. Y. Zhao, K. Gao, P.F. Huang, G.R. Liu Specified-time affine formation maneuver control of multiagent systems over directed networks. IEEE Trans. Autom. Control **69**(3), 1936–1943 (2024)
31. J.W. Demmel, *Applied Numerical Linear Algebra* (SIAM, Philadelphia, 1997)

Part V
Conclusions

Chapter 11
Conclusions and Challenges

11.1 Conclusions

This monograph focuses on multi-fixed-wing UAV systems, addressing key issues related to cooperative guidance and control under complex environments, diverse task requirements, and practical system constraints. The research has theoretical and technical advancements in areas such as rapid system convergence, control performance constraints, event-triggered strategies, and safety control. These contributions provide effective solutions to achieving cooperative control of multi-UAVs under task-oriented multi-constraint conditions. The contributions and achievements of this monograph are summarized as the following aspects:

1. To enhance the fixed-time convergence of six-degree-of-freedom multi-UAV system subject to actuator faults and state constraints, a distributed fixed-time fault-tolerant control scheme is proposed. To reduce the adverse influences caused by parameter uncertainties and actuator faults, the bound estimation approach and well-defined smooth functions are introduced, such that the tracking errors of velocity and attitude stably converge to the residual sets around origin within fixed time.
2. To achieve the fixed-time synchronization tracking of six-degree-of-freedom multi-UAV system in the presence of actuator faults and saturation, a new adaptive-gain-based control design with fixed-time sliding mode surface was proposed. Moreover, model uncertainties and external disturbances could be effectively handled by using a bound estimation method and smooth functions.
3. To address the specified-time synchronization tracking problem of six-degree-of-freedom multi-UAV system with actuator faults, the novel specified-time performance functions are designed in such a way that the desired performance bounds can be imposed on velocity and attitude tracking errors. Then a distributed specified-time control scheme is constructed with adaptive robustifying terms to enhance the fault-tolerant ability and compensate the modeling uncertainties.

M. Lv et al., *Unmanned Aerial Vehicles Formation Control and Safety Guarantee*,
Cognitive Intelligence and Robotics, https://doi.org/10.1007/978-981-96-5829-9_11

4. To further reduce the signal transmission burden of six-degree-of-freedom multi-UAV system subject to actuator faults and lumped disturbances, an event-triggered cooperative synchronization fault-tolerant control methodology is investigated. By integrating an event-triggering mechanism, the time instants of updating control signals are determined. The actuator faults and the network induced errors are handled via the bound estimation approach and some well-defined smooth functions.
5. To further reduce the communication burden of six-degree-of-freedom multi-UAV system in the presence of actuator faults and lumped disturbances, an event-based fault-tolerant coordinated control is utilized such that the consensus tracking of velocity and attitude is achieved. By introducing a dynamic event-triggered communication mechanism, the connected communications are scheduled while avoiding the unnecessary information exchanges among UAVs. Regarding the actuator faults and lumped disturbances, the disturbance observer technique is incorporated into the control scheme to increase the composite approximation capability.
6. To address the problem of resource wastage in six-degree-of-freedom multi-UAV inter-communications, a dynamic self-triggered communication mechanism is proposed, which not only reduces the communication frequency of UAVs but also does not require continuous monitoring of state information. In addition, a time-varying constrained bounded function is proposed, which effectively relaxes the restrictions on the initial state of the system.
7. To ensure the performance robustness and optimality of multi-UAV system in the face of strong disturbances, a feedforward-feedback learning-based optimal control methodology is established to facilitate cooperative formation tracking. By leveraging backstepping-based feedback control, the problem of formation tracking is transformed into an equivalent optimal regulation problem. Consequently, a learning-based feedforward control scheme is devised, wherein the critic-only echo state network is employed to approximate the optimal feedforward control policies.
8. To address the resilient fixed-time security control problem for multiple UAVs under asynchronous DoS attacks, a novel resilient fixed-time observer is introduced to accurately estimate the leader's reference in spite of the attack signals. Then, an attack-oriented fixed-time security consensus controller is proposed to ensure the fast and stable convergence of the networked UAVs.

11.2 Open Problems and Challenges

The monograph has explored the cooperative control issues of multi-UAV systems under task-oriented multi-constraints, focusing on speed and attitude control and formation control, yielding several preliminary results. However, there remain shortcomings due to the depth of study and research duration, warranting further investigation in the following areas:

1. A critical challenge in multi-UAV formation control is coordinating to prevent potential collisions between UAVs and ensuring timely evasive actions against both stationary and moving obstacles in the flight environment. Given the nonholonomic constraints and maximum/minimum linear velocity constraints of fixed-wing UAVs, conventional obstacle avoidance algorithms may not be directly applicable. Therefore, advancing research into collision avoidance and obstacle avoidance strategies within multi-UAV formations is a vital area for future exploration. This could involve developing new algorithms that account for the specific dynamics and constraints of UAV systems.
2. The current design of cooperative controllers primarily considers limited communication resources. However, in nonideal network environments, communication among UAVs can suffer from delays or packet loss, which directly affects the performance and stability of cooperative control. Investigating cooperative control strategies that can effectively operate under these conditions holds practical significance. Future research could focus on developing robust communication protocols and control algorithms that maintain system performance despite network uncertainties, ensuring reliable coordination among UAVs in real-time applications.
3. The complete control of multi-UAV systems encompasses three key phases: navigation, guidance, and control. While this monograph has concentrated on guidance and control aspects, it has not thoroughly addressed navigation issues. Effective navigation involves planning reasonable flight trajectories based on mission requirements and environmental constraints, which are influenced by multiple factors. Further research is needed to develop comprehensive navigation strategies that integrate mission objectives with dynamic environmental conditions, ensuring optimal path planning and execution for multi-UAV systems.

MIX
Papier aus verantwortungsvollen Quellen
Paper from responsible sources
FSC® C105338

If you have any concerns about our products,
you can contact us on
ProductSafety@springernature.com

In case Publisher is established outside the EU,
the EU authorized representative is:
Springer Nature Customer Service Center GmbH
Europaplatz 3, 69115 Heidelberg, Germany

Printed by Libri Plureos GmbH
in Hamburg, Germany